Soft, Hard, and Hybrid Janus Structures

Synthesis, Self-Assembly, and Applications

Soft, Hard, and Hybrid Janus Structures

Synthesis, Self-Assembly, and Applications

Editors

Zhiqun Lin
Georgia Institute of Technology, USA

Bo Li
University of Illinois at Urbana-Champaign, USA

World Scientific

NEW JERSEY • LONDON • SINGAPORE • BEIJING • SHANGHAI • HONG KONG • TAIPEI • CHENNAI • TOKYO

Published by

World Scientific Publishing Europe Ltd.

57 Shelton Street, Covent Garden, London WC2H 9HE

Head office: 5 Toh Tuck Link, Singapore 596224

USA office: 27 Warren Street, Suite 401-402, Hackensack, NJ 07601

Library of Congress Cataloging-in-Publication Data
Names: Li, Bo (Writer on chemical engineering), editor. | Lin, Zhiqun, 1972– editor.
Title: Soft, hard, and hybrid Janus structures : synthesis, self-assembly, and applications / edited by Bo Li (University of Illinois at Urbana-Champaign), Zhiqun Lin (Georgia Institute of Technology).
Description: New Jersey : World Scientific, 2017. | Includes bibliographical references.
Identifiers: LCCN 2016055756 | ISBN 9781786343123 (hc : alk. paper)
Subjects: LCSH: Curves on surfaces. | Surfaces (Physics) | Nanostructured materials.
Classification: LCC QA643 .S64 2017 | DDC 620.1/15--dc23
LC record available at https://lccn.loc.gov/2016055756

British Library Cataloguing-in-Publication Data
A catalogue record for this book is available from the British Library.

Desk Editors: Chandrima Maitra/Mary Simpson/Shi Ying Koe

Typeset by Stallion Press
Email: enquiries@stallionpress.com

Printed in Singapore

Preface

By chemically designing particle from the "bottom", colloidal particles have enabled future devices with advanced functionalities through self-assembly. Janus particles (JPs), which possess two distinct surface chemical composition within one particle, have shown unique properties not only due to the coupling effect between two composing parts, but also due to the orientational interactions between particles, making them exhibit complicated but intriguing collective properties and self-assembled structure.

This book summarizes recent progress on biphasic JPs. As a new category of materials, the synthesis method is always the primary focus, which is also the major part of this book, in Chapters 1–6. In general, two types of strategies are employed for JP preparation: site-specific surface engineering on a colloidal particle core or by joining two distinct materials together. The former strategy often leads to soft–hard hybrid JP while the latter yields all-inorganic or all-organic JPs. It should also be noted that the length scale of JP determines the synthetic routes, that is, Janus nanoparticle (JNP) is synthesized though a totally different pathway compared to Janus microparticle. Consequently, the applications of the JP of different sizes vary greatly.

Instead of being considered as individual units, JPs are utilized as building blocks for assembly to achieve further functionalities. To this end, the self-assembly behavior of JP is discussed in Chapter 7. In the last three chapters, the applications of JP are reviewed, focusing on different research areas. Same as any other nanomaterials, the structure of this book follows conventional research flow: synthesis — self-assembly — application. We hope this book will benefit all researchers working in the area of JPs or particles with anisotropic structure/properties.

Chapter 1 covers the synthesis method of JPs with interesting morphologies, such as hollow, non-spherical, sheets/disks, rods/cylinders, rings and cone-like shape. It should be noted that there does not exist a universal method for Janus synthesis, and each shape may require a specific synthetic route. However, among various methods such as emulsion polymerization, protect-modification upon 3D interface (Pickering emulsion) and 2D interface and self-assembly of copolymers, emulsion polymerization is relatively more versatile and with excellent scalability.

Chapter 2 reviews the main methods of synthesis of JPs, and synthesis of both polymeric and hybrid JNPs with their diameter in the nanometer range. Moreover, a synthetic approach based on reversible-addition fragmentation chain transfer (RAFT) in seeded emulsion polymerization is discussed that yields typical particle size within tens of nanometers.

Chapter 3 provides a glimpse into the synthesis of all-inorganic multicomponent nanocrystals by illustrating progress achieved in the wet-chemical development of last-generation breeds of so-called hybrid or heterostructured nanocrystals (HNCs) in asymmetric non-core/shell geometries. The distinct material modules are interconnected in heterodimer, heterooligomer and anisotropic multidomain architectures via heteroepitaxial bonding interfaces of limited extension. In particular, this chapter focuses on the HNCs incorporating at least one magnetic material component combined with semiconductors and/or plasmonic metals, due to its potential for enhanced, unconventional magnetic behavior and even new properties and capabilities.

Chapter 4 reviews the fabrication methods on magnetic JPs, such as the miniemulsion/solvent evaporation, shrinkage-gelation and microfluidics-based synthesis. Also in this chapter, the applications of magnetic JPs are discussed, such as swimming micromotors, which will be reviewed in detail in Chapter 8, electronic display and biological applications.

Chapter 5 specifically introduces the synthetic routes to Fe_3O_4@ polymer magnetic JPs, including solvent evaporation-induced phase separation, miniemulsion polymerization, assembly and *in situ* growing of particles, Pickering emulsion, seed emulsion polymerization, microfluidics-based synthesis and so on. At the end of this chapter, the applications are briefly discussed.

Chapter 6 summarizes recent progress of synthesis, self-assembly and applications of amphiphilic JNPs and triblock JNP (TJNP) analogs formed by the phase separation of ligands on the inorganic nanoparticle (NP) surface. Importantly, JNPs with inorganic NP core and phase separating polymer ligand possess hybrid functionality, which can be considered as organic molecular analogs but with unique multiscale interactions through self-assembly. Moreover, the strategies for assembling JNP/TJNP amphiphilies and their biological applications are also included in this chapter.

Chapter 7 is devoted to the recent numerical and experimental studies of the self-assembly of Janus ellipsoids to examine the role of particle aspect ratio and patch orientation in dictating associated cluster morphologies and properties. Based on computer simulation, the process and structural and thermodynamic consequences of self-assembly of oblate, prolate spheroidal particles and patchy ellipsoids are investigated.

Chapter 8 describes how catalytic Janus swimmers can be used to generate autonomous motion within fluids at small scales. First, synthesis methods of catalytic Janus swimmers are briefly discussed. Then, the propulsion mechanism was explored. Moreover, this chapter emphasizes recent characterization method for analyzing catalytic Janus swimmers' motion and motion control strategies. The potential applications and future directions of catalytic Janus swimmers are also mentioned at the end of the chapter.

Chapter 9 focuses particularly on the biomedical applications of JPs. Due to the desire of keeping different functions (e.g., cell targeting, molecular sensing and *in vivo* imaging) spatially separated while combining them in the same structural unit, JPs are considered to be the key to solving this challenge. In this chapter, current progress in synergistic drug delivery, cell targeting, multimodal bioimaging, biosensing, remote control of biological functions, self-propelled therapeutic motors and, more importantly, the combinations of these capacities, are systematically reviewed.

Chapter 10 highlights the recent developments in the employment of JPs as novel building blocks for active multifunctional materials. The stabilization efficiency of the JPs can be combined with the ability of interface structuring/functionalizing at the active interfaces, while JPs can be utilized as advanced building blocks to construct materials with controlled chemical and topographical heterogeneity as well as wettability and adhesion at the surface. The applicability and rational design of multifunctional materials based on the JPs is reviewed in this chapter, closing with the future perspectives and vision.

We are pleased to present the collection of reviews by established leaders and emerging researchers. The synthesis of JPs is just emerging. There is still lack of a universal and facile route to large-scale synthesis of JNPs at low cost, which restrains many applications at the device scale. The theoretical study of self-assembly of JPs has brought important insights but the experimental control over JP self-assembly remains challenging, especially when hierarchical self-assembly exists. We believe that future development of the synthesis methods and finely control over self-assembly of JPs will lead to applications in new areas.

The editors thank all the authors who have contributed to this book, and the help provided by Beibei Jiang at Georgia Institute of Technology during the editing process of the book. We also thank Mary Simpson of World Scientific Publishing for her remarkable patience and ensuring a finished product that we are proud of.

About the Editors

Dr. Bo Li is currently a postdoc with Prof. Charles M. Schroeder in Department of Chemical & Biomolecular Engineering at University of Illinois at Urbana-Champaign. He received both his B.S. and Master's degrees in Materials Science and Engineering from Xi'an Jiaotong University in 2008 and 2010, respectively. He received his Ph.D. in Materials Science and Engineering from Georgia Institute of Technology in 2015 with Prof. Zhiqun Lin. His research interests include self-assembly of nanomaterials, nanomaterial synthesis using organic templates (e.g., DNA, self-assembled block-copolymer micelles, unimolecular star-like block-copolymers) or by conventional wet-chemistry strategies, theoretical modeling, and numerical simulation. He is particularly interested in realization of artificial neuron and artificial brain structure using "bottom-up" self-assembly.

Dr. Zhiqun Lin is currently a Professor of Materials Science and Engineering at Georgia Institute of Technology. He received the B.S. degree in Materials Chemistry from Xiamen University, Fujian, China in 1995, Master's degree in Macromolecular Science from Fudan University, Shanghai, China in 1998, and the Ph.D. degree in Polymer Science and Engineering from University of Massachusetts at Amherst in 2002. He did his postdoctoral research at University of Illinois at Urbana-Champaign. He joined the Department of Materials Science and Engineering at Iowa State University as an Assistant Professor in 2004 and was promoted to Associate Professor in 2010. He moved to Georgia Institute of Technology in 2011, and became a Professor in 2014. His research interests include polymer-based nanocomposites, block copolymers, polymer blends, conjugated polymers, quantum dots (rods, tetrapods, and wires), functional nanocrystals (metallic, magnetic, semiconducting, ferroelectric, multiferroic, upconversion, and thermoelectric) of different architectures (plain, core/shell, hollow, and Janus), solar cells (perovskite solar cells, organic–inorganic hybrid solar cells, and dye-sensitized solar cells), lithium-ion batteries, hydrogen generation, hierarchically structured and assembled materials, and surface and interfacial properties. He has published more than 190 peer-reviewed journal articles (with an h-index of 55), 10 book chapters, and 2 books. Currently, he serves as an Associate Editor for *Journal of Materials Chemistry A*, and an editorial advisory board member for *Nanoscale*. He is a recipient of Frank J. Padden Jr. Award in Polymer Physics from American Physical Society, an NSF Career Award, a 3M Non-Tenured Faculty Award, and an invited participant at the National Academy of Engineering's 2010 US Frontiers of Engineering Symposium. He became a Fellow of Royal Society of Chemistry in 2014 and a Japan Society for Promotion of Science (JSPS) Fellow in 2015. More information on his research can be found at http://nanofm.mse.gatech.edu/.

Contents

Chapter 1

Design, Synthesis, and Application of Anisotropic Janus Particles*

Chengliang Zhang†, *Fuxin Liang, and Zhenzhong Yang*‡

State Key Laboratory of Polymer Physics and Chemistry,
Institute of Chemistry, Chinese Academy of Sciences,
Beijing 100190, China
†zhangcl@iccas.ac.cn
‡yangzz@iccas.ac.cn

Abstract

In this chapter, we comprehensively introduce the recent developments on anisotropic Janus particles, including hollow structure, non-spherical, sheets/disks, rods/cylinders, ring, etc. Also, we systematically compare different methodologies and present some emerging application examples. At last, we conclude some issues and trending for next research as well.

Keywords: Janus particles, anisotropic, composite, microstructure.

---

*The authors thank NSF of China, Chinese Academy of Sciences and The Ministry of Science and Technology for financial support.

1.1. Introduction

The physical structure (shape and size) and the distribution of chemical composition are two essential defining properties of Janus particles. In last decade, many efforts have been devoted to the precise control of the structure and composition of Janus particles for achieving excellent properties. Janus particles with various shapes and different compositions have been prepared, some of which have demonstrated important and interesting applications. As shown in Fig. 1, the representative shapes of Janus particles include spherical, non-spherical, sheets/disks, rods/cylinders, etc. The particle size varies from micro to nanoscale. The composition contains polymer, inorganic, metal or hybrid materials.

Inspired by the initial research of French scientists [1,2] and encouraged by Gennes' lecture [3], researchers chose all kinds of spheres as research objects, for example, monodispersed polystyrene (PS) spheres and silica spheres. So, at the first research stage, the Janus particles displayed only chemical asymmetry, not structural asymmetry. Compared with

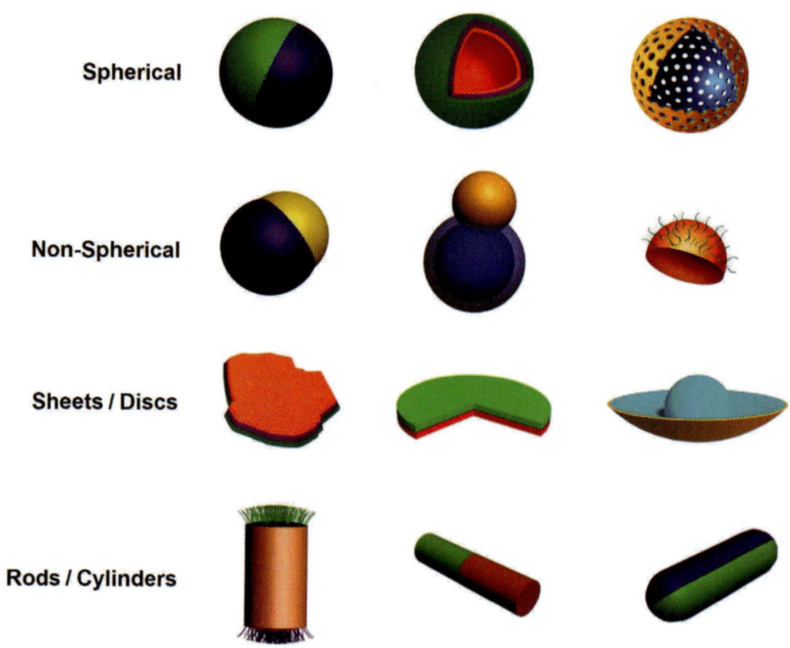

Figure 1. Schemes of Janus particles with various shapes.

spheres, anisotropic particles have more interesting and amazing properties due to their higher length or diameter or aspect ratio and concomitant structural and chemical distribution. In this chapter, we will focus on Janus particles with anisotropic shapes, especially non-spherical, sheets/disks, rods/cylinders, and other interesting morphologies, and review the synthesis methods and emerging applications of anisotropic Janus particles. Finally, we will present a summary and emerging trends as well.

1.2. Janus Hollow Spheres

Spherical Janus particle is the most extensively investigated of this type of particles. The most popular research idea is to compartmentalize two compositions on the exterior surface of the template particles by using a protective interface to separate two different areas, followed by modification of the two half surfaces, respectively. The strategy is versatile. Therefore the compositions of two hemispheres are tunable, for example, polymer/polymer [4], polymer/inorganic [5], polymer/metal (metallic compound) [6], polymer/protein [7], inorganic/organic ligand complex [8] and silica/metal [9]; furthermore, the diameter of Janus spheres can be changed from the nanometer to the micrometer scale by choosing the template particles with different sizes.

Janus hollow spheres are a special form of spherical Janus particles, and their shell is asymmetric with two different compositions compartmentalized onto interior and exterior surfaces although their outermost surface is centrosymmetric. In 2011, we reported a facile method of large-scale preparation of inorganic Janus hollow spheres with hydrophilic outside surface and hydrophobic inside surface by a self-assembled sol–gel process [10]. As shown in Fig. 2(a), the emulsion is first formed with three different silanes in oil phase. Then, the sol–gel process will be triggered by changing the pH value and temperature. Along with the reaction, the amine- and phenyl-terminated sols can self-assemble at the interface, facing the aqueous phase, and the oil phase, respectively. After the sol–gel process is completed, the oil core is removed by dissolution. Thus the Janus hollow spheres with holes on the shell can be achieved due to osmotic pressure, which is conductive to mass transport. Although the Janus hollow spheres are overall centrosymmetric, the shell is asymmetric with two different compositions compartmentalized onto interior and

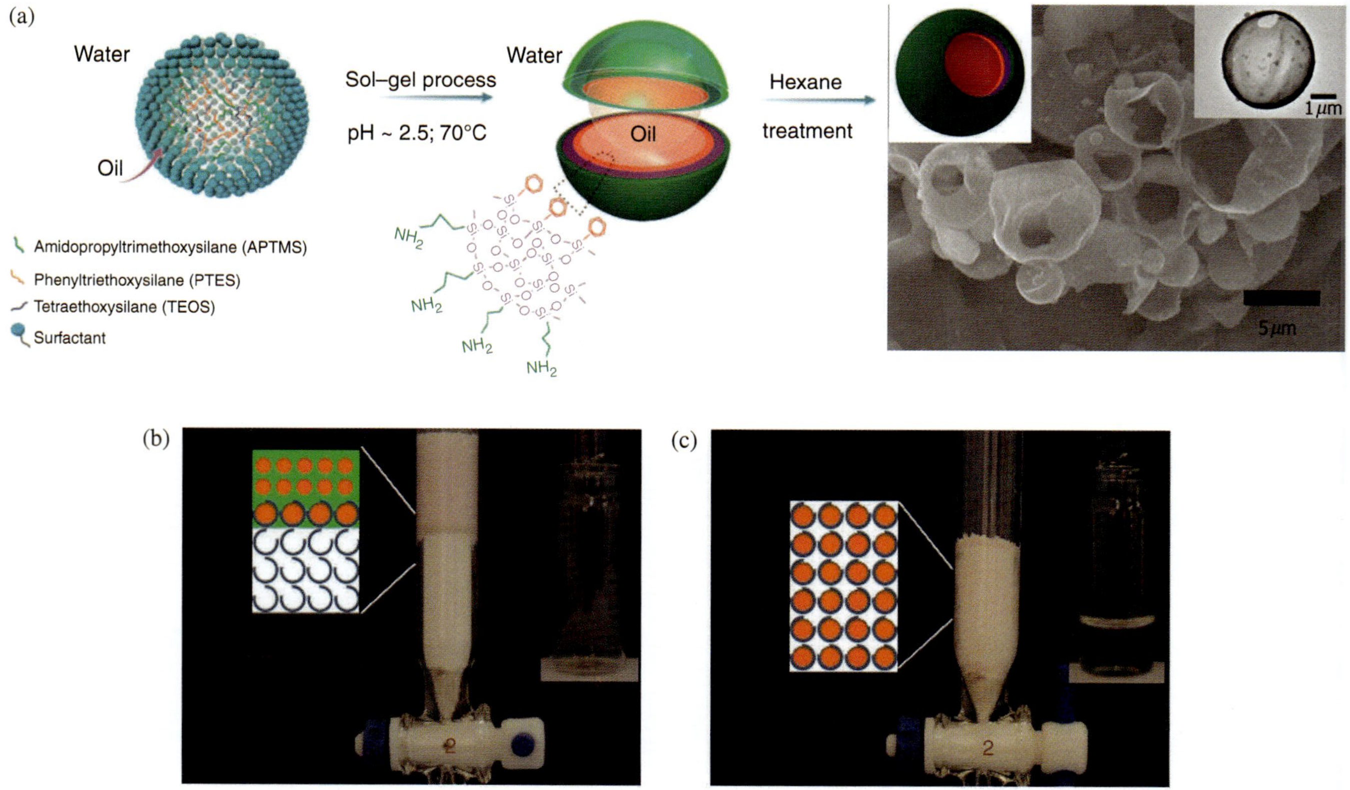

Figure 2. (a) SEM and inset TEM images of inorganic Janus hollow spheres. (b) The toluene–water emulsion was added to the chromatography stacked with Janus hollow spheres. (c) Water eluted while toluene was absorbed. Reprinted with permission from Ref. [10]. Copyright (2011) Royal Society of Chemistry.

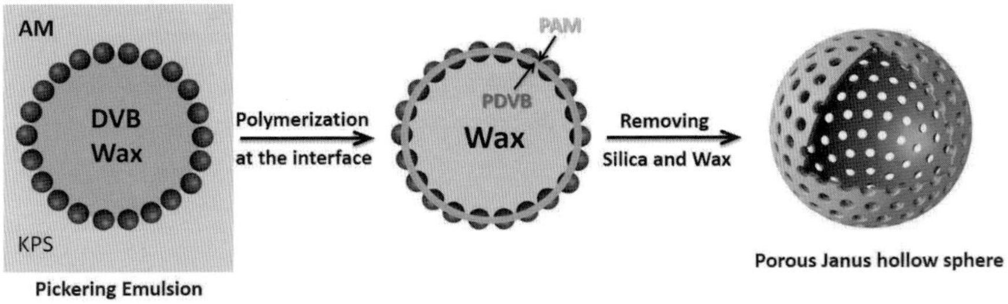

Figure 3. Schematic synthesis of porous Janus hollow sphere. Reprinted with permission from Ref. [11]. Copyright (2012) Elsevier.

exterior surfaces. Also, we present an application to demonstrate the properties of Janus hollow spheres. The as-prepared Janus hollow spheres are stacked into a column for separation. When the toluene–water emulsion passes through the column, toluene is absorbed because the hydrophobic cavity is favorable to loading toluene inside, as clear water is eluted (Figs. 2(b) and 2(c)).

Pickering emulsion is a classic and effective strategy for preparing Janus particles using 3D interface as the protecting layer. We observed that the particles also can be used as templates to create uniform pores. So we combined Pickering emulsion and interfacial polymerization to prepare porous Janus hollow spheres [11]. As shown in Fig. 3, SiO_2 particles, wax, oil-soluble monomer and water-soluble monomer are formed through Pickering emulsion. After polymerization at the interface, the amphiphilic copolymer film is formed at the gap between the particles. After removing SiO_2 particles and wax, the porous Janus hollow spheres will be formed. The pore size and the shell thickness can be modified by changing the monomer content. In comparison with other porous materials, the advanced property of such Janus particles is to preferentially and easily capture lipophilic molecules and even particles as well inside the cavity.

Further, for preparing a robust shell with hard templates (SiO_2 particles) free, we proposed another strategy for preparing porous Janus hollow spheres based on the sol–gel process at the emulsion interface and polymer grafting onto the interior surface (Fig. 4) [12]. The binary surfactants instead of SiO_2 particles are used to make the transverse channels

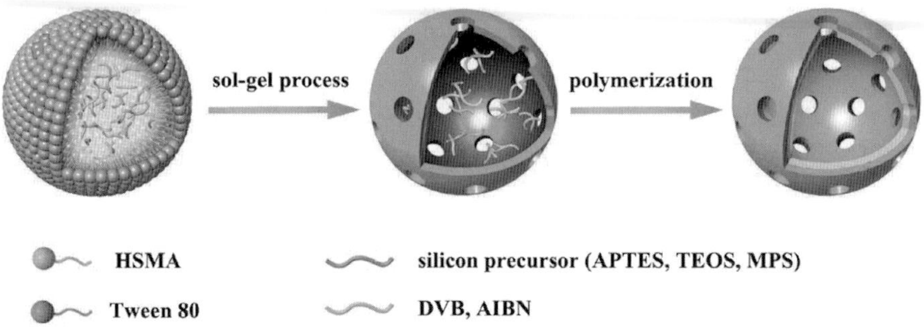

Figure 4. Schematic synthesis of the Janus cage with polymer–inorganic bilayers. Reprinted with permission from Ref. [12]. Copyright (2013) American Chemical Society.

across the shell due to phase separation. The resulting Janus hollow spheres are further functionalized by growing responsive polymers onto the interior surface for easy loading of the desired materials. In addition, magnetic nanoparticles can be integrated into the composite shell to make the Janus spheres work under a magnetic field.

For showing the special property of such porous Janus hollow spheres, we also presented two interesting examples in controlled loading/release. In the first example [13], we first prepared the porous Janus hollow spheres with the catalyst TiO_2 nanoparticles encapsulated inside the cavity (Fig. 5(a)). Then, such Janus hollow spheres were able to concentrate organic compounds from their aqueous surroundings due to their hydrophobic interior surface and transverse channels across the shell. Upon UV irradiation, the organic compounds can be easily decomposed. The method is expected to be used for other kinds of Janus microcontainers with different shell composition and catalytic species, for example enzymes.

In the second example [14], we fabricated bilayered porous Janus hollow spheres with a pH-responsive exterior layer and a biodegradable interior layer. Then, doxorubicin (DOX) was loaded into the cavity as a model drug. The results of *in vitro* release showed a faster rate at more acidic pH values between 7.4 and 5.5 due to the pH-responsive shell (Fig. 5(b)). Also, it is noted that as-prepared Janus hollow spheres can be favorably taken up by cells under tumor acidic conditions, i.e., pH 6.5, which makes it a suited candidate for targeted release.

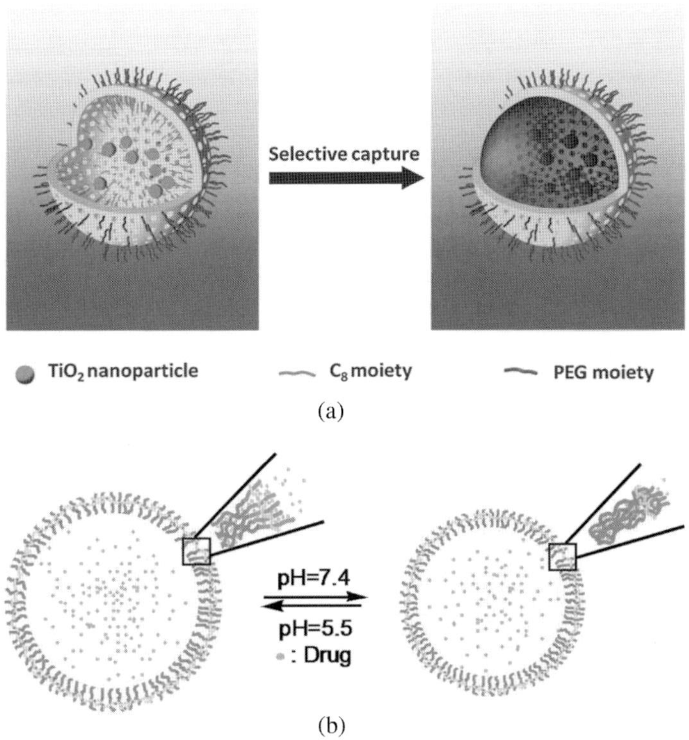

Figure 5. (a) Schematic of Janus microreactor. (b) Schematic of the drug release behavior of pH-responsive polymeric Janus hollow spheres at pH 7.4 and 5.5. Reprinted with permission from Refs. [13,14]. Copyright (2013) and (2015) Royal Society of Chemistry.

1.3. Non-spherical Janus Particles

The spherical Janus particles, solid spheres or hollow spheres, have an asymmetric chemical composition while the structure is overall centrosymmetric. In comparison with general spherical particles, Janus particles with asymmetric composition and structure also have very good properties and find more promising applications, such as the recently published example of self-assembly [15] and self-propulsion motor [16,17]. Actually, the classic emulsion polymerization is a the very effective method to control the particle structure, thereby creating various interesting shapes [18,19]. Figure 6 shows the morphological evolution of non-spherical Janus particles starting from spherical particles based on the emulsion method, and the typical shapes include acorn-like, snowman-like, and dumbbell-like.

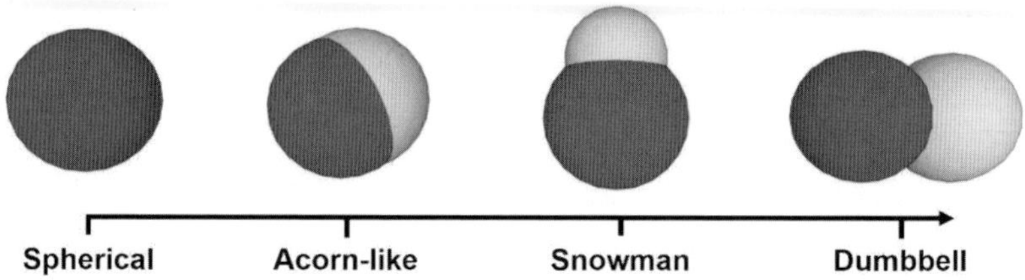

Figure 6. The morphological evolution of non-spherical Janus particles starting from spherical particle.

Phase separation of multiple component polymer materials is the key to controlling the microstructure and properties of materials, which can be induced by either polymerization or solvent evaporation. By controlling some critical parameters — for example, composition of monomers or polymers, cross-linking degree, and surfactant concentration — the resulting shape of the particles can be modified accordingly.

In the early-stage research, Okubo *et al.* [20–22] carried out a series of important experiments to show that the morphology of heterogeneous polymer composite particles could be changed from raspberry-like to snowman-like by seeded emulsion polymerization, depending upon the type and weight ratio of polymers. In 1990, Sheu *et al.* [18,19] reached a milestone in this field. They chose the polystyrene (PS) particles as a model, and systematically investigated the process of seeded emulsion polymerization. The non-spherical particles were formed by separation of the second-stage monomer from the cross-linked seed network. The resulting morphology can be modified by changing the degree of cross-linking. Also, they concluded that this structure follows a nucleation-and-growth mechanism of forming anisotropic structure by analysis of the thermodynamics of phase separation.

Reculusa *et al.* [23] developed this strategy and chose inorganic silica particles as a seed to prepare inorganic–polymer composite particles by seeded emulsion polymerization. The mechanism of forming anisotropic morphology is nucleation-on-surface, different from the process of phase separation of polymer/polymer, due to the hard silica seed particles. Actually, it is not easy to control such process of nucleation-on-surface, and more nodules would be formed on the silica surface as well, resulting

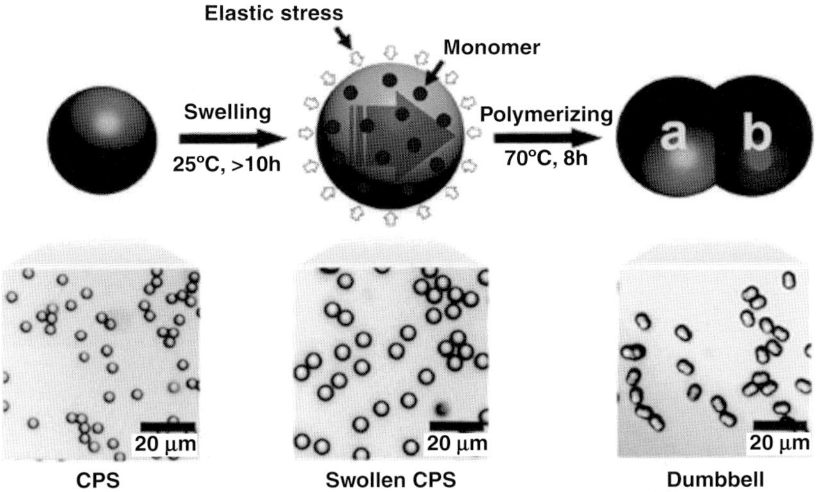

Figure 7. Schematic synthesis of anisotropic non-spherical dumbbell particles by seeded polymerization. Reprinted with permission from Ref. [25]. Copyright (2006) American Chemical Society.

in raspberry-like shape [24]. Also, many free polymer particles were formed in the continuous phase except on the surface.

Although the resulting particles prepared by above methods have anisotropic structure, e.g., snowman-like shape, they cannot be treated as Janus particles in the strict sense because their chemical compositions are not separated specifically. Further, Kim *et al.* [25] extended the seeded emulsion polymerization to synthesize polymeric Janus particles having chemical and structure asymmetry simultaneously. As shown in Fig. 7, the particles retained their spherical shape until polymerization. The phase separation caused by elastic stress gave rise to formation of a new bulb, resulting in dumbbell-like particles. By employing surface treatment, the amphiphilic Janus particles could be obtained, which may be used as a colloid surfactant to stabilize the emulsion [26]. Furthermore, by further controlling the thermodynamic and kinetic parameters in the process of phase separation, e.g., cross-linking density or reaction time, some interesting shapes would be formed, including rods and triangles [27].

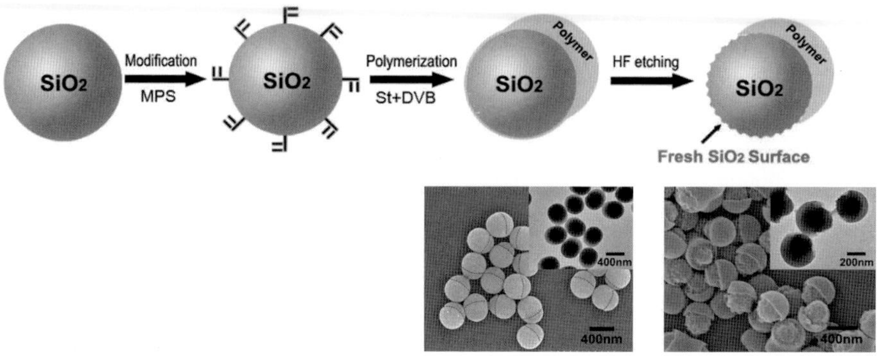

Figure 8. Schematic large-scale synthesis of submicrometer-sized Janus particles by seeded emulsion polymerization. Reprinted with permission from Ref. [28]. Copyright (2010) Royal Society of Chemistry.

Besides precisely controlling the composition and morphology to ensure the Janus structure, the yield is also an important issue for real application. Therefore, we developed two large-scale synthetic methods to prepare the submicrometer-sized Janus particles based on seeded emulsion polymerization. The first method [28] is shown in Fig. 8. First, the silica seed particles are introduced to the vinyl group on the surface by surface treatment with a special silane coupling agent. After polymerization of monomers on the surface, the anisotropic polymer/silica composite particles are formed due to the dewetting effect, which can also be treated as the phase separation of polymer on the inorganic particle surface. In addition, we investigated the evolution of morphology along with the surfactant and cross-linking degree. By using hydrofluoric acid (HF) to chemically etch the thin polymer part for generating a fresh silica surface, the real polymer/silica Janus particles are produced. The ratio of polymer to silica can be modified by changing concentrations of HF. Furthermore, the family of Janus particles can be enriched by further modification and growing other materials.

In the second method [29] (Fig. 9(a)), we choose the cross-linked polyacrylonitrile (PAN) hollow polymer particles as starting seeds, and then the monomer mixture of styrene/divinylbenzene (St/DVB) is mixed with the seed emulsion accompanied by polymerization at high temperature. Since

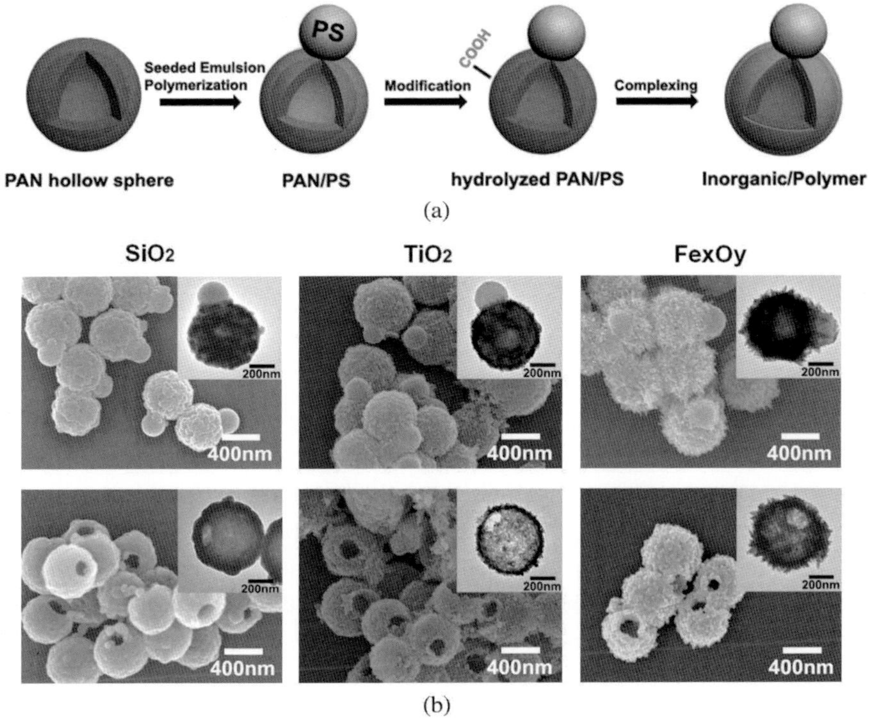

Figure 9. (a) Schematic large-scale synthesis of Janus particles by using polymer hollow spheres as seeds. (b) SEM and insert TEM images of various Janus hollow particles: SiO_2/PS; TiO_2/PS; Fe_xO_y/PS and the derived inorganic hollow spheres with a single hole after removing polymer by calcination. Reprinted with permission from Ref. [29]. Copyright (2010) American Chemical Society.

PAN and PS are typically immiscible, PS does separate from the PAN shell, especially such process can be driven by elastic-retractile force of the polymer network during cross-linking. Consequently, PAN/PS Janus polymer particles are formed. In addition, the important functional group, e.g., carboxylic acid (–COOH), is derived by selective hydrolysis of PAN, which is conducive to a further favorable growth of other materials therein to form Janus composite particles, such as SiO_2/polymer, TiO_2/polymer, Fe_xO_y/polymer, etc. By removing the polymer by calcination, the inorganic hollow particles with a single hole can be obtained (Fig. 9(b)), which will be an emerging functional material. The resulting SiO_2/PS Janus particles are amphiphilic because the SiO_2 part is hydrophilic and the PS part is hydrophobic. Therefore, such amphiphilic Janus particles serve as a solid surfactant to stabilize the oil/water emulsion. Meanwhile, the Janus particles have a preferential orientation at the oil/water interface, i.e., SiO_2 part faces the external water phase while PS part faces the inner oil phase.

Besides compositions, the particle size is another problem, especially the fabrication of nanosized particles. Therefore, we extend the above strategy to develop a facile approach to prepare the anisotropic Janus nanoparticles [30]. As shown in Fig. 10, first, the patchy composite spheres with many bulges are formed on the surface by seeded emulsion polymerization. The bulge size can be converted from submicrometer to nanometer by changing the reaction parameters, such as weight ratio of monomer to seed or pH value. Since the bulges are *in situ* grown from the seed sphere with their root part tightly embedded onto the surface, the root does not undergo any change while the exposed part, i.e., the bulges, is modified by changes in reaction parameters. Then, Janus particles can be obtained after removing the polymer seed spheres using a solvent.

Actually, based on above the strategy of protection-modification at the interface, we also have presented an effective method previously using the classic Pickering emulsion. The basic flow chart is shown in [31] (Fig. 11(a)): (1) modified SiO_2 particles are used to stabilize paraffin wax/water emulsion, namely Pickering emulsion; (2) the exposed part is etched by HF. The resulting particle shape can be varied along with the etching time, from spherical to mushroom-like. Furthermore, polymer can be grown on the corona part (i.e., unetched part) by emulsion polymerization. By increasing the amount of the monomer, the polymer/SiO_2 Janus

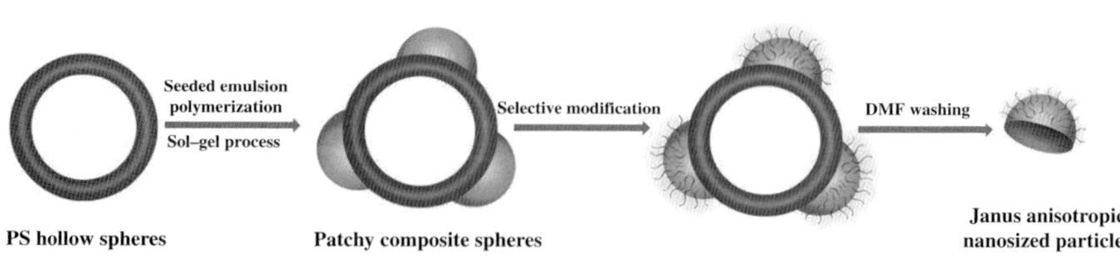

Figure 10. Schematic synthesis of Janus anisotropic nanoparticles. Reprinted with permission from Ref. [30]. Copyright (2013) American Chemical Society.

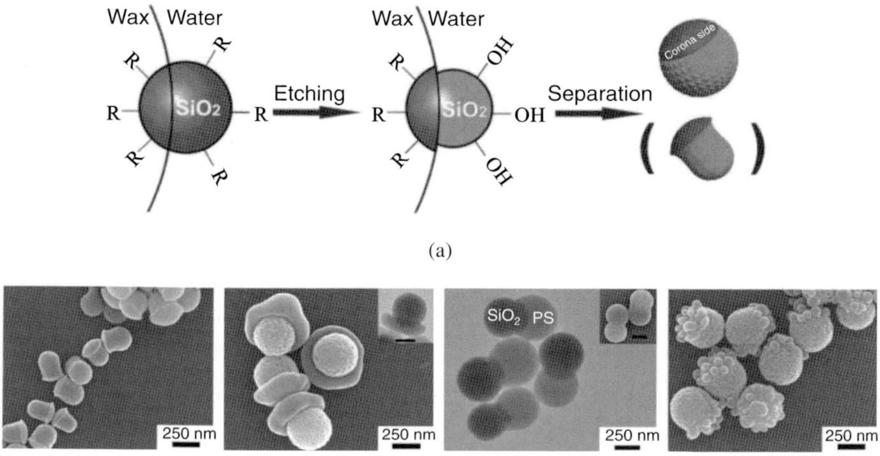

(a)

(b)

Figure 11. (a) Schematic synthesis of Janus particles by etching. (b) The derived typical Janus particles (from left to right): mushroom-like; with a cap grown on the corona side; dimer-like; flower-like. Reprinted with permission from Ref. [31]. Copyright (2009) Royal Society of Chemistry.

particles with various shapes would be obtained, from dimer-like to flower-like (Fig. 11(b)).

Except for the strategy of forming anisotropic shapes induced by phase separation, we also developed another method of making snowman-like Janus structure by emulsion swelling core–shell composite particles (Fig. 12) [32]. The as-prepared Janus particles can be further extended to other compositions, with improved physical properties and tuned Janus balance (the ratio of polymer/inorganic parts) by swelling polymerization. The resultant anisotropic Janus particles with a different ratio of hydrophilic/hydrophobic parts can be used as solid surfactants to stabilize water/oil or oil/water emulsion, respectively.

Lee *et al.* reported a method of preparing pH-responsive Janus particles based on phase separation in seeded emulsion polymerization [33]. The shapes of Janus particles can be tailored by changing the pH value of the solution. Also, the amphiphilic property will be changed accordingly due to the ratio change of hydrophilic/hydrophobic sides, namely Janus balance, which will induce the phase inversion of emulsion (Fig. 13).

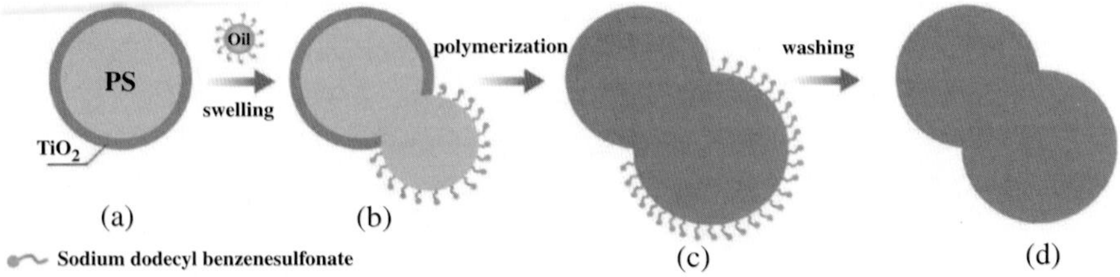

Figure 12. Schematic synthesis of Janus particles by emulsion swelling core–shell composite particles. Reprinted with permission from Ref. [32]. Copyright (2012) American Chemical Society.

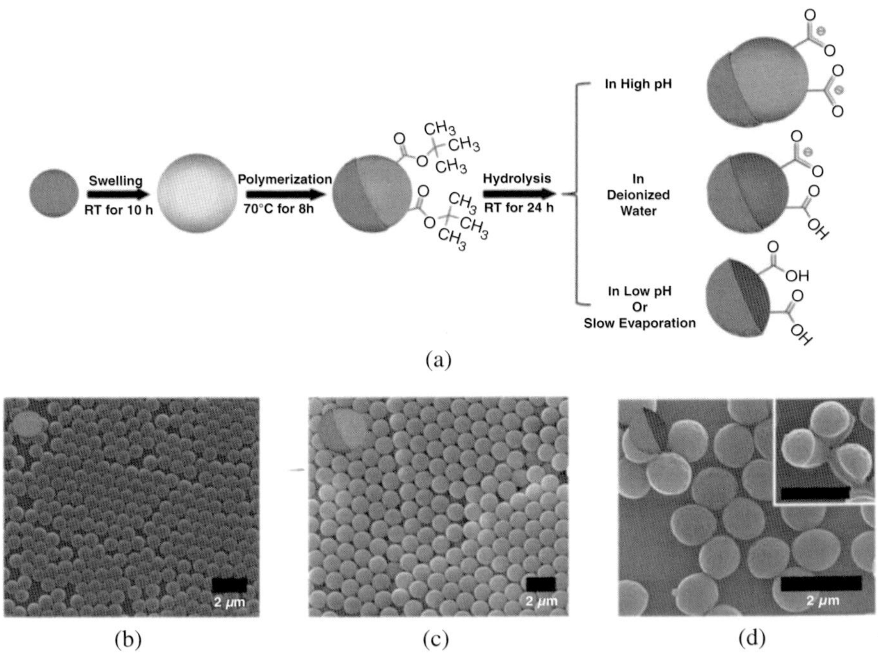

Figure 13. (a) Schematic illustration for the synthesis of amphiphilic Janus particles by seeded emulsion polymerization followed by acid hydrolysis, and SEM images of (b) PS seed particles, (c) composite particles after polymerization, (d) Janus particles after hydrolysis. Reprinted with permission from Ref. [33]. Copyright (2014) American Chemical Society.

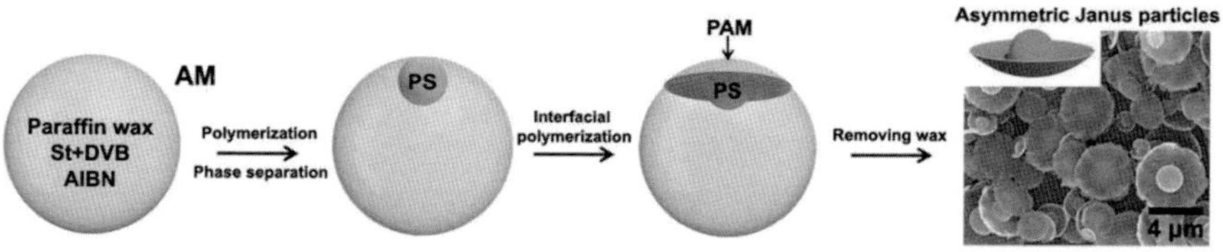

Figure 14. Schematic synthesis of UFO-like Janus particles. Reprinted with permission from Ref. [34]. Copyright (2011) American Chemical Society.

By controlling the interfacial tensions at the triple phase contact line, the spreading coefficients of polymer can be modified. Thus, the UFO-like Janus particles are prepared [34] (Fig. 14). The polymer particles are formed *in situ* in the internal oil phase by phase separation during emulsion polymerization, and migrate toward the interface and anchor there due to Pickering effect. By selective growth of the functional materials onto the desired sides, the composition and microstructure of the Janus particles can be controlled. The representative PS/PAM Janus particles are amphiphilic and can be used as solid surfactants to stabilize emulsions, which preferentially orientate at the interface. The method is general and scalable to derive a huge family of asymmetric Janus composite particles, which are important for exploring their practical applications.

1.4. Janus Sheets/Disks

Janus sheets/disks deserve more attention due to their highly anisotropic shape as well as chemistry. According to the axis of compartmentalized compositions, the Janus sheets/disks have two representative shapes as shown in Fig. 15. Compared to the spherical shape, the rotation of Janus sheets at the interface is greatly restricted, thereby making the emulsion more stable as the planar configuration facilitates their favorable orientation [35].

In 1980s, the first method of preparing inorganic Janus sheets was proposed by crushing the modified glass hollow spheres [36]. Although the size of as-prepared Janus sheets was big, generally on the order of micrometers, the preparation strategy involved an important decision, i.e.,

Figure 15. Schemes of Janus sheets/disks.

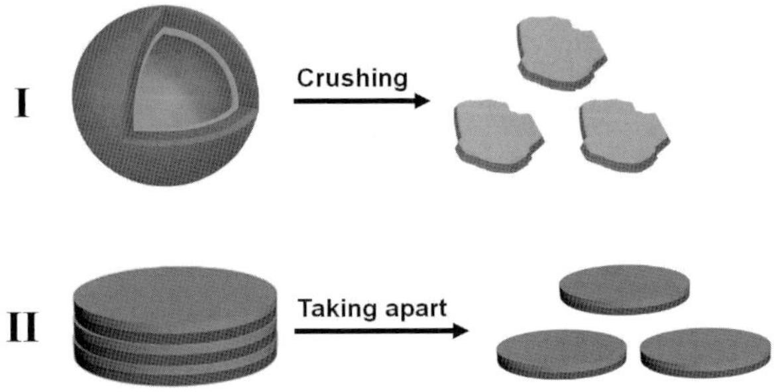

Figure 16. Schematic synthesis of Janus sheets/disks.

crushing the modified spheres (Fig. 16(a)), which had a crucial effect on further research.

In 2011, we developed a facile approach for large-scale production of inorganic Janus sheets by applying the crushing strategy [37]. First, silica Janus hollow spheres with hydrophilic surface on the outside and hydrophobic surface on the inside were fabricated by the self-assembled sol–gel process at an emulsion interface. Then, after crushing the as-prepared Janus hollow spheres, amphiphilic Janus sheets were achieved. The Janus sheets were used as solid surfactants to emulsify immiscible liquids to form the stabilized "dry" droplet because of their distinctive Janus property. If the as-formed emulsion flowed through a porous mica membrane, water eluted, whereas the toluene "dry" droplets were left and preserved in air rather than allowed to coalesce or collapse. Upon stirring, the "dry" oil droplets coalesced to release the oil, so that the Janus sheets can be collected for reuse (Fig. 17). Besides inorganic materials, the compositions of Janus sheets also can be extended to other materials, for example, polymer by favorable grafting onto the desired side [38]. Also, the functional polymer, e.g., pH-sensitive polymer PDEAEMA and thermal-responsive polymer PNIPAM, can be grafted on the two sides by

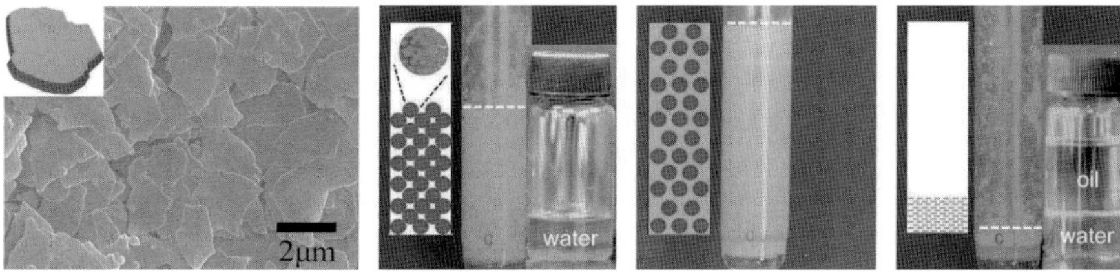

Figure 17. SEM image of inorganic Janus sheets and oil/water separation demo using as-prepared Janus sheets. Reprinted with permission from Ref. [37]. Copyright (2011) Wiley-VCH Verlag GmbH & Co. KGaA.

controlled polymerization, atom-transfer radical-polymerization (ATRP) and reversible addition-fragmentation chain transfer (RAFT). The as-prepared dual-responsive Janus polymer/inorganic sheets can be used to stabilize the oil/water emulsion. It is noted that such emulsion will be easily demulsified by adjusting pH or temperature. Further, the paramagnetic Fe_3O_4 nanoparticles were favorably grown onto one side to prevent the Janus sheets and emulsion from being manipulated by the magnetic field [39]. In the same way, the wettability of such dual-responsive sheets can be changed by modifying the temperature [40]. If a photo-responsive spiropyran-containing polymer (PSPMA) is grafted onto one side of silica sheets, the light-triggered Janus composite sheets can be achieved because the wettability on the PSPMA side will be changed by UV irradiation [41]. By selective treatment of the imidazolin-terminated side of Janus sheets, the ionic liquid functionalized Janus sheets are synthesized. The stability of emulsions stabilized by as-prepared Janus sheets are reversibly triggered by anion exchange [42]. Additionally, the composition of Janus sheets also can be extended to polymeric ionic liquid (PIL) by polymerization of ionic liquid monomers onto one side of silica sheets. Then, the Keggin-structured polyoxometalates (POMs), e.g., $H_3PW_{12}O_{40}$, can be introduced onto the PIL side based on anion exchange, thereby endowing the Janus sheets with the character of excellent catalytic degradation for dye [43].

Besides compositions, the size of Janus sheets also can be modified by choosing special monomers. Recently, we presented a method of large-scale preparation of Janus sheets with a molecular scale thickness

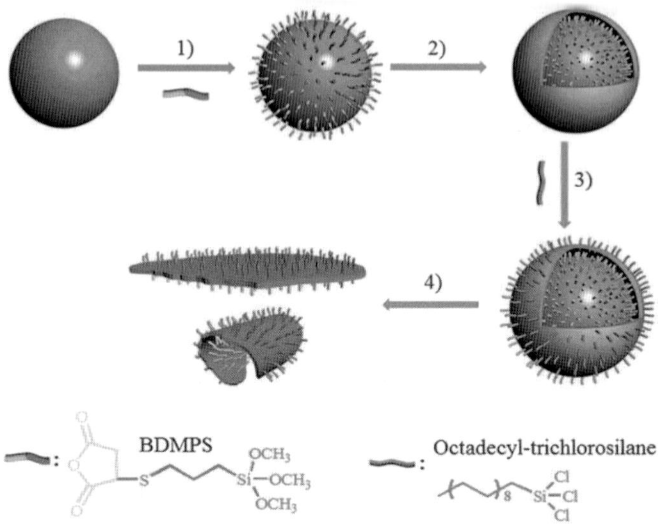

Figure 18. Schematic synthesis of flexible Janus sheets: (1) BDMPS self-assemble to monolayer on the template surface; (2) the inorganic shell is formed by sol–gel process; (3) use of octadecyl-trichlorosilane to modify the shell exterior surface to make a hydrophobic side; (4) Janus sheets are obtained after removing templates. Reprinted with permission from Ref. [44]. Copyright (2015) Royal Society of Chemistry.

(3.5 nm) by a sol–gel process of a self-assembled monolayer of an amphiphilic silane (e.g., 3-butyldianhydride mercaptopropyl trimethoxysilane (BDMPS)) at the oil/water interface (Fig. 18) [44]. The as-prepared Janus sheets are flexible due to their very thin thickness, which makes it feasible to wrap objects just with a single Janus sheet. If different silanes are chosen to selectively modify the exterior surface, the family of Janus sheets will be extended, for example, to producing amine/acid Janus sheets [45]. Also, we use reactive polymer surfactant instead of silane to self-assemble and cross-link at the oil/water interface, thus fabricating polymeric Janus sheets with pH- or magnetic-responsive character [46]. Further, we proposed a similar facile method to prepare pH-responsive Janus sheets by polymerization at the oil–water interface. The stability of emulsions stabilized by such Janus sheets can be easily achieved by maintaining the pH value [47].

The other important strategy of preparing Janus sheets/disks is to dismantle the self-assembled copolymer bulk (Fig. 16(b)), in three basic steps: (1) forming bulk by self-assembly of copolymer; (2) selectively cross-linking one domain; (3) dismantling the bulk by dissolution and sonication. Müller *et al.* did a pioneering work on this. They first reported a method of synthesizing disk-like Janus polymer particles with an inner cross-linked layer and two different outer sides, which can be used to stabilize oil/water emulsion [48]. By using a different copolymer, the composition of Janus disks will be easily modified, e.g., achieving organic/inorganic Janus objects [49].

For Janus disks, the key issue is how to prepare them with controlled uniform shape and size, especially at the nanoscale. Recently, we found that the diblock copolymer is an available candidate if the disk-stacked structure can be disassembled by some appropriate solvents [50]. As shown in Fig. 19, first, the PS-*b*-P4VP diblock copolymer is chosen due to its capacity of forming lamellar structures in bulk. Then, the disk-stacked particles are formed by emulsion droplet-confined assembly of PS-*b*-P4VP solution. The as-formed disk-stacked particles can be disassembled to P4VP/PS/P4VP sandwiched nanodisks by using ethanol to selectively dissolve P4VP domain while the PS domains are less influenced. After cross-linking both the P4VP layers with 1,5-diiodopentane, the nanodisks are antisolvent and the profile is kept. Further, if chloroform is used to dissolve the PS domain, the sandwiched nanodisk will be disassembled into two PS/P4VP Janus nanodisks, which provides a new building block to construct the complex superstructure.

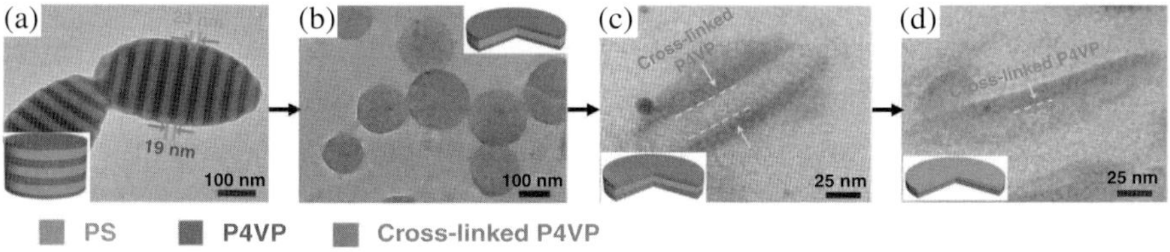

Figure 19. Schematic synthesis of the Janus nanodisks: (a) disk-stacked particles formed by self-assembly of diblock copolymer; (b) obtaining sandwiched nanodisks by selectively disassembling the P4VP layer; (c) cross-linking the P4VP layers; (d) obtaining Janus nanodisks by selectively disassembling PS intermediate layer. Reprinted with permission from Ref. [50]. Copyright (2014) Wiley-VCH Verlag GmbH & Co. KGaA.

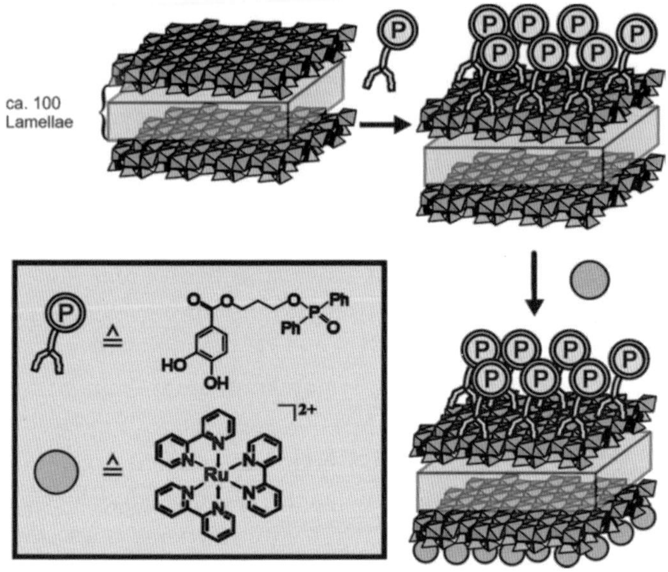

Figure 20. Schematic fabrication of amphiphilic Janus platelets by selective modification of natural kaolinite. Reprinted with permission from Ref. [51]. Copyright (2012) Wiley-VCH Verlag GmbH & Co. KGaA.

Besides polymers, the inorganic materials with plate-stacked structure, e.g., natural minerals, also can be chosen as target objects to fabricate Janus platelets by selective modification. Breu *et al.* reported a method of large-scale fabrication of amphiphilic Janus platelets by selective modification of natural mineral kaolinite (Fig. 20) [51]. Kaolinite is a dioctahedral-layered silicate with two opposite external surfaces, i.e., $Al_2(OH)_4$ octahedral surface and SiO_4 tetrahedral surface. If two surfaces are modified by catechol and $[Ru(bpy)_3]^{2+}$, respectively, the amphiphilic Janus platelets will be obtained, thereby providing an important potential solid emulsifier due to the cheap raw materials.

1.5. Janus Rods/Cylinders

In comparison to spherical shapes, the rods/cylinders are 1D in shape with a high length/diameter (L/D) aspect ratio. They serve as new building

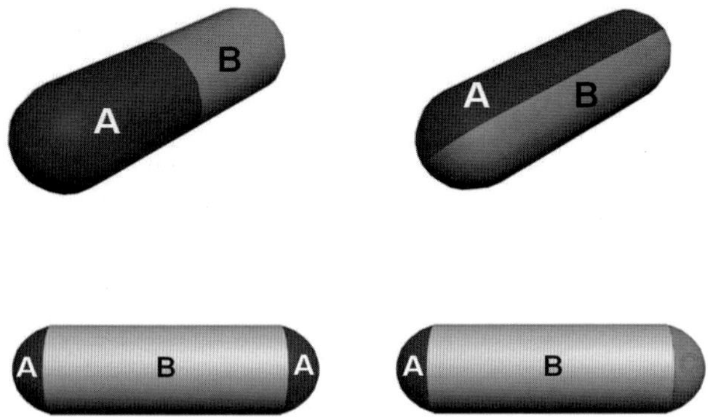

Figure 21. Schemes of Janus rods/cylinders.

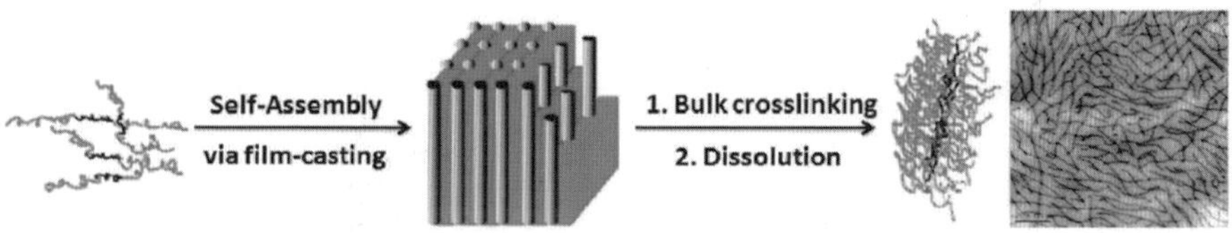

Figure 22. Schematic synthesis of Janus cylinders. Reprinted with permission from Ref. [52]. Copyright (2015) American Chemical Society.

blocks to construct novel and interesting superstructures. The representative shapes of Janus rods/cylinders are shown in Figs. 21 and 22. Müller *et al.* prepared Janus cylinders by self-assembly of block copolymer with the compartmentalized line parallel or perpendicular to the cylinder axis, respectively [52]. The as-prepared Janus cylinders can form hierarchical superstructures from fibers to fibrillar networks upon concentration of the solution [53]. Also, they studied the self-assembly of Janus cylinders at liquid/liquid interfaces. They found that the Janus cylinders can significantly enhance surface activity in comparison to homogeneous cylindrical copolymer particles although such Janus cylinders are slightly amphipathic [54]. Further, the Janus concept also can be extended to A–B–A and A–B–C structures except for the general A–B structure. For example, Manners *et al.* reported an ingenious method of preparing such Janus cylinders. The monodispersed cylindrical seed micelles are first created.

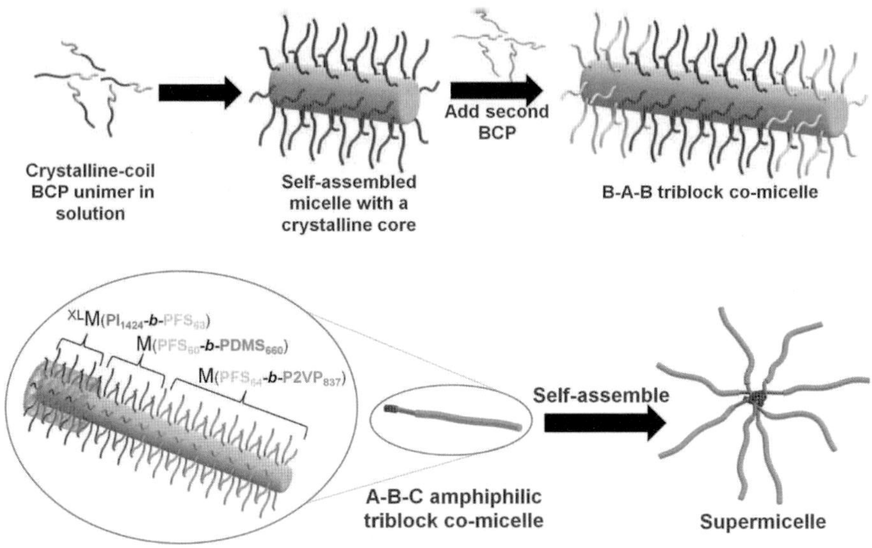

Figure 23. Schematic synthesis of A–B–A and A–B–C Janus cylinders. Reprinted with permission from Ref. [55]. Copyright (2012) American Association for the Advancement of Sciences.

Upon addition of other block copolymers, the new segments will grow at two ends of as-formed seed micelles to form A–B–A or A–B–C Janus cylinders. Further, the as-prepared A–B–C Janus cylinders can self-assemble into hierarchical star-shaped superstructure [55] (Fig. 23).

Although the self-assembly approach can be used to fabricate nanosized Janus cylinders, smaller amount of species in compositions, low yield, and polydispersion restrict further applications of the Janus materials because the self-assembly processes are determined by thermodynamics. Then, the strategy of topo-modification is developed to prepare the Janus cylinders with modifiable compositions and shapes due to the diversity of cylindrical templates. For example, an arrow head gold nanorods can be used as a template to generate Janus nanorods with A–B–A structure by coating with a bilayer of cetyltrimethyl ammonium bromide (CTAB) along the wall, and the two arrowhead edges are coated with thiol-terminated PS molecules [56]. In addition, silica rods can also be used as templates to prepare the Janus rods using a multistep process involving electric field

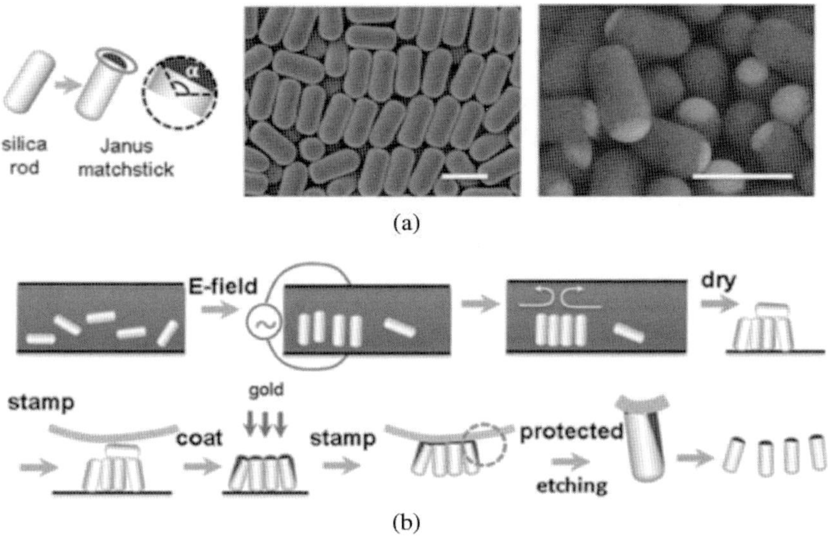

Figure 24. Schematic synthesis of silica Janus rods. Reprinted with permission from Ref. [57]. Copyright (2012) American Chemical Society.

alignment and crystallization, microcontact printing, and selective metallization (Fig. 24) [57]. The resulting silica–Au Janus rods can self-assemble into multipods (bi-, tri-, and tetra-pods). In the case of carbon nanotubes (CNTs) as templates, Janus CNTs are achieved by dissymmetric grafting with different polymers [58] or deposition of metals [59].

1.6. Others

1.6.1. Janus Rings

Rings are especially intriguing due to their promising performance in optical and electrical fields [60,61]. In 2010, we first reported Janus rings with two compositions, distributed onto interior and exterior surfaces, which are fabricated by combinational template synthesis and skiving microprocess method (Fig. 25) [62]. Polymer nanotubes and their arrays are polymerized from the pore surface via ATRP inside the porous anodic alumina oxide (AAO) membrane templates. By a selective partial

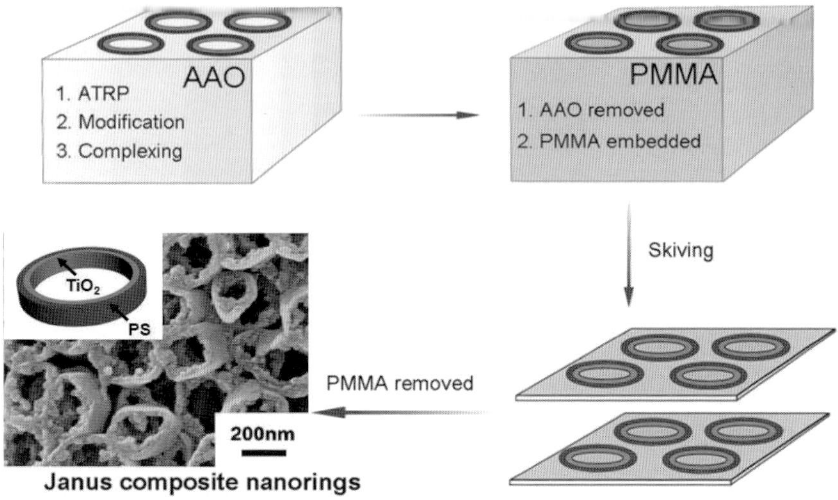

Figure 25. Schematic illustration of preparing TiO_2/PS Janus rings. Reprinted with permission from Ref. [62]. Copyright (2010) Elsevier.

modification of the interior surface of the polymer nanotubes followed by a favorable growth of other materials, double-shelled composite nanotubes and their arrays are synthesized. The corresponding Janus composite nanorings are obtained by skiving. The characteristic dimension, microstructure, and composition of these as-synthesized Janus composite nanorings are modifiable, such as TiO_2/PS, PANi/PS, etc.

1.6.2. Strawberry-like Janus Particles

The anisotropic structure can endow Janus particles with special properties. Recently, we presented a successful example of achieving a robust superhydrophobic coating by Janus particles (Fig. 26) [63]. First, the strawberry-like Janus particles are prepared by self-organized sol–gel process at a patchy interface. The resulting Janus particles have PS grafted on the rough side and imidazolin grafted on the flat side. Then, the as-prepared particles can self-organize into a layer on substrates. Nanoscale roughness on the hydrophobic PS side determines the superhydrophobic performance. In comparison to previous hydrophobic coating, it is observed that the imidazolin group on the flat side can make a robust

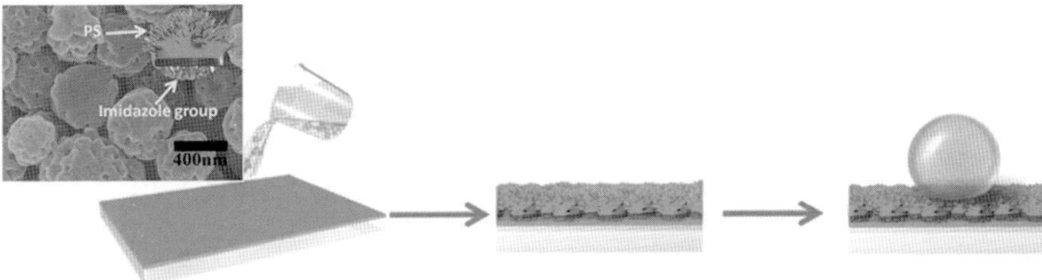

Figure 26. Schematic fabrication of the robust superhydrophobic coating by strawberry-like Janus particles. Reprinted with permission from Ref. [63]. Copyright (2015) Nature Publishing Group.

coating by covalent binding onto substrates via cations initiating the cross-linking of the intermediate epoxy resins. The resulting coating can tolerate organic solvents and high water flushing speeds. If the hydrophobic side is smooth, then the coating is highly adhesive to water. This procedure can be used to fabricate unique coatings on a diverse range of substrates with varied compositions and shapes.

1.6.3. Cone-like Janus Particles

The surfactant is like a "magical hand" and can be used to fabricate emulsion droplets of various shapes due to the presence of interfacial tension. For the binary surfactant system, more variable factors endow them with more potential to generate amazing shapes. Recently, we proposed a method for a large-scale synthesis of cone-like PS composite Janus particles by tuning the ratio of binary surfactants (Tween-80 and HSMA) through interfacial polymerization (Fig. 27) [64]. We found that the different ratios of Tween-80 to HSMA affect the outward convex tension at the triple-phase contact line, thus determining the formation of a patchy interface after phase separation. So the cone shape is achieved by squeezing the deformable PS particles at a certain ratio. The cone side can be selectively modified to achieve Janus structure while the bottom flat side is protected onto the paraffin sphere surface. Robust composite coatings are fabricated using the cone-like Janus particles. Also, the wettability of the coating can be adjusted from highly adhesive for water to

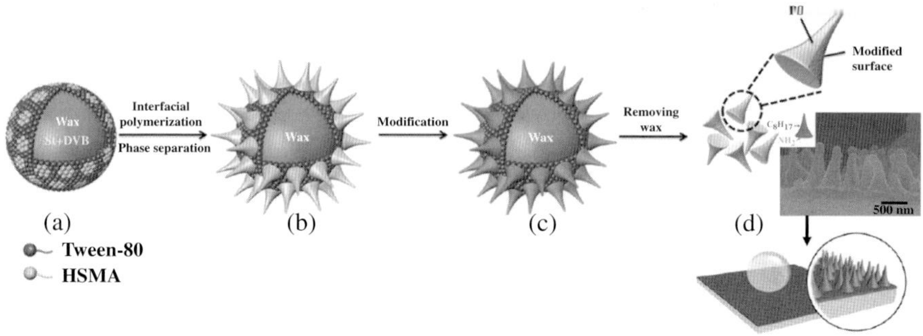

Figure 27. Schematic synthesis of cone-like Janus particles and the resulting hydrophobic coating. Reprinted with permission from Ref. [64]. Copyright (2015) American Chemical Society.

superhydrophobic by simply changing the roughness of the coating, which is achieved by using different Janus particles with varied size distribution.

1.7. Conclusion and Outlook

In this chapter, we summarized the recent research on anisotropic Janus particles, including hollow structure, non-spherical, sheets/disks, rods/cylinders, ring, etc. Also, we systematically introduced the methodology and some examples of their application, e.g., as solid surfactant to stabilize oil/water emulsion. None of methods or strategies can be applied globally, and each method is of limited scope. The main methods include emulsion polymerization, protect-modification upon 3D interface (pickering emulsion) and 2D interface, and self-assembly of copolymers, of which emulsion polymerization is the versatile method due to its capacity of forming various anisotropic structures and capability of a facile scale-up. For future research, we point to some important topics and trends as follows:

(1) It is still crucial important to precisely control the size and compositions of Janus particles, and the urgent need is to develop a method

to prepare the nanosized polymer Janus particles with uniform shape.

(2) Using biotechnology, the biomaterial Janus particles should be prepared and explored for application in medicine.

(3) The focus should be on a special application field, and promotion of deeper research, e.g., solid surfactant, nanomotor, separation materials.

References

[1] Casagrande, C.; Veyssié, M. *C. R. Acad. Sci. (Paris) II*, **306**, 1423 (**1988**).

[2] Casagrande, C.; Fabre, P.; Raphaël, E.; Veyssié, M. *Europhys. Lett.*, **9**, 251 (**1989**).

[3] de Gennes, P. G. *Science*, **256**, 495 (**1992**).

[4] Liu, B.; Wei, W.; Qu, X. Z.; Yang, Z. Z. *Angew. Chem. Int. Ed.*, **47**, 3973 (**2008**).

[5] Anderson, K. D.; Luo, M.; Jakubiak, R.; Naik, R. R.; Bunning, T. J.; Tsukruk, V. V. *Chem. Mater.*, **22**, 3259 (**2010**).

[6] Paunov, V. N.; Cayre, O. J. *Adv. Mater.*, **16**, 788 (**2004**).

[7] Tang, J. L.; Schoenwald, K.; Potter, D.; White, D.; Sulchek, T. *Langmuir*, **28**, 10033 (**2012**).

[8] Hong, L.; Jiang, S.; Granick, S. *Langmuir*, **22**, 9495 (**2006**).

[9] Petit, L.; Sellier, E.; Duguet, E.; Ravaine, S.; Mingotaud, C. *J. Mater., Chem.* **10**, 253 (**2000**).

[10] Liang, F. X.; Liu, J. G.; Zhang, C. L.; Qu, X. Z.; Li, J. L.; Yang, Z. Z. *Chem. Commun.*, **47**, 1231 (**2011**).

[11] Li, J.; Wang, Y. H.; Zhang, C. L.; Liang, F. X.; Qu, X. Z.; Li, J. L.; Wang, Q.; Qiu, D.; Yang, Z. Z. *Polymer*, **53**, 3712 (**2012**).

[12] Chen, Y.; Yang, H. L.; Zhang, C. L.; Wang, Q.; Qu, X. Z.; Li, J. L.; Liang, F. X.; Yang, Z. Z. *Macromolecules*, **46**, 4126 (**2013**).

[13] Zhao, L. L.; Zhu, L. J.; Chen, Y.; Wang, Q.; Li, J. L.; Zhang, C. L.; Liang, F. X.; Qu, X. Z.; Yang, Z. Z. *Chem. Commun.*, **49**, 6161 (**2013**).

[14] Zhao, Z. G.; Zhu, F. Y.; Qu, X. Z.; Wu, Q. H.; Wang, Q.; Zhang, G. L.; Liang, F. X. *Polym. Chem.*, **6**, 4144 (**2015**).

[15] Chen, Q.; Yan, J.; Zhang, J.; Bae, S. C.; Granick, S. *Langmuir*, **28**, 13555 (**2012**).

[16] Gao, W.; Wang, J. *ACS Nano*, **8**, 3170 (**2014**).

[17] Wang, J.; Gao, W. *ACS Nano*, **6**, 5745 (**2012**).

[18] Sheu, H.; El-Aasser, M.; Vanderhoff, J. *J. Polym. Sci. Part A: Polym. Chem.*, **28**, 629 (**1990**).
[19] Sheu, H.; El-Aasser, M.; Vanderhoff, J. *J. Polym. Sci., Part A: Polym. Chem.*, **28**, 653 (**1990**).
[20] Okubo, M.; Katsuta, Y.; Matsumoto, T. *J. Polym. Sci., Polym. Lett. Ed.*, **18**, 481 (**1980**).
[21] Okubo, M.; Ando, M.; Yamada, A.; Katsuta, Y.; Matsumoto, T. *J. Polym. Sci., Polym. Lett. Ed.*, **19**, 143 (**1981**).
[22] Okubo, M.; Katsuta, Y.; Matsumoto, T. *J. Polym. Sci., Polym. Lett. Ed.*, **20**, 45 (**1982**).
[23] Reculusa, S.; Poncet-Legrand, C.; Perro, A.; Duguet, E.; Bourgeat-Lami, E.; Mingotaud, C.; Ravaine, S. *Chem. Mater.*, **17**, 3338 (**2005**).
[24] Reculusa, S.; Poncet-Legrand, C.; Ravaine, S.; Mingotaud, C.; Duguet, E.; Bourgeat-Lami, E. *Chem. Mater.*, **14**, 2354 (**2002**).
[25] Kim, J. W.; Larsen, R. J.; Weitz, D. A. *J. Am. Chem. Soc.*, **128**, 14374 (**2006**).
[26] Kim, J. W.; Lee, D.; Shum, H. C.; Weitz, D. A. *Adv. Mater.*, **20**, 3239 (**2008**).
[27] Kim, J. W.; Larsen, R. J.; Weitz, D. A. *Adv. Mater.*, **19**, 2005 (**2007**).
[28] Zhang, C. L.; Liu, B.; Tang, C.; Liu, J. G.; Qu, X. Z.; Li, J. L.; Yang, Z. Z. *Chem. Commun.*, **46**, 4610 (**2010**).
[29] Tang, C.; Zhang, C. L.; Liu, J. G.; Qu, X. Z.; Li, J. L.; Yang, Z. Z. *Macromolecules*, **43**, 5114 (**2010**).
[30] Tang, C.; Zhang, C. L.; Sun, Y. J.; Liang, F. X.; Wang, Q.; Li, J. L.; Qu, X. Z.; Yang, Z. Z. *Macromolecules*, **46**, 188 (**2013**).
[31] Liu, B.; Zhang, C. L.; Liu, J. G.; Qu, X. Z.; Yang, Z. Z. *Chem. Commun.*, **45**, 3871 (**2009**).
[32] Liu, B.; Liu, J. G.; Liang, F. X.; Wang, Q.; Zhang, C. L.; Qu, X. Z.; Li, J. L.; Qiu, D.; Yang, Z. Z. *Macromolecules*, **45**, 5176 (**2012**).
[33] Tu, F. Q.; Lee, D. *J. Am. Chem. Soc.*, **136**, 9999 (**2014**).
[34] Wang, Y. H.; Zhang, C. L.; Tang, C.; Li, J.; Shen, K.; Liu, J. G.; Qu, X. Z.; Li, J. L.; Wang, Q.; Yang, Z. Z. *Macromolecules*, **44**, 3787 (**2011**).
[35] Nonomura, Y.; Komura, S.; Tsujii, K. *Langmuir*, **20**, 11821 (**2004**).
[36] Grüning, B.; Holtschmidt, U.; Koerner, G.; Rossmy, G. US Patent 4715986 (**1987**).
[37] Liang, F. X.; Shen, K.; Qu, X. Z.; Zhang, C. L.; Wang, Q.; Li, J. L.; Liu, J. G.; Yang, Z. Z. *Angew. Chem. Int. Ed.*, **50**, 2379 (**2011**).
[38] Chen, Y.; Liang, F. X.; Yang, H. L.; Zhang, C. L.; Wang, Q.; Qu, X. Z.; Li, J. L.; Cai, Y. L.; Qiu, D.; Yang, Z. Z. *Macromolecules*, **45**, 1460 (**2012**).

[39] Yang, H. L.; Liang, F. X.; Wang, X.; Chen, Y.; Zhang, C. L.; Wang, Q.; Qu, X. Z.; Li, J. L.; Wu, D. C.; Yang, Z. Z. *Macromolecules*, **46**, 2754 (**2013**).
[40] Zhao, Z. G.; Liang, F. X.; Zhang, G. L.; Ji, X. Y.; Wang, Q.; Qu, X. Z.; Song, X. M.; Yang, Z. Z. *Macromolecules*, **48**, 3598 (**2015**).
[41] Cao, Z. Q.; Wang, G. J.; Chen, Y.; Liang, F. X.; Yang, Z. Z. *Macromolecules*, **48**, 7256 (**2015**).
[42] Ji, X. Y.; Zhang, Q.; Liang, F. X.; Chen, Q. N.; Qu, X. Z.; Zhang, C. L.; Wang, Q.; Li, J. L.; Song, X. M.; Yang, Z. Z. *Chem. Commun.*, **50**, 5706 (**2014**).
[43] Ji, X. Y.; Zhang, Q.; Qu, X. Z.; Wang, Q.; Song, X. M.; Liang, F. X.; Yang, Z. Z. *RSC Adv.*, **5**, 21877 (**2015**).
[44] Liu, Y. J.; Liang, F. X.; Wang, Q.; Qu, X. Z.; Yang, Z. Z. *Chem. Commun.*, **51**, 3562 (**2015**).
[45] Liu, Y. J.; Wang, Q.; Qu, X. Z.; Liang, F. X.; Yang, Z. Z. *Sci. China Mater.*, **58**, 126 (**2015**).
[46] Wang, Q. G.; Liu, Y. J.; Qu, X. Z.; Wang, Q.; Liang, F. X.; Yang, Z. Z. *Colloid Polym. Sci.*, **293**, 2609 (**2015**).
[47] Zhou, P.; Wang, Q.; Zhang, C. L.; Liang, F. X.; Qu, X. Z.; Li, J. L.; Yang, Z. Z. *Chinese Chem. Lett.*, **26**, 657 (**2015**).
[48] Walther, A.; André, X.; Drechsler, M.; Abetz, V.; Müller, A. H. E. *J. Am. Chem. Soc.*, **129**, 6187 (**2007**).
[49] Zhang, K.; Gao, L.; Chen, Y. M. *Polymer*, **51**, 2809 (**2010**).
[50] Deng, R. H.; Liang, F. X.; Zhou, P.; Zhang, C. L.; Qu, X. Z.; Wang, Q.; Li, J. L.; Zhu, J. T.; Yang, Z. Z. *Adv. Mater.*, **26**, 4469 (**2014**).
[51] Hirsemann, D.; Shylesh, S.; Souza, R. A. D.; Diar-Bakerly, B.; Biersack, B.; Mueller, D. N.; Martin, M.; Schobert, R.; Breu, J. *Angew. Chem. Int. Ed.*, **51** 1348 (**2012**).
[52] Liu, Y.; Abetz, V.; Müller, A. H. E. *Macromolecules*, **36**, 7894 (**2003**).
[53] Walther, A.; Drechsler, M.; Rosenfeldt, S.; Harnau, L.; Ballauff, M.; Abetz, V.; Müller, A. H. E. *J. Am. Chem. Soc.*, **131**, 4720 (**2009**).
[54] Ruhland, T. M.; Gröschel, A. H.; Walther, A.; Müller, A. H. E. *Langmuir*, **27**, 9807 (**2011**).
[55] Rupar, P. A.; Chabanne, L.; Winnik, M. A.; Manners, I. *Science*, **337**, 559 (**2012**).
[56] Nie. Z. H.; Fava, D.; Kumacheva, E.; Zou, S.; Walker, G. C.; Rubinstein, M. *Nature Mater.*, **6**, 609 (**2007**).
[57] Chaudhary, K.; Chen, Q.; Juárez, J. J.; Granick, S.; Lewis, J. A. *J. Am. Chem. Soc.*, **134**, 12901 (**2012**).

[58] Priftis, D.; Sakellariou, G.; Baskaran, D.; Mays, J. W.; Hadjichristidis, N. *Soft Matter*, **5**, 4272 (**2009**).
[59] Loget, G. L.; Lapeyre, V.; Garrigue, P.; Warakulwit, C.; Limtrakul, J.; Delville, M. H.; Kuhn, A. *Chem. Mater.*, **23**, 2595 (**2011**).
[60] Kong, X. Y.; Ding, Y.; Yang, R.; Wang Z. L. *Science*, **303**, 1348 (**2004**).
[61] Aizpurua, J.; Hanarp, P.; Sutherland, D. S.; Käll, M.; Bryant, G.W.; Abajo, F. J. G. *Phys. Rev. Lett.*, **90**, 057401 (**2003**).
[62] Zhou, Q.; Li, J.; Zhang, C. L.; Qu, X. Z.; Liu, J. G.; Li, J. L.; Yang, Z. Z. *Polymer*, **51**, 3606 (**2010**).
[63] Yang, H. L.; Liang, F. X.; Chen, Y.; Wang, Q.; Qu, X. Z.; Yang, Z. Z. *NPG Asia Materials*, **7**, e176 (**2014**).
[64] Zhao, H.; Liang, F. X.; Qu, X. Z.; Wang, Q.; Yang, Z. Z. *Macromolecules*, **48**, 700 (**2015**).

Chapter 2

Synthesis and Applications of Polymeric Janus Nanoparticles

Hien T. T. Duong[*,†], *Duc Nguyen*[*], *Chiara Neto*[*,‡], *and Brian S. Hawkett*[*,§]

[*]Key Centre for Polymers and Colloids, School of Chemistry, The University of Sydney, NSW 2006, Australia
[†]Now at the Initiative for Biomedical Materials & Devices (IBMD), School of Mathematical and Physical Sciences, Faculty of Science, University of Technology Sydney, NSW 2006 Australia
[‡]chiara.neto@sydney.edu.au
[§]brian.hawkett@sydney.edu.au

Abstract

The properties of Janus particles (JPs) have been extensively studied in recent years, but their practical applications are still hampered by limitations in the synthetic approach, especially for particles on the nanoscale. Simple and versatile technologies for the well-defined and scalable synthesis of JPs are highly sought after for broad applications. This chapter reviews the main existing approaches for the synthesis of JPs, with particular emphasis on seeded emulsion polymerization. Emulsion-based methods have proven to be effective for the large-scale synthesis of multifunctional JPs with nanometer size range. Important applications of JPs are also reviewed and discussed.

Keywords: Janus nanoparticles, self assembly, block polymer, amphiphilic.

2.1. Introduction

Research into Janus particles (JPs) has received great attention over the past decade. In his Nobel lecture in 1991, Pierre-Gilles de Gennes suggested that asymmetric colloidal particles with different chemical compositions on the two lobes could exhibit a special behavior at interfaces, and named them "Janus grains" [1]. Due to their asymmetry, JPs have the ability to offer more advanced chemical and physical properties compared to that of their symmetric homogeneous counterparts and in particular may behave as amphiphilic surfactants. However, for years after de Gennes' Nobel lecture, research into JPs was still slow, as evidenced by the limited number of publications in the field. The main limitations to the advancement of the field were the difficulty in synthesizing well-defined JPs and also in characterizing them, especially in the case of nanometer-sized JPs. From around 2005, the significance of JPs in a wide range of applications has become clear, including in surfactants, electrochemistry, catalysis, electronics, sensors, optics, superhydrophobic textiles, and nanomedicine and this has driven research to pursue different methods for the fabrication of JPs. In the past 10 years, JPs have become a hot topic of research, as evidenced by the exponential growth in the number of publications in this area. Our latest search using the keyword "Janus particles" on Web of Science shows 1,608 published research papers on this topic, which is more than 16 times higher than in 2010.

Several excellent reviews have been published in recent years addressing the most common synthetic methods leading to JPs [2–5]. This chapter will review and discuss the main methods for the synthesis of JPs, and in particular, we will focus on the synthesis of both polymeric and hybrid Janus nanoparticles (JNPs), with diameter in the nanoscale range. We will showcase a synthetic approach based on reversible-addition fragmentation chain transfer (RAFT) in seeded emulsion polymerization [6–8], which our group has been among the first to employ, together with the group of El-Aasser [9,10]. More recently other groups have been successful in similar approaches [11–16]. However, our group remains the first to be able to synthesize nanoscale polymer JNPs by emulsion polymerization, with typical particle size in the tens-to-one-hundred nanometers. The properties and potential use of JNPs in a wide range of applications, derived from their asymmetric structure, will be reviewed and discussed.

2.2. Preparation Methods

In this section, we will describe the main synthetic approaches to date for the preparation of JNPs. The synthesis of uniform JPs with well-defined geometry and size remains a major challenge, especially for the particles with diameter below 100 nm. From around 2005, the advancement in synthetic tools, including controlled polymerization techniques, has allowed different methods for synthesizing JPs to emerge. The methods most commonly used to synthesize JPs to date are grouped into three main categories: masking, self-assembly and phase separation, as shown in Fig. 1 [4]. In what follows, we will provide a critical assessment of these methods. When discussing the phase separation method, we will focus on the synthesis of polymeric JP from seeded emulsion polymerization.

2.2.1. Masking

Masking is one of the first techniques that were used to synthesize JPs. The method involves the modification of only one side of the particle surface, while the other side is protected or masked. In 1988, Casagrande *et al.* first introduced the term "Janus bead" and used the masking method to describe glass spherical particles with diameter of 40–50 μm with one hydrophilic

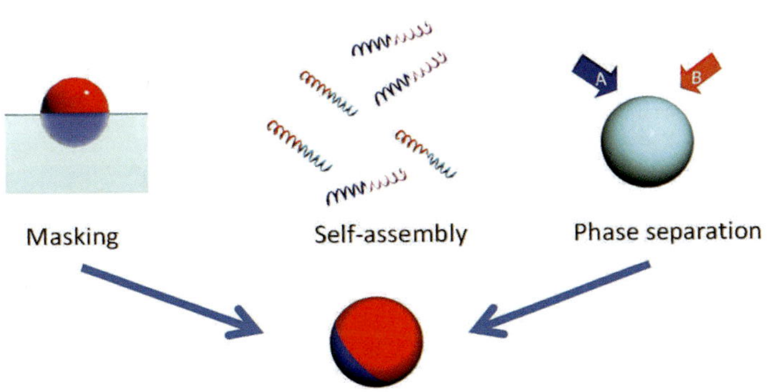

Figure 1. Schematic diagram illustrating three common strategies for the preparation of JPs: masking, self-assembly, and phase separation.

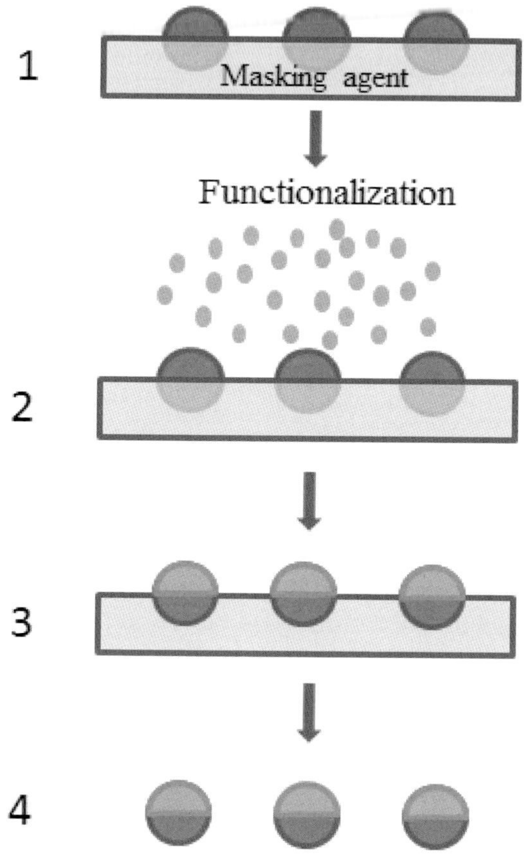

Figure 2. Schematic view of the synthesis of NPs via masking method.

hemisphere and the other hydrophobic [16]. The authors investigated the use of the prepared Janus beads for their potential stabilization of oil–water emulsions and observed a different behavior at oil–water interfaces compared to the precursor glass beads [17]. In general, the synthesis of JPs via masking process involves four steps, as shown in Fig. 2: (1) homogeneous particles are trapped at the interface between two immiscible phases and one side of the particle is protected; (2) the other half of the particle at the interface is modified with desired properties; (3) the masking agent is removed; and (4) heterogeneous JPs are obtained. In this section, we will review separately three examples of JPs prepared using a masking process with different immiscible phases, including liquid–liquid, liquid–wax hybrid, liquid–solid, and gas–liquid and gas–solid interfaces.

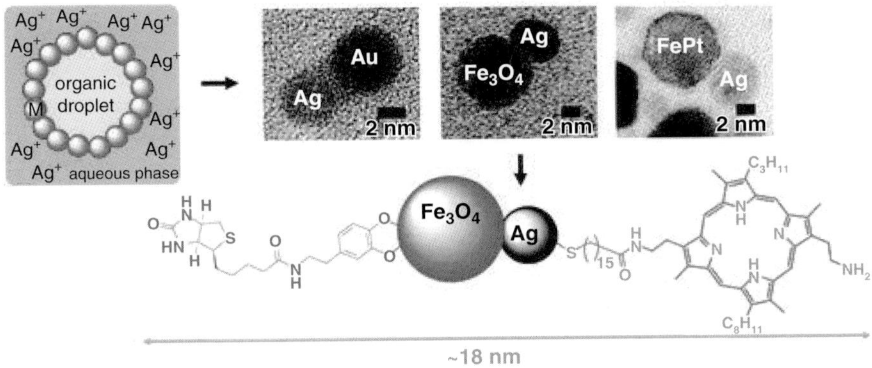

Figure 3. Synthetic scheme for functional heterodimers of NPs formed at a liquid–liquid interface. Reproduced with permission from Ref. [18]. Copyright (2005) American Chemical Society.

2.2.1.1. *Masking at a liquid–liquid interface*

Asymmetric nanoparticles (NPs) were obtained at the interface between two immiscible fluids. Gu *et al.* employed this strategy to prepare F_3O_4–Ag, FePt–Ag or Au–Ag hydrophobic Janus nanostructures (NPs) [18]. As shown in Fig. 3, these heterodimers made of two distinct nanospheres were prepared in two steps. First, oil-in-water Pickering emulsions were formed stabilized by F_3O_4, FePt or Au solid seed NPs. Then catalytic reduction of a noble metal such as silver ions in the aqueous phase by Fe(II) on the seeds took place at the surface of the oil droplets. The prepared JNPs could be functionalized through selective post-modification with different functional molecules to render them hydrophilic, which is useful for different biomedical applications including protein binding, bioimaging, and pathogen detection (Fig. 3) [18]. It should be noted that this approach has been only reported for nucleation of silver ions using aforementioned different types of NPs as seeds.

2.2.1.2. *Masking at a liquid–wax hybrid interface*

The Granick group prepared JPs using Pickering emulsion of paraffin wax droplets stabilized by silica particles at a temperature where the wax is molten [19]. The use of paraffin wax droplets as the dispersed oil phase is an interesting idea. The wax was easily solidified by cooling below its

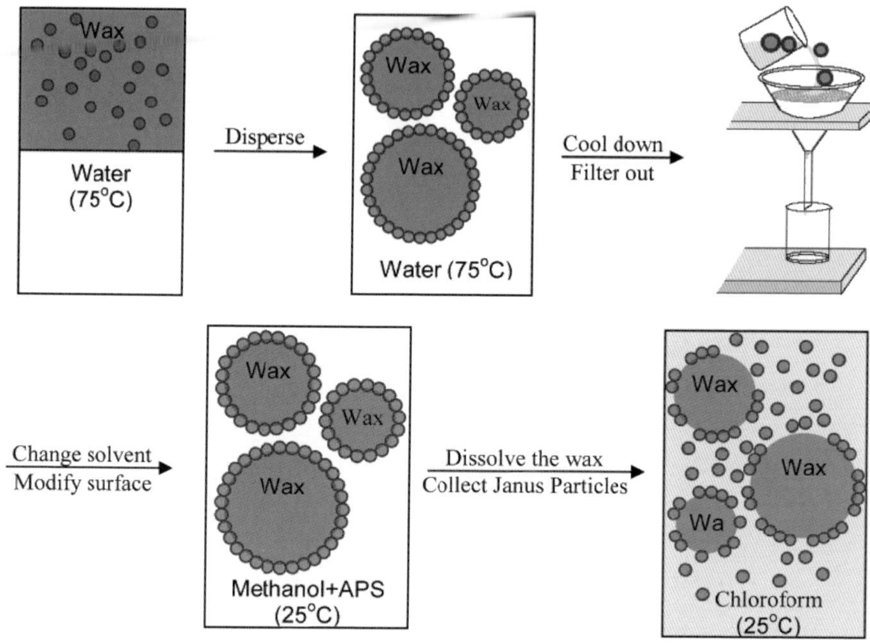

Figure 4. Schematic synthesis of JPs by functionalizing silica particles adsorbed onto an emulsion of paraffin wax in water. Reproduced with permission from Ref. [19]. Copyright (2006) American Chemical Society.

melting temperature, thus locking the particles at the wax–water droplet interface. These stable wax droplets were washed and the exposed surface of silica particles was chemically functionalized with silanes in organic solvents such as methanol. The JPs were then collected after adding chloroform to dissolve the wax (Fig. 4). This method is relatively simple and cost-effective to produce a relatively large quantity of JPs with distinct properties on each side. The main drawback of this method is that the particles could detach from the interface of the wax droplets during the chemical functionalization in the presence of organic solvents, thus leading to mixtures of Janus and homogeneous particles. The technique was later modified to a solvent-free method by the same group, in which the exposed surface of the particles was functionalized with silanes via vapor deposition [20]. Cui *et al.* used polydimethylsiloxane (PDMS) inert polymer film to create a liquid–solid interface. The exposed surface of the

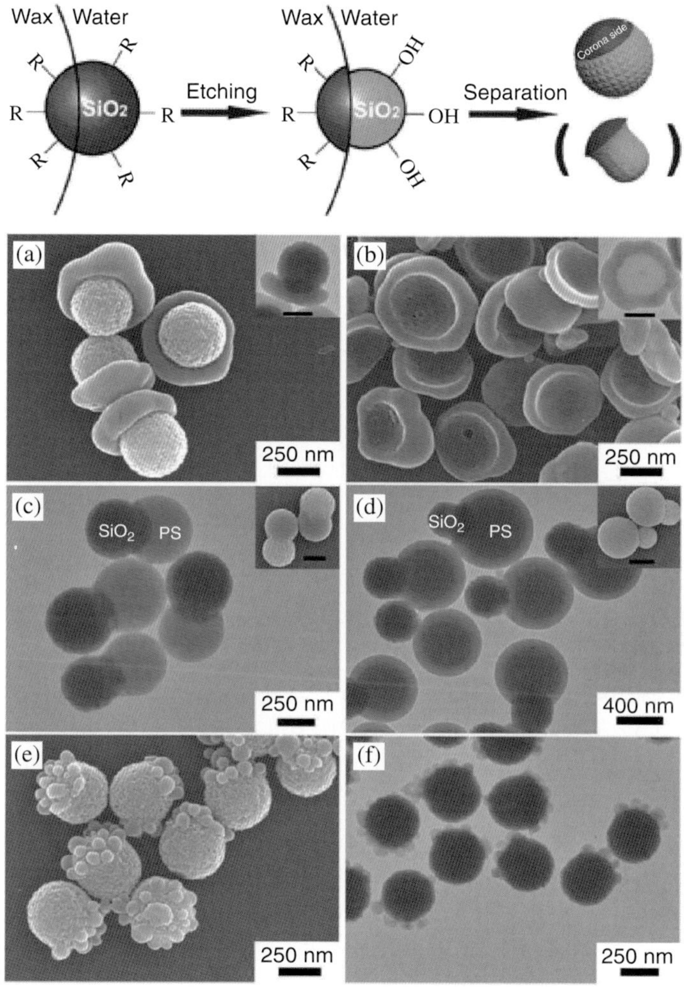

Figure 5. Scheme for synthesis of non-spherical JPs by asymmetric wet-etching. SEM and inset TEM micrographs of (a) Janus SiO_2/PS composite colloids with a PS cap grown on the corona side; (b) PS nanocaps after removing SiO_2 from (a); (c) and (d) Janus SiO_2/PS dimers with increased PS volume fraction (the dark area corresponding to silica and the gray area corresponding to PS); (e) and (f) Janus SiO_2/PS composite colloids with PS nanoflowers grown on the corona side. Reproduced with permission from Ref. [22]. Copyright Royal Society of Chemistry.

particles could be modified with metal, oxides, or alloys such as silver, gold, nickel and titania [21].

This method was also employed by other groups to produce Janus silica particles with high yield. Liu *et al.* prepared acorn- and mushroom-shaped silica particles by introducing polymerizable vinyl groups onto the surface of the silica particles followed by the etching of the exposed surface with ammonium fluoride (Fig. 5) [22]. The vinyl groups on the protected hemisphere were intact and the polymer cap was formed by polymerization after removal of the wax. Perro *et al.* reported the synthesis of Janus silica particles on a large scale by modifying the exposed surface of silica particles with amine groups as evidenced by the adsorption with gold NPs on one hemisphere [23].

He *et al.* prepared Janus $Cu_2(OH)_2CO_3$/CuS microspheres via a Pickering emulsion route by using styrene monomer as the oil phase instead of wax [24]. During the polymerization of styrene, PS microspheres were formed and covered by a layer of $Cu_2(OH)_2CO_3$ particles. The exposed surface was then treated with thioacetamide solution, allowing the conversion of $Cu_2(OH)_2CO_3$ to CuS. The PS core was then removed by treating with toluene at elevated temperature and the Janus $Cu_2(OH)_2CO_3$/CuS microspheres were obtained. Pardhy *et al.* employed both styrene-based and wax-based double Pickering emulsion approach to prepare Laponite clay Janus colloids [25]. In the first step, a Pickering emulsion with styrene as the oil phase was stabilized by Laponite particles. The stable colloids of Laponite-armored PS were then used as solid stabilizers for a second Pickering emulsion with wax as the oil phase. One side of the colloids was trapped in the wax and the exposed side was chemically modified by cation-exchange method, allowing the formation of JP with anisotropic surface charge [25].

2.2.1.3. *Masking at a liquid–solid interface*

In this method, homogeneous particles were immobilized on a solid substrate instead of being trapped at the interface between two fluids. Takei and Shimizu thermally deposited gold onto only one side of latex microspheres facing the evaporation source for subsequent chemical modification with a thiol compound in solution, to form dipolar JPs, which could be useful in applications ranging from microscopic probe to biomedicine [26]. One more example of the liquid–solid interface method was reported by Sardar *el al.* [27]. Citrate-stabilized gold NPs were immobilized onto a silanized glass surface bearing amine groups by displacing citrate groups. The exposed gold surface was functionalized with a thiol compound bearing hydroxyl group while the masked side of the gold NPs attached to the silane layer was modified with thiols bearing carboxylic or amine group, forming Janus gold NPs as shown in Fig. 6. Gold heterodimers were synthesized by the coupling reaction of these asymmetrically functionalized NPs.

2.2.1.4. *Masking at a gas–liquid interface*

A good example for this method is the work of Pradhan *et al.* who prepared Janus gold NPs with hydrophobic character on one side and hydrophilic on the other by using the Langmuir technique (Fig. 7) [28]. In this study, hydrophobic alkane thiolate-coated gold NPs were dispersed on the water surface at low pressure. Upon mechanical compression, a compact monolayer of hydrophobic gold NPs was formed on the water surface. At high pressure, the interfacial mobility of the NPs was restricted, which favored the ligand exchange reactions at the particle surface by introducing a hydrophilic thiol ligand such as 3-mercaptopropane-1,2-diol, resulting in the formation of amphiphilic JNPs.

2.2.1.5. *Masking at a gas–solid interface*

The Whitesides group employed this multistep gas–solid strategy to prepare hollow metallic hemispheres, which could be used as metallic foams, heterogeneous catalysts, magnetic filters, and superhydrophobic surfaces [29]. In the first step, a layer of silica colloids drop-casted on a glass slide was coated with titanium or nickel by electron-beam evaporation. After deposition of a thin film of gold or platinum, metal-coated silica particles were detached from the glass slide surface by sonication, followed by treatment with hydrogen fluoride (HF) aqueous solution to dissolve the silica and metallic half-shells were obtained (Fig. 8) [29].

In summary, although masking is a multistep process, it is probably one of the most versatile techniques to prepare JPs made of different

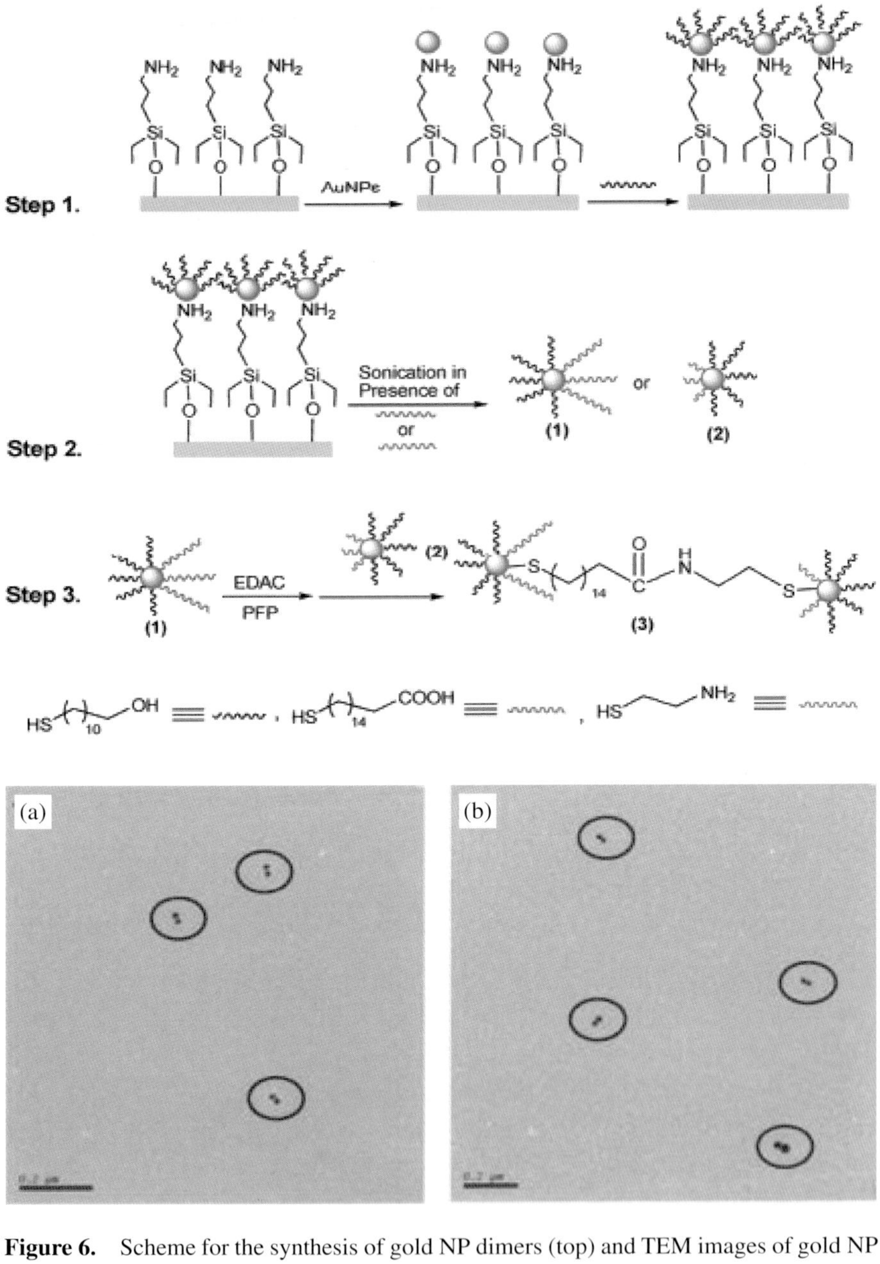

Figure 6. Scheme for the synthesis of gold NP dimers (top) and TEM images of gold NP dimers synthesized via coupling of 16–16 nm (a) and 30–30 nm (b) particles (bottom). Scale bar is 200 nm. Reproduced with permission from Ref. [27]. Copyright (2007) American Chemical Society.

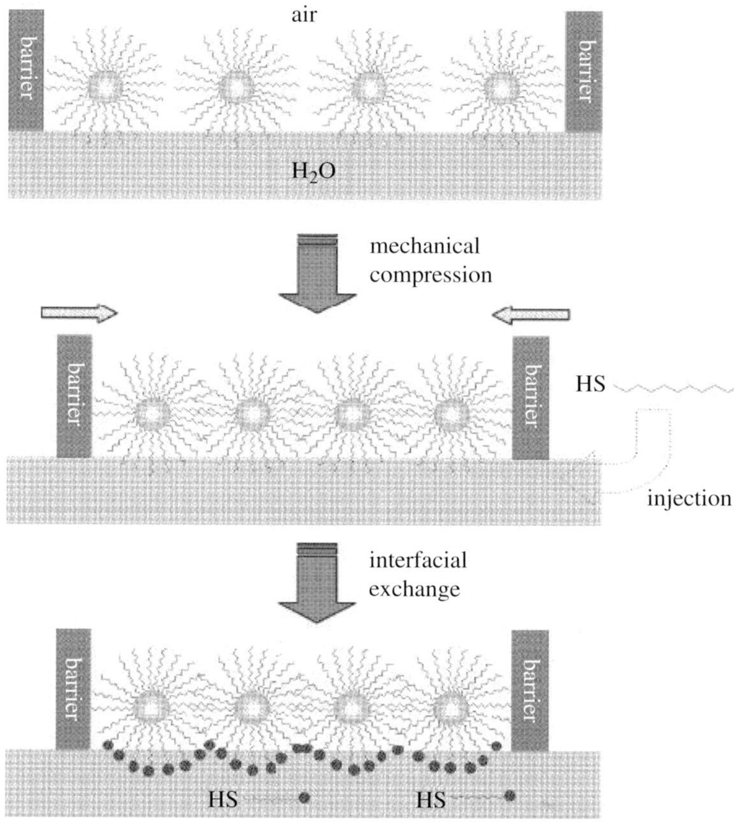

Figure 7. Scheme for preparation of JNPs using Langmuir technique. Reproduced with permission from Ref. [28]. @ Copyright (2007) John Wiley & Sons.

materials with a wide variety of functional groups. There is still a need to improve the method, which is even simpler and more accessible for large-scale synthesis of JPs.

2.2.2. Janus Micelles by the Self-Assembly of Block Copolymers

The self-assembly of amphiphilic block copolymers in solution into symmetric core–shell structures has been studied extensively in the last few decades, since Eisenberg *et al.* in the 1990s discovered the variety of morphologies that can be formed upon microphase separation of two

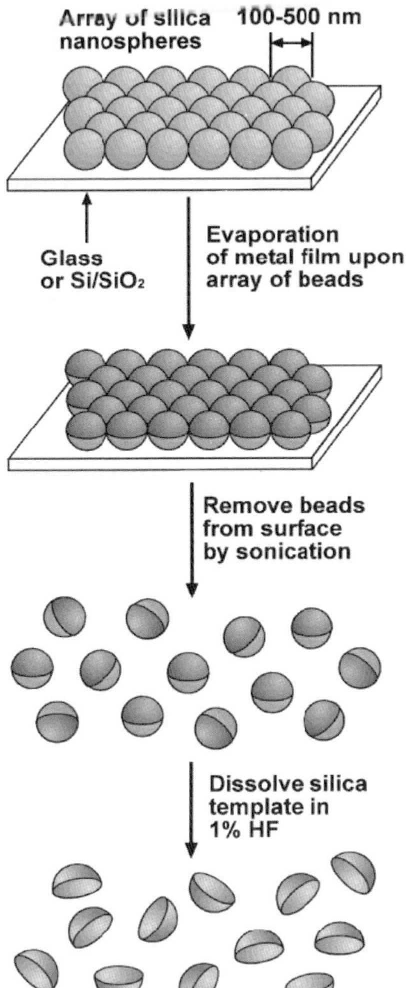

Figure 8. Scheme for fabricating metallic half-shells. Reproduced with permission from Ref. [29]. Copyright (2002) American Chemical Society.

incompatible blocks, including spheres, rods, and vesicles [30]. Rapid progress in this attractive research area takes advantage of advanced living/controlled polymerization techniques, which allow the synthesis of block copolymers with well-defined architecture, composition, and narrow molecular weight distribution. The preparation of JPs through the self-assembly of block copolymers is very challenging and requires

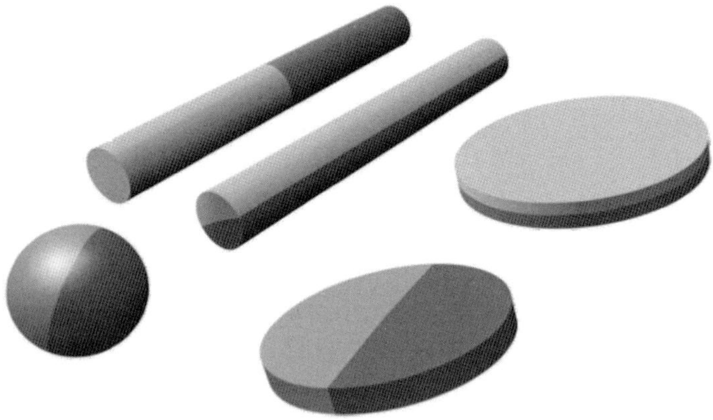

Figure 9. Spherical, cylindrical and disk-shaped Janus particles formed by synthetic and self-assembly approaches for nanoscale polymeric Janus particles. Reproduced with permission from Ref. [3] Copyright (2013) American Chemical Society.

extensive study of the thermodynamics of polymer mixtures, as well as the parameters affecting the self-assembly behavior, such as concentration, temperature, pH, and ionic strength. Different architectures of JPs have been prepared by using self-assembly approaches as depicted in Fig. 9. This area has been recently reviewed by the Müller group [3].

Both di-block and tri-block copolymers have been used for the direct synthesis of JPs. In the case of di-block copolymers, two combinations can allow for the preparation of Janus micelles. First, a combination of AB and BC diblock copolymers was used for the production of Janus nanoobjects, in which the B common segments is insoluble in a given solvent and forms the micellar core [31–33]. Even though sufficient incompatibility of the A and C corona blocks is required for the comicellization into mixed micelles, a too large incompatibility can lead to the formation of two separate AB and BC micelle populations [31,34]. Ma *et al.* prepared Janus micelles by comicellization of PNIPAM-b-P4VP and PEG-b-P4VP in aqueous solutions in the presence of bivalent sulphate ions, which help the cross-linking of positively charged P4VP block [35]. In another approach, Voets *et al.* mixed AB and CD diblock copolymers, poly(acrylic acid)-*block*-poly(acryl amide) (PAA-b-PAAm) and poly(2-methyl vinylpyridinium iodide)-*block*-poly(ethylene oxide) (P2MVP-b-PEO), leading to the forced coassembly of the two polymers due to electrostatic interactions between the positively charged P2MVP blocks and the negatively charged PAA blocks, while the PEO and PAAm blocks phase-separated on the corona of Janus micelles [36–39]. Wurm *et al.* prepared PEO-b-PPO diblock copolymers bearing a ruthenium end-group via ring-opening polymerization and observed the formation of Janus micelles and their assembly into higher ordered super-micelles [40].

In the case of ABC tri-block terpolymers, Janus micelles could be assembled with two outer soluble end-blocks, A and C, and an inner insoluble block, B. The Müller group pioneered the synthesis of JPs based on the self-assembly of tri-block copolymers with central polybutadiene block able to cross-link, as depicted in Fig. 10 [2,3,41]. The structures of Janus micelles formed within thin films were well-ordered, and consisted of a cross-linked core and a corona with two different hemispheres. This method was explored extensively by the same group and Janus morphologies such as spheres [41], cylinders [42], disks [43], and lamellae [44] were observed depending on the nature of terpolymers and relative length of blocks. The key for successful JP synthesis is the use of polybutadiene, allowing the cross-linking reaction after film casting of the terpolymer. JPs with different morphologies were obtained upon dissolution of the film with suitable solvent and sonication. The self-assembly of tri-block copolymer in solution was also employed by other groups for the preparation of JPs. Sfika *et al.* prepared pH-responsive Janus star-like micelles from double hydrophilic poly(2-vinylpyridineb-poly(methyl methacrylate)-*block*-(poly(acrylic acid) (P2VP-PMMA-PAA) terpolymer in aqueous dilute solutions [45]. Du and Armes reported the formation of JPs by dissolving directly a primary amine-based tri-block copolymer, poly(ethylene oxide)-*block*-poly(3-caprolactone)-*block*-poly(2-aminoethyl methacrylate) (PEO-b-PCL-b-PAMA) in water, with PCL phase separating in the core and the two hydrophilic PEO and PAMA blocks on the surface. The PCL core was selectively modified with tetramethyl orthosilicate, leading to the formation of Janus nanostructures with a silica core within a PAMA-rich hemisphere [46].

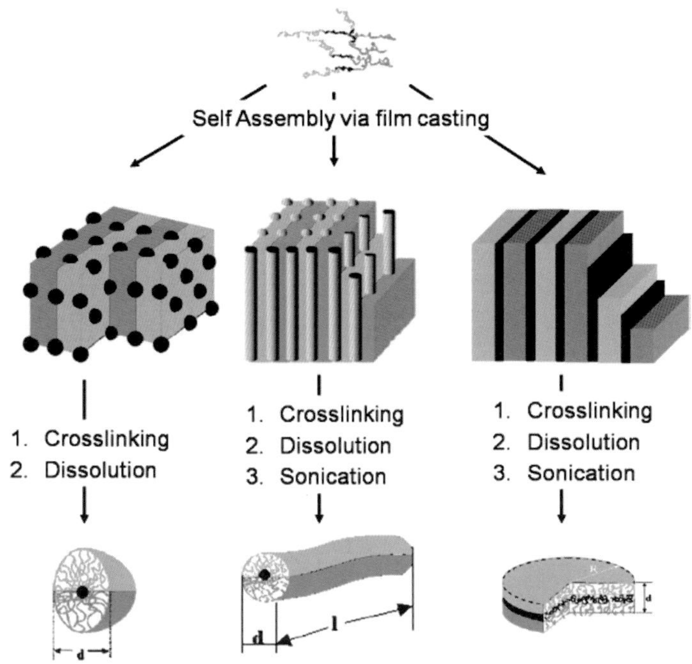

Figure 10. Scheme for the preparation of JNPs from terpolymers. The process starts with film casting of the terpolymer, followed by cross-linking of the intermediate block and finally by dissolution and sonication. Depending upon the structure and molecular weight of the terpolymer, film casting produces matrices with different morphologies: spherical, cylindrical, and disk-like. Reproduced from Ref. [2] with permission from the Royal Society of Chemistry.

2.2.3. Synthesis of Janus Nanoparticles from Phase Separation Method

In this method, the incompatibility of two polymers in a confined environment induces a powerful driving force for the phase separation of polymer mixtures, resulting in the formation of Janus structures. JNPs of two inorganic, two organic, or hybrid lobes can be prepared using the phase separation method as described in recent reviews [4,5]. This section will provide more detailed descriptions about JNPs synthesis using seeded emulsion polymerization method. Among the synthetic approaches for JNPs, emulsion-based method is the most effective method on a large scale. The seeded emulsion polymerization method starts with preformed solid seed particles dispersed in emulsions, followed by the swelling with additional monomer. The preformed seeds can be cross-linked or uncross-linked. Subsequent polymerization can induce the phase separation to obtain non-spherical JNPs with two chemically distinct parts.

2.2.3.1. *Formation of Janus nanoparticles from cross-linking seeds*

Seeded polymerization, which has been described by Sheu *et al.*, is one of the most flexible methods to prepare JPs with easy scalability [9]. This method involves the swelling of preformed cross-linked seed particles with monomer that during polymerization is expelled from the core to form a bulge in the JPs. It is a prerequisite that the swollen monomer and the seed matrix are incompatible, resulting in the phase separation of the newly formed polymer from the seed particles. In addition, the use of cross-linking agent is important for the monomer to be expelled onto the surface of the seed particles due to the high internal stress, especially at elevated temperature, which induces the formation of the Janus structure after free radical polymerization. The morphology of the anisotropic particle is governed by many different parameters including the degree of cross-linking of the seeds, the hydrophilicity of seed surface, the wettability of the monomer, size of the seeds, and monomer/polymer swelling ratio [6,9,47]. Formation of these particles followed the same mechanism discussed by Mock *et al.* [6] and Pham *et al.* [8] and presented in Fig. 11, where cross-linked polymer networks of the monomer-swollen seed particles shrunk on heating, expelling the polymerizing monomers. Depending on the wettability of the expelled monomers on the seed particles, different particle morphologies were formed. The wetting of the surface by the monomer was, in turn, influenced by stabilizing groups on the seed particle surface and compatibility between the extruded monomers and the seed polymer. Polymerization of the protruded monomers produced distinct bulges attached to the original seed particles.

The Weitz group prepared uniform JPs with chemical anisotropy in the micrometer range based on this technique, using cross-linked PS seeds

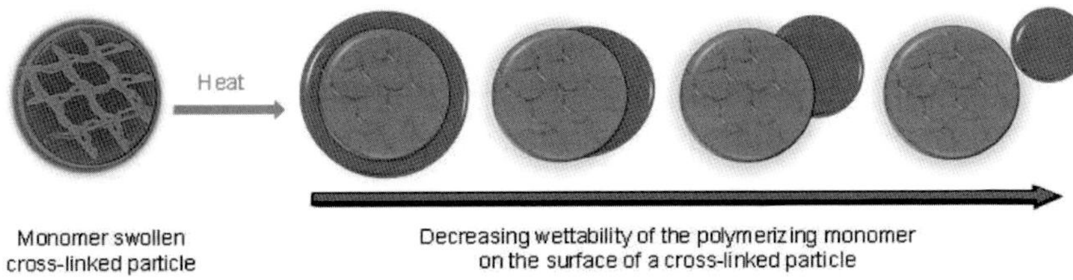

Figure 11. Schematic behavior of the monomer swollen cross-linked polymer seed on heating. The wetting ability of the monomer governs the morphology of the anisotropic particles formed during subsequent polymerization. In the reactions described here, complete detachment of a new particle is prevented by RAFT control making the new polymer continuous with the old. Reproduced with permission from Ref. [8]. Copyright (2015) Royal Society of Chemistry.

swollen with methyl methacrylate and butyl methacrylate, which are immiscible with PS [48]. This group also reported the synthesis of amphiphilic dumbbell JPs consisting of one hydrophilic poly (ethylene imine)-coated PS lobe and one hydrophobic PS lobe by employing surface modification. Interestingly, different shapes of anisotropic particles could be formed, including rod, cone, triangle, and diamond, depending on the cross-linking density gradients between the lobes [49]. This versatile strategy was employed by Tang *et al.* to prepare Janus polyacrylonitrile (PAN)/PS polymer colloids particles via styrene polymerization onto cross-linked PAN hollow spherical seeds [14]. Subsequent hydrolysis of PAN on one lobe leads to the formation of functional carboxylic acid, enabling a selective functionalization with magnetite and silica, as illustrated in Fig. 12. The prepared inorganic/organic composite Janus colloids can be used as solid surfactants.

By using the same immiscible pairs PS–PAN, Huang *et al.* synthesized raspberry-like PS–polyacrylonitrile (PAN) particles with multiple protrusions of PAN on the PS surface [50]. During the polymerization of AN monomer by γ-ray radiation, PAN polymer chains segregated from the cross-linked styrene-divinylbenzene-acrylic acid terpolymer (P(S-DVB-AA)) seed particles and multiple PAN lobes were formed onto the surface of the seeds due to the insolubility of AN monomer and PAN polymer (Fig. 13) [50].

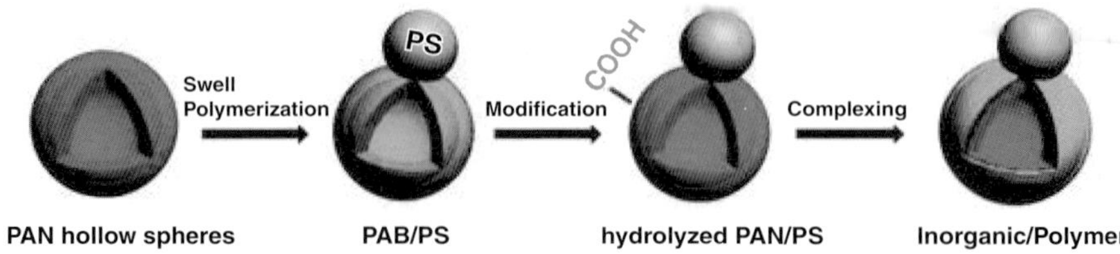

Figure 12. Schematic synthesis of mixed polymer/inorganic Janus colloids. Reproduced with permission from Ref. [14]. Copyright (2010) American Chemical Society.

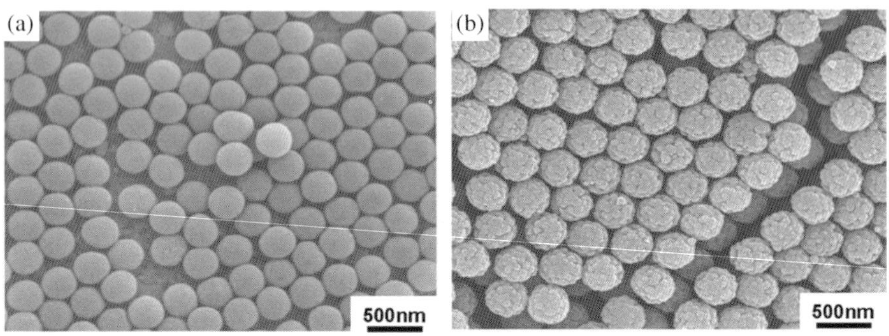

Figure 13. FESEM micrographs of (a) P(S-DVB-AA) particles and (b) PS–PAN raspberry-like particles. Reproduced with permission from Ref. 50. Copyright (2010) John Wiley & Sons.

In another study, the appearance of protrusions of PS onto highly cross-linked poly(*N*-isopropylacrylamide) (polyNIPAM), PS, and magnetite filled polyNIPAM as seed particles could be varied by varying the surfactant concentration, resulting in particles with multiple lobe particles. The roughness could be tuned by varying the experimental conditions such as initiator, surfactant, styrene concentrations, and also polymerization temperature (Fig. 14) [51].

The Okubo group employed a different approach for the preparation of spherical and "snowman-like" hydrophobic JPs by the solvent evaporation method [52,53]. Atom transfer radical polymerization (ATRP) was employed in combination with solvent evaporation method by this group as an elegant approach to prepare amphiphilic mushroom-like JPs in a two-step synthetic methodology. As shown in Fig. 15, in the first step spherical poly(methyl methacrylate) (PMMA)/poly(styrene-2-(2-bromoisobutyryloxy)ethyl methacrylate) (P(S-BIEM)) macroinitiator JPs were formed, due to the phase separation induced by solvent

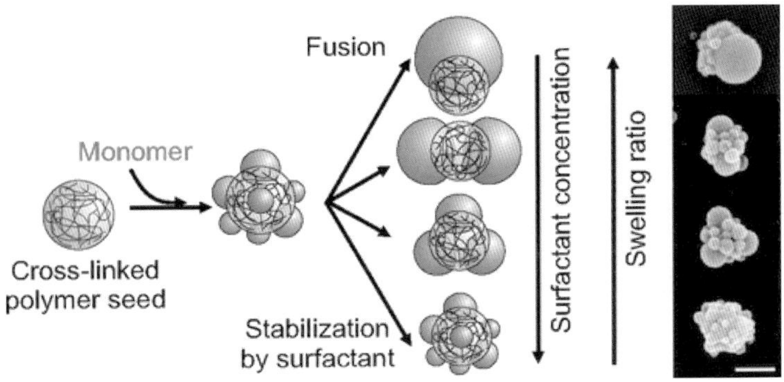

Figure 14. Schematic of the synthesis of multiple lobe particles, with systematic variation of several factors on the resulting patchy particles. Reproduced with permission from Ref. [51]. Copyright (2011) American Chemical Society.

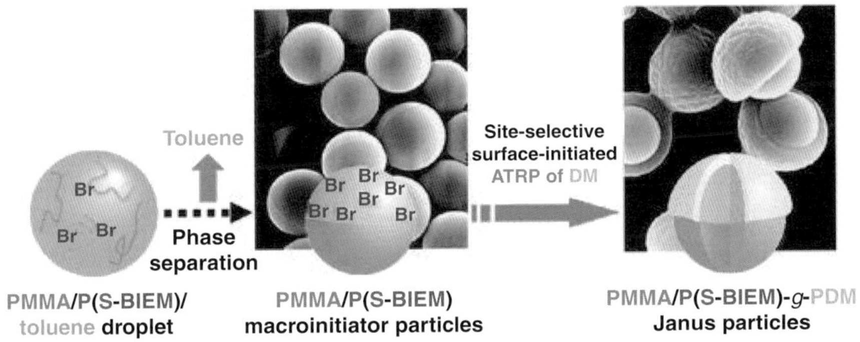

Figure 15. Preparation of "mushroom-like" JPs by site-selective surface-initiated atom transfer radical polymerization in aqueous dispersed systems. Reproduced with permission from Ref. [54]. Copyright (2010) American Chemical Society.

evaporation. Subsequently, 2-(dimethylamino)ethyl methacrylate (DM) monomer was further polymerized via surface-initiated ATRP mechanism to form pH-responsive mushroom-like PMMA/P(S-BIEM)-*graft*-poly(DM) [54]. The prepared mushroom-like JPs with amphiphilic character could be used as solid surfactants to stabilize oil-in-water emulsion. The stability of the emulsion was controlled by changing pH and temperature, which could lead to the adsorption/desorption of the stimuli-responsive JPs [55].

Our group recently reported the synthesis of polymeric JNPs based on this seeded emulsion approach using RAFT polymerization [8]. We

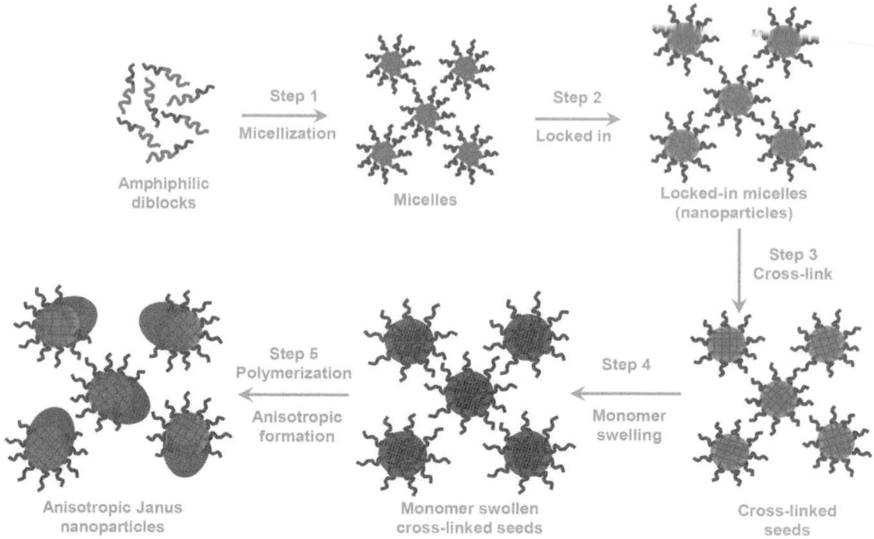

Figure 16. Schematic procedure for the synthesis of anisotropic JNPs. Reproduced from Ref. [8] with permission from the Royal Society of Chemistry.

pioneered the use of RAFT to mediate the emulsion polymerization processes [56,57]. The synthesis of polymer JNPs in this work consists of five steps, as described in Fig. 16. In what follows, the synthesis will be described in more detail.

Amphiphilic di-block copolymers containing hydrophobic block of PS and hydrophilic block of poly(acrylic acid) or poly(acrylamide) were first synthesized using RAFT polymerization as shown in Fig. 17. The prepared di-block copolymers were self-assembled into micelles in water with an average size of 20 nm after neutralization of the carboxylic acid functional groups (step 1). The formed micelles were swollen by monomer, which then underwent further free radical emulsion polymerization to extend the chain of hydrophobic PS blocks. At this stage, the initial micelles were "locked in" to NPs stabilized by the negative charges of polyacrylic acid or steric hydrophilic hairy layers of polyacrylamide blocks (step 2). The NPs were then cross-linked by divinyl benzene to generate uniform spherical cross-linked seeds (step 3). Seed NPs were swollen with monomers such as styrene or methyl methacrylate/butyl acrylate mixture, by simply mixing the monomers and particles overnight

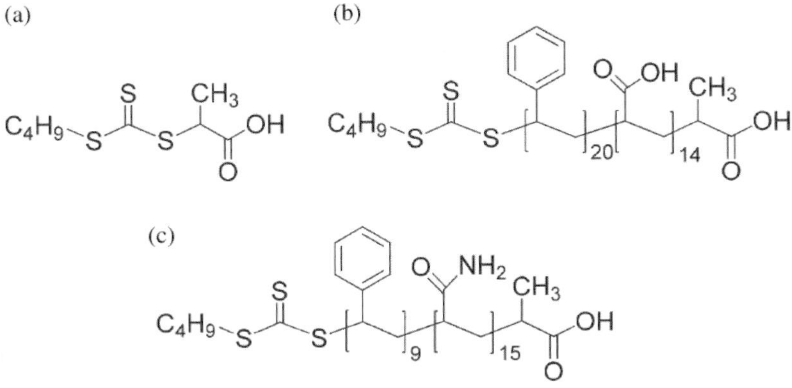

Figure 17. Structure of RAFT agent (a) and RAFT diblock of styrene and acrylic acid (b) or styrene acrylamide (c). Reproduced from Ref. [8] with permission from the Royal Society of Chemistry.

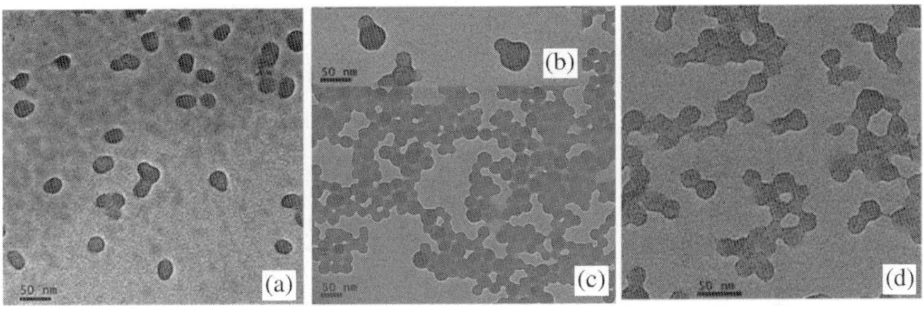

Figure 18. TEM micrographs: (a) ellipsoidal; (b) and (c) snowman-shaped; (d) dumb-bell-shaped polymer JNPs. Reproduced with permission from Ref. [8]. (2015) Royal Society of Chemistry.

under constant stirring (step 4). Polymeric JNPs, which are anisotropic in shape, chemical composition, and surface property were obtained after polymerizing monomer swollen cross-linked seed particles (step 5). It is interesting that the shape of JNPs was controlled by varying the charge density of the initiator and the composition of monomer swollen in the cross-linked seeds. While Janus ellipsoidal NPs were formed when V-501 was used as an initiator to grow the cross-linked seed particles, snowman-shaped anisotropic particles were obtained when KPS was used (Figs. 18(a)–18(c)). When styrene was replaced by a blend of MMA/BA monomers, we observed the formation of dumbbell shape (Figs. 18(d)).

The preparation of JPs by growing newly formed polymer onto the surface of preformed seeds involves several steps. An approach that can produce anisotropic particles directly is highly sought after.

In addition to polymeric JPs, polymer–inorganic hybrid particles were extensively studied. Ge *et al.* described the synthesis of asymmetric composite colloids consisting of two lobes: one lobe contains shell-cross-linked Fe_3O_4@SiO_2 PS composite colloids (Fig. 19(b)) and the other containing PS phase-separated after polymerization of swollen styrene at elevated temperature (Figs. 19(c) and 19(d) [58]. These uniform and well-defined anisotropic structures could be useful for novel devices and materials.

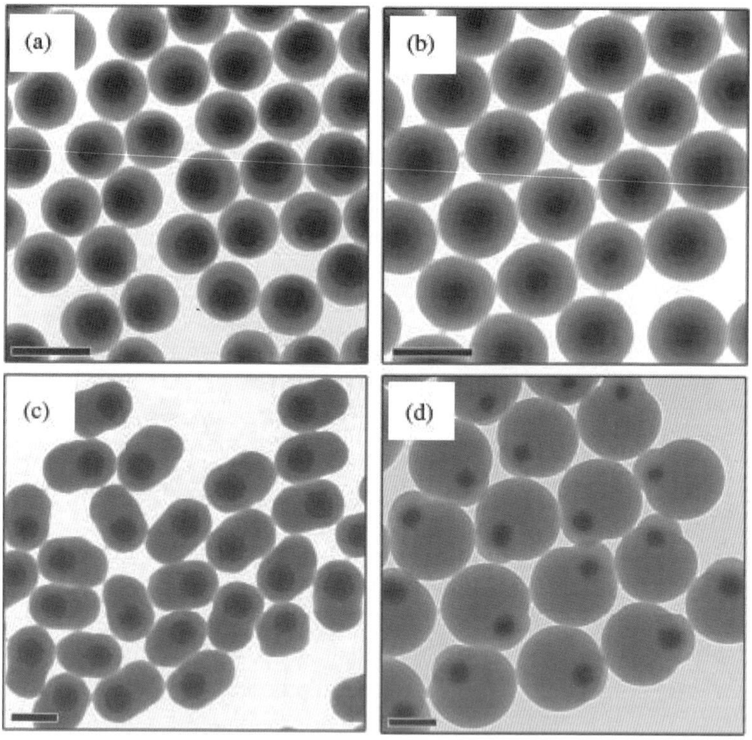

Figure 19. TEM micrographs of a series of Fe_3O_4@SiO_2 PS composite colloids with complex structure produced by emulsion polymerization of styrene using MPS-grafted Fe_3O_4/SiO_2 particles as seeds: (a) and (b) Spherical colloids produced in one-step emulsion polymerization (a) without, and (b) with DVB as cross-linker; (c) ellipsoids formed by swelling and phase separation in one-step emulsion polymerization; (d) doublets produced by separated steps of swelling and phase separation. All scale bars are 400 nm. Reproduced with permission from Ref. [58]. Copyright (2007) American Chemical Society.

2.2.3.2. *Formation of Janus nanoparticles without cross-linking seed*

From the previous section, it is clear that it is very common for the synthesis of JPs by seeded polymerization method that the seeds are cross-linked with difunctional reactive monomers such as divinylbenzene (DVB), which facilitates the polymer protrusion, resulting in the formation of anisotropic structures [47].

In another approach, Nagao *et al.* reported the successful synthesis of anisotropic polymer particles using surfactant free emulsion polymerization in the presence of the reactive silane coupling agent 3-methacryloxypropyl-trimethoxysilane (MPTMS) instead of difunctional DVB monomer (Fig. 20) [59]. This method allows the preparation of anisotropic polymer particles containing inorganic particles, such as silica, which can be removed by dissolving in hydrofluoric acid to form hollow polymer particles.

Janus hybrid colloidal particles with various metal NPs including Au and Ag were synthesized by Xia *et al.* using a simple, robust, and scalable approach [60]. The key for the successful synthesis of uniform metal–PS

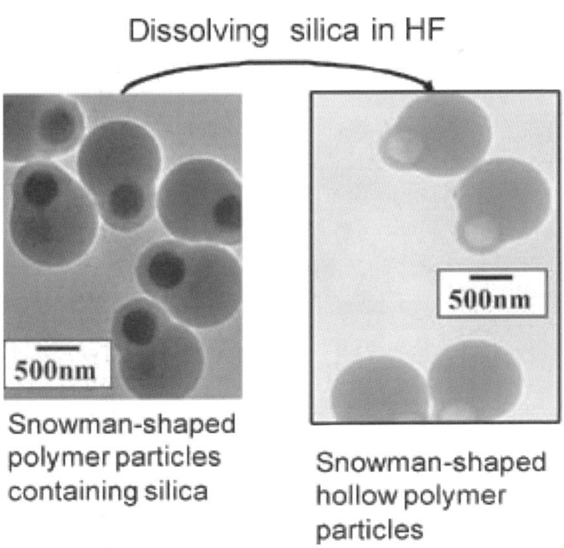

Figure 20. Synthesis of anisotropic polymer particles with soap-free emulsion polymerization in the presence of a reactive silane coupling agent. Reproduced with permission from Ref. [59]. Copyright (2008) John Wiley & Sons.

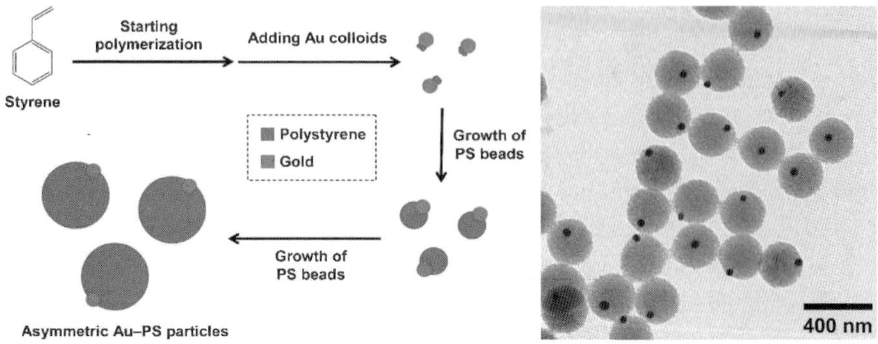

Figure 21. Schematic diagram for the synthesis of asymmetric Au–PS JNPs and their morphology. Reproduced with permission from Ref. [60]. Copyright (2009) American Chemical Society.

asymmetric particles with only one metal nanoparticle on its surface is the addition time point for metal nanoparticle colloids, which is only few minutes after the polymerization process (Fig. 21). The addition of metal NPs prior the polymerization can lead to the formation of PS beads, which contain aggregates of metal NPs instead of one single nanoparticle. The yield of hybrid particles is only 50% when metal NPs were added 30 min after polymerization. This is a robust strategy for the synthesis of hybrid JPs. Even though the type of metal particles and their size should not influence the yield, the amount of added styrene monomer was found to be crucial for achieving 100%.

Yu *et al.* reported an alternative strategy to produce anisotropic polymer particles with a snowman-like shape with two distinct lobes in a few minutes by the protrusion of uncross-linked PS core from a polyelectrolyte multilayer shell in a water and tetrahydrofuran (THF) mixture (Fig. 22) [61]. It was crucial to use mixed solvents to prevent the complete dissociation of PS in THF, resulting in the formation of a polyelectrolyte multilayer capsule. The process could be extended to different materials to generate both JPs with both structural and chemical contrast.

In a similar approach, Park *et al.* found that dumbbell-shaped NPs could easily be synthesized by emulsion polymerization without cross-linking of seed particles. When PS–poly(styrene-*co*-trimethoxysilylpropylacrylate (TMSPA)) core–shell particles were swollen with a monomer, the

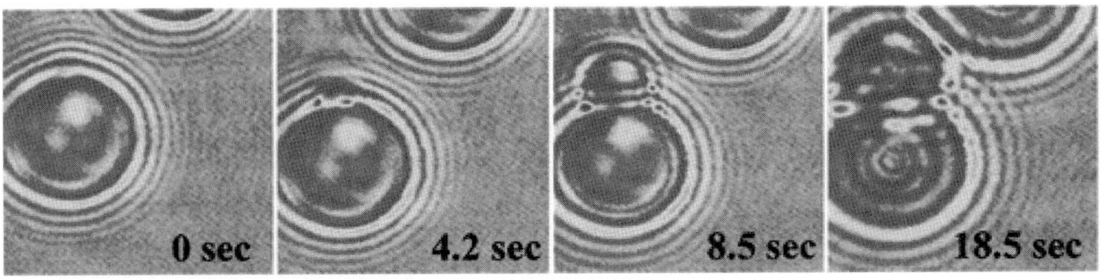

Figure 22. Genesis of anisotropic colloidal particles via protrusion of PS from polyelectrolyte multilayer encapsulation. Reproduced with permission from Ref. [61]. Copyright (2009) American Chemical Society.

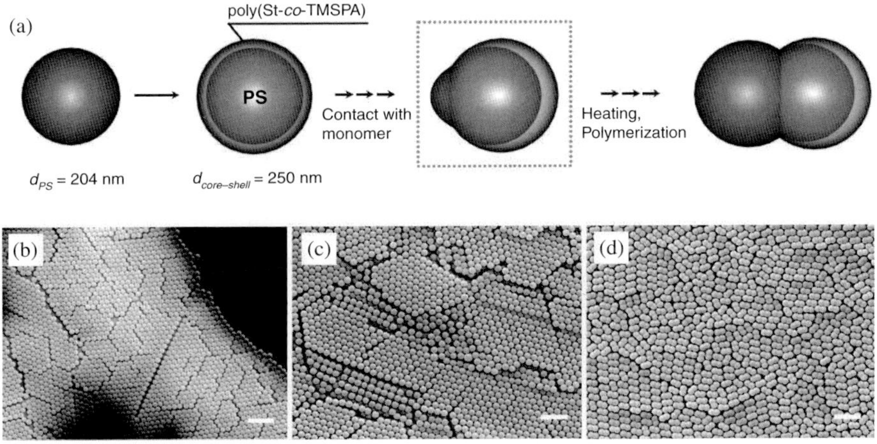

Figure 23. Synthesis of dumbbell-shaped polymer NPs: (a) Schematic representation of two-step seeded emulsion polymerization. (b)–(d) Scanning electron micrographs of (b) PS NPs, (c) PS–poly(styrene-*co*-TMSPA) core–shell nanoparticles, and (d) symmetric dumbbell-shaped. Reproduced with permission from Ref. [62]. Copyright (2010) American Chemical Society.

hydrophobic PS core was expelled onto the hydrophilic shell, forming another lobe of dumbbell-shaped polymer JPs (Fig. 23) [62].

In summary, among the approaches for the synthesis of JNPs, seeded emulsion-based method is the most efficient and promising in scalability, cost effectiveness, and varied composition, relevant to industrial application. The composition of JNPs can be easily controllable by post modification of each compartment with functional groups.

2.3. Properties of Janus Nanoparticles

Recent advancement in nanotechnology enabled the synthesis of JPs with various shapes other than just spherical, including ellipsoids, dumbbells, and snowman. Due to their asymmetric structure and chemical composition, JNPs possess a number of unique properties compared with their homogeneous counterparts. JNPs may have one lobe composed of hydrophilic groups and the other hydrophobic groups [63]. which gives JPs unique properties of both surface activity and amphiphilicity, compared to homogenous particles, which can only present surface activity [64]. Surface activity relates to the tendency to adsorb at interfaces, while amphiphilicity implies the two parts have different affinities for different chemical environments.

2.3.1. Interaction of Amphiphilic JPs at Interfaces

Based on theoretical prediction, Binks *et al.* showed amphiphilic JNPs adsorb at the water–oil interfaces with three times stronger energy than homogenous NPs with the same size and chemical nature [64,65]. Experimental data qualitatively confirmed Binks' theoretical prediction of the effect of JPs at liquid–liquid interfaces. The stronger attachment of JNPs at the water–oil interface can be attributed to their amphiphilic properties as interfacial activity was increased by increasing the amphiphilic character of the particles (Fig. 24) [66]. This property of JNPs makes them more advantageous in the stabilization of emulsions than homogeneous analogs [67].

It is noted that the geometry of JPs influence strongly on their surface activity [68–70]. Bon *et al.* studied the interfacial tension of Janus spheres, Janus cylinders, and Janus disks based on both experimental and simulation data. Interfacial activity of the JPs measured by pendant drop technique due to their adsorption at an oil–water interface was strongly influenced by JP shape (Fig. 25). Janus cylinders showed the maximum reduction in the equilibrium interfacial tension. Janus spheres showed a moderate reduction in surface activity. The addition of Janus disks resulted in the smallest reduction in interfacial tension. Simulation and experimental data also showed the change in JNP shape from sphere to cylinder and disk influenced not only the interfacial tension but also adsorption kinetics, packing behavior, and energy barriers [70].

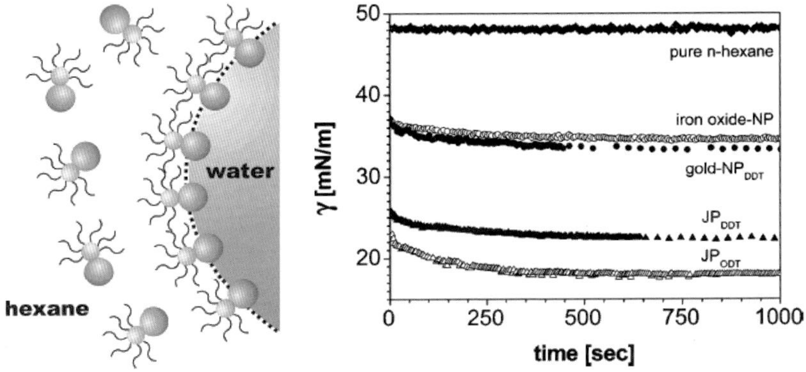

Figure 24. Schematic representation of JPs at the hexane–water interface (left) (light gray: gold part; dark gray: iron oxide part) and interfacial tension vs. time (right). NP: homogeneous NPs, JPs. The gold parts were modified using dodecanethiol (DDT) or octadecanethiol (ODT). Reproduced with permission from Ref. [66]. Copyright (2006) American Chemical Society.

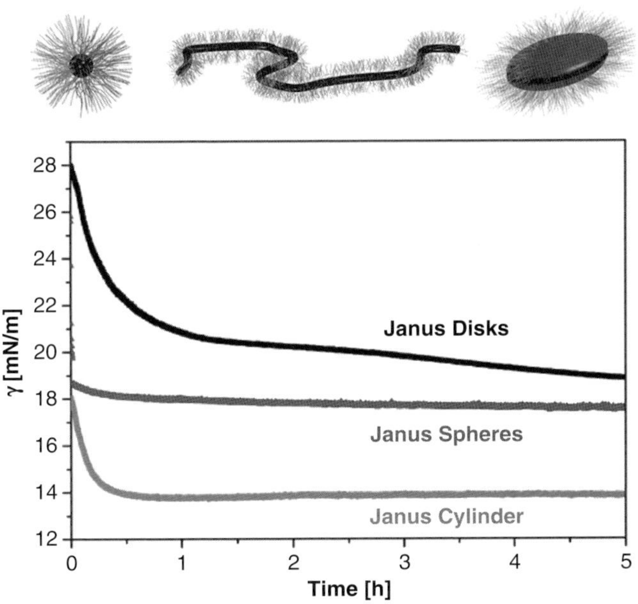

Figure 25. Influence of the JP shape on the interfacial tension at a water/toluene interface. Reproduced with permission from Ref. [70]. Copyright (2013) American Chemical Society.

2.3.2. Self-Assembly

JPs have the ability to self-assemble into complex hierarchical structures in both aqueous and organic solvents [41,43,69,71–76]. For example, spherical Janus micelles having one half of PS and other half of poly(methyl methacrylate) (PMMA) self-assemble into clusters in various organic solvents, such as tetrahydrofuran [41]. The Müller group observed the controlled self-assembly of their prepared Janus amphiphilic micelles into supramicelles with different shapes including cylinders [42,77], spherical [2] and disks [43,44] in water, with the PS side facing each other and the PMAA side exposed to the water above a critical concentration [3].

He *et al.* described a novel strategy for the synthesis of silica JPs with controlled asymmetric shapes including moon-like, dumbbell, and rod-like [78]. The surface of both lobes of the prepared JPs can be chemically modified to introduce amphiphilic features, allowing the spontaneous self-arrangement into various clusters such as hollow spherical or tube-like morphology (Fig. 26). These dense structures could be useful for a wide range of applications such as self-assembly, photonic crystals, and drug delivery.

Hong *et al.* studied the assembly of spherical particles with opposite electric charge, "zwitterionic" micron-sized JPs. The shapes of self-assembled clusters was characterized by epifluorescence microscopy and Monte Carlo computer simulations, which showed that charge asymmetry of individual JPs is preserved in the clusters [74].

The self-assembly of JPs can also be controlled by various parameters including pH, temperature, and external field. The Hatton group studied extensively the self-assembly behavior of prepared JPs composed of magnetite NPs coated on one lobe with poly(acrylic acid) (a pH-dependent polymer) and functionalized on the other lobe with either PS sodium sulfonate (a pH-independent polymer) or poly(*N*-isopropyl acrylamide) (temperature-dependent polymer) (Fig. 27). It was found that their prepared JNPs are monodispersed at high pH values and low temperatures, but can self-assemble at pH values less than 4 (for PSSNa) or at high temperatures than 31°C (for PNIPAM) to form stable dispersions of clusters of approximately 80–100 nm in size [79,80].

The formation of chain-like assemblies of aforementioned concentric and eccentric Janus magnetic NPs was observed when they were

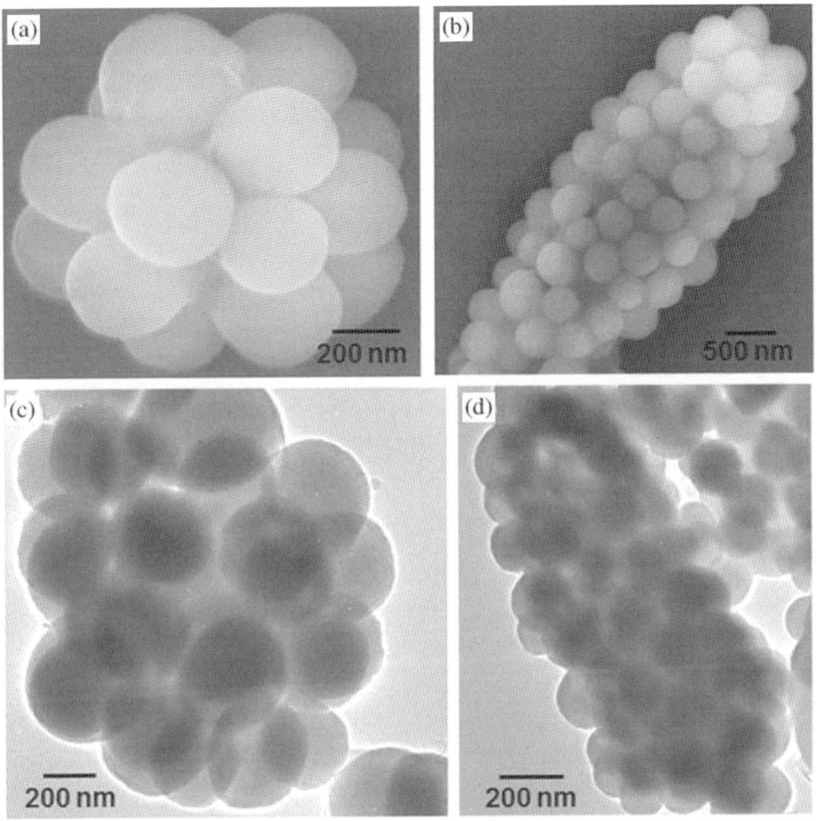

Figure 26. Self-assembly of amphiphilic JPs in water–ethanol (1:1, vol.) mixture. SEM and TEM micrographs of (a) and (c) hollow clusters; (b) and (d) tube-like self-assembled structures. Reproduced from Ref. [78], with permission from the Royal Society of Chemistry.

subjected to a magnetic field (Fig. 28) [58]. The morphology of the chain was linear for concentric particles, whereas zigzag chains were observed for eccentric ones. This observation was in good agreement with another magnetic JP system [81].

2.4. Applications of Janus Nanoparticles

JPs possess unique properties that make them "smart" materials in a wide range of applications including stabilizers for emulsion [82], catalysts in hydrogen peroxide decomposition [83], water-repellent fibers [84],

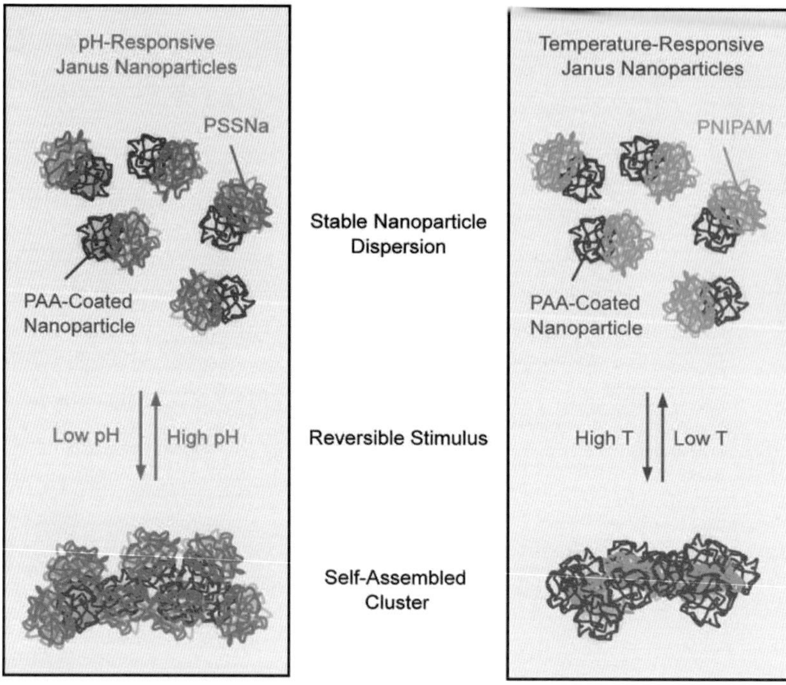

Figure 27. Schematic representation of the JNPs and their self-assembled structures on application of an appropriate stimulus. Note that the PSSNa polymers constitute the stabilizing shell of the clusters, while the PNIPAM chains collapse and form the hydrophobic core of the clusters. Reproduced with permission from Ref. [80]. Copyright (2008) American Chemical Society.

biomolecular sensing [85], imaging and magnetolytic therapy [86,87], electrotrics [88]. and drug delivery [5]. In this section, we will showcase the use of JNPs in emulsion stabilization and drug delivery.

2.4.1. Emulsion Stabilization

The chemical asymmetry of JPs can introduce amphiphilic properties that induce the particles to self-assemble [74,89], and allows them to be used to stabilize oil–water emulsions [11,66,70]. Emulsions stabilized by JPs are more stable than those prepared with homogeneous particles, because the higher adsorption energy of a Janus spherical particle to a fluid–fluid interface [64,90]. Due to the difference in hydrophobicity between the

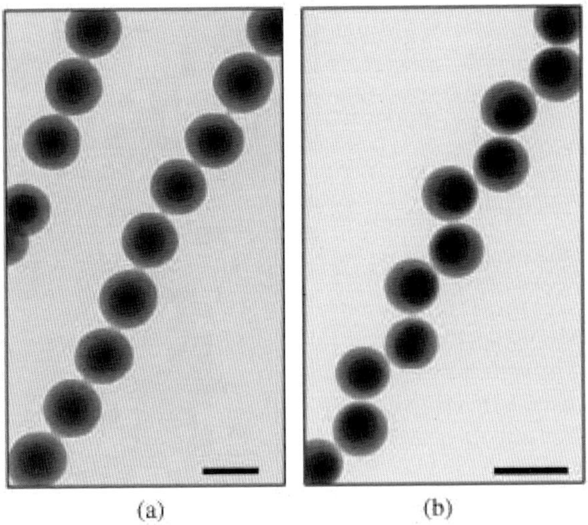

Figure 28. TEM micrographs of self-assembled chainlike structures of (a) concentric and (b) eccentric spherical colloids under an external magnetic field. Both scale bars are 400 nm. Reproduced with permission from Ref. [58]. Copyright (2007) American Chemical Society.

seed and the bulge, the JNPs synthesized by us using seeded emulsion polymerization approach showed good surface activity. We demonstrated the formation of polymer particles stabilized by ellipsoidal PS JPs in a surfactant-free *ab initio* free radical emulsion polymerization of styrene or BA/MMA [8]. The Janus NPs appear to play an important role in the nucleation and stabilization of the particles formed in an *ab initio* surfactant-free emulsion polymerization as well as particle size distribution as shown in Fig. 29.

2.4.2. Applications in Biological Sciences and Drug Delivery

There is an increasing interest in employing Janus nanostructures for biological sciences and drug delivery. The Lahann groups prepared JNPs obtained by electrodynamic jetting of two liquid solutions containing immiscible polymers and reported the good biocompatibility with cells [91–93]. The Ying group prepared biocompatible magnetic-quantum dot heterodimers coated with silica, which is potentially useful for specific imaging of live cells [94]. The use of cadmium-based quantum dots,

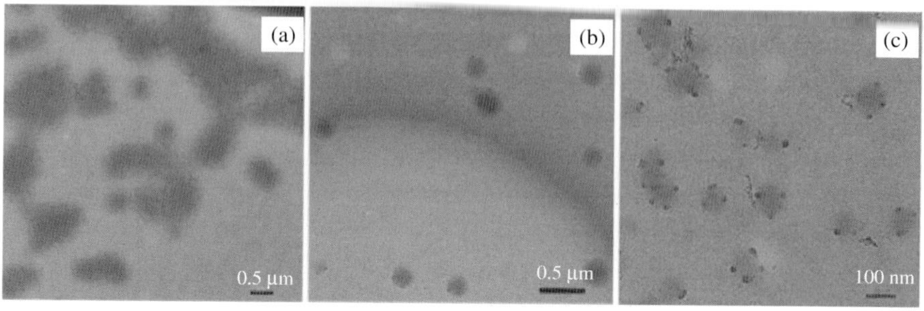

Figure 29. TEM micrographs of the RuO_4 stained surfactant-free poly(BA-*co*-MMA) particles produced by surfactant-free *ab initio* free radical emulsion polymerization: without Janus NPs (A); 1 wt.% of Janus JNPs and 5 wt.% Janus NPs (C). Reproduced from Ref. [8], with permission from the Royal Society of Chemistry.

however, raises some environmental and health concerns. Thus, Ying group developed hydrophobic Fe_3O_4–Ag heterodimer nanocomposites, which were subsequently functionalized with hydrophilic groups for live cells imaging with two-photon fluorescence microscopy [95]. Similarly, Sotiriou *et al.* prepared heterodimer Fe_2O_3–Ag NPs encapsulated by a thin silica layer rendering the particles hydrophilic and biocompatible, facilitating their use as a potential biomarkers (Fig. 30) [87].

Hu and Gao prepared PS–magnetite JNPs for dual applications of imaging and magnetolytic therapy [96]. The authors further demonstrated magnetic field modulated imaging and efficacy in cancer treatment. Wu *et al.* developed a new type of multifunctional cellular probe, Janus nanocoral composed of PS and gold lobes. The gold lobe was used for imaging and the PS lobe was functionalized with antibodies to target receptors for breast cancer cells, enabling the potential for both detecting and imaging. The PS lobe might be used as a reservoir for therapeutic agents, making this nanocoral system a useful candidate for biomedical applications [85]. Along this line, JPs are attractive candidates for drug delivery owing to their unique structure of having two separate compartments, which allows the transport of incompatible drug molecules on the same vehicle. Xie *et al.* prepared for the first time biocompatible JPs that were composed of the biodegradable polymer poly(lactic-*co*-glycolic acid) (PLGA) by using fluidic-based nanoprecipitation. These particles were investigated for the dual delivery of a hydrophobic (paclitaxel), and

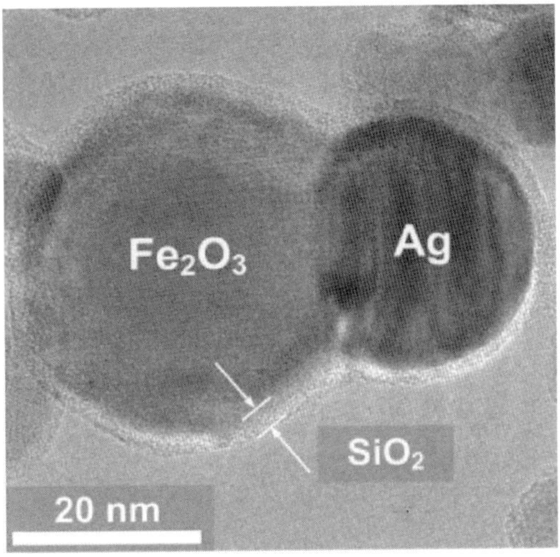

Figure 30. TEM micrograph of SiO_2-coated Ag–Fe_2O_3 sample. Reproduced with permission from Ref. [87]. Copyright (2011) American Chemical Society.

a hydrophilic (doxorubicin hydrochloride) anticancer drug [97]. The results indicated the efficient release of both drugs from the JPs (Fig. 31).

2.5. Conclusions and Future Perspectives

JPs have been shown to be smart materials presenting combined properties of different materials in one single unit. JPs have attracted great attention based on their asymmetric chemical composition and structure leading to their unique properties. Among the existing methods used for the synthesis of JPs, seeded emulsion polymerization is the most promising due to its low cost, scalability, and versatility. In addition, this method enables the synthesis of JPs at nanoscale length, which is essential for certain applications such as drug delivery. Efforts are continuing to further increase the simplicity and versatility of the preparation of JPs in large quantities with tunable chemistry, suitable for the industrial applications. The self-assembly of JPs into hierarchical structures should be investigated more extensively as the properties of these complex structures could be as exciting as those of JPs. In addition, high-end applications of JNPs such as painting, ceramics, and photonic materials need more attention from industry to transfer the

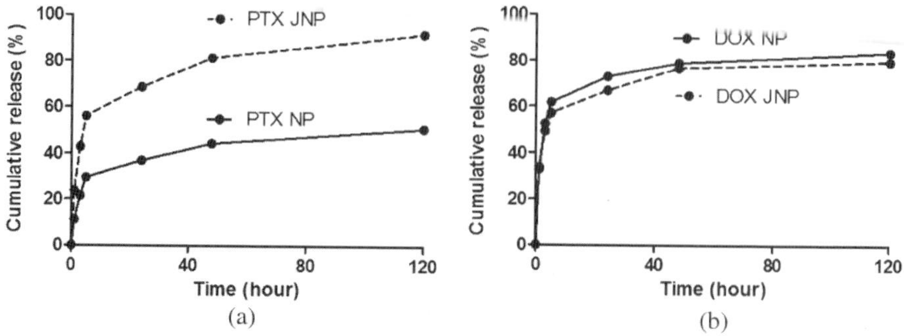

Figure 31. (a) Cumulative release of drugs from NPs. (b) Release profiles of (a) PTX and (b) DOX from JNPs and homogeneous NPs into phosphate buffer at 37°C. For each experiment, 2 g/ML of JPs and 1 mg/ML of homogeneous particles were used. DOX: Doxorubicin hydrochloride; PTX: Paclitaxel. Reproduced with permission from Ref. [97]. Copyright (2012) American Chemical Society.

laboratory scale to advanced technologies. For biomedical applications, the cell uptake mechanism of JNPs is currently poorly understood, because existing studies have only focused on particles with uniform surface functionality. Therefore, there are still many exciting aspects of JNPs to be explored for the development of a new generation of this unique material in the solution of real-world problems.

Acknowledgments

The authors acknowledge the Australian Research Council and DuluxGroup Australia for funding.

References

[1] de Gennes, P. G. *Angew. Chem. Int. Ed.*, **31**, 842 (**1992**).
[2] Walther, A.; Müller, A. H. E. *Soft Matter*, **4**, 663 (**2008**).
[3] Walther, A.; Müller, A. H. E. *Chem. Rev.*, **113**, 5194 (**2013**).
[4] Lattuada, M.; Hatton, T. A. *Nano Today*, **6**, 286 (**2011**).
[5] Tran, L.-T.-C.; Lesieur, S.; Faivre, V. *Expert Opin. Drug Delivery*, **11**, 1061 (**2014**).
[6] Mock, E. B.; de Bruyn, H.; Hawkett, B. S.; Gilbert, R. G.; Zukoski, C. F. *Langmuir*, **22**, 4037 (**2006**).

[7] Ganeva, D. E.; Sprong, E.; de Bruyn, H.; Warr, G. G.; Such, C. H.; Hawkett, B. S. *Macromolecules,* **40**, 6181 (**2007**).
[8] Pham, B. T. T.; Such, C. H.; Hawkett, B. S. *Polym. Chem.,* **6**, 426 (**2015**).
[9] Sheu, H. R.; El-Aasser, M. S.; Vanderhoff, J. W. *J. Polym. Sci., Part A: Polym. Chem.,* **28**, 629 (**1990**).
[10] Chen, Y. C.; Dimonie, V.; El-Aasser, M. S. **24**, 3779 (**1991**).
[11] Tu, F.; Lee, D. *J. Am. Chem. Soc.* **136**, 9999 (**2014**).
[12] Kim, J.-W.; Larsen, R. J.; Weitz, D. A. *J. Am. Chem. Soc.,* **128**, 14374 (**2006**).
[13] Kraft, D. J.; Vlug, W. S.; van Kats, C. M.; van Blaaderen, A.; Imhof, A.; Kegel, W. K. *J. Am. Chem. Soc.,* **131**, 1182 (**2008**).
[14] Tang, C.; Zhang, C.; Liu, J.; Qu, X.; Li, J.; Yang, Z. *Macromolecules,* **43**, 5114 (**2010**).
[15] van Ravensteijn, B. G. P.; Kamp, M.; van Blaaderen, A.; Kegel, W. K. *Chem. Mater.,* **25**, 4348 (**2013**).
[16] Casagrande, C.; Veyssie, M. *C. R. Acad. Sci., Ser. II,* **306**, 1423 (**1988**).
[17] Casagrande, C.; Fabre, P.; Raphaël, E.; Veyssié, M. *Europhys. Lett.,* **9**, 251 (**1989**).
[18] Gu, H. W.; Yang, Z. M.; Gao, J. H.; Chang, C. K.; Xu, B. *J. Am. Chem. Soc.,* **127**, 34 (**2005**).
[19] Hong, L.; Jiang, S.; Granick, S. *Langmuir,* **22**, 9495 (**2006**).
[20] Jiang, S.; Schultz, M. J.; Chen, Q.; Moore, J. S.; Granick, S. *Langmuir*, **24**, 10073 (**2008**).
[21] Cui, J.-Q.; Kretzschmar, I. *Langmuir*, **22**, 8281 (**2006**).
[22] Liu, B.; Zhang, C.; Liu, J.; Qu, X.; Yang, Z. *Chem. Commun.,* **2009**, 3871.
[23] Perro, A.; Meunier, F.; Schmitt, V.; Ravaine, S. *Colloids Surf., A* **332**, 57 (**2009**).
[24] He, Y. J.; Li, K. S. *J. Colloid Interface Sci.,* **306**, 296 (**2007**).
[25] Pardhy, N. P.; Budhlall, B. M. *Langmuir,* **26**, 13130 (**2010**).
[26] Takei, H.; Shimizu, N. *Langmuir,* **13**, 1865 (**1997**).
[27] Sardar, R.; Heap, T. B.; Shumaker-Parry, J. S. *J. Am. Chem. Soc.,* **129**, 5356 (**2007**).
[28] Pradhan, S.; Xu, L.; Chen, S. *Adv. Funct. Mater.,* **17**, 2385 (**2007**).
[29] Love, J. C.; Gates, B. D.; Wolfe, D. B.; Paul, K. E.; Whitesides, G. M. *Nano Lett.,* **2**, 891 (**2002**).
[30] Zhang, L. F.; Eisenberg, A. *Science,* **268**, 1728 (**1995**).
[31] Palyulin, V. V.; Potemkin, I. I. *Macromolecules,* **41**, 4459 (**2008**).
[32] Hu, J.; Liu, G. *Macromolecules,* **38**, 8058 (**2005**).
[33] Cheng, L.; Zhang, G.; Zhu, L.; Chen, D.; Jiang, M. *Angew. Chem. Int. Ed.,* **47**, 10171 (**2008**).

[34] Halperin, A. *J. Phys. France,* **49**, 131 (**1988**).
[35] Ma, R.; Wang, B.; Xu, Y.; An, Y.; Zhang, W.; Li, G.; Shi, L. *Macromol. Rapid Commun.,* **28**, 1062 (**2007**).
[36] Voets, I. K.; de Keizer, A.; de Waard, P.; Frederik, P. M.; Bomans, P. H. H.; Schmalz, H.; Walther, A.; King, S. M.; Leermakers, F. A. M.; Cohen Stuart, M. A. *Angew. Chem. Int. Ed.,* **45**, 6673 (**2006**).
[37] Voets, I. K.; Fokkink, R.; Hellweg, T.; King, S. M.; Waard, P. d.; Keizer, A. d.; Cohen Stuart, M. A. *Soft Matter,* **5**, 999 (**2009**).
[38] Voets, I. K.; Fokkink, R.; de Keizer, A.; May, R. P.; de Waard, P.; Cohen Stuart, M. A. *Langmuir,* **24**, 12221 (**2008**).
[39] Voets, I. K.; de Keizer, A.; Cohen Stuart, M. A.; de Waard, P. *Macromolecules,* **39**, 5952 (**2006**).
[40] Wurm, F.; König, H. M.; Hilf, S.; Kilbinger, A. F. M. *J. Am. Chem. Soc.,* **130**, 5876 (**2008**).
[41] Erhardt, R.; Böker, A.; Zettl, H.; Kaya, H.; Pyckhout-Hintzen, W.; Krausch, G.; Abetz, V.; Müller, A. H. E. *Macromolecules,* **34**, 1069 (**2001**).
[42] Liu; Abetz, V.; Müller, A. H. E. *Macromolecules,* **36**, 7894 (**2003**).
[43] Walther, A.; André, X.; Drechsler, M.; Abetz, V.; Müller, A. H. E. *J. Am. Chem. Soc.,* **129**, 6187 (**2007**).
[44] Walther, A.; Drechsler, M.; Muller, A. H. E. *Soft Matter* **5**, 385 (**2009**).
[45] Sfika, V.; Tsitsilianis, C.; Kiriy, A.; Gorodyska, G.; Stamm, M. *Macromolecules,* **37**, 9551 (**2004**).
[46] Du, J.; Armes, S. P. *Soft Matter,* **6**, 4851 (**2010**).
[47] Sheu, H. R.; El-Aasser, M. S.; Vanderhoff, J. W. *J. Polym. Sci., Part A: Polym. Chem.,* **28**, 653 (**1990**).
[48] Kim, J.-W.; Larsen, R. J.; Weitz, D. A. *J. Am. Chem. Soc.,* **128**, 14374 (**2006**).
[49] Kim, J. W.; Larsen, R. J.; Weitz, D. A. *Adv. Mater.,* **19**, 2005 (**2007**).
[50] Huang, H.; Liu, H. *J. Polym. Sci., Part A: Polym. Chem.,* **48**, 5198 (**2010**).
[51] Kraft, D. J.; Hilhorst, J.; Heinen, M. A. P.; Hoogenraad, M. J.; Luigjes, B.; Kegel, W. K. *J. Phys. Chem., B* **115**, 7175 (**2011**).
[52] Saito, N.; Kagari, Y.; Okubo, M. *Langmuir,* **22**, 9397 (**2006**).
[53] Tanaka, T.; Nakatsuru, R.; Kagari, Y.; Saito, N.; Okubo, M. *Langmuir,* **24**, 12267 (**2008**).
[54] Tanaka, T.; Okayama, M.; Kitayama, Y.; Kagawa, Y.; Okubo, M. *Langmuir,* **26**, 7843 (**2010**).
[55] Tanaka, T.; Okayama, M.; Minami, H.; Okubo, M. *Langmuir,* **26**, 11732 (**2010**).

[56] Ferguson, C. J.; Hughes, R. J.; Pham, B. T. T.; Hawkett, B. S.; Gilbert, R. G.; Serelis, A. K.; Such, C. H. *Macromolecules,* **35**, 9243 (**2002**).
[57] Ferguson, C. J.; Hughes, R. J.; Nguyen, D.; Pham, B. T. T.; Gilbert, R. G.; Serelis, A. K.; Such, C. H.; Hawkett, B. S. *Macromolecules,* **38**, 2191 (**2005**).
[58] Ge, J.; Hu, Y.; Zhang, T.; Yin, Y. *J. Am. Chem. Soc.* **129**, 8974 (**2007**).
[59] Nagao, D.; Hashimoto, M.; Hayasaka, K.; Konno, M. *Macromol. Rapid Commun.,* **29**, 1484 (**2008**).
[60] Ohnuma, A.; Cho, E. C.; Camargo, P. H. C.; Au, L.; Ohtani, B.; Xia, Y. *J. Am. Chem. Soc.,* **131**, 1352 (**2009**).
[61] Yu, H. K.; Mao, Z.; Wang, D. *J. Am. Chem. Soc.,* **131**, 6366 (**2009**).
[62] Park, J.-G.; Forster, J. D.; Dufresne, E. R. *J. Am. Chem. Soc.,* **132**, 5960 (**2010**).
[63] Granick, S.; Shan, J.; Qian, C. *Physics Today,* **62**, 68 (**2009**).
[64] Binks, B. P.; Fletcher, P. D. I. *Langmuir,* **17**, 4708 (**2001**).
[65] Aveyard, R. *Soft Matter,* **8**, 5233 (**2012**).
[66] Glaser, N.; Adams, D. J.; Boker, A.; Krausch, G. *Langmuir,* **22**, 5227 (**2006**).
[67] Kumar, A.; Park, B. J.; Tu, F.; Lee, D. *Soft Matter,* **9**, 6604 (**2013**).
[68] Park, B. J.; Lee, D. *ACS Nano,* **6**, 782 (**2012**).
[69] Ruhland, T. M.; Gröschel, A. H.; Walther, A.; Müller, A. H. E. *Langmuir,* **27**, 9807 (**2011**).
[70] Ruhland, T. M.; Gröschel, A. H.; Ballard, N.; Skelhon, T. S.; Walther, A.; Müller, A. H. E.; Bon, S. A. F. *Langmuir,* **29**, 1388 (**2013**).
[71] Nie, Z.; Li, W.; Seo, M.; Xu, S.; Kumacheva, E. *J. Am. Chem. Soc.,* **128**, 9408 (**2006**).
[72] Erhardt, R.; Zhang, M.; Böker, A.; Zettl, H.; Abetz, C.; Frederik, P.; Krausch, G.; Abetz, V.; Müller, A. H. E. *J. Am. Chem. Soc.,* **125**, 3260 (**2003**).
[73] Dendukuri, D.; Hatton, T. A.; Doyle, P. S. *Langmuir,* **23**, 4669 (**2007**).
[74] Hong, L.; Cacciuto, A.; Luijten, E.; Granick, S. *Nano Lett.,* **6**, 2510 (**2006**).
[75] Li, Z.-W.; Lu, Z.-Y.; Sun, Z.-Y.; An, L.-J. *Soft Matter,* **8**, 6693 (**2012**).
[76] Rocklin, D. Z.; Mao, X. *Soft Matter* **10**, 7569 (**2014**).
[77] Walther, A.; Drechsler, M.; Rosenfeldt, S.; Harnau, L.; Ballauff, M.; Abetz, V.; Müller, A. H. E. *J. Am. Chem. Soc.,* **131**, 4720 (**2009**).
[78] He, J.; Hourwitz, M. J.; Liu, Y.; Perez, M. T.; Nie, Z. *Chem. Commun.,* **47**, 12450 (**2011**).
[79] Lattuada, M.; Hatton, T. A. *J. Am. Chem. Soc.,* **129**, 12878 (**2007**).

[80] Isojima, T.; Lattuada, M.; Vander Sande, J. D.; Hatton, T. A. *Acs Nano,* **2**, 1799 (**2008**).
[81] Isojima, T.; Suh, S. K.; Sande, J. B. V.; Hatton, T. A. *Langmuir,* **25**, 8292 (**2009**).
[82] Walther, A.; Hoffmann, M.; Mueller, A. H. E. *Angew. Chem. Int. Ed.,* **47**, 711 (**2008**).
[83] Valadares, L. F.; Tao, Y.-G.; Zacharia, N. S.; Kitaev, V.; Galembeck, F.; Kapral, R.; Ozin, G. A. *Small,* **6**, 565 (**2010**).
[84] Synytska, A.; Khanum, R.; Ionov, L.; Cherif, C.; Bellmann, C. *ACS Appl. Mater. Interfaces,* **3**, 1216 (**2011**).
[85] Wu, L. Y.; Ross, B. M.; Hong, S.; Lee, L. P. *Small,* **6**, 503 (**2010**).
[86] Hu, S.-H.; Gao, X. *J. Am. Chem. Soc.,* **132**, 7234 (**2010**).
[87] Sotiriou, G. A.; Hirt, A. M.; Lozach, P.-Y.; Teleki, A.; Krumeich, F.; Pratsinis, S. E. *Chem. Mater.,* **23**, 1985 (**2011**).
[88] Nisisako, T.; Torii, T.; Takahashi, T.; Takizawa, Y. *Adv. Mater.,* **18**, 1152 (**2006**).
[89] Chen, Q.; Whitmer, J. K.; Jiang, S.; Bae, S. C.; Luijten, E.; Granick, S. *Science,* **331**, 199 (**2011**).
[90] Tu, F.; Park, B. J.; Lee, D. *Langmuir,* **29**, 12679 (**2013**).
[91] Yoshida, M.; Roh, K. H.; Mandal, S.; Bhaskar, S.; Lim, D. W.; Nandivada, H.; Deng, X. P.; Lahann, J. *Adv. Mater.,* **21**, 4920 (**2009**).
[92] Yoshida, M.; Roh, K.-H.; Lahann, J. *Biomaterials,* **28**, 2446 (**2007**).
[93] Bhaskar, S.; Gibson, C. T.; Yoshida, M.; Nandivada, H.; Deng, X.; Voelcker, N. H.; Lahann, J. *Small,* **7**, 812 (**2011**).
[94] Selvan, S. T.; Patra, P. K.; Ang, C. Y.; Ying, J. Y. *Angew. Chem. Int. Ed.,* **46**, 2448 (**2007**).
[95] Jiang, J.; Gu, H.; Shao, H.; Devlin, E.; Papaefthymiou, G. C.; Ying, J. Y. *Adv. Mater.,* **20**, 4403 (**2008**).
[96] Hu, S. H.; Gao, X. H. *J. Am. Chem. Soc.,* **132**, 7234 (**2010**).
[97] Xie, H.; She, Z.-G.; Wang, S.; Sharma, G.; Smith, J. W. *Langmuir,* **28**, 4459 (**2012**).

Chapter 3

Magnetically Active Asymmetric Nanoheterostructures Based on Colloidal All-Inorganic Multicomponent Nanocrystals

P. Davide Cozzoli[*,†,‡], *Concetta Nobile*[*], *Riccardo Scarfiello*[*,†], *Luigi Carbone*[*], *and Angela Fiore*[*,†]

[*]CNR NANOTEC — Institute of Nanotechnology, Sede di Lecce, c/o Campus Ecotekne, via Monteroni, 73100 Lecce, Italy
[†]Department of Mathematics and Physics "E. De Giorgi", University of Salento, c/o Campus Ecotekne, via Monteroni, 73100 Lecce, Italy
[‡]davide.cozzoli@unisalento.it

Abstract

Colloidal inorganic nanocrystals (NCs) constitute an important class of advanced nanomaterials owing to the flexibility with which their dimensionality-dependent physical–chemical properties can be controlled by engineering their compositional, structural, and geometric features in the synthesis stage and the versatility with which they can be exploited in disparate technological fields, spanning from optoelectronics, energy conversion/production to catalysis, and biomedicine.

In recent years, building upon knowledge acquired on the thermodynamic and kinetic processes that underlie NC evolution in liquid media, synthetic nanochemistry research has made tremendous advances, opening new possibilities for designing, creating, and mastering increasingly complex NC-based assemblies, in which sections of different materials are grouped together into free-standing, easily processable multifunctional nanocomposite systems. This chapter will provide an overview of this fast-growing research field by illustrating progress achieved in the wet-chemical development of last-generation breeds of so-called hybrid or heterostructured nanocrystals (HNCs) in asymmetric non-core/shell geometries, in which distinct material modules are interconnected in heterodimer, heterooligomer, and anisotropic multidomain architectures via heteroepitaxial bonding interfaces of limited extension. The focus will be on HNCs that incorporate at least one magnetic material component combined with semiconductors and/or plasmonic metals, which hold potential for generating enhanced, unconventional magnetic behavior, on one side, and diversified or even new properties and capabilities, on the other side. Various synthetic strategies, all based on the manipulation of seeded-growth techniques, will be described and rationally interpreted within the framework of the currently understood mechanisms of colloidal heteroepitaxy.

Keywords: Colloidal synthesis, nanocrystals, heterostructures, epitaxy, seeded growth, topological control, surface energy, strain, magnetism, multifunctionality.

3.1. Introduction

The discovery of the dimensionality dependence of the chemical–physical behavior of nanoscale matter has laid the basis for the current scientific revolution of nanoscience and nanotechnology, stimulating the development of approaches for the controllable fabrication and in-depth investigation of nanostructures. In this realm, colloidal inorganic nanocrystals (NCs), solution free-standing crystalline nanoparticles (<100–150 nm) entirely grown in liquid media, stand apart from other classes of nanomaterials owing to the precision and flexibility with which their crystal structure, size, shape, and surface functionalities, hence their properties, can be tailored in the fabrication stage, and to the versatility with which they can be exploited in artificial mesoscopic materials, innovative processes, and devices [1–5]. Disparate technological applications have already been demonstrated in various fields, including optoelectronics, catalysis, energy conversion and production, sensing, environmental remediation, and biomedicine, and some of them have also been introduced into the market [4–7].

The wet-chemistry route, which allows manipulating the thermodynamic parameters and kinetic processes underlying NC formation through adjustment of temperature and the liquid growth environment, has proven to be particularly effective for delivering a wealth of semiconductor, metal, and oxide NCs with precisely engineered quality features across diverse size-morphological regimes [1–5,8–17]. Recently, to meet the ever-growing demand for "smart" nanostructured systems capable of exhibiting enhanced or even entirely new properties as well as diversified capabilities for multitask applications, synthetic nanochemistry research has progressed tremendously, enabling access to new breeds of complex multicomponent NCs, broadly referred to as hybrid or heterostructured nanocrystals (HNCs). These are sophisticated free-standing multidomain nanoparticles, characterized by a spatially controlled distribution of their composition, crystal structure, and geometry, which embody two or more different material modules permanently connected through epitaxial heterojunctions. Among the most advanced prototypes reported so far are onion-like HNCs with concentric or eccentric core/shell and yolk/shell topologies, on one side, and domain-segregated heterodimers, heterooligomer, and asymmetric-shaped HNCs, in which geometry-defined material modules are arranged precisely in space, on the other side [4,5,18–29]. Rational design and synthetic development of HNCs combine together to define a smart paradigm in nanochemistry, whereby the elaboration of artificial hybrid nanostructures with an increasingly high level of structural–architectural sophistication and compositional richness is envisioned as an avenue toward overcoming technological potential limitations of conventional single-material nanoparticles. The creation of HNCs allows anticipating achievement of enhanced and/or diversified, or entirely unprecedented properties and functionalities by combining control over the structural and geometric features of their constituent modules, on one hand, with suitable engineering of their 3D connectivity, on the otherside.

HNCs promise to deliver unexpectedly new, broad, and exciting scenarios in both fundamental science and future technology. Primarily, robust, liquid-phase processable HNCs made up of bound inorganic sections, each featuring distinctive optoelectronic, magnetic, and/or chemical properties, represent appealing multifunctional nanoplatforms on which new processes and applications can be designed and manipulated. These encompass, for example, the possibility to assemble hybrid "superstructures", perform cooperative catalytic conversions, install an anisotropic surface distribution of functional molecules, or elaborate multimodal techniques for biomedical diagnostics and therapeutics [4,5,18–24,26–29]. Additionally, the electronic communication that is naturally established in HNCs across their interconnected material domains may underpin exchange-coupling interactions between non-homologous moieties, leading to modulated, synergistically enhanced, or even entirely unprecedented properties and functionalities, an opportunity otherwise prohibited by their isolated material components and their physical-mixture counterparts. For example, HNCs based on semiconductors and/or noble metals may exhibit anomalous absorption/emission and/or conductivity due to modifications in electronic structure, confinement, recombination, separation, and relocation dynamics of photostimulated charge carriers, and/or to plasmon-to-exciton coupling [24,26–28]. In the case of HNCs incorporating magnetic phases and plasmonic metals, abnormally modified or mutually switchable magnetic and optical responses may reflect the synergistic interplay of magnetic, magneto-optical, and surface plasmon resonance properties through various exchange-coupling mechanisms [5,18,20,21,30,31]. Electronic-structure changes or hybridization via the heterointerfaces and the programmable distribution and destination of charge carriers across potential barriers of tunable heights and widths have important implications for the (photo)catalytic activity and chemical-sensing capabilities of HNCs [28,32–35]. These prerogatives clearly indicate that establishment of effective bonding junctions among dissimilar nanoscale materials may be deliberately exploited to engineer the chemical–physical behavior and functionalities of modular HNCs constructed from appropriate material associations [18,20,21,24,25,28,30].

The tailored synthesis of colloidal HNCs with predetermined topologies requires a high level of creativity in architectural design and exquisite synthetic ingenuity. Indeed, the formation of multimaterial architectures occurs at a critical thermodynamic–kinetic crossover. At this point, the delicate dynamics that governs the structural and geometric evolution of individual components inevitably interplays with complex heteroepitaxial deposition or transformative growth pathways, which are, in turn, heavily affected by facet-dependent reactivity, interfacial strain, solid-state atomic diffusivity, and/or amenability to lattice-ion exchange and/or incorporation. Given the current limited availability of general mechanistic guidelines, control over such complex dynamics remains very challenging, and has to be optimized for specific material associations against the particular reaction conditions applied.

This chapter will provide a glimpse into this fast-growing research field by illustrating progress achieved in the wet-chemical development and characterization of last-generation breeds of HNCs in asymmetric non-core/shell geometries, in which distinct material modules are interconnected in heterodimer, heterooligomer, and anisotropic multidomain architectures via heteroepitaxial bonding interfaces of limited extension. The focus will be on HNCs that entail at least one magnetic material component combined with semiconductors and/or plasmonic metals, which hold potential for generating enhanced, unconventional magnetic behavior, on one hand, and diversified or even new properties and capabilities, on the other. For a rich selection of material associations, various synthetic strategies, all based on the manipulation of seeded-growth techniques, will be described and rationally interpreted within the framework of the currently understood mechanisms of colloidal heteroepitaxy. The unique properties and functionalities offered by such generations of complex nanomaterials will also be briefly highlighted.

3.2. Basic Concepts in the Synthesis of Colloidal Heterostructures

3.2.1. Synthesis of Single-Material Nanocrystals

Colloidal NCs are generated upon reaction of molecular precursors in a liquid medium that usually contains selected coordinating solvents and stabilizing agents (e.g., ligands, polymers, surfactants, or self-assembled soft nanotemplates such as micelles). The synthesis is triggered at a suitable temperature, on reaching which highly reactive "monomers" species, the smallest atomic constituent units of the target material lattice, are released and, above a critical supersaturation threshold, they condense, thus initiating the nucleation of NCs and feeding their subsequent growth. The organic stabilizers play several key roles during NC evolution: they can (i) form complexes with the monomers, thus dictating their actual chemical potential in the solution; (ii) dynamically adsorb onto/desorb from the surface of the growing NCs, guaranteeing controllable incorporation of monomers and steady growth; (iii) act as size- and shape-regulating agents; and (iv) prevent the NCs from irreversible aggregation and ensure their solubility [4,5,8,10].

Judicious adjustment of temperature, and the type and the relative concentrations of precursors and organic stabilizers influence numerous thermodynamic conditions and kinetic processes underlying NC formation in liquid media, including solution supersaturation, reactant and monomer diffusion rate, lattice crystallization, solid-state in-lattice ion diffusivity, relative polymorph stability, and crystallographic-direction-dependent lattice growth rate [4,5,8]. Experimental and theoretical insights have indicated that, in order to produce ideally monodisperse NCs, a discrete, burst-like nucleation event must be combined with a diffusion-controlled growth, and these two reaction stages should be temporally separated [9,10,36]. Within this framework, systematically tuned sizes and narrow size variances can be achieved by balancing the relative consumption of monomers between nucleation and the growth stages across an appropriate time evolution of monomer concentration. Depending on the particular material target, this dynamics may be realized by applying appropriate reactant delivery techniques (e.g., a primary swift "hot-injection" combined with secondary slow additions of extra reactants), by manipulating the unique reactivity of the system (e.g., a "delayed" nucleation event followed by rapid autocatalytic growth), or deliberate promotion of digestive ripening (to promote growth of the larger, more stable nanoparticles at the cost of the spontaneous dissolution of the smallest, unstable ones) [9,10,36,37]. Importantly, the dynamic binding of organic stabilizers can significantly affect the relative stability of the surface facets enclosing the growing NCs, thereby shaping their formation into non-spherical habits (e.g., cubes, polyhedrons, rods, wires, polypods). In particular, anisotropic lattice development and branching-out, most frequently observed in materials forming in reduced-symmetry crystal phases and/or featuring polytypisms, may be promoted within kinetically dominated growth regimes, and/or under conditions facilitating growth-symmetry breaking (e.g., in the presence of soft surfactant or polymer lamellar templates, of foreign particle catalysts, or when crystal-oriented attachment pathways are allowed, or growth may be conducted in external electric or magnetic fields or under microwave irradiation) [4,5,8,14–17,38–42].

3.2.2. Thermodynamics of Heterostructure Formation

From a thermodynamic viewpoint, the construction of an HNC may be formally viewed as a sequence of reaction steps, each involving the addition of one secondary material domain to a pre-existing (hetero) nanocrystal substrate and the concomitant formation of a corresponding bonding heterointerface. In general, the material modules that are grouped together are likely to be both chemically and structurally dissimilar (i.e., crystallize in different phases and/or with dissimilar lattice parameters). As a first approximation, the energy balance underlying colloidal heterodeposition can be considered to be analogous to that accounting for the epitaxial growth of thin-film heterostructures and strained quantum dots onto large-area oriented substrates, performed by vapor-phase techniques (e.g., molecular beam epitaxy, chemical vapor deposition, etc.). This is illustrated in Scheme 1. The sign of total Gibbs free surface energy change function, ΔG_S, which accounts for the heterogeneous deposition of a secondary material (2) over a preformed substrate of a different composition/structure (1), will essentially dictate the growth mode of the former [21,43]:

$$\Delta G_S = \gamma_1 - \gamma_2 + \gamma_{1,2}, \tag{1}$$

where γ_1 and γ_2 are the solid/solution interfacial energies of the bare surfaces (facets) of the substrate and of the secondary material through which the heterojunction will be attained, and $\gamma_{1,2}$ is the solid/solid interfacial energy associated with the formed heterointerface(s). Assuming that these parameters are not noticeably affected by the heterodeposition process itself, the terms γ_1 and γ_2 can be expected to be mainly influenced by

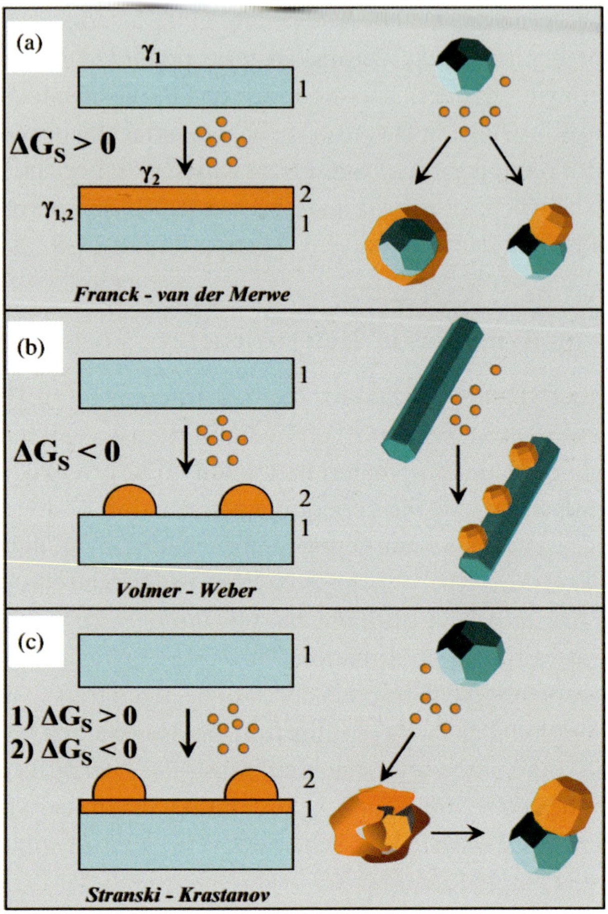

Scheme 1. Comparative sketches illustrating possible heterogeneous deposition modes for a secondary material (referred to as "2") that is deposited from the respective molecular precursors onto a preformed seed substrate of a different material (referred to as "1"): (a) Franck–van der Merwe; (b) Volmer–Weber; and (c) Stranski–Krastanov regimes. Adapted from Ref. [20] with permission. Copyright (2008) Wiley-VCH Verlag GmbH & Co. KGaA.

the binding or adsorption of solution species (e.g., surfactants, ligands, reactive monomers) and, to a lesser extent, by the extension and structure of all other facets exposed to the liquid environment, while $\gamma_{1,2}$ will strictly depend on the bonding strength and degree of structural similarity of the involved lattices at the interfacial region.

If the secondary material exposes lower energy surfaces (i.e., $\gamma_2 < \gamma_1$) and/or can crystallographically match with the substrate to a good extent (hence, $\gamma_{1,2}$ is small), then its deposition will likely proceed layer-by-layer, resulting in a continuous and uniform coverage ($\Delta G_S > 0$: *Frank–van der Merwe* mode in Scheme 1(a)). On the other hand, if the secondary material features higher energy surfaces (i.e., $\gamma_2 > \gamma_1$) and/or is significantly lattice-mismatched (hence, $\gamma_{1,2}$ is high), then it will tend to deposit as an array of island-to-droplet-like domains as a means of minimizing of the overall interfacial area (hence the interfacial misfit strain) shared with the seed substrate underneath ($\Delta G_S < 0$: *Volmer–Weber* mode in Scheme 1(b)). The mean interdomain distance would approximately scale with the extent of excess strain that has to be accommodated. An evolutionary mixed-deposition regime could also be observed (*Stranski–Krastinov* mode in Scheme 1c). In the early stages, the secondary material may grow according to a layer-by-layer mode ($\Delta G_S > 0$). Subsequently, as the deposited layer exceeds a threshold thickness and/or reaches a critical composition (for example, in those cases in which it chemically reacts with the substrate underneath, forming an alloy or solid solution), and/or an excess of thermal energy is provided (e.g., reaction temperature is increased), subsequent growth will proceed in the form of segregated domains protruding out of the initially deposited layer ($\Delta G_S < 0$) in response to an intensification of interfacial strain fields. Exceedingly high strain fields, combined with a strong cohesive energy of the secondary material, may also led to a complete dewetting of the initially deposited thin layer and its reshaping into discrete domains [44,45].

3.2.3. Liquid-Phase Epitaxy via Seeded Growth

The most widely exploited strategy to synthesize HNCs relies on the so-called *seeded growth* technique, which represents the solution-phase analog of vapor-phase heteropitaxial deposition. According to this approach, the liquid growth environment contains preformed NCs of a suitable material, which serve as primary "seeds" for accommodating secondary inorganic domains of different materials upon reaction of the respective molecular precursors. According to the classical nucleation theory (CNT) [2,5,8–10,20,21], the energy barrier, ΔG^*_{het}, which has to be surpassed for a foreign material to nucleate heterogeneously onto a pre-existing condensed phase (e.g., the seeds) is lower than the activation energy, ΔG^*_{hom}, required to trigger otherwise independent homogeneous nucleation of distinct crystal embryos, according to Eq. (2):

$$\Delta G^*_{het} = f(\theta)\Delta G^*_{hom}, \tag{2}$$

where the "wetting" function is given by

$$f(\theta) = \frac{(2+\cos\theta)(1-\cos\theta)^2}{4} \tag{3}$$

with $0 < f(\theta) < 1$; the contact angle θ depends on the surface-tension equilibrium Eq. (4) attained at the three-boundary seed/"droplet"/solution region (which will depend on nature of the materials, extension, and structure of the exposed surfaces, geometry of the substrate seed and of secondary domain deposited thereon, and on energy balance described by Eq. (1)) [2,21,43,46]:

$$\gamma_1 = \gamma_2 \cos\theta + \gamma_{1,2}. \tag{4}$$

Note that the barrier for the growth of a heterogeneously nucleated domain, ΔG^*_{growth}, is far lower than both ΔG^*_{hom} and ΔG^*_{het} and corresponds to the limiting case of complete wetting ($f(\theta) \rightarrow 0$ for $\theta \rightarrow 0$). In an equivalent way, heterogeneous nucleation can be understood as requiring a much lower chemical potential of solution monomers (proportional to their concentration) to be triggered, relative to homogenous nucleation:

$$\Delta\mu_{het} < \Delta\mu_{hom}. \tag{5}$$

The deposition regimes predictable on the basis of Eq. (1) can be safely transcribed to the context of a seeded-growth synthesis, whereby the overall energy change explaining the preference for a given topological configuration depends on how the surface and interfacial energy terms

Eq. (1) are competitively balanced with each other. For instance, on a NC seed with well-defined facets, a secondary material can either form an ubiquitous shell (leading to an onion-like HNC) or grow as a discrete module (leading to a heterodimer habit), if the condition for a *Frank– van der Merwe* regime is either met for all facets exposed, or selectively for just one of them, respectively (Scheme 1(a)). On the other hand, under circumstances favoring a *Volmer–Weber* regime, one or more sufficiently extended facets of the original seeds may accommodate multiple domains of the foreign material (Scheme 1(b)). In the intermediate case of a *Stranski–Krastinov* or *dewetting* regime, a transformation from metastable architectures, where the secondary material has initially formed an ubiquitous thin coverage on some or all facets of the seed, to a phase-segregated heterodimer/oligomer heterostructure could be observed as a convenient pathway toward lowering of interfacial strain (Scheme 1(c)).

Finally, it is important to remark that the creation of nanoscale hetero-interfaces in a colloidal solution can benefit from the binding of organic stabilizers or other solution species, which can significantly influence the surface energy terms (i.e., γ_1 and γ_2), to the point that even large values of $\gamma_{1,2}$ may be significantly compensated for. This opportunity lays the basis for the high synthetic flexibility of colloidal epitaxy routes, which can pave the way to "exotic" HNCs made of rather structurally dissimilar materials that would otherwise be difficult to combine into nanoheterostructures.

3.3. Non-Core/Shell Nanocrystal Heterostructures

The configuration in which HNCs have most frequently been engineered is the so-called core/shell topology. In such systems, an inner NC "core" is evenly enwrapped within a "shell" made of one or more layers of other materials, which ultimately governs or mediates all interactions with the external environment. Semiconductors, metals, and oxides arranged in centrosymmetric onion-like or eccentric core/shell configurations share large connecting heterointerfaces, across which direct electronic communication and hybridization may lead not only to chemical–physical properties distinct from those inherent to the individual components (e.g., increased photoluminescence, emission over spectral ranges prohibited for the individual material components, enhanced or modified plasmon absorption, increased magnetic anisotropy, etc.), but also to exchange coupling of non-homologous properties (e.g., exciton–plasmon coupling, magnetic exchange bias, etc.), depending on the specific material association.

HNCs with coherently grown epitaxial interfaces may be achieved when the core and shell materials are characterized by similar crystal structures and closely matching lattice parameters, which permit misfit strain to be kept at acceptably low levels and prevents the generation of dislocations or other defects as long as the coating thickness is sufficiently small. However, various circumstances may occur under which the requirements of lattice compatibility can be fairly less prohibitive [5,21,27]. For example, pathways enabling plastic strain relaxation may be available when the deposition proceeds non-epitaxially, for example, when the shell grows polycrystalline or even partially amorphous. Misfit strain constraints could easily be circumvented under kinetically driven conditions, albeit at the cost of incorporation of a large density of crystal defects at the interfacial regions and/or within the shell. In such cases, the core/shell interface may ultimately entail a number of small-area coherent heterojunctions at which dissimilar crystallographic relationships locally hold between the relevant lattices. Alternatively, the ligand environment may allow the interfacial energy to be efficiently counterbalanced by a proportional decrease in surface energy associated with the outermost exposed shell surfaces.

Over the past decade, however, a high level of synthetic diversity has been achieved by colloidal seeded-growth routes, allowing access to magnetic HNCs with asymmetric non-core/shell topologies. The available broad library comprises nanoarchitectures featuring a spatially asymmetric distribution of their composition and crystal structure. These are heterodimer and heterooligomer type HNCs that incorporate distinct size- and shape-controlled domains of dissimilar materials interconnected through one or multiple heterointerfaces of limited extension. Distinct from their core/shell counterparts, these asymmetric nanoheterostructures can be regarded as the inorganic analogs of complex organic molecules equipped with a number of functional moieties [25]. While grouping the properties of the component materials and allowing their electronic interactions, these HNCs offer a diversified set of surface platforms onto which a topologically controlled distribution of functional moieties may eventually be anchored [5,18,20,21,24,25].

In seeded-growth synthesis, various circumstances can make HNC growth switch from core@shell to a phase-segregated development regime. Under thermodynamically controlled conditions, the topology will be dictated by the ultimate surface energy balance accompanying the deposition event or heterostructuring process Eq. (1). For example, materials that do not form alloys and/or are strongly lattice-uncorrelated can evolve into oligomer-type heterostructures as a pathway toward minimizing the overall interfacial strain at a proportionally smaller cost of increased surface energy (associated to the multiple material surfaces exposed). Equally, small bonding junctions may be induced to form among pre-existing NCs as a means of alleviating the high surface energy, which would otherwise characterize their physical mixtures, for example, in the case of ineffective stabilization by the surface ligands. Other growth circumstances favoring the formation of non-core@shell HNCs may concern the introduction of seeds characterized by site-preferential accessibility or chemical reactivity that could arise from varying degrees of lattice matching at the exposed facets, or from kinetically overdriven deposition conditions [5,18,20,21,24,25].

3.3.1. Heterodimers and Heterooligomers Based on Nearly Isotropic-Shaped Material Domains

A variety of heterodimers and heterooligomers that are composed of two or more isotropically shaped modules can readily be prepared, within the framework of seeded-growth based techniques, by exploitation of different mechanisms of heterostructuring, as sketched in Scheme 2: (a) direct heterogeneous nucleation; (b) non-epitaxial deposition followed by thermally driven coalescence–crystallization and/or solid-state atomic diffusion; (c) reactions at liquid/liquid interfaces; (d) self-regulated homogeneous–heterogeneous nucleation; and (e) crystallographically induced attachment. Representative examples that illustrate the level of architectural precision and sophistication achievable are collected in Figures 1–5.

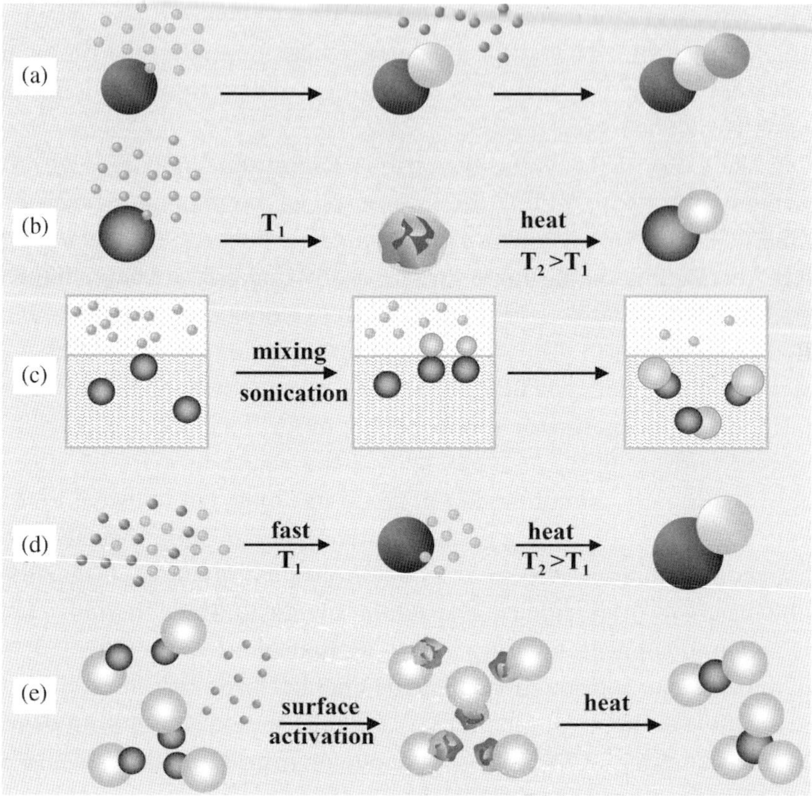

Scheme 2. Sketch of mechanisms underlying formation of oligomer-type HNCs: (a) direct heterogeneous nucleation; (b) and (c) non-epitaxial deposition followed by thermally driven coalescence/crystallization and/or solid-state atomic diffusion; (c) reactions at liquid/liquid interfaces; (d) self-regulated homogeneous–heterogeneous nucleation; (e) crystallographically induced attachment of preformed heterodimers. Adapted from Ref. [20] with permission. Copyright (2008) Wiley-VCH Verlag GmbH & Co. KGaA.

3.3.1.1. *Direct heterogeneous nucleation*

Direct heterogeneous nucleation (Scheme 2(a)) is among the most frequently exploited pathway toward constructing HNCs in the form of binary and ternary oligomers composed of diverse associations of magnetic, metal, and materials. Such HNCs have been envisioned as key elements on which new technological solutions, especially in areas (e.g., sensing, imaging, and therapy in biomedicine) where multifunctionality and multitasking are in great demand, may be founded. For example, it has been proposed that distinct material modules of an HNC can be used as platforms for the site-specific anchoring of selected biomolecules or ligands [18,47–49]. In addition, while a metal or semiconductor domain can enable optical detection, a magnetic module can be utilized for complementary purposes, such as for magnetic resonance imaging (MRI), optical imaging, and magnetic separation [18,47–54]. The existence of bonding heterointerfaces through which dissimilar materials can electronically communicate has clearly been recognized to impact on the magnetic [30,54–60], optical [61–63], transport [55], magneto-optical [31,64] and (electro)catalytic [65–73] and energy-storing properties of appropriately engineered HNCs [74].

Thermal decomposition of metalorganic precursors in the presence of preformed Me^1, Me^2, Me^3, Fe_3O_4, or FePt seeds in non-coordinating solvents, such as octadecene (ODE) or phenyl ether-containing oleic acid (OLAC), oleyl amine (OLAM), and/or tri-*n*-octyl phosphine (TOP) surfactants at 200–300°C has enabled access to heterodimer HNCs made of two nearly spherical and/or cubic-shaped domains epitaxially interconnected, such as of Me^1–Fe_3O_4 (Me^1 = Au, AuAg, PtPd, AuPd, AuPt, Pt, Ni, Cu), Me^2–CoO (Me^2 = Au, AuAg, Pt), Me^3–MnO (Me^3 = Au, Ag), Fe_3O_4–PbS, Fe_3O_4–CdSe, FePt–YZ (YZ = $CoFe_2O_4$, PbSe, CdSe, In_2O_3, Fe_3O_4, MgO), and Fe_3O_4–MnO, respectively [52,53,55,59,67,68,71–73,75–87]. Depending on the geometric features of the two component domains, these heterostructures exhibit diverse morphological profiles, spanning from peanut-, dumbbell- to brick- or flower-like (Figs. 1(a)–1(c)). In most cases, the non-negligible difference in lattice parameters can be considered to be the main driving force that promotes segregation of the concerned materials into discrete domains, oriented relative to one another so as to guarantee coherent heterointerfaces with reduced extension and minimal strain. The formation mechanism was investigated in detail for the Au–Fe_3O_4 association, for which a fundamental role of the solvent in regulating the Fe_3O_4 nucleation sites on the Au seeds was identified [77,78]. The dumbbell-like configuration yielded through reactions performed in non-polar media was explained by invoking induction of charge polarization at those regions of the Au seeds where Fe_3O_4 had initially been deposited. Such polarization resulted in a corresponding electron density depletion at other locations,

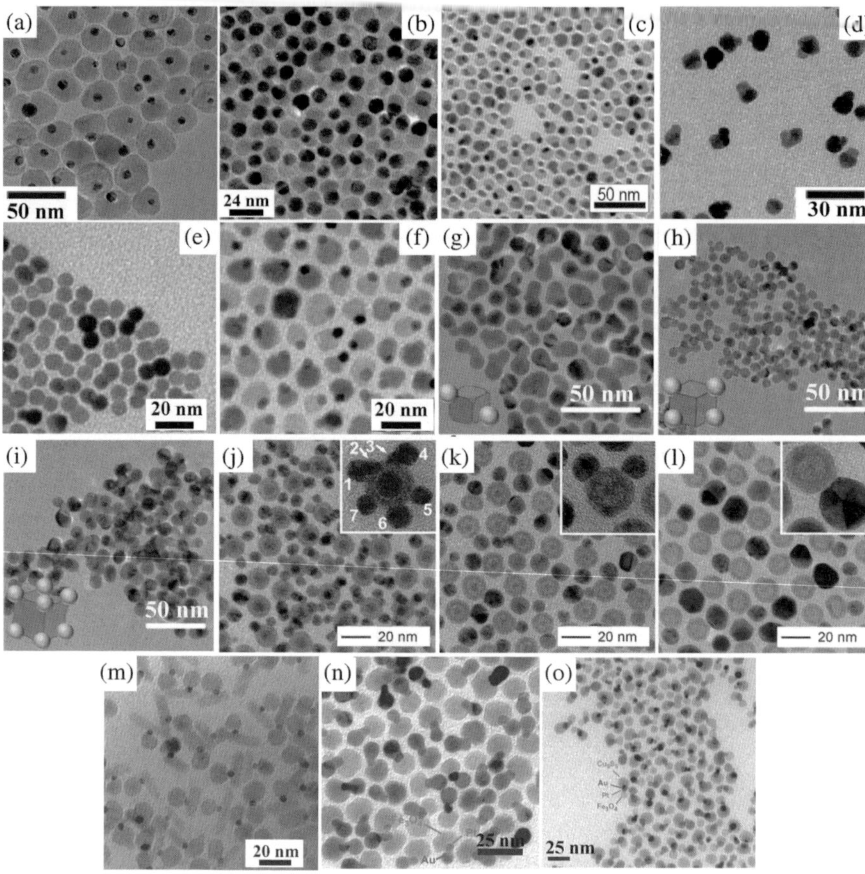

Figure 1. Examples of heterodimer and heterotrimer HNCs synthesized by direct epitaxial heterogeneous nucleation onto preformed seeds (cf. Scheme 2(a)). The panel shows galleries of low-magnification TEM images of (a) peanut-shaped Au–MnO HNCs. Adapted from Ref. [52] with permission. Copyright (2008) American Chemical Society); (b) dumbbell-like Au–Fe_3O_4 HNCs. Reproduced from Ref. [77] with permission. Copyright (2005) American Chemical Society; (c) peanut-shaped FePt–In_2O_3 heterodimer HNCs with cubic-shaped FePt domains. Reproduced from Ref. [83] with permission. Copyright (2021) American Chemical Society; (d) FePt–Au heterodimer HNCs made of cubic-shaped FePt and spherical-shaped Au domains (reproduced from Ref. [50] with permission, copyright American Chemical Society); (e) symmetric $CoPt_3$–Au heterodimer HNCs synthesized according to Ref. [100]; (f) peanut-shaped Pt–Fe_3O_4 HNCs. Adapted from Ref. [68] with permission. Copyright (2010) American Chemical Society; (g)–(i) trimeric and oligomeric FePt–Au HNCs grown starting from cubic FePt seeds. Reproduced with permission

where further nucleation events could, in fact, be inhibited. In contrast, in syntheses carried out in a more polar electron-donor solvent, any electron deficiency generated over the Au surface could be compensated for and leveled off by the medium molecules, which contributed to render the seed a more suitable ground either for the accommodation of multiple Fe_3O_4 "petals" or even for the attainment of uniform oxide coverage [68,77,78]. Recently, it has been discovered that the formation of AuPt–Fe_3O_4 heterodimers grown from AuPt seeds involved the initial evolution of AuPt–Fe_3O_4 core–shell HNCs (at relatively low temperature, 190°C), followed by partial dewetting of the Fe_3O_4 shell and its reshaping into a discrete spherical Fe_3O_4 domain attached aside (at much higher temperatures, 280–320°C) [72]. Along this process, the AuPt seed domain ultimately retained a thin discontinuous Fe_3O_4 shell on the hemisphere region that was diametrically opposite to the position of the major Fe_3O_4 domain. A similar mechanism relying on the dewetting of an unstable Fe_3O_4 shell formed on the metal seeds at intermediate stages under relatively mild conditions may be expected to be involved in the formation of other coinage-metal–Fe_3O_4 heterodimers. Actually, interesting "tug-of-war" etching–destabilization experiments, aimed at probing the chemical reactivity of these heterostructures [69,72,88], have showed that Au could be selectively leached out from Me–Fe_3O_4 (Me = Au, AuPt) heterodimers upon oxidation with I_2 at near room temperature. Depending on the topological profile of the starting heterodimers, either nearly spherical Fe_3O_4 NCs, each bearing a concavity, or peanut-/dumbbell-like solid-hollow homodimer Fe_3O_4 NCs were obtained. In all cases, the formed concave

Figure 1. (***Continued***) from Ref. [94]. Copyright (2013) Royal Society of Chemistry; (j–l) evolution from Fe@Fe_xO_y–Ag heterooligomer HNCs to lower-order hollow-Fe–Fe_xO_y–Ag heterooligomers to hollow-Fe_xO_y–Ag heterodimers via ripening and Kirkendall diffusion. Reproduced with permission from Ref. [105]. Copyright (2011) Wiley-VCH Verlag GmbH & Co. KGaA; (m) Au–Fe_3O_4–PbS heterotrimer HNCs obtained by nucleating a rod-shaped PbS section on the Au domain of Au–Fe_3O_4 heterodimer seeds. Reproduced from Ref. [75] with permission. Copyright (2006) American Chemical Society); (n,o) Ag–Pt–Fe_3O_4 and Cu_9S_5–Pt–Fe_3O_4 heterotrimer HNCs obtained by chemoselective nucleation of Au or Cu_9S_5 on the Pt domain of Pt–Fe_3O_4 heterodimer seeds. Reproduced from Ref. [117] with permission. Copyright (2012) Nature Publishing Group.

region or void corresponded to the volume that was originally occupied by the Au domain; in particular, when bimetallic AuPt seeds were involved, the nanocontainer section could encase a residual Pt domain [69,72,88]. The production of such exotic nanostructures with concave surfaces or extended cavities indirectly revealed that the metal (hemi)domain of these heterodimers, which had initially been assumed to be totally "naked", thus protruding out into the external liquid medium, could frequently accommodate a thin, porous, or discontinuous (hence, chemically accessible) shell of Fe_3O_4 instead. These findings corroborate the dewetting-driven mechanistic picture highlighted earlier.

In additional mechanistic investigations on other metal/metal oxide and metal/semiconductor systems, the transition from dumbbell- to flower-like geometry was achieved by increasing the temperature and/or relative precursor to seed proportions [51–53,75,76,78,82]. In some cases, the nature of ligands bound to the surface of the starting seeds or added during the seeding stage was found to be critical to driving the preference for a heterodimer topology over a core–shell one, which pointed to the influence of kinetic processes on topology selection [51–53,55,76–78,82,84].

In an attempt to rationalize the formation of peanut-shaped FePt–In_2O_3 heterodimer HNCs [83], the crystallographic relationships between two materials were investigated on the basis of the coincidence site lattice theory (CSLT) [89–92]. Upon studying the degree of matching between points of the FePt and In_2O_3 lattices, and the frequency at which this correspondence occurred along the relevant heterojunctions, several couples of facets of the two materials were identified to be capable of forming satisfactorily lattice-matched heterointerfaces. An empirical law (bonding energy criterion) for explaining the preference for the relative FePt–In_2O_3 lattice orientations that were observed experimentally in the heterodimers was proposed: instead of assuming that In_2O_3 overgrowth took place on the facets of the FePt seeds, where lattice mismatch could be minimized, the epitaxial deposition of In_2O_3 was considered to most favorably take place on the FePt crystal facets, in which the first atomic monolayer of the deposited In_2O_3 had the strongest chemical affinity for the seed [83].

By a reverse reaction scheme, heterodimers made of FePt–Au [50,75,76,93–99], $CoPt_3$–Au [100,101], and Fe_3O_4–Ag [51,102–106], were generated upon performing reduction of Au(I)–, Au(III)–, and Ag(I)–ligand complexes onto FePt, $CoPt_3$, and Fe_3O_4, seeds, respectively, with mild reducing agents (alkyl amines, alkyl diols, Ar/H_2 atmosphere) at moderate temperatures (<120°C) (Figs. 1(d)–1(f)). With the introduction of an extra reaction step, γ-Fe_2O_3–Cu_2O heterodimer HNCs were produced from parent γ-Fe_2O_3–Cu heterodimers upon post-synthesis air oxidation of their Cu sections [107].

The properties of these magnetic/metal/semiconductor HNCs clearly diverge from those of the individual components alone. For example, the luminescence of semiconductor modules is severely quenched due to the metal contact promoting electron transfer, thereby decreasing the probability of radiative electron–hole recombination [24,26,28]. The plasmon resonance features of coinage-metal sections results in being largely modified depending on the dielectric characteristics of the proximal environment (solution + attached domain of another material) [75,76,95–98,101,102,108–110]. In addition, the relevant magnetic parameters of magnetic domains are often found to deviate from these of the otherwise isolated components [50,81,93,94,104,111], indirectly revealing an influence of interfacial electron communication on the electronic and magnetic behavior [51,75,100–103,108].

Mechanisms of HNC formation have been elaborated on the basis of experimental evidence and of dedicated experiments. The possibility to tune the geometric parameters of $CoPt_3$–Au, FePt–Au, and Fe_3O_4–Ag heterodimers can be rationalized on considering that the $CoPt_3$ and Fe_3O_4 seeds utilized in the reported cases could act as redox active heterogeneous catalysts with varying degrees of reactivity toward reduction of the relevant metal–complex precursor, depending on their size and shape (faceting). The seed to precursor concentration ratio and temperature regulate the size to which the secondary Au and Ag domains could ultimately be grown by dictating the rate and extent of secondary metal-ion addition at their surfaces [51,94,100,104–106]. When interfacial strain is not prohibitive, smaller-sized seeds may accommodate a continuous thick shell of the foreign materials as a means of reducing their high surface energy [112]. On the other hand, when the seeds expose crystallographically and/or chemically inequivalent facets and/or the emergent interfacial strain, being highly dependent on surface curvature, may not be alleviated, heterodimer-type topologies are preferred over core–shell ones [51,94,101]. A recent detailed study on the formation of Au–Pt–alloy heterodimer HNCs [113] has unexpectedly revealed that the heteroepitaxial deposition of Au on the Py-alloy seeds proceeds through the initial formation of a thin Au shell that exerts high stress onto the seed underneath. In the later reaction stage, the intensification of lattice strain and its subsequent relaxation accompany Au dewetting. Thus, the Au/Pt-alloy heterostructures undergo a topological transition from a core@shell to a dumbbell-like configuration.

When relatively large-faceted NCs enclosed by stable facets with comparable surface structure and/or reactivity are utilized as seeds, growth of multiple domains of the secondary components may occur at equivalent facets or in proximity of surface sites (defects, edges grooves), where nucleation may be kinetically favored and/or interfacial strain may be minimized. These circumstances may lead to heterooligomer HNCs with binary composition, namely heterostructures comprising flower-like assemblies of multiple domains of the secondary materials arranged around a single central core that essentially retains the geometry and structure of the original seeds [82,90,91,105,106,112,114,115]. The use of appropriate surface-binding surfactants [106] or delicate adjustment of reducing conditions [94] have been proposed as valuable strategies for either accentuating or leveling off the reactivity of the different seed facets exposed, thereby modulating the frequency of heterogeneous nucleation events (Figs. 1(g)–1(i)). In another interesting case study [105], during heterogeneous Ag deposition on spherical Fe@Fe_xO_y seeds in organic media, an evolution from high-order multiply Ag-decorated Fe@Fe_xO_y–Ag heterooligomers to lower-order hollow-Fe–Fe_xO_y–Ag heterooligomers and, finally, to hollow-Fe_xO_y@Ag heterodimers was captured (Figs. 1(j)–1(l)). Across this sequence, the average Ag domain and Fe_xO_y shell size increased, while the inner Fe cores of the seeds shrunk until their complete disappearance. The time evolution of the density and size of the Ag domains was found to be consistent with the classical LaMer nucleation and Ostwald ripening mechanisms, according to which the growth of nanoparticles is initiated from a temporally limited nucleation burst, followed by subsequent growth of the larger nanoparticles within the population upon addition of the monomers derived from the dissolution of the smaller, unstable nanoparticles. Parallel to this process, the Fe@Fe_xO_y substrate seeds underwent progressive oxidation and Kirkendall diffusion processes [11–13] (most likely driven by trace O_2, Ag^+ and/or NO_3^- ions), which eventually generated hollow Fe_xO_y domains with thick walls [105].

Conceptually mimicking the total-synthesis approach used to construct complex organic molecules bearing multiple functional moieties, elaborate multicomponent heterooligomer HNCs with progressively higher nuclearity order have been engineered by performing additional chemoselective heterogeneous nucleation steps on heterostructured seeds (Scheme 2(a)). For this purpose, well-known chemical pathways for growing the target materials were rationally combined and applied under appropriate conditions. For example, Fe_3O_4–Au–PbSe heterotrimers (Fig. 1(m)) were synthesized by directing deposition of PbSe onto binary Fe_3O_4–Au dumbbell-shaped heterodimer seeds upon reaction of Pb/Se–surfactant complexes [75,116]. Interestingly, the tertiary PbSe component nanostructures could grow out of the Au domains anisotropically, via a solution–liquid–solid (SLS) growth mechanism,[5] developing into nanorod sections (Fig. 1(c)) that eventually detached from the heterodimer seeds and transferred into the solution [116]. Similarly, linear Fe_3O_4–Pt–Me (Me = Au, Ag, Ni, Pd) and Fe_3O_4–Pt–Me_xS_y (Me = Pb, Cu) heterotrimers (Figs. 1(n) and 1(o)) were constructed by site-specific deposition of the desired tertiary metal or metal sulfide component on the Pt domain of Fe_3O_4–Pt heterodimer seeds, a process that was not accompanied by formation of other isomer side products [117]. This reaction dynamics is intriguing in that it can be regarded as a NC-related analog of regiospecificity in molecular systems, in which out of several products that may be formed with different spatial arrangements of their functional moieties, only one is ultimately obtained. On the basis of competently designed control experiments, the observed chemoselectivity was preliminarily rationalized as correlating with an electron enrichment of the Pt domains in the Fe_3O_4–Pt heterodimer seeds due to charge transfer from the adjacent Fe_3O_4 domain (a process not achievable in isolated Pt nanoparticles, in fact). Further microscopic investigation of chemoselectivity in the formation of Ag–Pt–Fe_3O_4 heterotrimer HNCs [118] revealed an initial indiscriminate Ag nucleation onto both the Pt and Fe_3O_4 surfaces of Fe_3O_4–Pt seeds, followed by surface diffusion and coalescence of Ag onto the Pt surface to yield the final Fe_3O_4–Pt–Ag heterotrimer product. The size of the Ag domain of Fe_3O_4–Pt–Ag correlated with the overall surface area of the Fe_3O_4–Pt seeds, which was consistent with a mechanism of Ag coalescence through a surface-mediated process. Additionally, small iron oxide islands on the Pt surface of the Fe_3O_4–Pt seeds (deposited during their synthesis) were identified as defining the morphology of the Ag domain [118]. To obtain other ternary HNC isomers, the concept of a solid-state protecting group was introduced [119]. A thin amorphous iron oxide shell was installed onto the Pt domain of preformed Fe_3O_4–Pt heterodimers to serve as a solid-state protecting group that isolated the Pt moiety, thus redirecting the nucleation of a third domain of Ag or Au to an otherwise disfavored site, namely the Fe_3O_4 domain. This strategy thus allowed producing the distinct and otherwise inaccessible Ag–Fe_3O_4–Pt and Au–Fe_3O_4–Pt heterotrimer isomers, respectively.

Occasionally, heterogeneous nucleation processes have been identified to interplay and/or compete with redox replacement, cation exchange, and the Kirkendall reaction pathways, leading to heterostructures that, however, may embody an irregular distribution of their chemical composition or contain voids in one of the component domains [99,108,120,121]. In the realm of magnetic-based HNCs, an exception is provided by the outcome of the reaction of pre-synthesized metallic α-Fe NCs with nickel acetylacetonate [122]. Mechanistic investigations into this system indicated occurrence of galvanic replacement of Fe in the α-Fe NCs for Ni, followed by rapid oxidation of both Ni and Fe to yield multidomain alloyed-Fe_xNi_y–Me_3O_4 (Me = Fe, Ni) heterostructures.

3.3.1.2. *Post-deposition coalescence–crystallization and solid-state diffusion*

Detailed investigations on heterostructures based on either FePt or γ-Fe_2O_3 and metal chalcogenides of the type MeX (Me = Cd, Zn, Hg; X = S, Se) have revealed how interfacial strain emerging upon post-deposition coalescence/crystallization and solid-state diffusion may jointly impact their topological evolution [76,89–91,95,123,124]. Examples of heterodimer and heterooligomer HNCs derived from these pathways are reported in Fig. 2. Upon reacting either γ-Fe_2O_3 or FePt seeds with

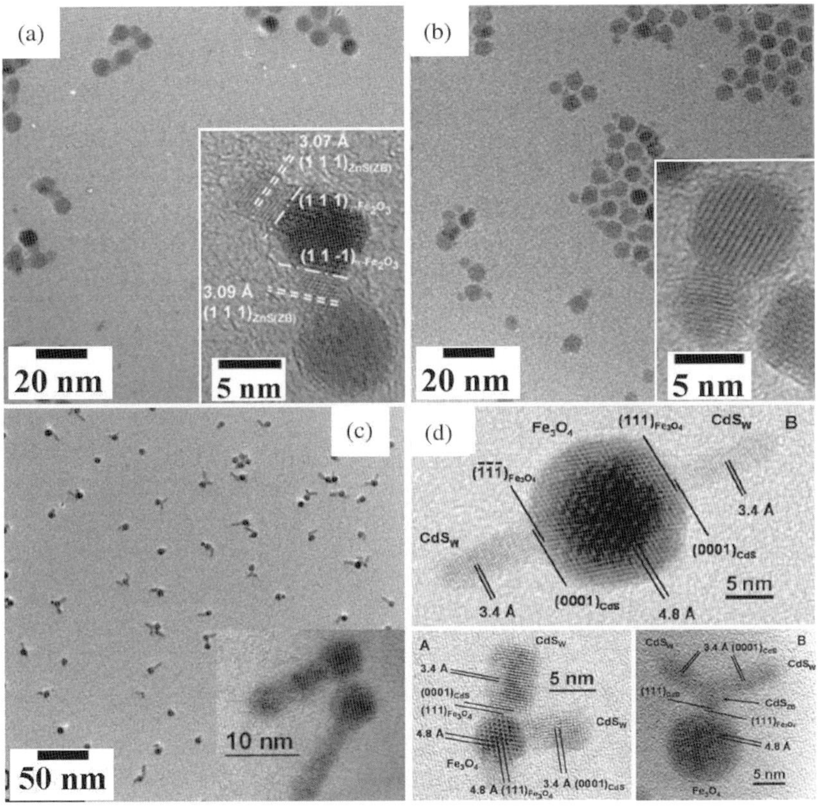

Figure 2. Examples of heterooligomer HNCs synthesized by postdeposition coalescence-crystallization and/or solid-state atomic diffusion (cf. Scheme 2(b)). The panels show low-magnification TEM galleries and selected HRTEM images of: (a) and (b) Fe_3O_4–ZnS heterodimers and heterooligomer HNCs. Adapted from Ref. [89] with permission. Copyright (2005) American Chemical Society; (c) and (d) Fe_3O_4–CdS heterooligomer HNCs with one or multiple shaped CdS sections departing from a single Fe_3O_4 seed. Reproduced from Ref. [91] with permission. Copyright (2009) American Chemical Society.

suitable organometallic precursors that were added sequentially at low temperature, a highly defective and amorphous MeX layer was initially deposited. Upon prolonged heating at 280°C, the amorphous MeX shell gradually crystallized, causing induction of strain across the lattice-mismatched seed/MeX heterojunctions. Over time, the shell coalesced, dewetted, and segregated into a discrete MeX grain aside of the γ-Fe_2O_3 seed [76,89,90,95,123,124] (Scheme 2(b)). Such evolution was explained on the basis of energy minimization arguments. The large junction tension in the annealed core@shell nanostructure could be greatly relieved upon reduction of the interfacial area between the two materials. This energy gain could therefore be large enough to compensate for the proportionally smaller increase in the overall surface energy that eventually accompanied formation of a non-core/shell heterostructure. In the case of the γ-Fe_2O_3–MeX system, the observed topological evolution was rationalized on the basis of the CSLT theory [89–92]. Indeed, the mean number of MeX domains that could be accommodated on each γ-Fe_2O_3 seed correlated with the degree of lattice match achievable at the γ-Fe_2O_3/MeX heterointerfaces and with the seed size [89–91] (Figs. 2(a) and 2(b)). Adjustment of the ligand environment and of the growth kinetics regime could facilitate installation of either one or multiple CdS sections that developed anisotropically out of γ-Fe_2O_3 seeds [91] (Figs. 2(c) and 2(d)). These magnetic-oxide/semiconductor HNCs could still exhibit appreciable fluorescent emission from the semiconductor MeX domains, while retaining the typical superparamagnetic behavior of magnetic modules [76,89,90,95,123,124], which suggests their potential exploitation as bifunctional probes for dual-mode bioimaging [47,95].

3.3.1.3. *Reactions at liquid/liquid interfaces between immiscible solvents*

An interesting technique to synthesize magneto-plasmonic heterodimer HNCs relies on performing the deposition of a secondary metal component on seeds located at the liquid/liquid interface between immiscible liquid phases [125,126] (Scheme 2(c)). Examples of HNCs derived by this strategy are shown in Fig. 3. In the reported procedure, an aqueous solution containing an Au(III) or Ag(I) salt is put in contact with an immiscible organic phase that contains surfactant-capped γ-Fe_2O_3/Fe_3O_4 or FePt seeds dissolved in a non-polar solvent, such as dichlorobenzene, dichloromethane, hexane, or DOE. Upon ultrasonic irradiation under inert atmosphere, an emulsion forms, which consists of a continuous aqueous phase containing organic microdroplets stabilized by the hydrophobic-capped seeds that have self-assembled at the organic/water interfaces

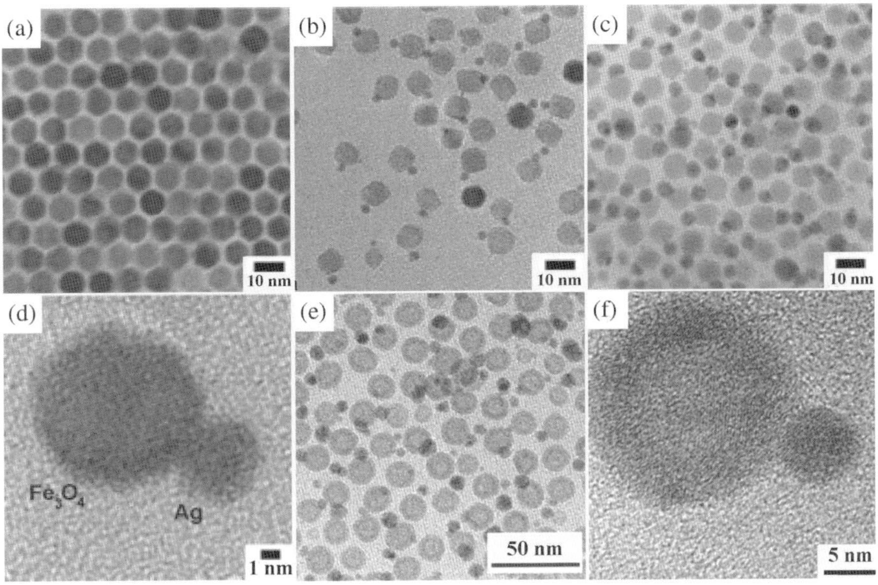

Figure 3. Examples of heterodimer HNCs synthesized at liquid/liquid interfaces (cf. Scheme 2(c)). The panels report variable-magnification, low-resolution TEM images of: (a) Starting Fe_3O_4 seeds; (b)–(d) Fe_3O_4–Ag heterodimers thereof with different Ag domain size. Reproduced from Ref. [125] with permission. Copyright (2005) American Chemical Society; (e) and (f) Fe_3O_4–Ag heterodimers made of one solid Ag domain and one hollow Fe_3O_4 domain. Adapted from Ref. [126] with permission. Copyright (2010) American Chemical Society.

("colloidosomes") [125]. Under these conditions, the seeds provide catalytically active sites onto which the Ag^+ or $AuCl_4^-$ ions are reduced to the respective metals during the ultrasonication. As the seeds are only partially exposed to the aqueous phase, metal deposition is spatially restricted to a small region of their surface and proceeds self-catalytically, thus allowing the formation of only one metal domain on each seed (Figs. 3(a)–3(d)). The "colloidosome"-based approach has been extended to the synthesis of solid–Ag/hollow–γ-Fe_2O_3 heterodimers starting from hollow γ-Fe_2O_3 seeds prepared by manipulating the Kirkendall mechanism [11–13,29,126] (Figs. 3(e) and 3(f)). These heterodimers were used to accommodate a site-differential surface distribution of biomolecules to be exploited for biomedical applications [11,47,126].

3.3.1.4. *Self-regulated homogeneous and heterogeneous nucleation*

A few reports have documented the synthesis of heterodimer HNCs by one-pot approaches, in which no separate seed preparation step is involved. In these cases, the reagents required to build up the heterostructures are introduced together into the same surfactant medium since the beginning of the synthesis. Conditions may be serendipitously discovered under which both the homogeneous and heterogeneous nucleation processes that underlie the formation different material modules self-govern, in such a way that they are triggered at distinct times and then proceed consecutively, with negligible temporal overlapping. Examples of HNCs enabled by such mechanisms are illustrated in Fig. 4.

Co_9S_8–PdS_x and Cu_2S–In_2S_3 heterodimers with acorn-, bottle- and larva-like shapes were produced by copyrolysis of the corresponding cobalt and copper carboxylate precursors in the presence of alkylthiols as the sulfur source [127,128] (Figs. 4(a)–4(c)). During the reaction, selective homogeneous nucleation and growth of one material did occur first, which was followed by the nucleation and growth of the second component that developed through an interface of graded composition. As these metal sulfides are only partially miscible and feature large interfacial energy, a solid-solution cannot be generated, whereas phase–segregation of two materials into distinct domains takes place instead (Scheme 2(d)). Similarly, Au–Ni HNCs were obtained upon heating $HAuCl_4$ and $Ni(NO_3)_2$ in octadecyl amine (ODA) at 120°C, during which initially generated Au seeds triggered the otherwise kinetically hindered reduction of Ni^{2+}–ODA complexes at their surface [129].

Another interesting case is represented by the evolution of FePt–Fe_xO_y heterodimers for which domain size tunability could be easily achieved (Fig. 4(d)). The two material modules formed sequentially in two steps during the reaction of platinum acetylacetonate and $Fe(CO)_5$ in a hot OLAM/OLAC/ODE mixtures [130]. Initially, homogeneous nucleation and growth of FePt NCs was allowed to take place at lower temperature ($T \approx 180$–200°C); then, a thin polycrystalline Fe_xO_y shell was deposited onto the *in situ* generated FePt seeds upon prolonged heating at $T \approx 295$°C. The shell rapidly dewetted out and evolved into a separate cubic spinel

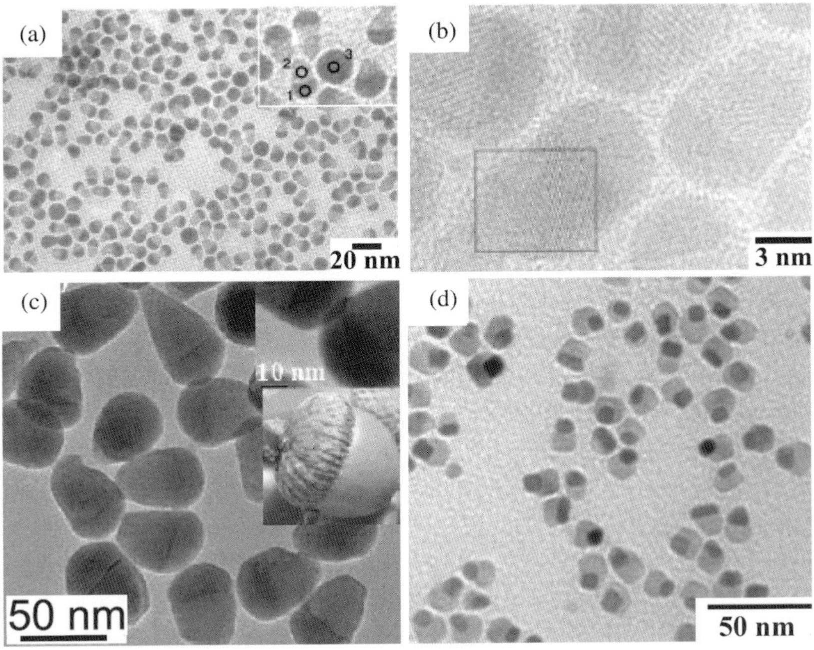

Figure 4. Examples of HNCs synthesized by self-controlled nucleation-growth mechanisms (cf. Scheme 2(d)). The panels show low-magnification TEM and HRTEM images of: (a) and (b) acorn-shaped Co_9S_8–PdS_x heterodimers. Reproduced from Ref. [127] with permission. Copyright (2004) American Chemical Society; (c) acorn-shaped Cu_2S–In_2S_3 heterodimers. Reproduced from Ref. [176] with permission. Copyright (2006) American Chemical Society; (d) peanut-shaped FePt–Fe_3O_4 heterodimers (synthesized according to Ref. [130]).

Fe_3O_4@γ-Fe_2O_3 core@shell domain [131] attached aside, a process that allowed alleviating the intervening interfacial strain. Since each reaction stage was selectively activated under distinct thermal conditions, then regulation of both the temperature and the heating time guaranteed that the two material sections of the HNCs formed at distinct stages of the synthesis course. As a consequence of the magnetic exchange coupling holding between the two soft and hard materials, the HNCs exhibited tunable single-phase-like magnetic behavior and superior performance as MRI contrast agents, not otherwise achievable by their individual components [130,131].

3.3.1.5. *Oriented fusion of preformed HNCs*

A total-synthesis framework with the potential to access increasingly complex HNCs envisions utilization of small low-order HNCs as multifunctional inorganic precursors provided with reactive moieties that can be combined to create additional bonding heterointerfaces, hence, to form larger, higher-order HNCs. Examples of HNCs obtained by exploitation of this concepts are shown in Fig. 5. Ternary Fe_3O_4–Au–Fe_3O_4 HNCs, in which an Au section bridging two Fe_3O_4 domains, were obtained by inducing welding of the Au domains that belonged to distinct peanut-shaped Au–Fe_3O_4 heterodimers with the assistance of added elemental sulfur [75] (Scheme 2(e)). Due to its high affinity to Au surfaces, S was assumed to adsorb on the Au portions of the heterodimers and to facilitate displacement of the capping surfactants thereon, thus making the

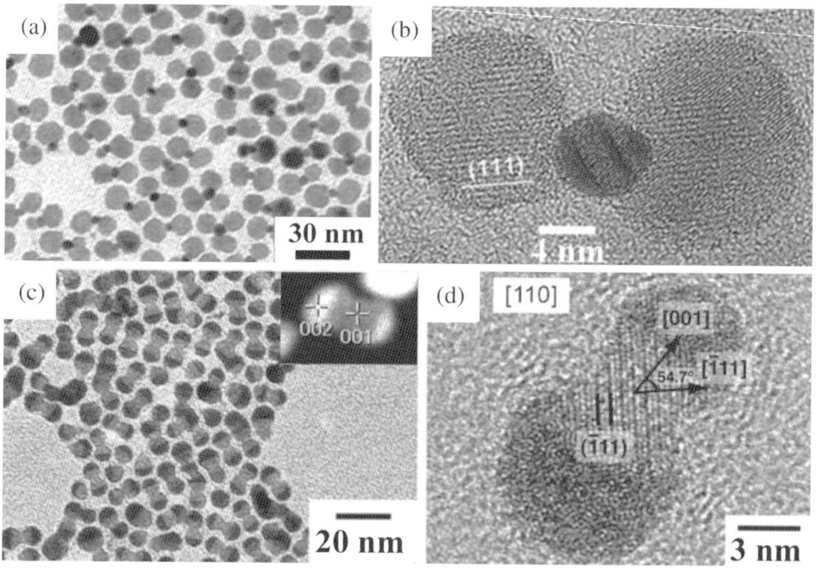

Figure 5. Examples of ternary HNCs synthesized by induced fusion of preformed heterodimer seeds (cf. Scheme 2(e)). The panels show low-resolution TEM and selected HRTEM images of: (a) and (b) Ternary Fe_3O_4–Au–Fe_3O_4 HNCs with an intermediate Au connecting domain. Reproduced from Ref. [75] with permission. Copyright (2006) American Chemical Society. (c) and (d) Ternary PdS_x–Co_9S_8–PdS_x HNCs with a Co_9S_8 bridging section. Reproduced from Ref. [132]. Copyright (2007) Wiley-VCH Verlag GmbH & Co. KGaA.

Au sections prone to reducing their total surface energy upon fusing with each other (Figs. 5(a) and 5(b)). Extension of this strategy allowed access to higher-order oligomers with linear or variably branched topology. For example, Fe_3O_4–Pt–Au–Au–Pt–Fe_3O_4 HNCs were synthesized upon S-participated selective coalescence of preformed Au–Pt–Fe_3O_4 heterotrimers through their Au domains [117].

In another approach, peanut-shaped PdS_x–Co_9S_8–PdS_x HNCs, in which a central Co_9S_8 section connected two PdS_x domains (Figs. 5(c) and 5(d)), were synthesized by reacting PdS_x seeds with cobalt acetate and octadecanethiol precursors in DOE at 230°C [128,132]. As demonstrated by time monitoring of the reaction course, the ternary PdS_x–Co_9S_8–PdS_x HNCs evolved upon crystal-oriented coalescence of acorn-shaped PdS_x–Co_9S_8 heterodimers that had been generated in the earlier reaction stages. The process appeared to be promoted by the weak organic passivation of the Co_9S_8 domains, which made them inclined to weld with each other.

3.3.2. Heterostructures Based on Anisotropically Shaped Material Modules

Magnetic HNCs with a spatially asymmetric arrangement of their component modules have been created by manipulating heterogeneous nucleation/growth reactions on anisotropically shaped seeds, such as nanorods, nanowires, and branched NCs, which most commonly crystallize in non-centrosymmetric phases [5,8]. In addition to offering peculiar shape-dependent properties, anisotropic seeds normally exhibit a pronounced facet-dependent reactivity, thus representing interesting platforms over which pathways to site-selective implantation of secondary domains can be studied and rationalized. On a thermodynamic basis, the preference for secondary material deposition to take place at certain locations of asymmetrically shaped seeds should correspond to the topological configuration allowed by the overall surface-interface energy balance accompanying the formation of all the involved heterointerfaces Eq. (1) [5,8,24,28,133]. For example, selective heterogeneous nucleation may be viewed as being energetically convenient when it permits elimination of some unstable facets of the seeds (e.g., high-index facets at the apexes of nanorods) at the proportionally smaller cost of the interfacial strain energy spent in the formation of the heterojunctions. However, when nucleation and growth occur under kinetically controlled regimes, and/or other chemical or physical transformative pathways come into play (e.g., intraparticle ripening, atomic diffusion, ion exchange), understanding of the mechanisms underlying topology selection may be less straightforward [5,8,20,24,28,133–139].

The following paragraphs describe major synthetic strategies that have been developed to construct magnetically responsive HNCs incorporating anisotropic (linear or branched) material sections. The main growth mechanisms that have been exploited, sketched in Scheme 3, include: (i) site-specific heterogeneous nucleation/growth; (ii) surfactant-controlled site-selective deposition; and (iii) strain-driven heteroepitaxial growth. Examples that demonstrate the degree of synthetic control achievable can be found in Figs. 6–8.

3.3.2.1. *Site-specific heterogeneous nucleation/growth*

The most developed breeds of anisotropically shaped HNCs are represented by nanoheterostructures based on the archetypal II–VI metal–chalcogenide compounds of the type MeX (Me = Cd, Zn, Mn; X = S, Se) and transition metal oxides (ZnO, TiO_2) for which a high level of synthetic capabilities have been matured over the past two decades [5,8,21,26,28,133,140]. Nanoplatelets, nanowires, nanorods, or polypods of these materials are commonly trapped in low-symmetric crystal structure (e.g., hexagonal, tetragonal, orthorhombic) that preferentially extend along or perpendicular to their axis of higher symmetry. Shape anisotropy has important consequences on the seeding behavior of these NCs when used in seeded-growth syntheses. The facets at the apex, edges, and longitudinal/basal sidewalls generally feature distinct atomic arrangement, hence dissimilar chemical reactivity. In nanorods/wires, the lack of a plane of symmetry perpendicular to their major elongation axis implies that the two terminal basal facets at the extremities are crystallographically, hence chemically, non-equivalent. Arrowing at the nanorod termini may further exacerbate such dissimilarities [91,134,136–138,141–146]. In line with the mechanism of their anisotropic growth, the longitudinal sidewalls

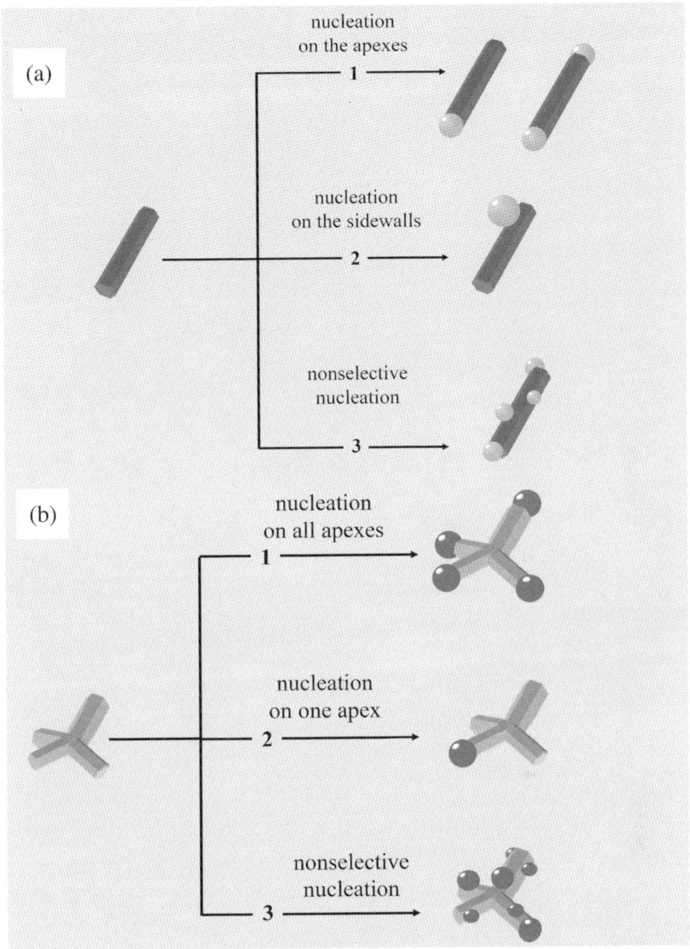

Scheme 3. Sketch of possible pathways leading to NC heterostructure grown from anisotropically shaped seeds: functionalization of a nanorod (a) or of a tetrapod (b) with spherical domains of a different material at selected locations. Adapted from Ref. [19] with permission. Copyright (2007) Bentham Science Publisher.

and the two tips of nanorods/nanowires can be expected to exhibit significantly different propensity to accommodate secondary material domains in seeded-growth strategies (Scheme 3(a), path 1). Furthermore, depending on the specific case, nanorods for which the lattice deviates from the

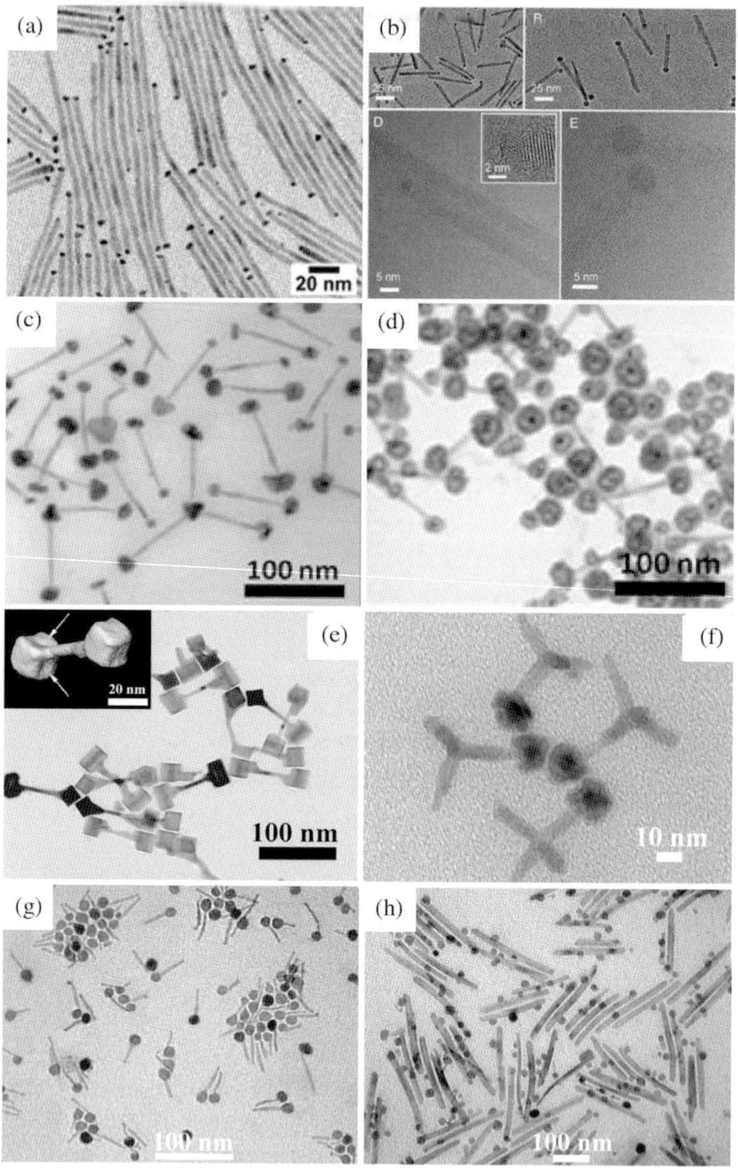

Figure 6. Examples of anisotropic HNCs synthesized by site-selective heterogeneous deposition on preformed nanorods seeds (cf. Scheme 3). The panels depict low-resolution TEM images of: (a) Nanodumbbells of NiPt–tipped CdS nanorods. Reproduced from Ref. [153] with permission. Copyright (2008) American Chemical Society; (b) nanodumbbells of Ni-tipped CdS nanorods. Reproduced from Ref. [155] with permission. Copyright (2016) American Chemical Society; (c) nanodumbbells of Pt–Co–tipped CdS nanorods obtained

ideal bulk crystallographic structure or which have been heterostructured into an inner core@shell architecture can hold an intrinsic dipole moment or accommodate defective or strained near subsurface regions, which may influence their seeding capabilities [28,133,141,143,144,147]. Demonstration of these concepts is illustrated by the fabrication of HNCs with matchstick-like or dumbbell-like topologies, in which either one or both apexes, or the longitudinal sidewalls of nanorod seeds are decorated with domains of other materials [26,28,91,97,133,134,136–139,141–144,146,148–158]. Representative examples are reported in Fig. 6.

At the relevant bonding heterointerfaces, the mismatched lattice arrangement can break the crystal periodicity to a varying extent, giving rise to noticeable interfacial strain, which may, in turn, affect their electronic–band structure [26,28,133,140] as well as the type and extent of exchange coupling between different magnetic phases or between the optical and magnetic functionalities [135–137,143,158,159] Misfit strain may be alleviated: (i) through induced interfacial defects, such as dislocations, stacking faults, or interfacial lattice adaptation (coherency strain) [140,160] that may be easily induced under the high-temperature conditions under which colloidal synthesis is typically carried out; and/or (ii) through gradual variation in the chemical composition and crystal-lattice parameters of the heterojunction regions, due to interdiffusion of the constituent atomic species [154,161]. Experimental results [135,143,154,158,159,161] and thermodynamic modeling [140,162] have indicated that the temperature

Figure 6. ***(Continued)*** upon Co deposition on the tips of preformed Pt-tipped CdS nanorods and (d) nanodumbbells of corresponding Pt@Co_xO_y–tipped CdS nanorods with Pt@Co_xO_y yolk–shell termini obtained upon Co oxidation of their parent Pt–Co-tipped CdS nanorods. Reproduced from Ref. [156] with permission. Copyright (2012) American Chemical Society; (e) nanodumbbells of Co nanorods bearing concave Fe cubes at the apexes, along with (see inset) a tomographic reconstruction of a single heterostructure, where arrows indicate concave surfaces on the terminal Fe cubes. Reproduced from Ref. [158] with permission. Copyright (2014) American Chemical Society; (f) CdSe–CdS core–shell tetrapods asymmetrically decorated with a single Au–Co core–shell tip domain. Reproduced with permission from Ref. [157]. Copyright (2016) Wiley-VCH Verlag GmbH & Co. KGaA; (g) nanomatchsticks made of single-γ-Fe_2O_3-tipped brookite TiO_2 nanorods. Reproduced from Ref. [146] with permission. Copyright American Chemical Society; (h) heterooligomers made of multiply γ-Fe_2O_3-decorated brookite TiO_2 nanorods. Adapted from Ref. [149] with permission. Copyright (2009) Royal Society of Chemistry.

and the particular chemical pathways being operative in the synthesis can be decisive in determining the attainment of either uniformly alloyed, graded, or abrupt interfaces.

As for their core@shell counterparts, in anisotropic HNCs based on associations of dissimilar semiconductors (magnetic) metals and metal oxides, the energy level structure across the interface can be purposely engineered upon proper heterostructure design, so as to promote confinement or spatial separation of the photogenerated charge carriers [24,26,28,133,140]. Interestingly, the overall magnetic anisotropy (and the magnetic behavior in general) may also be strongly affected, even when magnetic modules are connected with non-magnetic ones [135–137,143,158,159]. Such hybrid nanoarchitectures hold promise as key elements for the realization of optoelectronic devices and efficient photo(electro)catalytic platforms [24,26,28,133], and spintronic and magnetic recording systems [135–137,143,158,159].

Intriguing breeds of HNCs obtained by the heterogeneous nucleation/growth are represented by semiconductor–metal heterostructures, whereby anisotropic semiconductor NCs have been exploited as underlying substrates for accommodating various metal domains (Scheme 3(a), path 1). Standard pyrolytic organometallic routes were suitably combined and utilized to synthesize CdX–Me and Me–CdX–Me heterostructures based on CdX (X = Se, S) nanorods functionalized with magnetic and non-magnetic domains (Me = Au, Pt, PtNi, PtCo, Co, Ni) [97,137,148,152–155] (Figs. 6(a) and 6(b)). For these systems, a high degree of selectivity for metal nucleation at both seed apexes was readily achievable [152,153,155,163] (Scheme 3(a), path 1 or 3), whereas indiscriminate deposition at random sites was observed when large aspect-ratio nanorods/nanowires were used as starting seeds, or the synthesis was carried out at exceedingly high reactant-to-seed concentration ratios. However, when delicate organometallic chemistry was manipulated, the choice of coordinating ligands and adjustment of their content in the synthesis media appeared to be critical in ensuring topological selectivity, which pointed to the interplay of kinetic processes [137,153,155,156]. As an effort toward increased structural and compositional complexity, secondary metal domains on preformed heterostructures were envisaged to be exploitable as highly reactive sites where exclusive deposition of a third component could be directed. This route has actually been proven to be feasible within the context of rationally designed multiseeding-step reaction schemes [156,157]. In one report, Pt-tipped Cd@CdS core@shell nanorods were exploited to direct the otherwise infeasible overgrowth of Co domains onto the Pt apexes, leading to corresponding Pt@Co–CdSe@CdS–Pt@Co nanodumbbells with Co@Pt core@shell terminal domains; the HNCs could be then converted into Pt@Co_xO_y–CdSe@CdS–Pt@Co_xO_y heterostructures in which the Pt@Co_xO_y yolk@shell termini derived from selective oxidation of the Co sections [156] (Figs. 6(c) and 6(d)). In another protocol [157], asymmetrically Au-tipped CdSe@CdS tetrapods were first fabricated by accommodating one single Au sphere at one of the four-arm termini of preformed Cd@CdS core@shell tetrapods through a photodeposition process that proceeded through the initial attainment of multiply Au-decorated tetrapod arms, followed by intraparticle ripening of the nucleated Au patches into a single Au tip (this topological evolution is analogous to the one underlying the transition of double-tipped Au–CdSe–Au nanodumbbells into their single-tipped Au–CdSe nanomatchstick counterparts [97]) (Scheme 3(b), paths 1, 2); subsequently, thermolysis of $Co_2(CO)_8$ in the presence of the Au–CdSe@CdS tetrapod heterostructures and carboxylic-acid-terminated polystyrene ligands in 1,2,4-trichlorobenzene at $T = 140°C$ enabled selective overgrowth of a Co shell around the previously implanted Au domains. As a result, CdSe@CdS core@shell tetrapods heterostructures asymmetrically decorated with a single Au@Co core@shell domain at one apex were obtained (Fig. 6(f)).

In all of the aforementioned cases, the strong electronic band coupling between the semiconductor and the magnetic-metal sections was corroborated by a significant luminescence quenching, which transcribed into increased charge-carrier separating capabilities [26,28,133,164]. However, localized surface states at the relevant interfaces were also invoked to account for the dramatically altered optoelectronic properties of such HNCs [26,28,133,152,164,165]. Depending on the reaction medium (e.g., presence of hole/electron scavengers) and the irradiation conditions, the metal section could either act a sink for the electrons photogenerated within the semiconductor or help shuttling them to the solution, which enable, for example, efficient charge retention or enhanced photocatalytic reduction yield [26,28,133,155,164,166,167]. The magnetic properties [137,152] were also found to be modified upon heterojunction attainment. Interestingly, selective functionalization with magnetic metal domains was proven to be a valuable strategy toward enhancing the propensity of heterostructures to self-assemble through deliberate exploitation of tip-to-tip dipolar magnetic interactions [156,157].

Other relevant HNC prototypes obtained by the selective heterogeneous nucleation/growth mechanism are represented by all-metallic anisotropic heterostructures. Tadpole-like HNCs made of one-sided Au-tipped FePt nanorods, periodically Au-decorated FePt nanorods, and Au–FePt necklace-like heterostructures were obtained upon reduction of Au–OLAM complex on preformed FePt nanorod seeds in organic media under controlled Ar or Ar/H_2 atmosphere at 25–80°C [93,111]. Under the specified reaction conditions, the regioselectivity for the possible destination sites of deposited Au was found to correlate with the characteristics of the reducing atmosphere [111] or with the reducibility of the starting Au molecular precursor [93], which indeed appeared to dictate the number of nucleation sites on the FePt seeds. To explain the formation of a periodic arrangements of Au domains over the thin, high-aspect-ratio FePt nanorod seeds [93], an intraparticle Ostwald ripening mechanism [97] was invoked. All these FePt–Au HNCs exhibited appealing electrocatalytic activity for oxygen reduction [93], catalytic activity for methanol oxidation, and better CO tolerance on the catalyst surface compared to commercial Pt catalysts [111].

In a very recent case study, selective tip growth at the apexes of anisotropic seeds was observed to result in shaped domains of the secondary material [158]. Anisole solutions containing an organometallic iron precursor, $Fe(N\text{–}(SiMe_3)_2)_{22}$, lauric acid (LA), and hexadecylamine (HDA) were reacted with HDA-capped Co nanorods at 150°C under H_2 for 24 h, resulting in dumbbell-like HNCs that carried large cubic-shaped Fe domains with concave facets, covering the termini of the starting seeds [158] (Fig. 6(e)). Under these conditions the growth of Fe started from the tips and continued on the lateral facets along the Co nanorod longitudinal axis thereon, forming triangular prismatic domains. The latter eventually evolved into concave cubes surrounding the whole nanorod perimeter in proximity of both apexes. Some cobalt atoms appeared to be etched away from the seed tips during Fe overgrowth, whereas a small amount of non-crystallized Fe was found deposited on the Co domains. Interestingly, the soft iron nanocube domains behaved as easy nucleation centers that induced the magnetization reversal of the entire nanoheterostructure, leading to a drastic modification of the overall effective magnetic anisotropy [158].

Various examples of oxide-based HNCs have been reported. A nonaqueous approach to magnetic–semiconductor γ-Fe_2O_3–TiO_2 HNCs with switchable heterooligomer to heterodimer topologies and tunable geometric parameters was developed [146,149] (Figs. 6(g) and 6(h)). The synthesis exploited brookite TiO_2 nanorods with either rectangular- or tapered-shape profiles, the latter terminating into arrow-like apexes with high-index facets [168] as substrate seeds for accommodating γ-Fe_2O_3 upon decomposition of $Fe(CO)_5$ in OLAM/OLAC/hexadecan-1,2-diol mixtures at 280–300°C. The TiO_2 seeds exhibited size- and shape-dependent anisotropic reactivity, which allowed producing HNCs individually made of a single TiO_2 section functionalized with either one or multiple spherical γ-Fe_2O_3 domains at distinct locations in a highly controlled manner (Scheme 3(a), paths 1, 3). The surface–interface energy balance associated with the formation of the different architectures, examined on the basis of comparative experimental strain mapping of individual HNCs and on CSLT-based analysis of TiO_2/γ-Fe_2O_3 heteroepitaxy relationships [92], did not provide a conclusive explanation regarding the mechanism underlying the topological selectivity observed under specified condition. Alternatively, the synthesis outcome was rationalized within the framework of a model assuming rapid establishment of diffusion-limited growth conditions for the foreign γ-Fe_2O_3 component. According to this model, γ-Fe_2O_3 overgrowth on the TiO_2 seeds could switch from a thermodynamically controlled (corresponding to a nonselective γ-Fe_2O_3 deposition on the nanorod seeds) to a kinetically driven regime (corresponding to selective γ-Fe_2O_3 nucleation and growth on the seed apex) when highly anisotropically reactive seeds (those with arrow-like terminations) were utilized under conditions (high precursor-to-seed ratios) favoring establishment of steep monomer gradients across the diffusion layer around each seed [146]. The as-synthesized γ-Fe_2O_3–TiO_2 HNCs displayed a rich scenario of photocatalytic properties or modified magnetic responses, which clearly diverged from those afforded by their individual components and physical mixture counterparts [146,149].

Intriguing branched α-Fe_2O_3–SnO_2 HNCs were built from crystallographic-oriented epitaxial SnO_2 growth on shaped α-Fe_2O_3 seeds upon hydrothermal dehydration of $Sn(OH)_6^{2-}$ species [169,170]. For instance, when sixfold symmetrical spindle-shaped α-Fe_2O_3 seeds were used, multiple SnO_2 domains indiscriminately nucleated onto each (110) facet, developing into small adjacent rods that progressively merged together laterally to decrease the overall surface energy. On the other hand, when overgrowth took place on cubic-shaped seeds, the secondary SnO_2 nucleated on the α-Fe_2O_3 facets slantwise at fixed angle of 65°, an arrangement that reduced interfacial lattice mismatch down to 1% and prevented the generation of misfit dislocations. A similar tendency toward branching was observed on employing hexahedron-shaped α-Fe_2O_3 seeds [169]. Compared to the bare α-Fe_2O_3 particles, branched α-Fe_2O_3–SnO_2 HNCs showed prominent photocatalytic activity toward organic dye degradation under both visible and UV light irradiation [170]. Worthy to mention are also hyperbranched α-Fe_2O_3–SnO_2 heterostructures with a sixfold symmetry, composed of a SnO_2 nanowire stem and α-Fe_2O_3 nanorod branches, which were generated by hydrothermal overgrowth of iron oxide on preformed SnO_2 nanowire substrates prepared by a vapor-transport deposition route [171]. Such architectures demonstrated superior performance as anode material for lithium-ion batteries as a result of the synergistic effect of the α-Fe_2O_3 and SnO_2 components arranged in branched architectures.

3.3.2.2. *Surfactant-controlled selective heterogeneous nucleation*

For several material associations, careful investigation of the conditions underlying trapping of HNCs in non-equivalent topologies has indicated that the site-dependent accessibility and chemical reactivity of anisotropically shaped seeds can be rationalized the basis of the frequently postulated mechanism of facet-preferential adhesion of surfactants or ligands (Scheme 3). Examples of elaborate HNCs, which demonstrate this pathway, are reported in Fig. 7.

A clear influence of the growth environment on the ultimate location of the secondary material domains was inferred in the synthesis of Au-decorated Co nanorods from a AuCl(tetrahydrothiophene) complex precursor in the presence of lauric LA and HDA (Fig. 7(a)). In this system,

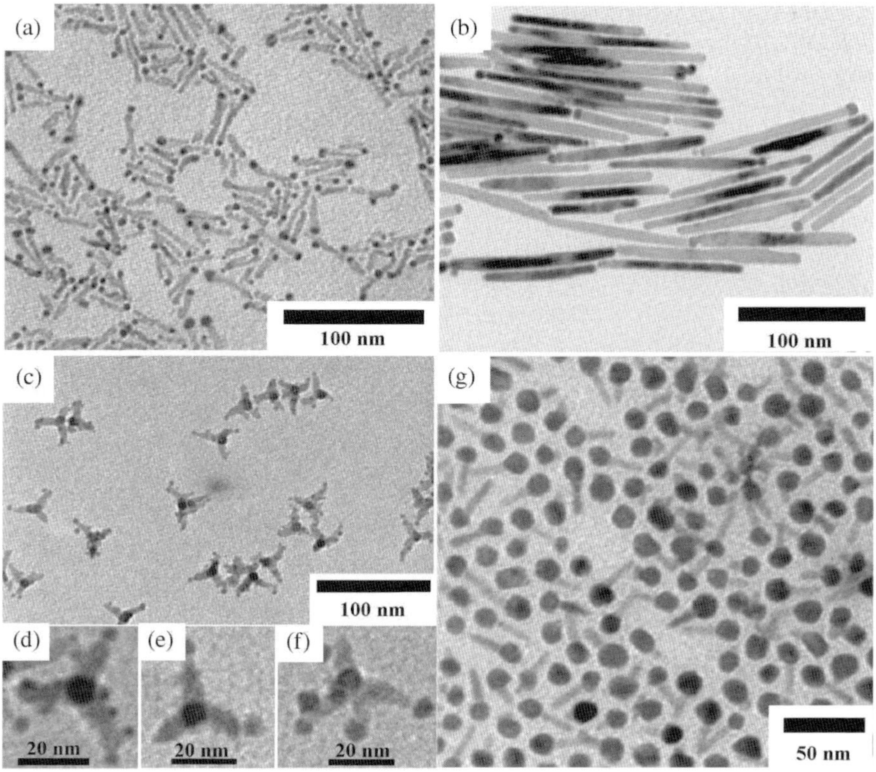

Figure 7. Examples of anisotropic HNCs synthesized by surfactant-controlled selective heterogeneous nucleation mechanism (cf. Scheme 4). The panel shows low-magnification TEM galleries of: (a) HNCs made of Cotipped anatase TiO_2 nanorods. Reproduced from Ref. [135] with permission. Copyright (2007) American Chemical Society; (b) dumbbell-like HNCs made of Au-tipped Co nanorods. Reproduced from Ref. [136] with permission. Copyright (2007) Wiley-VCH Verlag GmbH & Co. KGaA); (c)–(f) oligomer-type HNCs made of Codecorated γ-Fe_2O_3 tetrapods. Reproduced with permission from Ref. [159]. Copyright (2009) American Chemical Society; (g) matchstick-like HNCs made of eccentric CdSe–CdS nanorods tipped with a single Co head (synthesized according to Ref. [143]).

the Co seeds triggered the otherwise kinetically prohibited reduction of Au(I) to Au(0), while a dense protective coating of surface-adsorbed LA and HDA on the longitudinal sidewalls of the nanorods guaranteed that Au deposition was confined at their apexes [136].

Magnetic/semiconductor Codecorated TiO_2 nanorods were synthesized by manipulating the pyrolysis of $Co_2(CO)_8$ under assistance of octanoic acid (OCAC) and OLAM at 250–280°C [135] (Fig. 7(b)). The OCAC and OLAM concentrations along the synthesis course were judiciously time-programmed to regulate the accessibility of the reactive cobalt species to the different exposed TiO_2 nanorod facets, which enabled switching heterogeneous Co nucleation from a tip-preferential to a non-selective decoration regime, in which multiple metal domains formed indiscriminately also on the longitudinal sidewalls of the TiO_2 seeds (Scheme 3(a), paths 1 vs. 3). According to CSLT-based arguments [92], both types of TiO_2/Co HNCs were interpreted as being nearly equivalent in terms of interfacial strain that had to be accommodated, regardless of the different structure types and densities of heterojunction types. Hence, site-preferential Co overgrowth was rationalized as being a process that compensated for the increase in the overall surface energy caused by progressive depletion of the organic ligands on the respective seed facets at increasingly low surfactant concentration. As the two materials communicated through rather extended interfaces, proximity effects at the interface unpredictably influenced the magnetic anisotropy of the Co domains, despite them being connected to non-magnetic TiO_2 [135].

To achieve Co deposition on eccentric CdSe@CdS core@shell nanorods and γ-Fe_2O_3 tetrapods upon $Co_2(CO)_8$ decomposition, non-coordinating ODE was selected as the reaction medium, without any extra surfactants or ligands added in [143,159]. The seeds became activated toward heterogeneous nucleation due to partial displacement of their original capping molecules A technique that involved controllably slow $Co_2(CO)_8$ addition and delayed delivery of excess OLAC to a stage at which Co growth was reaching completion guaranteed the high-yield formation of heterostructures with high colloidal stability. In the first documented case, matchstick-like HNCs, made of single-Co-tipped CdSe@CdS nanorods, were obtained, whereby the Co domain formed on the apex that was diametrically opposed to the CdS region embedding the CdSe core [143] (Fig. 7(g)). This outcome was consistent with the growth mechanism of CdSe–CdS core–shell nanorods themselves, which involved asymmetric unidirectional development of the CdS shell from the CdSe seed cores [172] (Scheme 3a, path 1). Pertinent control syntheses, in which corresponding CdS-only nanorods were utilized as seeds, clarified that the inner core@shell structure of the CdSe@CdS core@shell seeds was an essential pre-requirement to ensuring that the apexes acted as the exclusive nucleation sites. The seeding behavior was tentatively attributed to electronic effects associated with the different inner dipole moments held by the CdSe@CdS seeds, with respect to their CdS-only counterparts. Interestingly, the CdSe@CdS–Co HNCs exhibited appreciable luminescence in spite of photoexcited electrons transferring from the semiconductor to the metal domain and showed anomalous ferromagnetic-like behavior at room temperature, most likely to be correlated with the structure of the heterojunctions formed. The second case pertains HNCs made of a γ-Fe_2O_3 tetrapod skeleton randomly decorated with multiple Co domains (Figs. 7(c)–7(f)) (Scheme 3b, path 3). In these branched heterostructures, FiM-FM exchange coupling resulted in a wealth of unique magnetic properties, including noticeable exchange bias, increased saturation magnetization, and coercivity, and enhanced thermal stability [159]. In a more recent report, HNCs with analogous topology, composed of multiply Codecorated CoO tetrapods, were generated through consecutive nucleation-growth events that took place at different temperatures during the thermal treatment of cobalt acetylacetonate and 1,2-dodecanediol reducing agent in ODE at 200°C and 320°C, respectively [173].

3.3.2.3. *Strain-driven heterostructure formation*

Interesting NC heterostructures have been created by profiting from the emergence of strain fields during heteroepitaxial growth. The cases illustrated in Fig. 8 show how strain may influence heteroepitaxial growth and the topology of the final heterostructures.

Asymmetric binary HNCs, each made of one rod-shaped TiO_2 section and one size-tunable γ-Fe_2O_3 spherical domain attached longitudinally, were obtained by decomposing $Fe(CO)_5$ on anatase TiO_2 nanorod seeds in a ternary mixture of OLAC, OLAM and 1,2-hexadecandiol at 240–300°C [174] (Figs. 8(a) and 8(b)). The γ-Fe_2O_3 deposition likely proceeded as a means of selectively eliminating high-energy edges on the longitudinal stepped sidewalls of the TiO_2 seeds (Scheme 3(a), path 2). However, the huge interfacial strain (8–11%) restricted the degree of "wetting" of the (011)/(101) facets underneath the edges, making the supporting TiO_2 section deformed and curved toward the γ-Fe_2O_3 sphere. The TiO_2 and

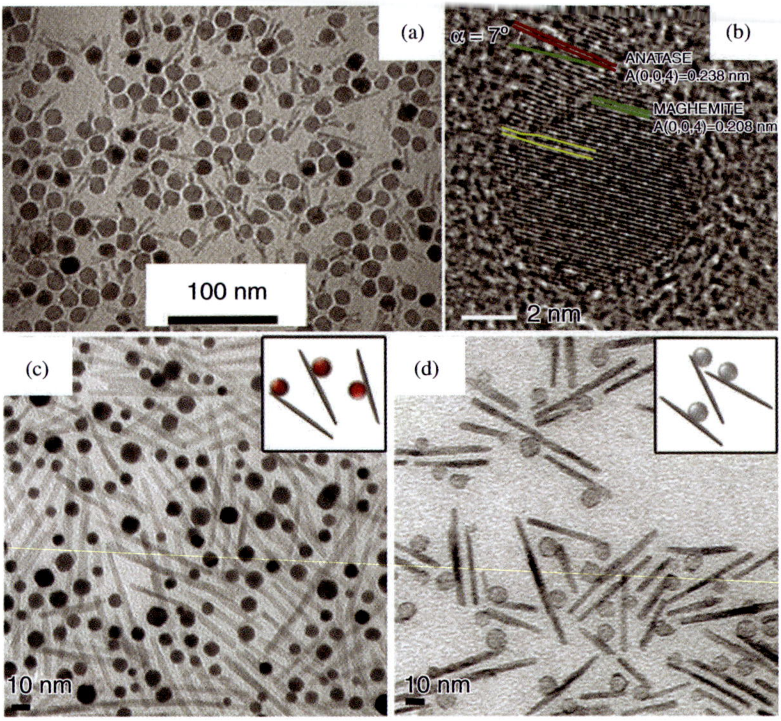

Figure 8. TEM and HRTEM examples of HNCs achieved by strain-driven heteroepitaxial growth (cf. Scheme 3). (a) and (b) Binary heterostructures made of single-γ-Fe_2O_3-functionalzied anatase TiO_2 nanorods. Reproduced with permission from Ref. [174]. Copyright (2006) American Chemical Society; (c) Binary heterostructures made of asymmetrically Au–Fe_xO_y-decorated Fe_xO_y nanorods and (d) corresponding all-Fe_xO_y nanostructures obtained after Au leaching. Reproduced with permission from Ref. [175]. Copyright (2016) Royal Society of Chemistry.

γ-Fe_2O_3 lattices were coincidently matched via a rather limited junction area at which the near-interface planes of the respective materials were locally bent. This curving allowed the interfacial strain to be relieved to a remarkable extent at a proportionally smaller cost of increased surface energy. These factors jointly accounted for the fact that reiterated deposition events were disfavored and that the ultimate size to which the γ-Fe_2O_3 domains could be grown was limited to a critical threshold value. Due to the small contact area shared with TiO_2, the magnetic properties of the HNCs essentially resembled those of isolated γ-Fe_2O_3 particles [174].

In another report, HNCs with analogous topologies, made of a spinel Fe_xO_y nanorod section asymmetrically decorated with a spherical Au@Fe_xO_y core@shell domain (Figs. 8(c) and 8(d)), were generated through an opposite reaction sequence [175]. Primary Au seeds were combined with $Fe(CO)_5$ in ODE-diluted mixtures of OLAM, OLAC, and dodecyldimethylammonium bromide (DDAB) at 300°C. A thin, nearly ubiquitous, yet discontinuous shell of spinel Fe_xO_y initially formed on the spherical Au seeds. However, upon prolonged annealing at high temperature, the combined effects of the strain increasing at the Au/Fe_xO_y interface and of the facet-selective binding of DDAB on Fe_xO_y promoted breaking of the Fe_xO_y growth symmetry, leading to tangential evolution of a Fe_xO_y nanorod out of each Au@Fe_xO_y core@shell intermediate. This mechanism was corroborated by topological and structural analyses of corresponding all-Fe_xO_y nanostructures that were derived from the Au@Fe_xO_y–Fe_xO_y nanorod heterostructures after Au had been leached out with I_2 (Fig. 8(d)).

3.4. Conclusions

The design and construction of modular HNCs combining NC domains of magnetic and non-magnetic materials represent a frontier research area in which nanochemistry approaches have progressed tremendously over the past years. The challenges posed by the wet-chemical synthesis of colloidal HNCs reflect the inherent difficulty of refining the ability to tailor specific nanoscale modules and of deepening the understanding of the intricately entangled thermodynamic conditions and kinetic processes that underlie their combination via heteroepitaxy pathways in liquid media. Studies and synthetic achievements documented so far suggest that an increased level of topological sophistication and regioselectivity in HNC engineering should be founded on a more detailed mechanistic knowledge of the solid-state and liquid-phase formation pathways that govern HNC evolution.

As of today, full realization of the technological potential of magnetic HNCs is hindered by the level of synthetic precision and property control with which these nanoheterostructures can be engineered to meet specific purposes. Unfortunately, the search for multifunctionality often conflicts with partial degradation of some native properties of the concerned component modules due to, for example, unfavorable changes in electronic structure and/or formation of defective interfaces. In many cases, it remains hard to unambiguously unravel and discriminate proximity effects from the emergence of new or abnormal magnetic behavior or other chemical–physical properties as a consequence of the electronic interactions and exchange coupling via the relevant heterojunctions, especially for associations of magnetic and non-magnetic materials. The support of theoretical investigations, which are still rather sparse, would be highly appreciated in an effort toward decoupling the roles played by heterointerfaces.

On the long term, it can be envisioned that the development of ingenious design and synthetic capabilities in the elaboration of magnetic HNCs will pave the way to both fundamental comprehension and practical utilization of novel unconventional properties and functionalities, particularly in the emerging fields of spintronics and magnetoplasmonics, green catalysis and biomedical theranostics.

Acknowledgment

The authors acknowledge financial support from the Apulia Regional Government through the project NANOAPULIA (cod. MDI6SR1 — CUP B38C14001140008).

References

[1] Burda, C.; Chen, X. B.; Narayanan, R.; El-Sayed, M. A. *Chem. Rev.*, **105**, 1025 (**2005**).

[2] *Advanced Wet-Chemical Synthetic Approaches to Inorganic Nanostructures;* P. D. Cozzoli Ed.; Transworld Research Network, Kerala, India, 2008.

[3] Niederberger, M.; Pinna, N. *Metal Oxide Nanoparticles in Organic Solvents: Synthesis, Formation, Assembly and Application;* Springer-Verlag, London, UK, 2009.

[4] Talapin, D. V.; Lee, J. S.; Kovalenko, M. V.; Shevchenko, E. V. *Chem. Rev.*, **110**, 389 (**2010**).

[5] Cozzoli, P. D.; Pellegrino, T.; Manna, L. *Chem. Soc. Rev.*, **35**, 1195 (**2006**).

[6] *Semiconductor Nanocrystal Quantum Dots: Synthesis, Assembly, Spectroscopy and Applications;* A. L. Rogach Ed.; Springer Wien New York, 2008.

[7] *Bio-Applications of Nanoparticles — Experimental Medicine and Biology. Vol.* **620**, W. C. W. Chan Ed.; Landes Bioscience and Springer Science+Business Media, LLC, New York, USA, 2007.

[8] Jun, Y.-w.; Choi, J.-s.; Cheon, J. *Angew. Chem. Int. Ed.*, **45**, 3414 (**2006**).

[9] Park, J.; Joo, J.; Kwon, S. G.; Jang, Y.; Hyeon, T. *Angew. Chem. In. Ed.*, **46**, 4630 (**2007**).

[10] Kwon, S. G.; Hyeon, T. *Acc. Chem. Res.*, **41**, 1696 (**2008**).

[11] An, K.; Hyeon, T. *Nano Today*, **4**, 359 (**2009**).

[12] Wang, W. S.; Dahl, M.; Yin, Y. D. *Chem. Mater.*, **25**, 1179 (**2013**).

[13] Zhang, Q.; Wang, W.; Goebl, J.; Yin, Y. *Nano Today*, **4**, 494 (**2009**).

[14] Baghbanzadeh, M.; Carbone, L.; Cozzoli, P. D.; Kappe, C. O. *Angew. Chem. Int. Ed.*, **50**, 11312 (**2011**).

[15] Bouet, C.; Tessier, M. D.; Ithurria, S.; Mahler, B.; Nadal, B.; Dubertret, B. *Chem. Mater.*, **25**, 1262 (**2013**).

[16] Yang, J.; Son, J. S.; Yu, J. H.; Joo, J.; Hyeon, T. *Chem. Mater.*, **25**, 1190 (**2013**).

[17] Lhuillier, E.; Pedetti, S.; Ithurria, S.; Nadal, B.; Heuclin, H.; Dubertret, B. *Acc. Chem. Res.*, **48**, 22 (**2015**).

[18] Jun, Y.-w.; Choi, J.-s.; Cheon, J. *Chem. Commun.*, 1203 (**2007**).

[19] Buonsanti, R.; Casavola, M.; Caputo, G.; Cozzoli, P. D. *Recent Pat. Nanotechnol.*, **1**, 224 (**2007**).

[20] Casavola, M.; Buonsanti, R.; Caputo, G.; Cozzoli, P. D. *Eur. J. Inorg. Chem.*, 837 (**2008**).

[21] Carbone, L.; Cozzoli, P. D. *Nano Today*, **5**, 449 (**2010**).

[22] Liu, J.; Qiao, S. Z.; Chen, J. S.; Lou, X. W.; Xing, X. R.; Lu, G. Q. *Chem. Commun.*, **47**, 12578 (**2011**).

[23] Purbia, R.; Paria, S. *Nanoscale*, **7**, 19789 (**2015**).

[24] de Mello Donegà, C. *Chem. Soc. Rev.*, **40**, 1512 (**2011**).

[25] Buck, M. R.; Schaak, R. E. *Angew. Chem. Int. Ed.*, **52**, 6154 (**2013**).

[26] Sitt, A.; Hadar, I.; Banin, U. *Nano Today*, **8**, 494 (**2013**).

[27] Melinon, P.; Begin-Colin, S.; Duvail, J. L.; Gauffre, F.; Boime, N. H.; Ledoux, G.; Plain, J.; Reiss, P.; Silly, F.; Warot-Fonrose, B. *Phys. Rep.*, **543**, 163 (**2014**).

[28] Banin, U.; Ben-Shahar, Y.; Vinokurov, K. *Chem. Mater.*, **26**, 97 (**2014**).

[29] Qi, J.; Lai, X.; Wang, J.; Tang, H.; Ren, H.; Yang, Y.; Jin, Q.; Zhang, L.; Yu, R.; Ma, G.; Su, Z.; Zhao, H.; Wang, D. *Chem. Soc. Rev.*, **44**, 6749 (**2015**).

[30] López-Ortega, A.; Estrader, M.; Salazar-Alvarez, G.; Roca, A. G.; Nogués, J. *Phys. Rep.*, **553**, 1 (**2015**).

[31] Armelles, G.; Cebollada, A.; García-Martín, A.; González, M. U. *Adv. Opt. Mater.*, **1**, 10 (**2013**).

[32] Chng, L. L.; Erathodiyil, N.; Ying, J. Y. *Acc. Chem. Res.*, **46**, 1825 (**2013**).

[33] Rawalekar, S.; Mokari, T. *Adv. Energy Mater.*, **3**, 12 (**2013**).

[34] Song, H. *Acc. Chem. Res.*, **48**, 491 (**2015**).

[35] Liao, F.; Lo, B. T. W.; Tsang, E. *Eur. J. Inorg. Chem.*, **2016**, 1924 (**2016**).

[36] Wang, Y.; He, J.; Liu, C.; Chong, W. H.; Chen, H. *Angew. Chem. Int. Ed.*, **54**, 2022 (**2015**).

[37] de Mello Donegà, C.; Liljeroth, P.; Vanmaekelbergh, D. *Small*, **1**, 1152 (**2005**).

[38] Zhang, J.; Huang, F.; Lin, Z. *Nanoscale*, **2**, 18 (**2010**).

[39] Li, H. B.; Kanaras, A. G.; Manna, L. *Acc. Chem. Res.*, **46**, 1387 (**2013**).

[40] Hu, S.; Wang, X. *Chem. Soc. Rev.*, **42**, 5577 (**2013**).

[41] Hu, L.; Zhang, R.; Chen, Q. *Nanoscale*, **6**, 14064 (**2014**).

[42] Wang, F.; Richards, V. N.; Shields, S. P.; Buhro, W. E. *Chem. Mater.*, **26**, 5 (**2014**).

[43] Markov, I. V., *Crystal Growth for Beginners: Fundamentals of Nucleation, Crystal Growth, and Epitaxy*, World Scientific: Singapore, 2003.

[44] Thompson, C. V. *Annual Rev. Mater. Res.*, **42**, 399 (**2012**).

[45] Gentili, D.; Foschi, G.; Valle, F.; Cavallini, M.; Biscarini, F. *Chem. Soc. Rev.*, **41**, 4430 (**2012**).

[46] Jerison, E. R.; Xu, Y.; Wilen, L. A.; Dufresne, E. R. *Phys. Rev. Lett.*, **106**, 186103 (**2011**).

[47] Gao, J.; Gu, H.; Xu, B. *Acc. Chem. Res.*, **42**, 1097 (**2009**).

[48] Bigall, N. C.; Parak, W. J.; Dorfs, D. *Nano Today*, **7**, 282 (**2012**).

[49] Lim, J.; Majetich, S. A. *Nano Today*, **8**, 98 (**2013**).

[50] Choi, J. S.; Jun, Y. W.; Yeon, S. I.; Kim, H. C.; Shin, J. S.; Cheon, J. *J. Am. Chem. Soc.*, **128**, 15982 (**2006**).

[51] Jiang, J.; Gu, H.; Shao, H.; Devlin, E.; Papaefthymiou, G. C.; Ying, J. Y. *Adv. Mater.*, **20**, 4403 (**2008**).

[52] Choi, S.-H.; Na, H. B.; Park, Y. I.; An, K.; Kwon, S. G.; Jang, Y.; Park, M.-h.; Moon, J.; Son, J. S.; Song, I. C.; Moon, W. K.; Hyeon, T. *J. Am. Chem. Soc.*, **130**, 15573 (**2008**).

[53] Schladt, T. D.; Shukoor, M. I.; Schneider, K.; Tahir, M. N.; Natalio, F.; Ament., I.; Becker, J.; Jochum, F. D.; Weber, S.; Köhler, O.; Theato, P.; Schreiber, L. M.; Sönnichsen, C.; Heinz, C. S.; Müller, W. E. G.; Tremel, W. *Angew. Chem. Int. Ed.*, **49**, 3976 (**2010**).

[54] Xu, C.; Xie, J.; Ho, D.; Wang, C.; Kohler, N.; Walsh, E. G.; Morgan, Jeffrey R.; Chin, Y. E.; Sun, S. *Angew. Chem. Int. Ed.*, **47**, 173 (**2008**).

[55] Lee, J.-S.; Bodnarchuk, M. I.; Shevchenko, E. V.; Talapin, D. V. *J. Am. Chem. Soc.*, **132**, 6382 (**2010**).
[56] Umut, E.; Pineider, F.; Arosio, P.; Sangregorio, C.; Corti, M.; Tabak, F.; Lascialfari, A.; Ghigna, P. *J. Magn. Magn. Mater.*, **324**, 2373 (**2012**).
[57] Velasco, V.; Munoz, L.; Mazario, E.; Menendez, N.; Herrasti, P.; Hernando, A.; Crespo, P. *J. Phys. D- Appl. Phys.*, **48**, 035502 (**2015**).
[58] Pineider, F.; de Julián Fernández, C.; Videtta, V.; Carlino, E.; al Hourani, A.; Wilhelm, F.; Rogalev, A.; Cozzoli, P. D.; Ghigna, P.; Sangregorio, C. *ACS Nano.*, **7**, 857 (**2013**).
[59] Kim, M.; Song, H. *J. Mater. Chem. C*, **2**, 4997 (**2014**).
[60] Schick, I.; Lorenz, S.; Gehrig, D.; Tenzer, S.; Storck, W.; Fischer, K.; Strand, D.; Laquai, F.; Tremel, W. *Beilstein J. Nanotechnol.*, **5**, 2346 (**2014**).
[61] Comin, A.; Korobchevskaya, K.; George, C.; Diaspro, A.; Manna, L. *Nano Lett.*, **12**, 921 (**2012**).
[62] Korobchevskaya, K.; George, C.; Diaspro, A.; Manna, L.; Cingolani, R.; Comin, A. *Appl. Phys. Lett.*, **99**, 011907 (**2011**).
[63] Levin, C. S.; Hofmann, C.; Ali, T. A.; Kelly, A. T.; Morosan, E.; Nordlander, P.; Whitmire, K. H.; Halas, N. J. *ACS Nano*, **3**, 1379 (**2009**).
[64] Li, Y.; Zhang, Q.; Nurmikko, A. V.; Sun, S. *Nano Lett.*, **5**, 1689 (**2005**).
[65] Yin, H.; Wang, C.; Zhu, H.; Overbury, S. H.; Sun, S.; Dai, S. *Chem. Commun.*, 4357 (**2008**).
[66] Wu, B.; Zhang, H.; Chen, C.; Lin, S.; Zheng, N. *Nano Res.*, **2**, 975 (**2009**).
[67] Wang, C.; Daimon, H.; Sun, S. *Nano Lett.*, **9**, 1493 (**2009**).
[68] Wang, C.; Yin, H.; Dai, S.; Sun, S. *Chem. Mater.*, **22**, 3277 (**2010**).
[69] Lee, Y.; Garcia, M. A.; Frey Huls, N. A.; Sun, S. *Angew. Chem. Int. Ed.*, **49**, 1271 (**2010**).
[70] Chen, S.; Si, R.; Taylor, E.; Janzen, J.; Chen, J. *J. Phys. Chem. C*, **116**, 12969 (**2012**).
[71] Sun, X.; Guo, S.; Liu, Y.; Sun, S. *Nano Lett.*, **12**, 4859 (**2012**).
[72] George, C.; Dorfs, D.; Bertoni, G.; Falqui, A.; Genovese, A.; Pellegrino, T.; Roig, A.; Quarta, A.; Comparelli, R.; Curri, M. L.; Cingolani, R.; Manna, L. *J. Am. Chem. Soc.*, **133**, 2205 (**2011**).
[73] George, C.; Genovese, A.; Casu, A.; Prato, M.; Povia, M.; Manna, L.; Montanari, T. *Nano Lett.*, **13**, 752 (**2013**).
[74] Liu, S.; Guo, S.; Sun, S.; You, X.-Z. *Nanoscale*, **7**, 4890 (**2015**).
[75] Shi, W.; Zeng, H.; Sahoo, Y.; Ohulchanskyy, T. Y.; Ding, Y.; Wang, Z. L.; Swihart, M.; Prasad, P. N. *Nano Lett.*, **6**, 875 (**2006**).
[76] He, S. L.; Zhang, H. W.; Delikanli, S.; Qin, Y. L.; Swihart, M. T.; Zeng, H. *J. Phys. Chem. C*, **113**, 87 (**2009**).

[77] Yu, H.; Chen, M.; Rice, P. M.; Wang, S. X.; White, R. L.; Sun, S. *Nano Lett.*, **5**, 379 (**2005**).
[78] Wei, Y.; Klajn, R.; Pinchuk, A. O.; Grzybowski, B. A. *Small*, **4**, 1635 (**2008**).
[79] Lin, F. H.; Doong, R. A. *J. Phys. Chem. C*, **115**, 6591 (**2011**).
[80] Jang, Y.; Chung, J.; Kim, S.; Jun, S. W.; Kim, B. H.; Lee, D. W.; Kim, B. M.; Hyeon, T. *Phys. Chem. Chem. Phys.*, **13**, 2512 (**2011**).
[81] Leung, K. C.-F.; Xuan, S.; Zhu, X.; Wang, D.; Chak, C.-P.; Lee, S.-F.; Ho, W. K. W.; Chung, B. C. T. *Chem. Soc. Rev.*, **41**, 1911 (**2012**).
[82] Zhai, Y.; Han, L.; Wang, P.; Li, G.; Ren, W.; Liu, L.; Wang, E.; Dong, S. *ACS Nano*, **5**, 8562 (**2011**).
[83] Wu, H. M.; Chen, O.; Zhuang, J. Q.; Lynch, J.; LaMontagne, D.; Nagaoka, Y.; Cao, Y. C. *J. Am. Chem. Soc.*, **133**, 14327 (**2011**).
[84] Nakhjavan, B.; Tahir, M. N.; Natalio, F.; Gao, H.; Schneider, K.; Schladt, T.; Ament, I.; Branscheid, R.; Weber, S.; Kolb, U.; Sonnichsen, C.; Schreiber, L. M.; Tremel, W. *J. Mater. Chem.*, **21**, 8605 (**2011**).
[85] Liu, F.; Zhu, J.; Yang, W.; Dong, Y.; Hou, Y.; Zhang, C.; Yin, H.; Sun, S. *Angew. Chem. Int. Ed.*, **53**, 2176 (**2014**).
[86] Yang, W.; Lei, W.; Yu, Y.; Zhu, W.; George, T. A.; Li, X. Z.; Sellmyer, D. J.; Sun, S. *J. Mater. Chem. C*, **3**, 7075 (**2015**).
[87] Victor, V.; Laura, M.; Eva, M.; Nieves, M.; Pilar, H.; Antonio, H.; Patricia, C. *J. Phys. D: Appl. Phys.*, **48**, 035502 (**2015**).
[88] Wang, C.; Wei, Y.; Jiang, H.; Sun, S. *Nano Lett.*, **9**, 4544 (**2009**).
[89] Kwon, K. W.; Shim, M. *J. Am. Chem. Soc.,* **127**, 10269 (**2005**).
[90] Kwon, K.-W.; Lee, B. H.; Shim, M., *Chem Mater.*, **18**, 6357 (**2006**).
[91] McDaniel, H.; Shim, M. *ACS Nano*, **3**, 434 (**2009**).
[92] Randle, V., *The role of the coincidence site lattice in grain boundary engineering*. Woodhead Publishing Limited: Cambridge, (England, 1997).
[93] Mao, J.; Cao, X.; Zhen, J.; Shao, H.; Gu, H.; Lu, J.; Ying, J. Y. *J. Mater. Chem.*, **21**, 11478 (**2011**).
[94] Zhu, J.; Wu, J.; Liu, F.; Xing, R.; Zhang, C.; Yang, C.; Yin, H.; Hou, Y. *Nanoscale*, **5**, 9141 (**2013**).
[95] Selvan, S. T.; Patra, P. K.; Chung Yen Ang; Ying, J. Y. *Angew. Chem. Int. Ed.*, **46**, 2448 (**2007**).
[96] Franchini, I. R.; Bertoni, G.; Falqui, A.; Giannini, C.; Wang, L. W.; Manna, L. *J. Mater. Chem.*, **20**, 1357 (**2010**).
[97] Mokari, T.; Sztrum, C. G.; Salant, A.; Rabani, E.; Banin, U. *Nat. Mater.*, **4**, 855 (**2005**).

[98] Zeng, J.; Jianliu, H.; Liu, C.; Wu, C. H.; Lin, Y.; Wang, X.; Zhang, S.; Hou, J.; Xia, Y. *Adv. Mater.*, **22**, 1936 (**2010**).

[99] Wark, S. E.; Hsia, C.-H.; Son, D. H. *J. Am. Chem. Soc.*, **130**, 9550 (**2008**).

[100] Pellegrino, T.; Fiore, A.; Carlino, E.; Giannini, C.; Cozzoli, P. D.; Ciccarella, G.; Respaud, M.; Palmirotta, L.; Cingolani, R.; Manna, L. *J. Am. Chem. Soc.*, **128**, 6690 (**2006**).

[101] Pazos-Perez, N.; Gao, Y.; Hilgendorff, M.; Irsen, S.; Perez-Juste, J.; Spasova, M.; Farle, M.; Liz-Marzan, L. M.; Giersig, M. *Chem. Mater.*, **19**, 4415 (**2007**).

[102] Yang, J.; Ying, J. Y. *Nat. Mater.*, **8**, 683 (**2009**).

[103] Zhang, L.; Dou, Y. H.; Gu, H. C. *J. Coll. Interface Sci.*, **297**, 660 (**2006**).

[104] Huang, J. M.; Sun, Y. H.; Huang, S. S.; Yu, K.; Zhao, Q.; Peng, F.; Yu, H.; Wang, H. J.; Yang, J. *J. Mater. Chem.*, **21**, 17930 (**2011**).

[105] Peng, S.; Lei, C.; Ren, Y.; Cook, R. E.; Sun, Y. *Angew. Chem. Int. Ed.*, **50**, 3158 (**2011**).

[106] Mao, Y. W.; Yi, P. W.; Deng, Z. W.; Ge, J. P. *CrystEngComm.*, **15**, 3575 (**2013**).

[107] Mirtchev, P.; Liao, K.; Jaluague, E.; Qiao, Q.; Tian, Y.; Varela, M.; Burch, K. S.; Pennycook, S. J.; Perovic, D. D.; Ozin, G. *J. Mater. Chem. A*, **2**, 8525 (**2014**).

[108] Yang, J.; Peng, J.; Zhang, Q.; Peng, F.; Wang, H.; Yu, H. *Angew. Chem. Int. Ed.*, **48**, 3991 (**2009**).

[109] Yang, J.; Levina, L.; Sargent, E. H.; Kelley, S. O. *J. Mater. Chem.*, **16**, 4025 (**2006**).

[110] Yang, J.; Elim, H. I.; Zhang, Q.; Lee, J. Y.; Ji, W. *J. Am. Chem. Soc.*, **128**, 11921 (**2006**).

[111] Wu, J. J.; Hou, Y. L.; Gao, S. *Nano Res.*, **4**, 836 (**2011**).

[112] Lee, K. S.; Anisur, R. M.; Kim, K. W.; Kim, W. S.; Park, T. J.; Kang, E. J.; Lee, I. S. *Chem. Mater.*, **24**, 682 (**2012**).

[113] Kwon, S. G.; Krylova, G.; Phillips, P. J.; Klie, R. F.; Chattopadhyay, S.; Shibata, T.; Bunel, E. E.; Liu, Y.; Prakapenka, V. B.; Lee, B.; Shevchenko, E. V. *Nat. Mater.*, **14**, 215 (**2015**).

[114] Hou, X.; Zhang, X.; Chen, S.; Kang, H.; Tan, W. *Coll. Surface Sci. A*, **403**, 148 (**2012**).

[115] Yu, S. M.; Hachtel, J. A.; Chisholm, M. F.; Pantelides, S. T.; Laromaine, A.; Roig, A. *Nanoscale*, **7**, 14039 (**2015**).

[116] Shi, W.; Sahoo, Y.; Zeng, H.; Ding, Y.; Swihart, M. T.; Prasad, P. N. *Adv. Mater.*, **18**, 1889 (**2006**).

[117] Buck, M. R.; Bondi, J. F.; Schaak, R. E. *Nat. Chem.*, **4**, 37 (**2012**).
[118] Hodges, J. M.; Morse, J. R.; Williams, M. E.; Schaak, R. E. *J. Am. Chem. Soc.*, **137**, 15493 (**2015**).
[119] Hodges, J. M.; Biacchi, A. J.; Schaak, R. E. *ACS Nano*, **8**, 1047 (**2014**).
[120] Liu, Y. L.; Walker, A. R. H. *J. Phys. Chem. C*, **114**, 4264 (**2010**).
[121] Wang, D.; Li, X.; Li, H.; Li, L.; Hong, X.; Peng, Q.; Li, Y. *J. Mater. Chem. A*, **1**, 1587 (**2013**).
[122] Slaton, R. D.; Bae, I.-T.; Lutz, P. S.; Pathade, L.; Maye, M. M. *J. Mater. Chem. C.*, **3**, 6367 (**2015**).
[123] Zanella, M.; Falqui, A.; Kudera, S.; Manna, L.; Casula, M. F.; Parak, W. J. *J. Mater. Chem.*, **18**, 4311 (**2008**).
[124] Gu, H. W.; Zheng, R. K.; Zhang, X. X.; Xu, B. *J. Am. Chem. Soc.*, **126**, 5664 (**2004**).
[125] Gu, H. W.; Yang, Z. M.; Gao, J. H.; Chang, C. K.; Xu, B. *J. Am. Chem. Soc.*, **127**, 34 (**2005**).
[126] Pan, Y.; Gao, J. H.; Zhang, B.; Zhang, X. X.; Xu, B. *Langmuir*, **26**, 4184 (**2010**).
[127] Teranishi, T.; Inoue, Y.; Nakaya, M.; Oumi, Y.; Sano, T. *J. Am. Chem. Soc.*, **126**, 9914 (**2004**).
[128] Teranishi, T.; Inoue, Y.; Saruyama, M.; Nakaya, M.; Kanehara, M. *Chem. Lett.*, **36**, 490 (**2007**).
[129] Wang, D.; Li, Y. *J. Am. Chem. Soc.*, **132**, 6280 (**2010**).
[130] Figuerola, A.; Fiore, A.; Di Corato, R.; Falqui, A.; Giannini, C.; Micotti, E.; Lascialfari, A.; Corti, M.; Cingolani, R.; Pellegrino, T.; Cozzoli, P. D.; Manna, L. *J. Am. Chem. Soc.*, **130**, 1477 (**2008**).
[131] Nolle, D.; Goering, E.; Tietze, T.; Schutz, G.; Figuerola, A.; Manna, L. *New J. Phys.*, **11**, 033034 (**2009**).
[132] Teranishi, T.; Saruyama, M.; Nakaya, M.; Kanehara, M. *Angew. Chem. Int. Ed.*, **46**, 1713 (**2007**).
[133] Costi, R.; Saunders, A. E.; Banin, U. *Angew. Chem. Int. Ed.*, **49**, 4878 (**2010**).
[134] Kudera, S.; Carbone, L.; Casula, M. F.; Cingolani, R.; Falqui, A.; Snoeck, E.; Parak, W. J.; Manna, L. *Nano Lett.*, **5**, 445 (**2005**).
[135] Casavola, M.; Grillo, V.; Carlino, E.; Giannini, C.; Gozzo, F.; Fernandez Pinel, E.; Garcia, M. A.; Manna, L.; Cingolani, R.; Cozzoli, P. D. *Nano Lett.*, **7**, 1386 (**2007**).
[136] Wetz, F.; Soulantica, K.; Falqui, A.; Respaud, M.; Snoeck, E.; Chaudret, B. *Angew. Chem. Int. Ed.*, **46**, 7079 (**2007**).

[137] Maynadiè, J.; Salant, A.; Falqui, A.; Respaud, M.; Shaviv, E.; Banin, U.; Soulantica, K.; Chaudret, B. *Angew. Chem. Int. Ed.*, **48**, 1814 (**2009**).

[138] Carbone, L.; Kudera, S.; Giannini, C.; Ciccarella, G.; Cingolani, R.; Cozzoli, P. D.; Manna, L. *J. Mater. Chem.*, **16**, 3952 (**2006**).

[139] Chakrabortty, S.; Yang, Jie A.; Tan, Yee M.; Mishra, N.; Chan, Y. *Angew. Chem. Int. Ed.*, **49**, 2888 (**2010**).

[140] Xu, B.; Zhou, G.; Wang, X. *NPG Asia Mater.*, **7**, e164 (**2015**).

[141] Pacholski, C.; Kornowski, A.; Weller, H. *Angew. Chem. Int. Ed.*, **43**, 4774 (**2004**).

[142] Milliron, D. J.; Hughes, S. M.; Cui, Y.; Manna, L.; Li, J. B.; Wang, L. W.; Alivisatos, A. P. *Nature*, **430**, 190 (**2004**).

[143] Deka, S.; Falqui, A.; Bertoni, G.; Sangregorio, C.; Poneti, G.; Morello, G.; De Giorgi, M.; Giannini, C.; Cingolani, R.; Manna, L.; Cozzoli, P. D. *J. Am. Chem. Soc.*, **131**, 12817 (**2009**).

[144] Dukovic, G.; Merkle, M. G.; Nelson, J. H.; Hughes, S. M.; Alivisatos, A. P. *Adv. Mater.*, **20**, 4306 (**2008**).

[145] Cheng, C.; Yu, K. F.; Cai, Y.; Fung, K. K.; Wang, N. *J .Phys. Chem. C*, **111**, 16712 (**2007**).

[146] Buonsanti, R.; Grillo, V.; Carlino, E.; Giannini, C.; Gozzo, F.; Garcia-Hernandez, M.; Garcia, M. A.; Cingolani, R.; Cozzoli, P. D. *J. Am. Chem. Soc.*, **132**, 2437 (**2010**).

[147] Menagen, G.; Mocatta, D.; Salant, A.; Popov, I.; Dorfs, D.; Banin, U. *Chem. Mater.*, **20**, 6900 (**2008**).

[148] Mokari, T.; Rothenberg, E.; Popov, I.; Costi, R.; Banin, U. *Science*, **304**, 1787 (**2004**).

[149] Buonsanti, R.; Snoeck, E.; Giannini, C.; Gozzo, F.; Garcia-Hernandez, M.; Garcia, M. A.; Cingolani, R.; Cozzoli, P. D. *Phys. Chem. Chem. Phys.*, **11**, 3680 (**2009**).

[150] Jonathan E. Halpert; Venda J. Porter; John P. Zimmer; Bawendi, M. G. *J. Am. Chem. Soc.*, **128**, 12590 (**2006**).

[151] Xi, L. F.; Boothroyd, C.; Lam, Y. M. *Chem. Mater.*, **21**, 1465 (**2009**).

[152] Jen-La Plante, I.; Habas, S. E.; Yuhas, B. D.; Gargas, D. J.; Mokari, T. *Chem. Mater.*, **21**, 3662 (**2009**).

[153] Habas, S. E.; Yang, P.; Mokari, T. *J. Am. Chem. Soc.*, **130**, 3294 (**2008**).

[154] Yuhas, B. D.; Habas, S. E.; Fakra, S. C.; Mokari, T. *ACS Nano* **3**, 3369 (**2009**).

[155] Nakibli, Y.; Amirav, L. *Chem. Mater.*, **28**, 4524 (**2016**).

[156] Hill, L. J.; Bull, M. M.; Sung, Y.; Simmonds, A. G.; Dirlam, P. T.; Richey, N. E.; DeRosa, S. E.; Shim, I.-B.; Guin, D.; Costanzo, P. J.;

Pinna, N.; Willinger, M.-G.; Vogel, W.; Char, K.; Pyun, J. *ACS Nano*, **6**, 8632 (**2012**).

[157] Pavlopoulos, N. G.; Dubose, J. T.; Pinna, N.; Willinger, M.-G.; Char, K.; Pyun, J. *Angew. Chem. Int. Ed.*, **55**, 1787 (**2016**).

[158] Liakakos, N.; Gatel, C.; Blon, T.; Altantzis, T.; Lentijo-Mozo, S.; Garcia-Marcelot, C.; Lacroix, L.-M.; Respaud, M.; Bals, S.; Van Tendeloo, G.; Soulantica, K. *Nano Lett.*, **14**, 2747 (**2014**).

[159] Casavola, M.; Falqui, A.; Garcia, M. A.; Garcia-Hernandez, M.; Giannini, C.; Cingolani, R.; Cozzoli, P. D. *Nano Lett.*, **9**, 366 (**2009**).

[160] Chen, X. B.; Lou, Y. B.; Samia, A. C.; Burda, C. *Nano Lett.*, **3**, 799 (**2003**).

[161] Koo, B.; Korgel, B. A. *Nano Lett.*, **8**, 2490 (**2008**).

[162] Sadowski, T.; Ramprasad, R. *Appl. Phys. Lett.*, **96**, 101906 (**2010**).

[163] Wang, L.; Liu, X. Z.; Yang, M.; Fan, Y. J.; Zhan, J. H. *Eur. J. Inorg. Chem.*, 897 (**2009**).

[164] Costi, R.; Saunders, A. E.; Elmalem, E.; Salant, A.; Banin, U. *Nano Lett.*, **8**, 637 (**2008**).

[165] Steiner, D.; Mokari, T.; Banin, U.; Millo, O. *Phys. Rev. Lett.*, **95**, 056805 (**2005**).

[166] Amirav, L.; Alivisatos, A. P. *J. Phys. Chem. Lett.*, **1**, 1051 (**2010**).

[167] Elmalem, E.; Saunders, A. E.; Costi, R.; Salant, A.; Banin, U., Growth of Photocatalytic CdSe-Pt Nanorods and Nanonets. *Adv. Mater.*, **20**, 4312 (**2008**).

[168] Buonsanti, R.; Grillo, V.; Carlino, E.; Giannini, C.; Kipp, T.; Cingolani, R.; Cozzoli, P. D. *J. Am. Chem. Soc.*, **130**, 11223 (**2008**).

[169] Niu, M.; Cheng, Y.; Wang, Y.; Cui, L.; Bao, F.; Zhou, L. *Cryst. Growth Des.*, **8**, 1727 (**2008**).

[170] Niu, M.; Huang, F.; Cui, L.; Huang, P.; Yu, Y.; Wang, Y. *ACS Nano*, **4**, 681 (**2010**).

[171] Zhou, W.; Cheng, C.; Liu, J.; Tay, Y. Y.; Jiang, J.; Jia, X.; Zhang, J.; Gong, H.; Hng, H. H.; Yu, T.; Fan, H. J. *Adv. Func. Mater.*, **21**, 2439 (**2011**).

[172] Carbone, L.; Nobile, C.; Giorgi, M. D.; Sala, F. D.; Morello, G.; Pompa, P.; Hytch, M.; Snoeck, E.; Fiore, A.; Franchini, I. R.; Nadasan, M.; Silvestre, A. F.; Chiodo, L.; Kudera, S.; Cingolani, R.; Krahne, R.; Manna, L. *Nano Lett.*, **7**, 2942 (**2007**).

[173] Deng, X.; Yang, D.; Tan, G.; Li, X.; Zhang, J.; Liu, Q.; Zhang, H.; Mellors, N. J.; Xue, D.; Peng, Y. *Nanoscale*, **6**, 13710 (**2014**).

[174] Buonsanti, R.; Grillo, V.; Carlino, E.; Giannini, C.; Curri, M. L.; Innocenti, C.; Sangregorio, C.; Achterhold, K.; Parak, F. G.; Agostiano, A.; Cozzoli, P. D *J. Am. Chem. Soc.*, **128**, 16953 (**2006**).

[175] George, C.; Genovese, A.; Qiao, F.; Korobchevskaya, K.; Comin, A.; Falqui, A.; Marras, S.; Roig, A.; Zhang, Y.; Krahne, R.; Manna, L. *Nanoscale*, **3**, 4647 (**2011**).
[176] Choi, S.-H.; Kim, E.-G.; Hyeon, T. *J. Am. Chem. Soc.*, **128**, 2520 (**2006**).

Chapter 4

Magnetic Anisotropic Particles: Synthesis and Applications

Xian Jun Loh[*,†,‡,||] *and Boon Mian Teo*[§,¶,**]

[*]Institute of Materials Research and Engineering, A*STAR, 2 Fusionopolis Way, Innovis, Singapore
[†]Department of Materials Science and Engineering, National University of Singapore, 9 Engineering Drive 1, Singapore 117576, Singapore
[‡]Singapore Eye Research Institute, 11 Third Hospital Avenue, Singapore 168751, Singapore
[§]Institute of Biomedical Engineering (IBME), University of Oxford, UK
[¶]Interdisciplinary Nanoscience Center (iNANO), Aarhus University, Denmark
[||]lohxj@imre.a-star.edu.sg
[**]boon.teo@eng.ox.ac.uk, bteo@inano.au.dk

Abstract

Recent breakthroughs in nanomaterial synthesis have led to continuous development of an expanding library of novel, complex, and shape-controlled structures with unique properties. Such anisotropic Janus particles (JPs) are colloids that provide asymmetry and can impart different chemical and physical properties and directionality within a single particle. The fabrication of such asymmetric particles has attracted tremendous interest amongst research scientists and this interest arises from their fascinating properties and the exciting range of potential application opportunities. JPs are typically divided into three categories, namely, polymeric, inorganic, and polymeric-inorganic. Each of this kind of JPs can be of a range of shapes such as spherical, snowman-shaped, rod-shaped, or cylindrical. Herein, this chapter focuses mainly on JPs with a magnetic component, with emphasis on the fabrication methods and also the unique properties and applications of magnetic JPs in recent years.

Keywords: Janus particles, magnetic field, iron oxide, nanoswimmers, biomedical applications, pickering surfactants.

4.1. Introduction

The research field of anisotropy colloidal systems (also famously known as Janus particles, JPs) and their potential applications have been booming since P. G. de Gennes introduced the concept of JPs in his Nobel Prize address in 1991 [1]. The term "Janus" comes from the ancient mythological Roman God, depicted as having twin faces looking to the past and the future, to creation and destruction. The term Janus is used to describe colloids that are asymmetric and can impart different chemical and physical properties within a single particle. Topics in this area of research in Janus colloids include different synthetic strategies, chemical composition-property relationships, self-assembly behaviors, and many of these JPs have already been realized in several exciting applications.

Many natural systems possess asymmetry, which is used for adaptation and mimicry by living organisms. It is interesting to note that the scales on pine cones have a structure that is Janus-like and a bilayer. They are able to open in dry environment and close in humid environments [2]. There are many other types of natural asymmetric Janus-like systems with sizes ranging from nanometers to macroscale. They include lipids, enzymes (Janus kinases), plant leaf surfaces, and varying claw size of fiddler crabs.

Colloidal JPs, possessing changing color, charge, and other physical properties at different sides of the particle, are very fascinating illustrations of synthetic asymmetric systems. As a result of their biphasic properties, these particles are unique compared to other types of particles. Actually, these particles can be seen as macroscopic equivalents of surfactants. They are able to organize at interfaces similarly. JPs show substantial prospects for various uses. To illustrate, Janus beads were shown to be very efficient as catalysts [3], as drug delivery vehicles [4], as modular blocks for the construction of self-assembled materials [5], as optical [6], and rheological probes [7], including basic components for designing electronic paper devices [8], as a compatibilizer for emulsions and blends [9]. Studies of the characteristics of JPs at interfaces is especially an exciting topic. JPs possess greater adsorption energy than homogeneous particles. Additionally, JPs can separate at interfaces and stabilize emulsions as surfactants do, or behave as bendable superhydrophobic barricade on the water surface. Their separation can be accurately adjusted by changing the ratio between two particle parts.

At first, efforts were focused on the synthetic strategies for fabricating novel JPs. Such fabrication methods include utilizing masking techniques [10], microfluidic devices [11], and electrohydrodynamic jetting [12], or techniques exploiting phase separation in confined space, lithography patterning [13], the pickering emulsion method [14], or methods based on the phase separation techniques [15]. JPs of different sizes, ranging from microscopic to nanoscale length, have also designed and engineered. The interested reader is referred to several well-written review articles [16] dedicated to the synthetic strategies and challenges in the literature. While efforts continue on the designing and engineering of anisotropic particles, researchers are also realizing the novel and attractive properties that such asymmetric colloids possess because of their different chemical composition and interesting shapes. For example, the amphiphilic nature of these particles makes them useful as colloidal surfaces in water-based emulsions [17] and Janus colloids coated with different chemical groups can perform an autonomous motion with the right choice of fuel and be utilized as artificial micro/nanomotors [18].

In this chapter, we introduce the interested reader to an in-depth overview of the thriving field of anisotropic magnetic Janus particles (MJPs). We aim to provide a comprehensive discussion of the different synthesis methods and self-assembly of MJPs with emphasis on the challenges of the different methods of their synthesis. For the second part of the chapter, we also highlight some of the recent advances in important proof-of-concept studies in the field of MJPs for a range of applications. In this regard, we hope to inspire the reader with novel ideas in colloidal synthesis and also to stimulate the birth of new concepts and exciting applications in the field of magnetic anisotropy colloids.

4.2. Fabrication Techniques of Janus Particles

Submicron hybrid magnetic polystyrene (PS) based JPs containing magnetic nanoparticles (MNPs) on one side were effectively fabricated using the miniemulsion/solvent evaporation technique [19]. Nanosized droplets of styrene (St) monomer in the presence of PS and MNPs were created in an aqueous continuous phase. Following the evaporation of St monomers, PS and MNPs precipitate to form spherical nanoparticles. When the MNPs:PS weight ratio was 1:1, the formation of Janus-like particles with MNPs located on one side was promoted as a result of the increased PS concentration during evaporation of the monomer solvent. In another report, anisotropic Janus magnetic polymeric nanoparticles were successfully prepared by mini emulsion polymerization of St and acrylic acid (AA) monomers in the presence of MNPs coated with oleic acid (OA). Upon the initiation of polymerization, phase separation between the polymer matrix and iron oxide nanoparticles coated with OA occurs in each particle, resulting in the formation of anisotropic magnetic particles with Janus morphology with an average hydrodynamic diameter of 250 nm. Through the precise control of the St, AA, MNPs, AIBN, and SDS content, the morphology of the Janus magnetic polymer nanoparticles could be well controlled. The fabrication of anisotropic magnetic hybrid microparticles by emulsion polymerization of styrene in the presence of (O/W) magnetic emulsion as a seed was also reported [20]. Emulsion polymerization of the monomer-swollen ferrofluid droplets was carried out to prepare anisotropic magnetic PS particles with an average hydrodynamic diameter of around 300 nm. As the PS phase and the ferrofluid droplets phase are incompatible, phase separation occurs, leading to the formation

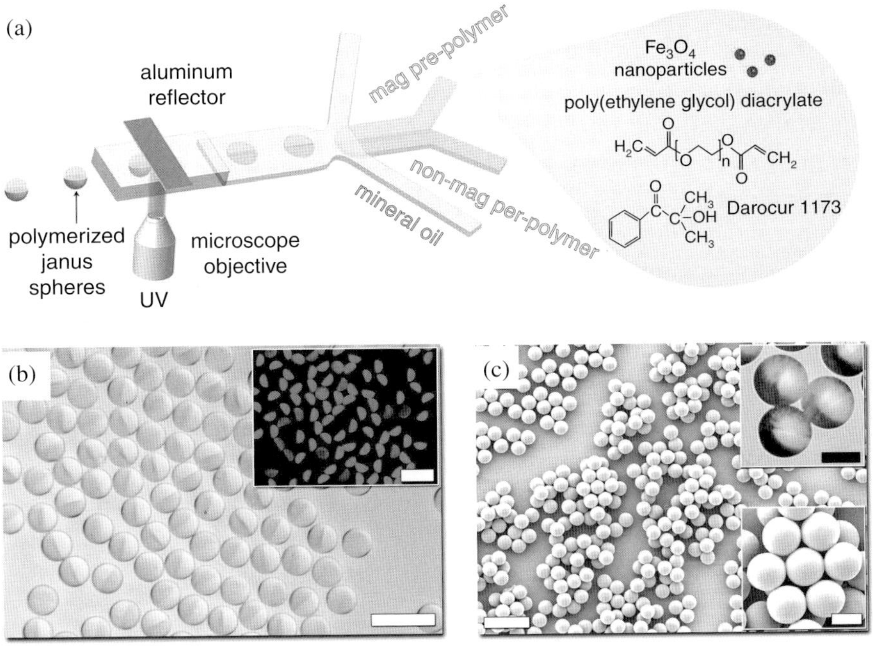

Figure 1. (a) Schematic of JP synthesis in a flow-focusing microfluidic device. (b) DIC and corresponding fluorescent (inset) images of MJPs generated from coflowing streams of polymer, one containing MNPs, and the other containing rhodamine B. The scale bars are 100 μm wide. (c) SEM and DIC (upper right inset) images of dried JPs. The scale bars are 100 μm wide and 25 μm wide for the insets. Reproduced from Ref. [21].

of Janus-type particle. The presence of the biphasic structure confirms that polymerization of styrene and AIBN in a magnetic emulsion could be a well-controlled process for Janus-like magnetic polymer particle synthesis. Doyle *et al.* reported the microfluidic synthesis of spherical Janus hydrogel particles with superparamagnetic properties and chemical anisotropy (Fig. 1) [21]. The particles assemble into stable chain-like microstructures under an external magnetic field. This is the first report on the synthesis of a uniform superparamagnetic JP and it also demonstrates bottom-up, field-driven assembly of hydrogels with controllable spatial distribution of biochemical payloads. With controllable compositions, these multifunctional microparticles hold promise for engineering complex mesoscale assemblies with heterogeneous geometries and biophysiochemical properties. Kim *et al.* reported a new magnetic nanocomposite material system and *in situ* fabrication process that is not shape limited and allows the programming of heterogeneous magnetic anisotropy at the microscale (Fig. 2) [22]. The combination of the self-assembling behavior of superparamagnetic nanoparticles, which have stronger magnetization than that of general paramagnetic materials, with a spatially modulated photopatterning fabrication technique was demonstrated. Through the tuning of the nanoparticle assembly and fixing of the assembled state using photopolymerization, microactuators were fabricated in which all the parts move in different directions under a homogeneous magnetic field. Polymeric nanocomposite actuators, capable of 2D and 3D complex actuations, were fabricated. This technique greatly simplifies the manufacturing process and has led to design rules for designing novel and complex microcomponents using a nanocomposite material with engineered magnetic anisotropy.

Another paper reports a novel technique for fabricating monodisperse Janus hydrogel beads with magnetic anisotropy smaller than the microchannel based on coagulation of Fe_3O_4 nanoparticles and a technique known as the shrinkage-gelation technique [23]. This technique avoids the limits of the minimum biphasic droplet size due to fear of clogging, and the problem of a blurring boundary on fabricated JPs by convective mixing across the interface inside the biphasic droplets. This technique is proposed for potential applications such as for the further development of electronic paper. Jeong *et al.* reported a simple method using microfluidic droplet-generation technique to prepare MJPs by using a solvent evaporation-induced phase separation and preferential partitioning of MNPs in the polymer blends [24]. Non-aqueous emulsion droplets of the polymer blends and nanoparticles solution coalesced to form JPs after solvent evaporation. The stabilizing polymer of the nanoparticles, which is compatible only with one of the polymer blends to be phase-separated, plays a key role in the anisotropic positioning of the nanoparticles in the JPs. Using this phase separation-based method and microfluidics, excellent control over the size, size distribution, and morphology of the particles is achieved. Another method for the

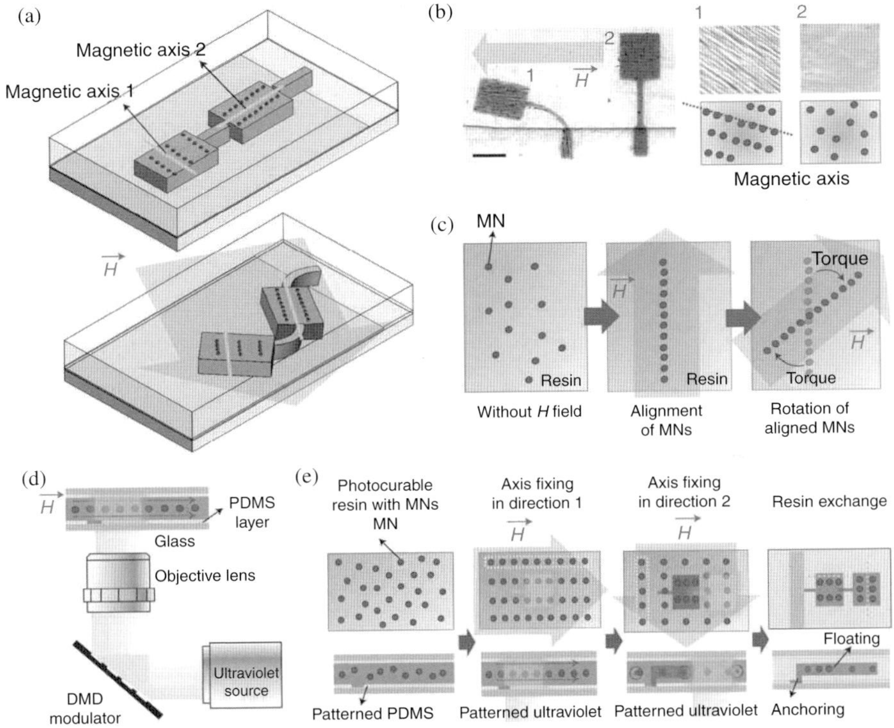

Figure 2. (a) The movement of a magnetic actuator. The magnetic actuator possesses two different magnetic axes; therefore it actuates in a zig–zag conformation when the magnetic field is applied. (b) Actuation of a simple magnetic cantilever. Under a homogeneous magnetic field, the magnetic cantilever, which contains self-assembled MNPs, only bends toward the field line. (Scale bar 50 μm). (c) MNPs self-assembly. Randomly dispersed MNPs are self-assembled along the uniform magnetic field line to minimize magnetic dipole interaction. When the magnetic field direction is changed, the assembled nanoparticles rotate along the field line. (d) Maskless lithography setup. The setup is composed of an UV light source, a digital micromirror device (DMD) modulator and objective lenses. The light is patterned through a DMD modulator, and focused on the microfluidic channel, polymerizing the resin. This process enables fabrication of microstructures with various shapes as well as the ability to fix the MNPs in the structure. (e) Magnetic axis fixing process. A mixture of photocurable resin and magnetic nanoparticles is injected into the microfluidic channel. A magnetic field is applied in direction 1, and the microstructure embedding with aligned MNPs in direction 1 is fabricated using maskless lithography. A magnetic field is applied in direction 2, and another microstructure embedded with aligned magnetic nanoparticles in direction 2 is attached to the previous structure using the same process. Finally, the resin is exchanged and the fabrication process is completed. The completed actuator is anchored at one end and free-floating at the other end. Reproduced from Ref. [22].

fabrication of such particles involves the flame synthetic route, which was shown to be very effective in the preparation of iron oxide/silica composite particles with a novel Janus structure [25]. The highly uniform particles could be dispersed very well in water, allowing options for further manipulation of the JPs to form self-assembled structures. These particles can be utilized as partially sacrificial templates in the preparation of hollow capsules with magnetic cores and soft PVP/Au nanoparticle shells. With regard to the direct synthesis of two chemically or physically dissimilar phases contained within one particle, polymerization can lead to the formation of two incompatible polymer segment blocks constrained within one macromolecular structure. A selection of macromolecules possessing Janus characteristics with two dissimilar covalently linked nanostructured units has been shown. Wooley *et al.* showed the synthesis of strongly dipolar monodisperse dendrimers [26]. The compounds possess high-dipole moments by the intelligent design of electron-withdrawing cyano groups and electron-donating benzyloxy groups at segmentally opposed regions of the chain ends of the dendrimers. Another report focused on comb copolymers of PS-*b*-PEO having a poly(chloroethyl vinyl ether)-*b*-poly(hydroxy ethyl vinyl ether) backbone with PS chains on one side and poly(ethylene oxide) chains on the other, i.e., (PCEVE-g-PS)-*b*-(POHEVE-g-PEO) [27]. By grafting polystyryllithium onto the reactive chloro functional group of the poly(chloroethyl vinyl ether) first by block polymerization and by grafting polyethylene oxide from the hydroxyl functions of the second block, the comb polymer was prepared. Through this, densely grafted amphiphilic Janus-type copolymers with high molar masses and narrow polydispersity were obtained. Unimolecular rodlike nanoobjects with distinct PS and PEO domains were observed. The polymer combs are able to self-assemble into different morphologies. When polymer deposits are made from a selective solvent of the PEO comb block, hyperbranched micelles formed in the solution retain their structure on the solid substrate, yielding well-defined spherical objects. In another example, Fréchet *et al.* synthesized dendronized block copolymers with non-centrosymmetric morphologies [28]. The covalent stabilization of Janus micelles is also an important approach for the synthesis of Janus unimolecular nanomaterials. Müller *et al.* showed that non-centrosymmetric nanoscopic particles, cylinders, and disks can be prepared from Janus micelles of triblock copolymers by selective cross-linking of the short middle blocks [29]. Additionally, Janus unimolecular nanoparticles can also be fabricated by polymerization or modification in the presence of asymmetrical templates [30]. However, it is challenging to come up with simple approaches for the fabrication of unimolecular Janus nanomaterials and to ascertain their non-centrosymmetric nanostructures by standard visualization techniques.

Berger *et al.* demonstrated a fresh useful idea for the fabrication of stimuli-responsive bicomponent JPs [5]. These particles are decorated by polymers with different polarity and charge. Sequential "grafting from" and "grafting to" methods were used to make these responsive JPs. Polymer was first grafted on one side of silica particles using the "grafting from" method by surface-initiated ATRP. The second polymer was anchored using the "grafting to" method in melt by reaction of the chain-end carboxylic group and functional groups on the other side of the particle surface (Fig. 3(A)). It was shown that the prepared JPs have sharp interfaces between the constituent phases. The bicomponent JPs responded to changes in pH and this led to the creation of hierarchically structured aggregates from the JPs. Fluorescence microscopy (Fig. 3(B)) established the Janus character of the acquired particles. Fluorescent microscopy images showed that one side of the particles possess more green fluorescence while the other one is more red-fluorescent, demonstrating that the polymers were grafted to the opposite sides of the particles. When viewed from the top, a combination of "green" and "red" fluorescent signals produced by the grafted polymers is observed.

The same group also reported a novel hairy hybrid Janus catalyst made up of a silica core and the hydrophilic PAA and hydrophobic PS on its opposite sides (Fig. 4) [31]. They showed simple and yet selective functionalization of amphiphilic JPs with Ag and Au nanoparticles. The catalytic species were localized selectively in the PAA-covered hemisphere of the JPs. The integrated metallic nanoparticles in the particles showed significant interfacial activity and efficient stabilization of water–oil emulsions that are pH-tunable. Also, the catalytic activity of Ag and Au was demonstrated by applying benchmark reactions such as the reduction of methylene blue, eosin Y, and 4-nitrophenol with extremely low amounts of catalyst being used. These particles bring in advantages as they are able to stabilize emulsions and reverse the stability of the emulsion by activating the responsive properties of the JPs, and the particles can be recovered after reaction and reused.

A key trial in nanoparticle functionalization is the fabrication of polymer-grafted Janus nanoparticles with diameter of less than 100 nm. Recently, a cyclic synthesis protocol using a reversible click reaction and "grafting to" strategies to synthesize such nanoparticles. The cyclic

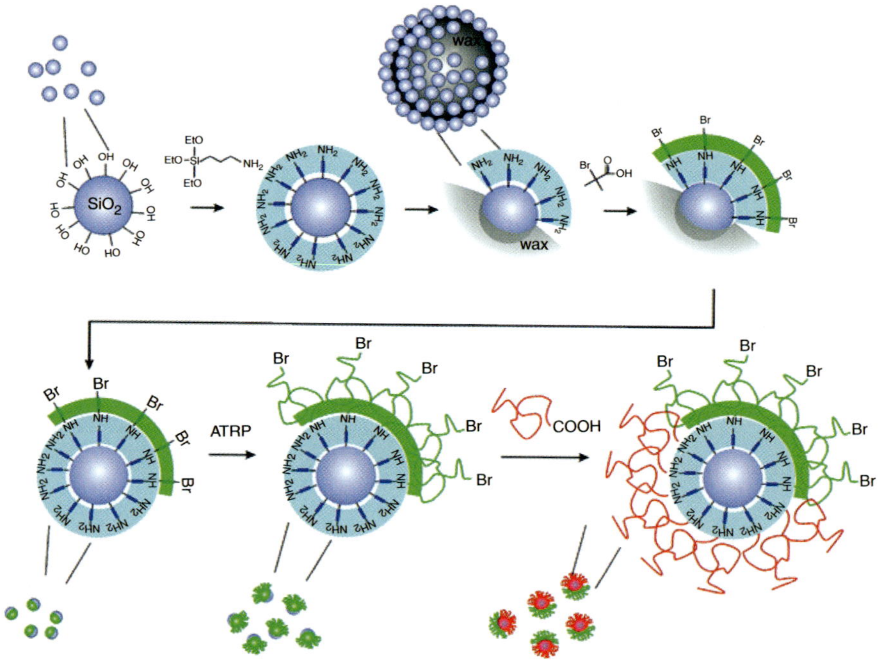

(A)

Figure 3. (Figure on facing page) (A) Scheme of the synthesis of bicomponent JPs by "grafting from" and "grafting to" approaches. The bare silica particles are coated by APS, assembled around wax colloidosomes and selectively modified by ATRP initiator at one side (upper panel). The first polymer (PtBA or PNIPAAm) is grafted by surface-initiated ATRP. The carboxyl-terminated second polymer (P2VP) is grafted to free amino groups on silica particles by the "grafting to" approach. (B) Fluorescence microscopy images of PtBA-PVP JPs obtained using FITC (a, green) and TRITC (b, red) filter sets. Image (c) is a color combination obtained from the images (a) and (b). Intensity profiles are given for numbered particles in images (a) and (b). Adapted with permission from American Chemical Society Ref. [5].

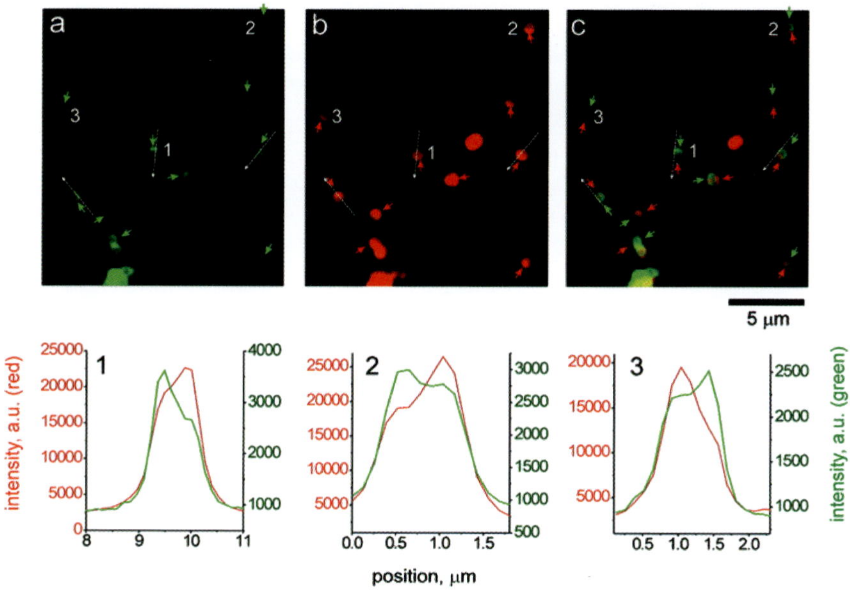

(B)

Figure 3. (*Continued*)

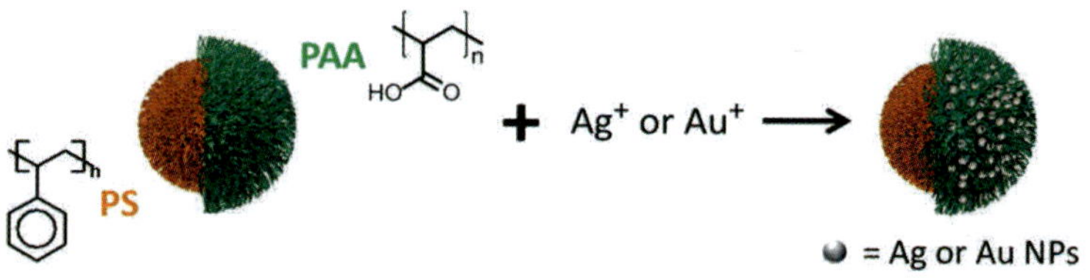

Figure 4. Scheme of the selective NP immobilization onto PAA/PS-JP for catalytic applications. Reproduced from Ref. [31].

synthesis route is depicted in Fig. 5. The first step involves the functionalization of the surfaces of the particles by azido and alkynyl groups. Then the smaller alkyne-functionalized particles were attached to the surface of the larger azido functionalized particles via the copper-mediated click reaction. The free unreacted smaller particles were removed in a repeated centrifugation–dispersion process. The next step involves the addition of azido-capped poly (methyl methacrylate) (N_3-PMMA) to the suspension of the particle complexes, and polymers were grafted to the smaller

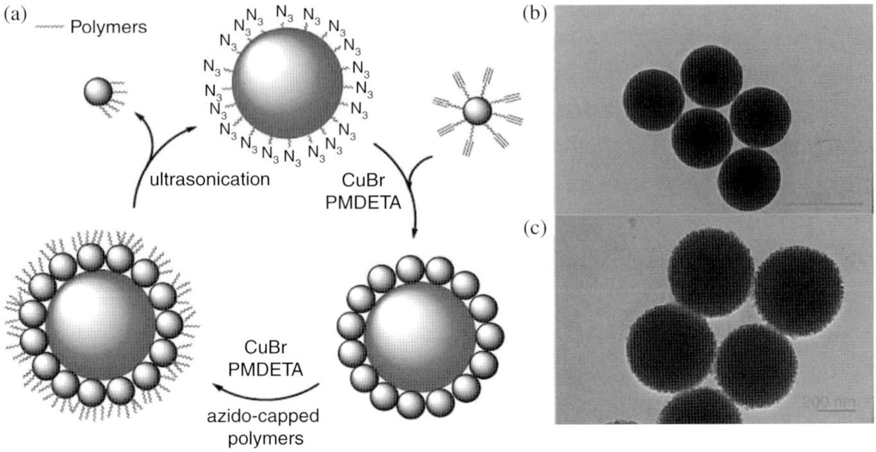

Figure 5. (a) Schematic illustration of the cyclic synthesis route for polymer-grafted Janus silica NPs by combining the reversible click reaction and grafting to strategies. (b) TEM image of azido-functionalized 500 nm particles. (c) TEM image of 500 nm particles with 15 nm NPs attached. Reproduced from Ref. [32].

particles by a click reaction again using a grafting to strategy. Finally, the PMMA-grafted Janus NPs were released under sonication, and the larger particles could be recycled for another round of attachment.

Nie *et al.* showed a microfluidics-based synthesis of JPs and three-phase particles [33]. The particles possessed a clear interface between the constituent phases and the structure and size distribution of microbeads could be accurately controlled. The amphiphilic JPs with various volume fractions of the constituent phases form clusters with different agglomeration degrees. Selective functionalization of the JPs was achieved by introducing functional moieties in one of the monomer phases or by conducting asymmetric bioconjugation of JPs with covalently attached bovine serum albumin after microfluidic synthesis. Figure 6 shows the process of synthesis. M1 contains methacryloxypropyl dimethylsiloxane and M2 contains a mixture of pentaerythritol triacrylate (45.0 wt.%), poly (ethylene glycol) diacrylate (45.0 wt.%), and AA (5.0 wt.%). Both M1 and M2 contained 4 ± 1 wt.% of 1-hydroxycyclohexyl phenyl ketone as the photoinitiator.

Erhardt *et al.* reported the fabrication and behavior of Janus micelles containing a polybutadiene (PB) core and a corona with a poly(methacrylic

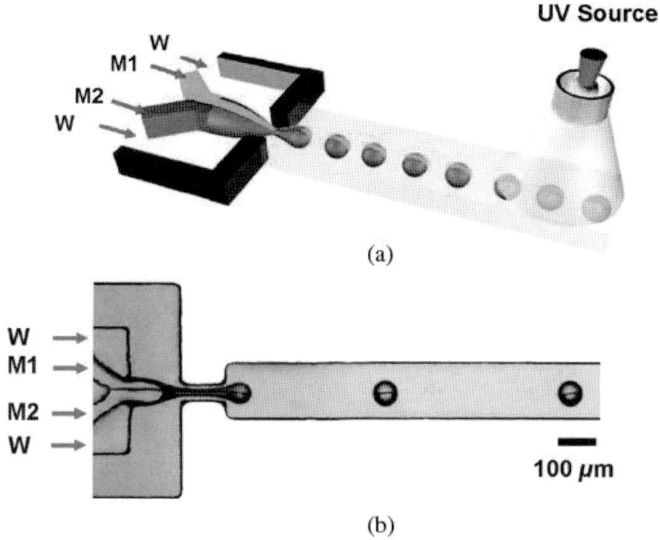

Figure 6. (a) Schematic of generation of Janus droplets from immiscible monomers M1 and M2, emulsified in an aqueous solution of SDS (W). The droplets are irradiated with UV light in the downstream channel. (b) Optical microscopy image of formation of Janus droplets. Flow rates of M1, M2, and W are 0.02, 0.02, and 4 mL/h, respectively. The height of the MFFD is 120 μm and the orifice width is 60 μm. The scale bar is 100 μm. Adapted with permission from American Chemical Society Ref. [33].

acid) (PMAA) and a PS hemisphere [29c]. The cross-linking of the middle block of the PS-*block*-PB-*block*-PMMA triblock copolymer in the bulk state, and subsequent alkaline hydrolysis of the PMMA ester groups led to the formation of the Janus micelles.

4.3. Applications of Magnetic Janus Particles

As a result of their non-symmetric nature, JPs and their assembled structures possess unique chemical and physical properties that are distinctly different from isotropic systems. Considerable efforts have been devoted to the understanding of these intriguing properties derived from their asymmetric structure and to a better understanding of the complex relationship between their structure and potential applications.

In the following sections, recent advances in using MJPs for a variety of applications will be reviewed. Particular attention will be given to new emerging applications and highlight the opportunities and challenges of this important field.

4.3.1. Magnetic Swimming Janus Micromotors

A myriad of autonomous biomolecular motors have evolved to perfection in nature to manufacture and transport biochemicals in the cytoplasm and provide motility of cells. Some excellent examples of locomotion in nature are motor proteins (dynein, kinesin, and myosin), which can sense, pick up, and deliver intracellular cargoes to specific targets [34]. The flagella biological motor in bacteria (*E. coli*), that is found in cell membranes, decomposes adenosine triphosphate for energy, providing motility to bacteria in liquids, and the spermatozoa of many organisms that swim to the ovum during the fertilization stage [35]. Such sophisticated biomotors have motivated scientists to fabricate artificial nanomotors with the ability to efficiently and precisely transport and deliver cargoes to targeted sites.

Anisotropic JPs can be propelled in a fluid when the right approach is taken into account for the negligible role of inertia at low Reynolds (Re) number. The Re number is defined according to the following equation:

$$\mathrm{Re} = \frac{UL_{\rho}}{\eta}, \tag{1}$$

where U and L are defined as the velocity and characteristic length, respectively, and ρ and η are the density and viscosity, respectively, of the fluid. The Re number is the ratio of inertial to viscous forces in a fluid. For small Re numbers, i.e., Re $\ll 1$, inertial forces become insignificant and viscous forces dominate. At low Re number, movement for tiny objects in liquids such as water becomes similar to movement in viscous liquids such as honey. As an example, the bacteria, *E. coli* has a Re number of 10^{-4} moving at typical speeds of ~20 μm s^{-1} in water [36]. At low Re number, the inertial terms in the Navier–Stokes equation can be considered negligible and gives a simplified linear expression, called Stokes equation and is defined as

$$\nabla p = \eta \nabla^2 U + f, \tag{2}$$

where p is the pressure, U is the flow velocity, and f is the external forces on the fluid. This equation assumes that Re = 0 but is valid for Re $\ll$ 1. This results in instantaneous and time-reversible fluid flow and flow parameters depend only on the time-dependent boundary conditions.

The autonomous locomotion of nanoscale particles represents a key challenge and research interest in the autonomy of artificial motors is enormous. Progress in this field toward the fabrication of Janus motors that can be propelled by different mechanisms, such as self electrophoresis [37], bubble propulsion [38], diffusiophoresis [39] and external stimuli such as light [40], magnetic [41], and ultrasound [42], have been tremendous. Systems that do not require external source of energy for their locomotion typically follow an uncontrolled trajectory. Among these propulsion procedures, magnetic propulsion is often a very popular technique and the most successful to date as it is non-invasive and offers easy control and navigation of the targeted motor. As the focus of this chapter is on MJPs, the interested reader is referred to the literature [43] for a full account of other propelling mechanisms. In the following sections, a brief introduction to magnetic field principles and recent advances in magnetic actuation of Janus colloids will be presented.

4.3.1.1. *Control by magnetic fields*

Magnetic fields are less invasive than other kinds of actuation and are accepted in the medical field such as in the widespread use of magnetic resonance imaging (MRI) [35b,44]. Typically, magnetic fields and magnetic field gradients can be produced using a strong permanent ferromagnet. However, the disadvantage of using such method in the manipulation of particles means that a change in the field direction or strength requires that the magnetic forces are moved and relocated in space. Alternatively, the use of devices containing electromagnets evaded this issue [36]. A typical configuration for such devices for producing homogeneous or gradient magnetic fields are two-stacked current-carrying coils that share a common axis and are separated by a gap. In the Helmholtz coil, the direction of the current in the two coils is the same. As a result, a homogenous field at the geometrical center of the coil system is generated. However, in the Maxwell coil configuration, the direction of the currents in the two coils is opposite, giving rise to magnetic gradient fields.

4.3.1.2. *Magnetic field principles*

In general, magnetic materials are subject to translational forces and rotational torques when exposed to an external magnetic field. The magnetic force as given in Eq. (3) depends on the non-uniformity of the magnetic field. As a result, net force is not exerted on the magnetic material in a homogeneous magnetic field.

$$\overrightarrow{F_m} = V_m(\overrightarrow{M} \cdot \nabla)\overrightarrow{B}. \tag{3}$$

Gradient fields are used to exert a translational force on magnetic materials. These materials can experience a force that pulls it in the direction of increasing field intensity. Such magnetic fields are typically generated with a Maxwell coil arrangement [45]. Materials exposed to a magnetic field also experience a torque and align their magnetization axis parallel to the external field.

$$\overrightarrow{T_m} = V_m \overrightarrow{M} \times \overrightarrow{B}, \tag{4}$$

where T_m [N m] is the torque that acts to align its magnetization M (A m^{-1}) with the external field B (T) and V_m (m^3) is the volume of the magnetic object.

As shown in Eq. (4), a Janus motor placed in the uniform magnetic field does not experience any force but only a torque until M is in collinear with B. The torque, T_m becomes zero and the Janus motor remains stationary. To generate a continuous actuation, the magnetic field has to exhibit a field gradient or undergo temporal change in the form of rotation, precession, oscillation, or on–off states. If the magnetic field exhibits a field gradient, it can produce a force to pull the motors, eliminating the need for an additional swim mechanism. However, if it undergoes a temporal change, different shapes and geometries of the motors can be considered with various actuation principles [46] (Fig. 7).

The design of magnetic motors is crucial as the type of material properties and shape can influence the motion mechanism. Ferromagnetic materials tend to show superparamagnetic properties when the size is reduced below a certain critical dimension. This is due to the absence of the motion of domain walls at the size below a critical dimension.

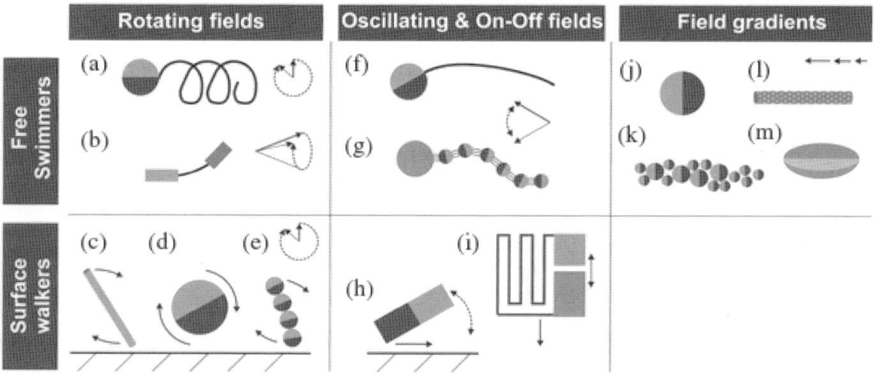

Figure 7. Different kinds of magnetic motors and their actuation methods. (a) Helical motor actuated by a rotating field. (b) Flexible nanorod actuated by a rotating or precession field. (c)–(e) Actuated with a rotating field, these motors tumble along the surface due to the fluidic drag imbalance. In this schematic, they would all tumble from left to right along the surface. (c) Nanorod; (d) permanent magnetic sphere; (e) self-assembled microbead chains. (f) and (g) In-plane oscillating fields actuate the head and tail, respectively. (h) Due to surface friction, this robot etches forward with every up-down oscillation. (l) Top view of a wireless resonant actuator, powered by on–off field oscillations. (j)–(m) Different types of nano and microrobots pulled by field-gradient forces. (j) Single permanent magnet; (k) multiple magnetic beads that group to attractive forces between them; (l) nickel or iron-filled carbon nanotubes and (m) soft-magnetic assembled microrobot. Adapted from Ref. [46] (2013) with permission. Copyright (2013) Royal Society of Chemistry.

Single- domain magnetic moments are formed below a certain critical size as shown in Table 1 [36,47]. The single domain size shown in Table 1 assumes the material in question is isotropic. Both magnetic field gradients and homogenous magnetic fields can be used to propel and control magnetically active Janus motors. However, there are several challenges involved and one important factor to consider is that the magnetic field tends to decay rapidly with the distance from the source. For magnetic propulsion experiments, the design of the magnetic setups is therefore a crucial parameter to consider.

Some early work on magnetic field control and manipulation of Janus motors was done by Mathieu *et al.* [48] and Martel *et al.* [49]. They demonstrated that nanoswimmers composed of ferromagnetic material can be successfully navigated under an external magnetic field. Others have exploited a clinical MRI system for magnetic propulsion of micro- and

Table 1. Single domain sizes of different magnetic spherical particles. Adapted from Ref. [36] with permission. Copyright (2011) Royal Society of Chemistry.

Material	Critical size (nm)
Co	70
Fe	14
Ni	55
Fe_3O_4	128
γ-Fe_2O_3	166

macrosized objects [35a,50]. Both the Helmholtz coils and electromagnetic coils have been used to generate magnetic fields for propulsion of ferromagnetic particles.

Rotational diffusion strongly influences the direction of Janus motors propelling in an aqueous H_2O_2 solution [3,51]. In one example, Palacci *et al.* showed that polymer spheres with a protruding hematite cube (Fig. 8(a)) can move in aqueous H_2O_2 solution under the action of blue-violet light as the light triggered the decomposition of H_2O_2 at the surface of the hematite particles and when facing this surface close to the substrate, the JPs started to slide above the generated osmotic flow (Figs. 8(b) and 8(c)) [52]. The authors were also able to guide clusters spontaneously form via phoretic attraction on a defined direction using a static external field. The external field also enhanced particle interactions, and the particles avoided reorientation effects due to rotational diffusion, which could lead to cluster disaggregation (Figs. 8(d)–8(i)).

Another successful study reported on demonstrations of autonomous nanomotors achieved in a nanorod composed of (platinum/nickel/gold/nickel//gold) (Pt/Ni/Au/Ni/Au) where the ferromagnetic nickel components were magnetized in a direction perpendicular to the rod axis (Fig. 9(i)) [53]. The orientation of the nanorod motors was perpendicular to the applied magnetic field direction and the rods were able to move in the direction of the Pt. The autonomous was entirely derived from the metal-catalyzed decomposition of H_2O_2 into oxygen and water while the

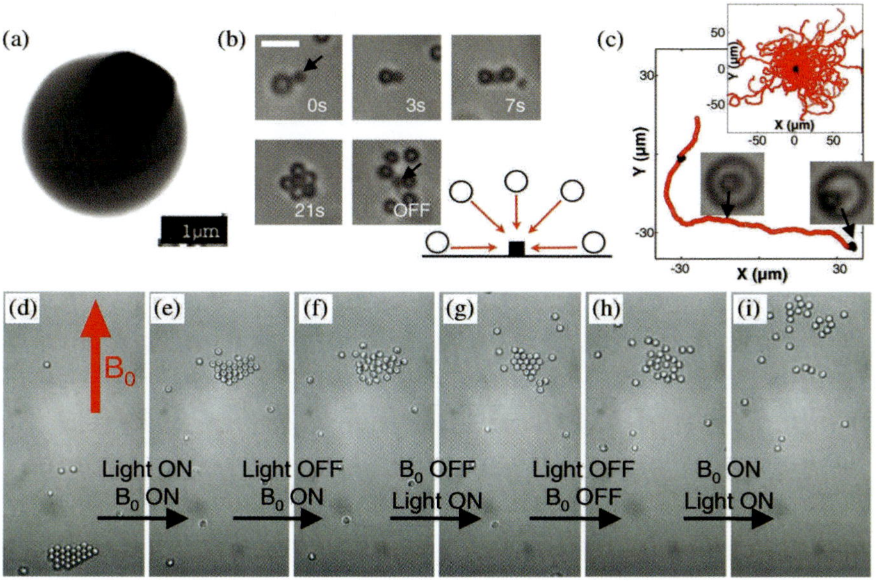

Figure 8. (a) Transmission electron microscopy (TEM) of the Janus colloid: a polymer colloidal sphere with protruding hematite cube (dark). (b) A hematite cube, indicated by an arrow, is immobilized on a surface and immersed in a solution of colloidal tracers. At $t = 0$ s, the blue light is switched on, triggering many particles with their origins aligned. (d)–(i) An external magnetic field $B_0 \sim 1\ mT$ was utilized for the decomposition of hydrogen peroxide on the hematite surface. The tracers are attracted to the hematite until they contact the cube. The attraction is isotropic with particles coming from all directions, thus discounting advective flow that must exhibit zero divergence. When the light is turned off, the attraction ceases and the tracers diffuse away. (c) A hematite cube protruding from a TPM polymer sphere moves on fixed glass substrate when exposed to blue light (red part of trace) and diffuses when the light is off (black part of trace). Initially, with no light, the hematite cube is oriented randomly (image, right) but rotates and faces downward toward the glass substrate when the light is turned on (image, left). The particle then surfs on the osmotic flow it induces between the substrate and itself. (Inset) A superposition of the trajectories orient the particles and direct their motion. The red arrow is the orientation of B_0. (d) and (e) The magnetic field is turned on, and the light is on; the crystal is self-propelled in the direction of the magnetic field, and crystal breakup is suppressed. (f) The light is turned off, and the magnetic field B0 is left; the crystal dissolves. (g) The magnetic field is turned off, and the light is turned on; particles collide, and the crystal is reformed. (h) The light is turned off, and the magnetic field remains off; the crystal dissolves. (i) The magnetic field is turned on first, and then the light is turned on. The particles all move in the field direction; they do not collide and do not crystallize. Adapted from Ref. [52] with permission. Copyright (2013) *Science*.

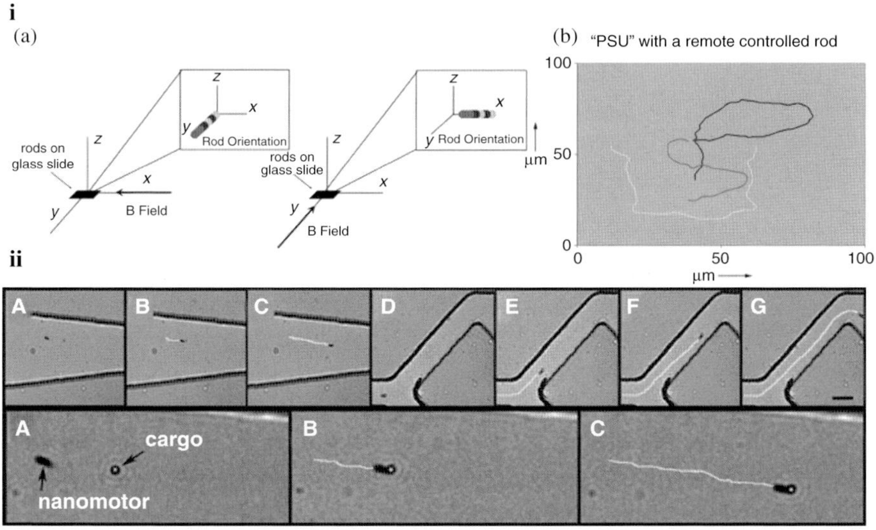

Figure 9. (i) (a) Rod orientation in an applied field, where the Cartesian coordinate axis to the left of the rod represents a sample of rods on the microscope stage and the arrow represents the location of the magnet in the *xy*-plane of the microscope stage. (b) The trajectory path of a Pt/Ni/Au/Ni/Au nanorod spelling the letters "PSU" in 5 wt.% H_2O_2. Adapted from Ref. [53] with permission. Copyright (2004) Wiley. (ii) Optical microscopy images of the dynamic loading of an Au/Ni/Au/Pt-CNT nanomotor with a 1.3 μm diameter magnetic microparticle cargo (A–C) and transport it through PDMS microchannels (D–G). Scale bar of panel G, 25 μm. Bottom images in panels A–C are magnified by 3.5 times that of the top images panels A–G. Adapted from Ref. [37] with permission. Copyright (2008) American Chemical Society.

magnetic field controlled the direction of the motion. The rods were manipulated along a trajectory, drawing the letters "PSU". In another publication, the same authors fabricated similar magnetic nanorods and demonstrated that their nanomotors can transport and magnetically steer PS microspheres [54]. Several investigations on the ability of nanomotors to carry and deliver a cargo have been conducted. In an early report by Wang, they also reported fabricated catalytic motors composed of Au/Ni/Au/Pt-carbon nanotube (CNT) and demonstrated that these motors can be used to pick up and deliver micron-sized PS beads coated with iron oxide in a microfluidic device (Fig. 9(ii)) [37b]. Their CNT-based motors can propel a relatively large cargo at high speeds through directed paths and junctions of a microchannel network and the nickel component of the motors offer controlled cargo manipulation, including enroute loading, dragging, and releasing.

In another study, Garcia-Gradilla *et al.* demonstrated a multifunctional nanomotor composed of three segments (Au–Ni–Au), which was propelled by ultrasound and guided by an external magnetic field using the Ni component [55]. The end of the Au component was made concave by sphere lithography technique to achieve shape asymmetry. They further modified the nanomotor Au surface by functionalizing lectin and antiprotein A antibody bioreceptors to facilitate capture and transport of *E. coli* and *Staphylococcus aureus* bacteria in complex media. The practical use of these fuel-free motors was successfully demonstrated by the isolation and separation of the biological targets *E. coli* and *S. aureus* using ConA and antiprotein A antibody functionalized micromotors.

In another investigation by Wang's group on the micromotor-based approach for the rapid screening, detection, and destruction of anthrax spore simulants, they functionalized their motors with anti-*B. globigii* antibody functionalized motors and the motors moved in a mixed contaminated solution of *B. globigii, S. aureus*, and *E. coli* via magnetically guided propulsion (Fig. 10(i)) [56]. These motors showed no interaction with excess of other bacterial counterparts and could concentrate multiple spores at their surface. Figure 10(ii) shows the "on-the-fly" spore detection protocol, which relies on the movement of anti-*B. globigii* antibody-functionalized micromotors in a contaminated solution to recognize, capture, and transport single and multiple spores.

The first work on helical micromotors mimicking bacteria propulsion, the artificial bacterial flagellum (ABF), was reported in 2007 by Bell *et al.* [57] and further characterized by Zhang in 2009 [41a,58]. The ABF they described consists of a rigid helical tail made of indium–gallium–arsenic/gallium–arsenic (InGaAs/GaAs) or indium–gallium–arsenic/gallium–arsenic/chromium (InGaAs/GaAs/Cr) and a ferromagnetic metal head containing chromium/nickel/gold (Cr/Ni/Au) layers obtained by a self-scrolling technique. The size of the helical tail measured at 2.8 mm in diameter and 30–100 mm in length. Upon exposing a rotating magnetic field, the magnetically induced rotation of the head was converted into translational motion, which is dependent on the chirality of the helix. The rotation of the magnetic head of the ABF was responsible for the wireless

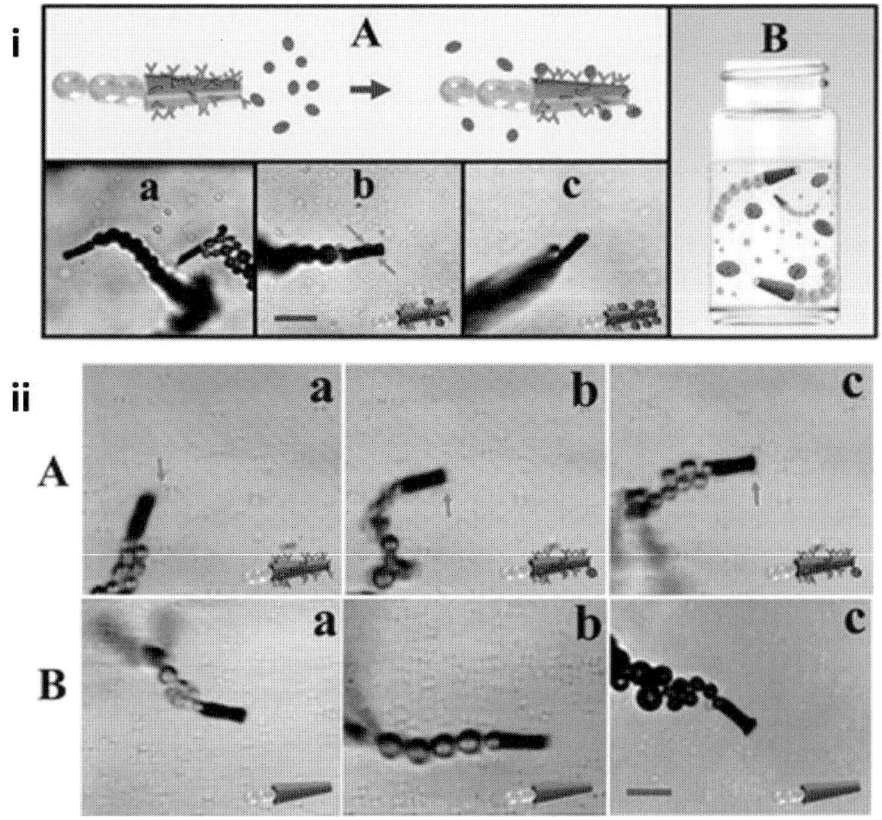

Figure 10. (i) (A) Schematic of the functionalized motors capturing and transporting *B. globigii* spores for their further destruction. (B) Sketch of micromotors swimming in a spore-containing solution for the accelerated destruction of the spores. (a)–(c) Time-lapse images illustrating the magnetically guided propulsion of anti-*B. globigii* antibody micromotors in a spore-containing aqueous solution (a) and functionalized micromotors transporting two target (b) and multiple target (c) spores. Spores, highlighted with an arrow. Scale bar, 20 μm. (ii) Micromotors capturing and transporting *B. globigii* spores. (A) Time-lapse of functionalized micromotors approaching (a), contacting (b) and carrying (c) *B. globigii* spores present at 9.5×10^5 CFU mL^{-1} (level 1×) in the solution. (B) Negative controls: time-lapse images of unmodified micromotor contacting but not capturing the spores at 5× (a), as well as modified micromotor bypassing both S. aureus cells at 10× (b) and *E. coli* at 1× (c). Conditions: PBS solution pH 7.4, containing 3.75% H_2O_2 and 5% NaCh. Spores, highlighted with light gray arrows. Scale bar, 20 μm. Adapted from Ref. [56] with permission. Copyright (2015) Royal Society of Chemistry.

motion and control of the motors in viscous and low Re number environment. The ABFs can be easily steered by changing the direction of the rotating magnetic field. The same authors later investigated the relationship between the velocity of ABF and the strength/rotation frequency of the applied magnetic field [41a]. They found that the maximum speed of their ABF motors was able to reach a value of 18 μm s^{-1}, which was comparable to the speed of bacteria [59].

They further demonstrated that their helical motors have considerable promise for biomedical applications by successfully functionalized titanium coated ABFs with temperature-sensitive dipalmitoyl phosphatidylcholine (DPPC)-based liposomes. The ability to load both hydrophilic and hydrophobic drugs and remotely controlled single-cell drug delivery was also demonstrated (Fig. 11). These liposome-functionalized ABFs displayed corkscrew swimming in 3D with micrometer positioning precision by exposure to an external rotating magnetic field [60]. They also demonstrated temperature triggered release of calcein from ABFs [61]. The decrease in the fluorescent intensity of ABFs upon temperature increase from 37°C to 41°C and the fluorescence intensity of the background increased accordingly indicated the significant release of entrapped calcein from the DPPC/monostearoyl phosphatidylcholine (MSPC) liposomes at 41°C. Very recently, they also surface-functionalized their magnetic micromotors with lipoplexes loaded with pDNA and remotely maneuvred these ABFs to HEK 293 cells *in vitro* [62]. The cells in contact with the ABFs were successfully transfected by the transported *p*DNA. These ABFs have been envisaged to be used in several applications such as sensors actuators, cell biology, and drug delivery purposes.

Fischer *et al.* also described a simple shadow growth method of producing large number of ABFs made up of ferromagnetic cobalt-coated silicon dioxide helices [41d]. The propellers are typically 200–300 nm in width and about 1–2 μm long, driven by a homogeneous magnetic field. At frequencies around 150 Hz, the helicals can move at speed of approximately 40 μm s^{-1}, which is comparable to the speed of a bacteria.

Wang's group recently reported a novel surface patterning technique for manufacturing microsized ABFs. Rotating magnetic fields of different frequencies were used to control the movement of the magnetic Au/$(Ag)_{flex}$/Ni nanowires. The nanomotors were functionalized with glucose

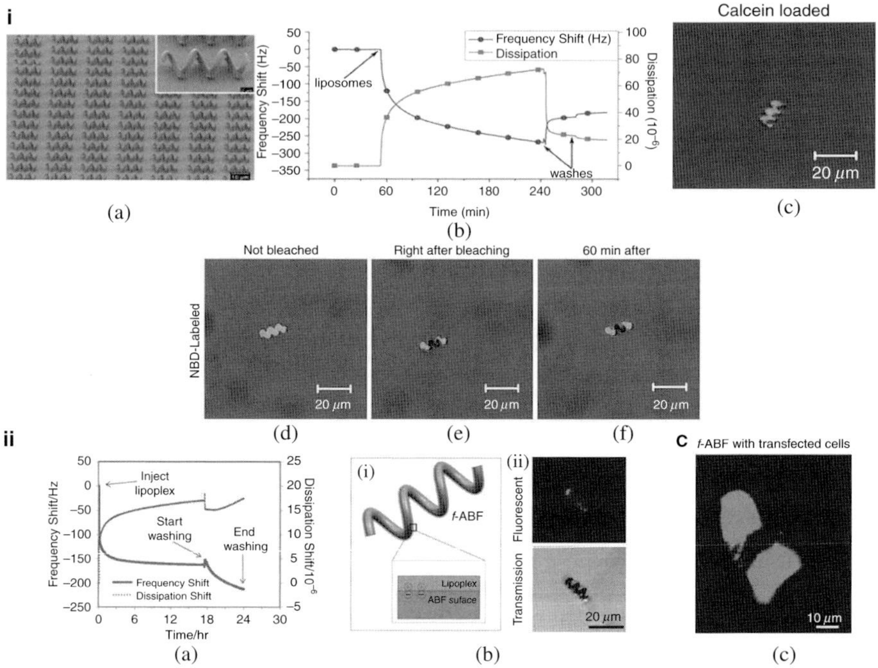

Figure 11. (i) (a) SEM image of a horizontal array of artificial bacterial flagella (ABFs), the inset shows a helical swimmer at a higher magnification. (b) Normalized resonance frequency shift and dissipation shift of the 15 MHz detection frequency for the adsorption of DOPE/DOTAP (3:1) liposomes on titanium-coated quartz crystal microbalance chips. (c) A representative image of ABFs coated with liposomes loaded with 50 mM calcein. (d) An ABF coated with NBD-labeled liposomes before bleaching, (e) after bleaching the center of the swimming microrobot, (f) and 60 min following the bleaching. Since no recovery was observed, it can be concluded that the liposomes are intact on the surface of the helical swimmers. Adapted from Ref. [60] with permission. Copyright (2014) Wiley. (ii) (a) QCM-D measurement of the frequency and dissipation response to the adsorption of the lipoplex on a TiO_2 crystal. The HEPES sodium buffer was used for washing. Only the third overtones are shown for the sake of clarity. (b) Functionalization of ABFs with the lipoplex. (i) The schematic of the functionalization of ABFs with lipoplex. (ii) The fluorescent and transmission images of a fluorescent-ABF by CLSM. The pDNA was marked using green fluorescence (fluorescein), as shown as the white spot. (c) *p*DNA transfection and protein expression by the cells that contacted fluorescent-ABFs. Adapted from Ref. [62] with permission. Copyright (2015) Wiley.

oxidase (GOx) on their surface. The catalytic cycle of GOx enzyme induced the growth of Au helical microstructures *via* the reduction of $AuCl_4-$ in the presence of hydrogen peroxide, which was produced as a result of a reaction between GO*x* and glucose. They demonstrated that their magnetic helical nanowires can be guided to induce direct "surface writing" or surface patterning of helical Au microstructures [63].

More recently, they described a novel magneto-acoustic hybrid fuel-free nanomotor constructed from bisegment nanomechanical elements, where the concave Ni-coated Au nanorod segments (essential for the ultrasound actuation), are connected to Ni-coated palladium (Pd) nano-helical magnetic segments [42c] (Fig. 12). They showed that their new class of nanomotors display efficient magnetic and acoustic propulsions in various untreated biofluids, including cell culture medium, serum, and blood and also displayed a distinct biomimetic collective behavior in response to alternating the external field.

Another example of nanomotors based on nanowires includes a flexible Ni–Ag nanomotor prepared by the templated electrodeposition method for the targeted delivery of doxorubicin encapsulated in the magnetic poly(D,L-lactic-*co*-glycolic acid) (PLGA) microparticles [64]. The microparticle and cargo loading onto the nanowire was performed via the magnetic interaction between the microparticle and the Ni component. They demonstrated the delivery of drug-loaded PLGA particles to HeLA cancer cells in biological media and the transport of the drug nanomotors through a microchannel from the pick up zone to the released microwell was also illustrated.

Chen *et al.* also 3D printed large-scale production of microscopic fish composed of poly(ethylene glycol) diacrylate (PEGDA)-based hydrogels containing functional nanoparticles such as iron oxide (for magnetic manipulation), platinum (for chemical propulsion) and polydiacetylene (PDA) nanoparticles(for toxin neutralization) (Figs. 13a–13c) [65]. The thickness of the microfish is approximately 30 μm and the length of the microfish is 120 μm. They optimized the shape of the fish to achieve a maximum speed of 780 $\mu m\ s^{-1}$ in a 15% peroxide solution (Fig. 13d). They further explored a potential application of the 3D-printed microfish, with the toxin-neutralizing nanoparticles encapsulated within the microfish for its use in detoxification applications (Fig. 13e).

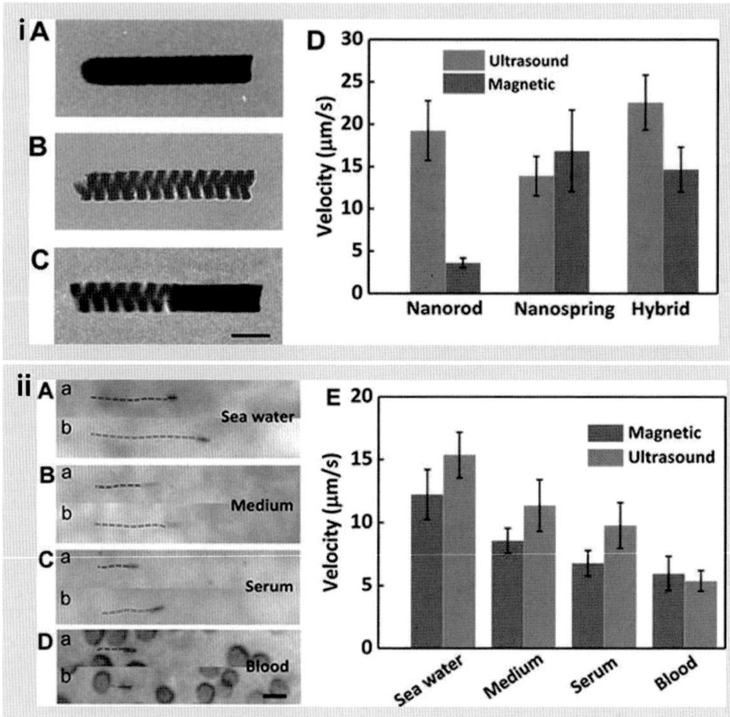

Figure 12. (i) TEM image of (A) Ni-coated Au nanorod, (B) Ni-coated Pd nanohelix, and (C) hybrid bisegment nanomotor. Scale bar: 500 nm. (D) Quantitative velocity evaluation of the motion of the three motor designs (shown in panels (A)–(C)). Ultrasound transducer power, 6 V; rotational magnetic field, 200 Hz. (ii) Movement of hybrid nanomotors in various media. Images illustrating the 2 s tracking lines of magnetic (dark gray) and ultrasound (light gray) propulsions of a hybrid nanomotor in (A) seawater, (B) cell culture medium, (C) serum, and (D) blood (50% in PBS buffer); the black circles are fresh red blood cells collected from six-week-old male ICR mice and anticoagulated with ethylenediamine tetraacetate. Scale bar: 10 μm. (E) Quantitative data of the velocity of the hybrid nanomotor in different modes and media, using an ultrasound transducer voltage of 4 V or a magnetic rotational frequency of 150 Hz. Adapted from Ref. [48c] with permission. Copyright (2015) American Chemical Society.

Very recently, Ghosh *et al.* reported on the first successful passage of magnetic nanomotors based on ferrite coatings in human blood [66]. They also showed that their nanomotors are cytocompatible with mouse myoblast cells.

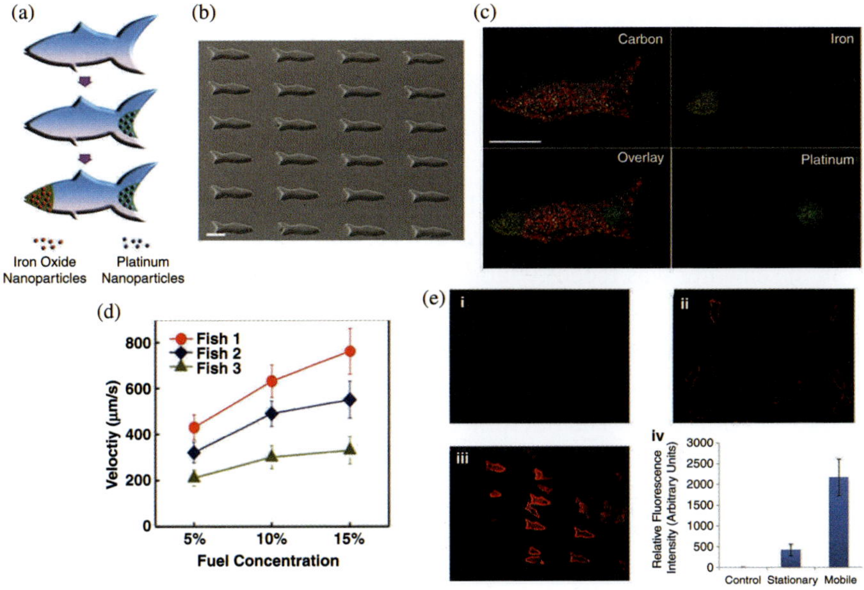

Figure 13. (a) Schematic illustration of the process of functionalizing a microfish for guided catalytic propulsion. Pt nanoparticles are first loaded into the tail of the fish for propulsion via catalytic decomposition of H_2O_2. Second, Fe_3O_4 nanoparticles are loaded into the head of the fish for magnetic control. (b) SEM images of a uniform array of microfish. (c) EDX spectroscopy images illustrating the iron-oxide head and platinum tail with respect to the PEGDA microfish body. Scale bar, 50 μm. (d) Speed profiles of microfish with different shape designs and different Pt nanoparticle concentrations at 5%, 10%, and 15% H_2O_2. (e) Fluorescent images demonstrating the detoxification capability of the microfish containing encapsulated PDA nanoparticles. (i) Control group of microfish incubated in 5% H_2O_2 without melittin toxin. (ii) Stationary microfish incubated in 2.5 mg mL^{-1} melittin solution (5% H_2O_2). (iii) Mobile microfish incubated in 2.5 mg mL^{-1} melittin solution (5% H_2O_2). (iv) Relative fluorescence intensity measurements indicating the amount of melittin absorbed by the microfish. Statistical analysis indicates that there is significant difference between any two conditions among control, stationary or mobile, $p < 0.05$. Adapted from Ref. [65] with permission. Copyright (2015) Wiley.

4.3.1.3. *Janus motors for environmental remediation*

Environmental sustainability is a core and long-term problem that our society faces, which needs to be addressed seriously. Several studies reported in the literature have shown that nanotechnology has been extremely useful in addressing environmental issues and the use of motors for environmental remediation have only been realized very recently [43b]. The proof-of-concept investigations have demonstrated fascinating potential application in a diverse range of environmental sustainability — the use of motors in water quality screening [67], removal of oil spills [68], and accelerated decontamination process [69,70].

Janus micromotors that utilize magnesium water reaction for propulsion in seawater was recently fabricated for practical environmental applications by Wang's group [71]. The micromotors were made up of biodegradable and environmental friendly magnesium (Mg) microparticles coated with Ni–Au bilayer patch for magnetic manipulation and surface modification. The micromotors were modified with self-assembled monolayers of long-chain alkanethiol to imbue the super hydrophobicity of the surface of the motors essential for "on the fly" collection of oil droplets (Fig. 14). Thus, the practicality of these micromotors in environmental remediation application was demonstrated. The authors showed that the motors could capture and transport oil droplets from contaminated seawater. The micromotors moved at a speed of 90 $\mu m\ s^{-1}$ and the speed dropped to 44 $\mu m\ s^{-1}$ upon collection of the oil droplets, reflecting the larger drag force (Fig. 14).

Zhao *et al.* designed a millimeter-sized polymer capsule motor based on Marangoni effects with nickel powder incorporated within the motors for guided motion under an external magnetic field. The motor capsule was made up of sodium dodecyl sulfate (SDS)/polysulfone (PSf) and could display long-range interaction with oil droplets and could be used to clean up oil contaminant from the water–air interface.

Although significant progress of the nanomotor design and the understanding of the ability of magnetic motors to pick up, transport, and delivery cargo has been achieved, several key challenges remain the focus of research in the coming years. Biomedical applications of such magnetic motors in the human body mainly focus on targeted drug delivery. Therefore attention should be given to fabrication of more biocompatible magnetic motors. Protective coating is necessary to avoid etching of the magnetic material by blood components. Targeting ligands or responsive to tumor microenvironments are also desirable to allow precise autonomous release of the therapeutic cargo to the specific tumor site. The size

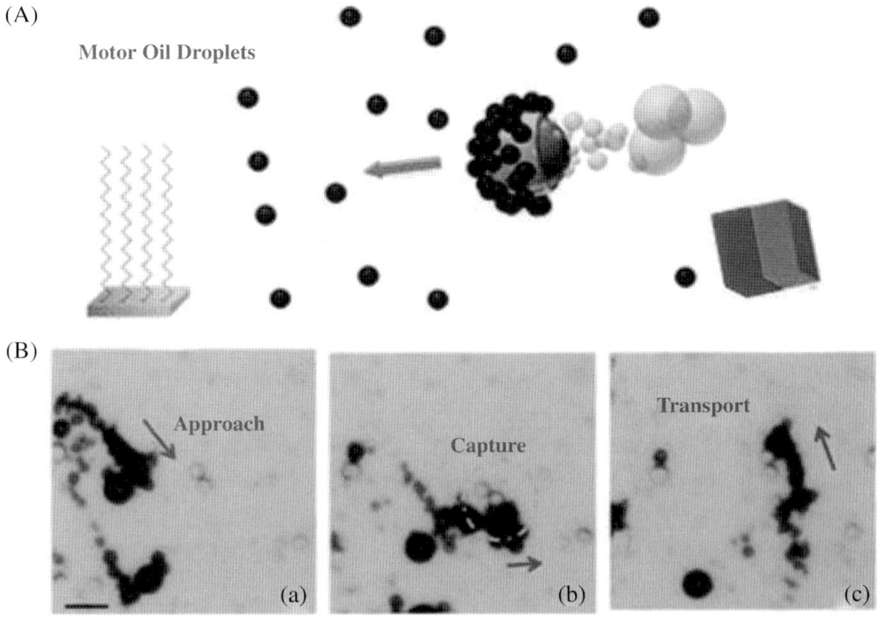

Figure 14. (A) Schematic diagram of hydrophobic seawater driven alkanethiol-modified Mg micromotor for environmental oil remediation. (B) Time lapse images of a Mg Janus micromotor approach (a), capture (b) and transport (c) the oil droplet in seawater. Scale bar, 50 μm. Adapted from Ref. [71] with permission. Copyright (2013) Royal Society of Chemistry.

of the motors should also be reduced to nanometer scale for drug delivery purposes, while taking into account the efficient and controlled nanoscale propulsion under low Re number.

For environmental sustainability, attention should be given to move these exciting proofs of concept to larger-scale studies toward the coverage of larger contaminated areas and wastewater. This will involve collaborative efforts among environmental scientists and engineers for translating such research activity into practical pilot studies involving different sample sizes. Another key challenge is the potential toxicity of these motors to prevent adverse environmental contamination. Research efforts in designing environmentally friendly motors based on green motors are currently underway [71,72]. We anticipate that such developments in novel and exciting motors capable of performing multiple tasks of sensing and destroying toxic pollutants can aid in environmental cleanup and make significant contribution toward addressing environmental remediation challenges that our society faces.

4.3.2. Microrheology

JPs with asymmetric configuration have considerable anisotropy that can give rise to inherent or rotation orientation upon external manipulation.

The concept of modulated optical nanoprobes (MOONs) was first implemented by Kopelman *et al.* [7]. The earliest configuration of MOONs consists of a spherical and fluorescent microsphere whereby one hemisphere is coated with a metal. This metal cap serves to prevent light from exciting the embedded fluorophores and also inhibit light from escaping the fluorescent microbead. These half-coated fluorescent beads, when suspended in a liquid medium, blink constantly under a fluorescence microscope due to the rotational Brownian motion; as a result, they are called Brownian MOONs. Large micrometer beads have an internal structure with different types of moon-like presence with beads smaller than the diffraction limit; the blinking signal alone is let to be resolved and used as an indicator for the rotational motion.

Brownian MOONs can be used for microrheological measurements on viscous drag and torques on the scale of micrometers to tens of nanometers and time scales that are difficult to study with other methods. The authors further extended the Brownian MOONs to magnetically modified MOONs (MagMOONs) that can be forced to rotate and blink upon the application of an external magnetic field [73]. These nanoviscometers can determine the kinematic viscosity from 5×10^{-5} to 3×10^{-4} m^2/s in a rotational magnetic field. They still extended their study to develop a torque-based asynchronous magnetic bead rotation (AMBR) detector that can measure nanoscale growth dynamics of individual bacterial cells [74]. However, both techniques have their limitations: that the viscosity measurements are limited by the intensity or reflection of light from particles and require very clear solutions and a complicated rotational magnetic field setup.

Passive microrheology measures the viscoelastic properties of fluids by monitoring the Brownian motion of implanted colloidal probes [75]. This technique has several advantages over the classical rheology method, such as the improved sensitivity and the high range of frequency. However, it cannot determine highly viscous materials using thermal fluctuations. Active microrheology uses an external field to monitor periodic rotation of particles and the material deformation [76]. Active magnetic rheology typically employs paramagnetic or ferromagnetic particles subjected to strong and inhomogeneous magnetic fields to generate a magnetic force of approximately 10^3–10^2 pN [77].

The simplest type of anisotropic probe is a magnetic rod, which has been used to investigate the rheological properties of monolayers at water–air interface. Rods can easily be absorbed at the interface due to their large aspect ratio. Interfacial viscometers using macroscopic rods have been developed by several groups [78]; however, nanometer-sized probes is only a recent invention [79]. Reducing the probe size will dramatically increase the sensitivity of the measurement as it results in an increase in the Boussinesq number (B_0), which is defined as the ratio between the interface and sub-phase drag (Eq. (5)):

$$B_0 = \frac{\eta_s}{\eta b^a}, \tag{5}$$

where η_s and η_b is the surface and bulk viscosities respectively and a is the characteristic length associated with the size of the probe. Nanorods are more sensitive than macroscopic ones since reducing a increases B_0 and reduces the coupling between the interfacial flow and the subphase.

In a recent work published by Dhar *et al.* [80], they investigated surface microrheology at an air–albumin interface with their rotating magnetic rods. Their magnet nanorod was able to reach a sensitivity of $\eta_s \sim 10^{-9}$ Ns m^{-1}, which was three times more sensitive than the resolution of macroscopic rods [81]. Their magnetic nanorod was made of nickel, with aspect ratio of 3 μm in length and 300 nm in diameter and was used to measure the surface viscosity of albumin at different aging times. Using this technique, it was possible to demonstrate that the surface viscosity of albumin increased significantly with aging compared to the surface pressure, suggesting the annealing with time of the protein film.

Choi *et al.* [82] used ferromagnetic microbuttons to measure the surface viscosity of a monolayer of phospholipid (DPPC) with a resolution of $\eta_s \sim 10^{-8}$ Ns m^{-1}. These magnetic microbuttons, 20 μm in diameter, were fabricated using photolithography. An in-plane oscillatory magnetic field was used to rotate these probes.

Apart from using these anisotropic probes for surface viscosity measurements, they have also been utilized in rheological measurements of bulk viscoelasticity [83]. In contrast to isotropic particles, anisotropic probes have the benefit of a uniform magnetic field to be torqued and easy visualization. Magnetic rotational microrheology has been shown to successfully measure the viscoelastic properties of living cells [84].

For a detailed discussion on probe based active microrheology, the interested reader is referred to the literature [76b,85].

4.3.3. Biomedical Applications

As JPs composed of magnetic components exhibit superior magnetic responsiveness, they can be utilized in several biomedical applications, such as drug delivery, biological imaging, high-throughput immunoassays, probes, and remote manipulation of devices. Kim *et al.* fabricated MJPs containing iron oxide particles (hematite, α-Fe_2O_3) and monodispersed SiO_2 particles, which could be used for the remote manipulation [86]. They showed that in a microfluidic device, the motions of the particles were constrained by the channel walls and chambers; as a result, it is possible to control the movements of the particles and manipulate them to the desired chamber by applying a rotating magnetic field with an axis parallel to the substrate. Figure 15(a) shows a bright field image of the microfluidic device containing the three different types of particles: (1) untreated, (2) fluorescein isothiocyanate (FITC) treated, and (3) tetramethyl rhodamine isothiocyanate (TRITC) treated particles. The fluorescence image (inset) indicates that each particle can be identified. Figure 15(b) shows that the system reached a certain point in which one black particle was advancing ahead while two dark gray particles were moving in the right-hand channel. The magnetic field was able to manipulate the dark gray particles toward the right and the black and light gray particles in the main channel to align against the right wall of the channel. After a period of time, the black particle reached the middle chamber and the two dark gray particles at the corner of the right-hand channel. When an

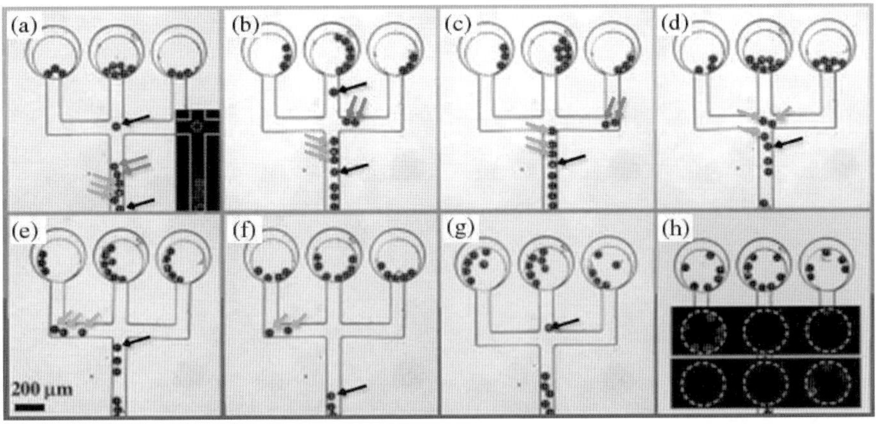

Figure 15. Optical images showing separation of MJPs in a microfluidic device. Insets of (a) and (h) show fluorescent microscope images. The black, dark gray, and light gray arrows denote the untreated, TRITC-treated, and FITC-treated MJPs. Adapted from Ref. [94] with permission. Copyright (2010) Wiley.

upward motion was applied, the two dark gray particles moved into the right-hand chamber (Fig. 15(d)). Similarly, the three light gray and one black particles were separated into the left and middle chambers, respectively (Figs. 15(e)–15(h)). The fluorescence microscopic images in the inset of Figure 15(h) showed the successful separation of the particles. Their study indicates that these magnetic JPs can be used in a broad range of biological systems, such as high-throughput immunoassays, biological probes, and microfluidic pumps and mixers.

In a promising work performed by Sun *et al.*, they synthesized novel Ag nanoparticle-decorated magnetic-silica Janus nanorods with different aspect ratio by changing the by changing the molar ratios of [TEOS]/[Fe_3O_4NP] (Figs. 16i(a)–16i(e)) [87]. Their magnetic Janus nanorods possessed high magnetic moments, strong affinity binding to bacteria, and highly effective and long-term antimicrobial activity against bacteria (Figs. 16(ii) and 16(iii)). Their Janus nanorods hold immense potential in a range of biomedical applications due to their excellent biocompatibility and non-hemolytic properties.

JPs with fluorophores and MNPs encapsulated within separate compartments have also showed magnetic field-modulated imaging, which have potential in an advanced technology in cancer cell therapeutics. In an

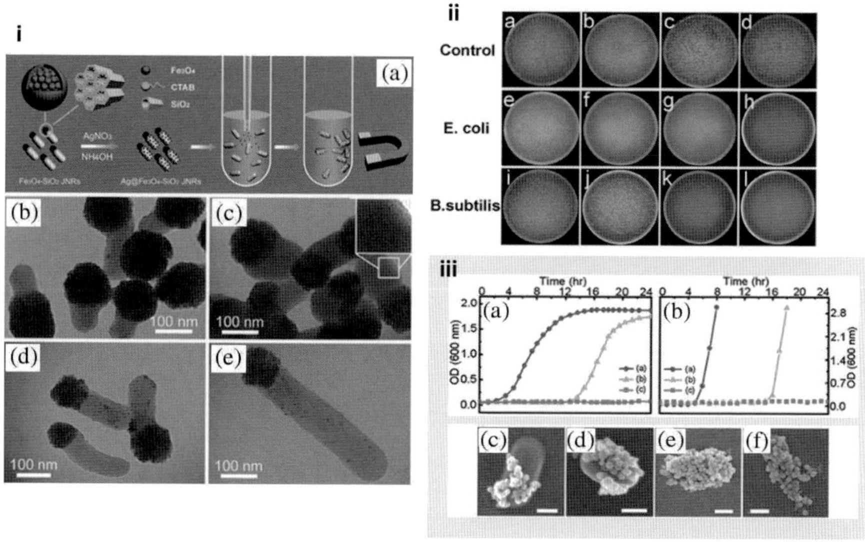

Figure 16. (i) (a) Schematic diagram for the synthesis of AgNPs@Fe_3O_4–SiO_2 Janus nanorods and their sterilization and separation process. TEM images AgNPs@Fe_3O_4–SiO_2 Janus nanorods with different lengths of 200 nm (b), 250 nm (c), 300 nm (d) and 500 nm (e). The inset in (c) is an enlarged image of the corresponding position. (ii) Antimicrobial activity of AgNPs@Fe_3O_4–SiO_2 Janus nanorods. Photographs of LB-agar plates coated with *E. coli* (a) and when supplemented with AgNPs@Fe_3O_4–SiO_2 Janus nanorods (b), and with *B. subtilis* (c) and when supplemented with AgNPs@Fe_3O_4–SiO_2 Janus nanorods (d). Photographs of LB-agar plates coated with *E. coli* (e)–(h) and *B. subtilis* (i)–(l) when supplemented with increasing concentrations of AgNPs–Fe_3O_4–SiO_2 Janus nanorods. (iii) Bacterial growth curve of *E. coli* (a) and *B. subtilis* (b) in LB liquid medium in the presence of AgNPs@Fe_3O_4–SiO_2 Janus nanorods. SEM images of AgNPs–Fe_3O_4–SiO_2 Janus nanorods deposition on bacteria for 30 min (c), 1 h (d) and 2 h (e), and the collapse of the bacterial membrane (f). Bar size = 0.5 μm. Adapted Ref. [87] with permission. Copyright (2012) Royal Society of Chemistry.

elegant research conducted by Gao *et al.* [88], they demonstrated that their magnetic nanocomposites can be efficiently attached to the cell surface using a magnetic field for imaging and therapy. Nanocomposite attachment to cells was achieved via non-specific adsorption promoted by magnetic manipulation. They investigated the therapeutic effect of their particles by exposing the cells to a spinning magnetic field, creating a mechanical force on the cell membrane. Upon exposure to the magnetic field, majority of the tumor cells were killed.

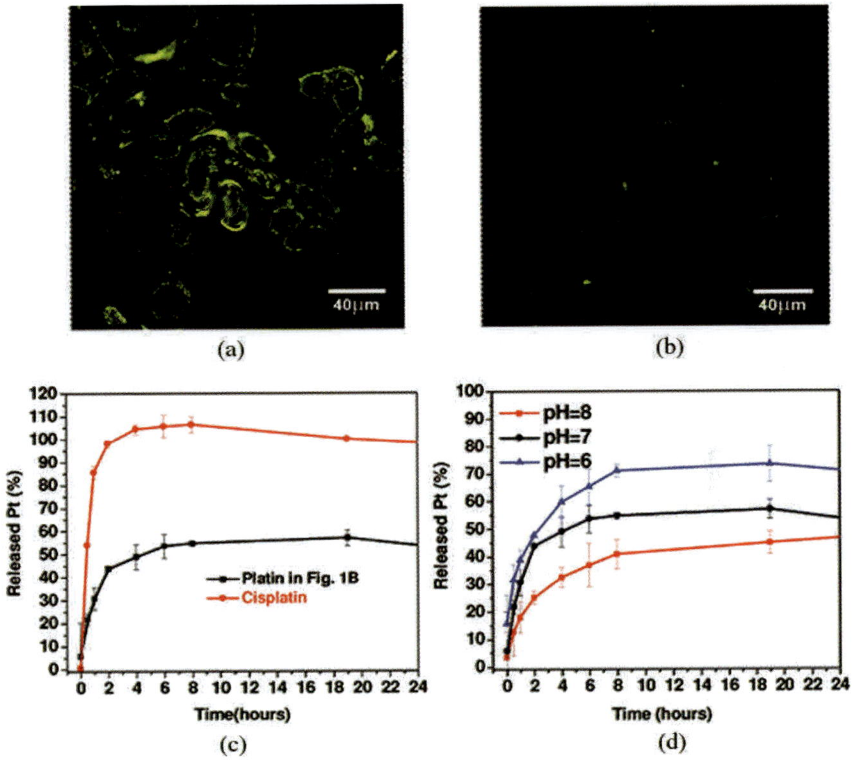

Figure 17. Reflection images of (a) Sk-Br3 and (b) MCF-7 cells after incubation with the same concentration of platin–Au–Fe_3O_4–Herceptin NPs. (c) Cisplatin and platin release curves at 37°C (pH 7). (d) pH-dependent Pt release from platin–Au–Fe_3O_4–Herceptin at 37°C. Adapted from Ref. [89] with permission. Copyright (2009) American Chemical Society.

Sun *et al.* designed dumbbell-shaped Au–Fe_3O_4 nanoparticles, which act as multifunctional carriers for target-specific platin delivery [89]. They showed that the release of platin under acidic conditions allows the nanoparticles to conjugate and become more toxic to the targeted tumor cells than free cisplatin (Fig. 17). The dumbbell-shaped Au–Fe_3O_4 particles have several advantages over single spherical particles, in that the presence of both Fe_3O_4 and gold improves the attachment of the antibody and platin complex and the combination of both components serves as both a magnetic and optical probe for tracing the platin complex in cells and biological systems.

The use of nanoparticles as contrast agents for *in vivo* bioimaging have increased the interest in multifunctional nanoparticles for theranostics, which was realized with MNPs for the first time by Czerlinski *et al.* in the late 1970s [90]. Since the discovery, superparamagnetic iron oxide nanoparticles are used in clinical applications [91] and several different formulations of MRI contrast agents such as Fe_3O_4, $MnFe_2O_4$, and MnO, have been the focus of research for their potential application as MRI imaging agents [92]. Inspired by the progress of spherical nanoparticles as contrast agents, anisotropic nanoparticles have become the focus of research for use as multimodal contrast agents [93]. For example, PEG-functionalized Au@MnO nanoflowers exhibit both optical and magnetic properties, which can be suitable for dual imaging [94]. Several other composite materials such as $Cu@Fe_3O_4$ and $Co@Fe_2O_3$, combining optical and magnetic properties, have also been used for simultaneous optical and magnetic imaging. Furthermore, studies have shown that the magnetic properties can also be enhanced due to the interaction at the nanointerface for the exceptionally large T2-relaxation times of $Co@Fe_2O_3$ in comparison to the single-component iron-based MRI agents [95]. Owing to their strong surface plasmon resonance, gold nanoparticles are the ideal candidates for optical imaging; additionally they exhibit strong X-ray absorption, which is used to increase the contrast in CT diagnostics [96]. Hence the combination of gold and iron oxide allows for simultaneous MRI and CT imaging [97].

Very recently, gold nanorods and $Au@MnO@SiO_2$ JPs (Fig. 18(i)) were reported to emit strong photoluminescence under two-photon excitation used for *in vitro* imaging (Fig. 18(ii)) [98]. Multiphoton microscopy is considered to be more superior than traditional microscopy as the fluorescence background is reduced based on the two-photon cross-section of most biomolecules, leading to less autofluorescence and enhanced penetration depth within biological samples. The effect of photobleaching can also be reduced by selective excitation of the focal volume [99].

4.3.4. Electronic Display Technology

Furthermore, JPs with asymmetric magnetic properties also allow the remote control of particle movement under the effect of magnetic field. This property could be exploited and considered as a promising strategy for

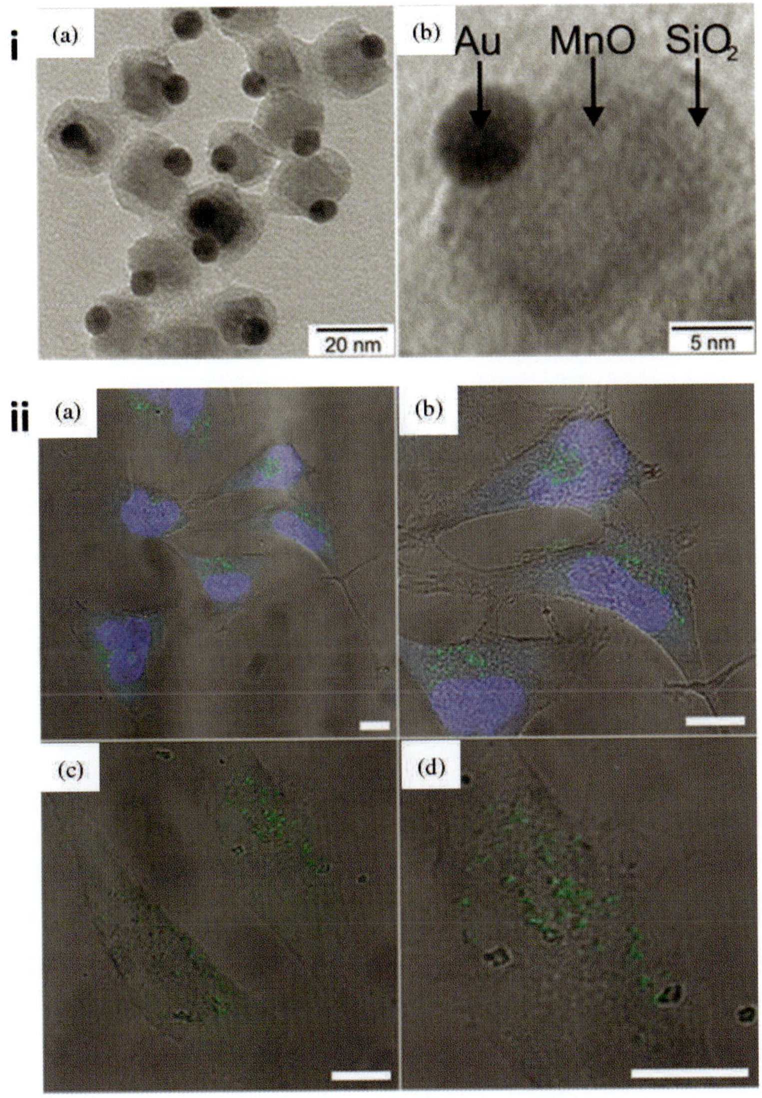

Figure 18. (i) (a) TEM bright field image of Au@MnO@SiO_2 nanoparticles, (b) TEM micrograph of a single Au@MnO@SiO_2 particle. (ii) (a,b) Confocal laser scanning microscopy images of HeLa cells coincubated with Au–MnO–SiO_2–Atto495 (green) for 24 h at 37°C. The cell nuclei (blue) were stained with DAPI. The sample was excited at 488 nm. (c),(d) Two-photon images of the same sample, excited with a two-photon laser at 970 nm, 30 mW. Scale: 10 μm. Adapted from Ref. [98] with permission. Copyright (2014) American Chemical Society.

engineering of electronic paper and displays. Bifunctional magnetic-fluorescent Janus supraballs made of poly(methyl methacrylate-*co*-2-hydroxyethyl methacrylate)/cadmium sulfide quantum dot polymer hybrids as one hemisphere and a mixture of modified Fe_3O_4 nanoparticles with poly(methyl methacrylate-*co*-2-hydroxyethyl methacrylate)/cadmium acrylate ionomers as the *other* hemisphere have been used as rotating beads in a magneto-driven fluorescent switch bead display with the orientation of the bead manipulated by an external field [100]. As shown in Figs. 19(a)–19(d), the Janus balls were introduced in the hole arrays and on varying

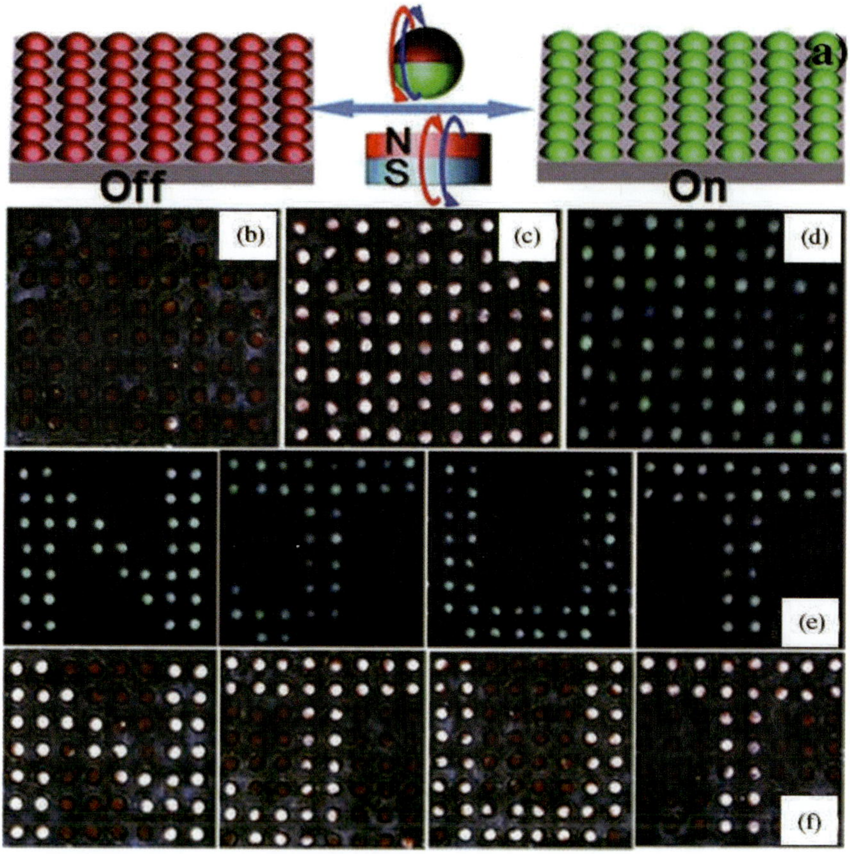

Figure 19. (a) Schematic representation of a fluorescent switch of JPs controlled by varying the direction of an external magnetic field. (b)–(f) Optical images of the magnetoresponsive bead display prepared from JPs: (b), (c), (f) under daylight and (d), (e) under UV irradiation. Adapted from Ref. [100] with permission. Copyright (2011) Wiley.

the direction of the external magnetic field, the alternation of the upward orientation of Fe_3O_4 nanoparticle hemispheres (OFF state) and QD–polymer hemispheres (ON state) allowed these Janus supraballs to be used as a fluorescent switch. When the Fe_3O_4 nanoparticle-hemispheres faced upward, the panel showed a reddish-brown color under daylight and black color under UV light. On the other hand, when the QD–polymer hemispheres orient upwardly, the panel displayed white color under daylight and bright blue emission under UV light. Interestingly, the magnetoresponsive rotation of these supraballs could realize free-writing on the flat panel using a magnetic needle. As shown in Figs. 19(e) and 19(f), the letters "N", "J", "U", and "T" successfully and freely displayed in the panel. The same

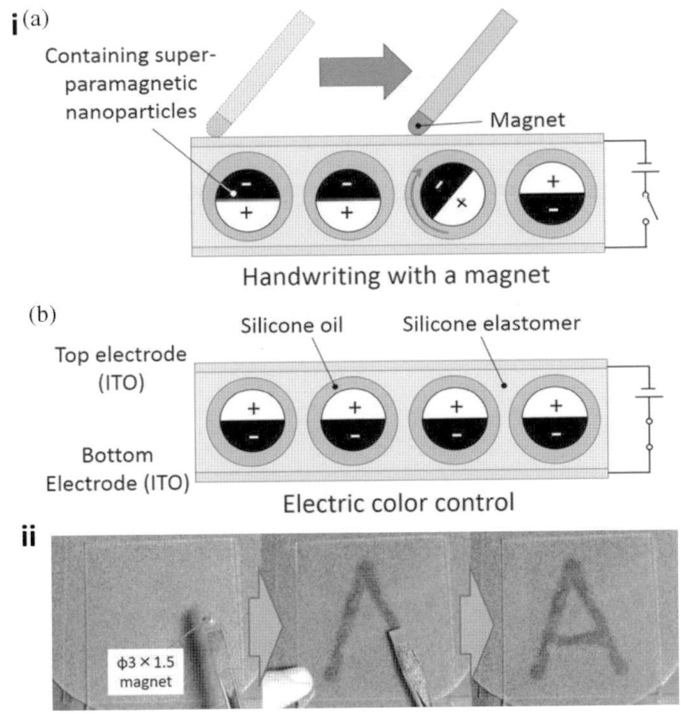

Figure 20. (i) Schematic illustration of dual-driven twisting ball display: (a) Handwriting with a magnet using a dual-driven twisting ball display. The black hemisphere of each JPs contains superparamagnetic nanoparticles. (b) Electric color control of the dual-driven twisting ball display. JPs rotate according to the direction of electric field because of their electric anisotropy. (ii) Magnetic handwriting with a $\phi\, 3 \times 1.5$ mm magnet with no voltage. Adapted from Ref. [8] with permission. Copyright (2015) American Institute of Physics.

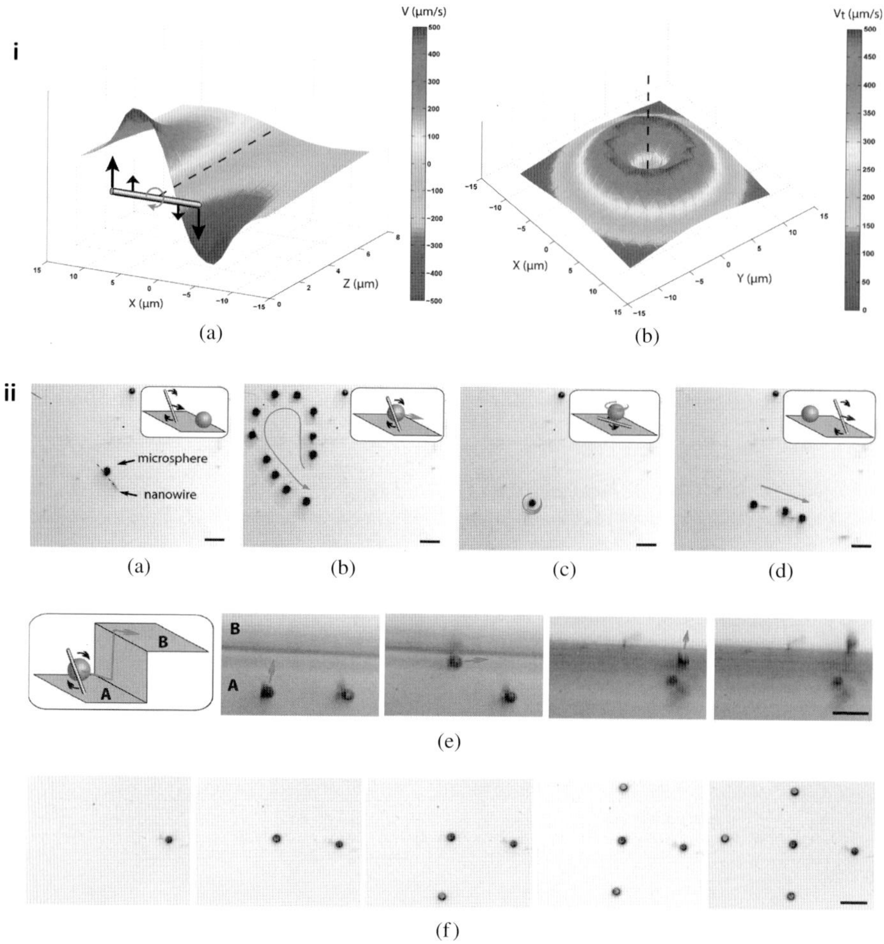

Figure 21. (i) Simulation of the tangential flow field induced by a rotating nanowire. The simulation uses the low Reynolds number assumption and is based on the method of fundamental solutions. (a) 2D flow velocity profile induced by the rotation of a 13 μm long nanowire with a frequency of 95 Hz in the plane formed by the nanowire and its rotational axis. The flow velocity is plotted from $Z = 1$ μm. The nanowire rotational axis, rotational direction, and velocity direction, are represented as a dashed line, gray arrow and black arrows, respectively. (b) 2D tangential flow velocity averaged over a full rotation of the nanowire in the plane normal to the rotational axis at $Z = 2$ μm. (ii) Manipulation of individual PS microspheres with a microvortex generated by a rotating nickel nanowire. (a) A 13 μm long nickel nanowire tumbles in the direction of a 6 μm diameter PS microsphere with an input field frequency $f = 23$ Hz. The dashed line represents the trapping trajectory. (b) The microsphere is trapped in the microvortex and translated with the rotating nanowire with $f = 25$ Hz. Images obtained during 10 s are superimposed.

authors also recently extended magnetic switching to photonic JPs possessing optical quality colloidal crystal in one hemisphere [101].

Very recently, researchers from Japan developed a versatile and cheap method of magnetic handwriting-enabled e-paper well suited for large displays like whiteboards. They fabricated JPs of 0.1 mm, with the black hemisphere containing pigments, magnetic particles and a charge control agent, and the other white hemisphere containing TiO_2 particles. The particles are sandwiched between two electrodes and by switching the direction of the voltage across the electrodes, the background display can alternate between black and white. A magnet pulling across the surface of the white display will attract the black hemisphere and the balls flip to face the magnetic [8] (Fig. 20). This technology can in the near future bring electronic paper closer to that of traditional paper.

4.3.5. Trapping and Sorting

A rodlike particle that is being torqued by an external rotating field can generate a hydrodynamic vertical flow. However, due to fluid flow at low Re, movement of tiny objects in liquid such as water become similar to movement in more viscous liquids like honey. Rotating anisotropic particles have applications in microscale trapping and sorting based on hydrodynamic or magnetic interactions. This concept was demonstrated by Petit *et al.*, where they synthesized 15 μm long and 200 nm wide rotating nickel nanowires by electrochemical deposition in an aluminum oxide template [102]. They also,

Figure 21. (*Continued*) (c) The nanowire rotates in a plane parallel to the horizontal wall and the microsphere is immobilized during 16 s. The input field frequency is progressively increased up to $f = 95$ Hz to induce fluidic trapping and then decreased to $f = 59$ Hz before the release of the microsphere. (d) Release of the microsphere after changing the rotational plane from horizontal to vertical orientation. Images obtained during 1 s are superimposed. (e) The microsphere is driven over a 100 μm step with $f = 28$ Hz. Surfaces "A" and "B" are the bottom and upper surfaces of the step, respectively. A lateral drift was observed during the vertical translation, attributed to surface roughness of the wall. (f) Sequential pick-and-place manipulation of five microspheres to form a cross-like pattern. Insets illustrate the different configurations with black and gray arrows corresponding to the velocities of the nanowire and the microsphere, respectively. The magnetic field strength is 3 mT in all manipulation tests. Scale bars are 20 μm. Adapted from Ref. [102] with permission. Copyright (2012) American Chemical Society.

as a proof of concept, demonstrated precise manipulation of single microspheres and microorganisms using microvortices at low Re number generated by the rotation of their magnetic nanowires and self-assembled magnetic bead doublets in fluid. Figure 21(a) and 21(b) show the low-velocity profile generated by the microvortices and was characterized by a local minima close to the center of the nanowire. They further demonstrated this concept to transport, trap and release 6 μm size PS microsphere.

4.4. Amphiphilic Properties and Pickering Surfactants

The most important feature of any JPs is their amphiphilic properties. They can be utilized as emulsion-stabilizing agents as molecular surfactants. Several researchers have investigated the behavior of JPs at the air–water interface and the application of Janus micelles as stabilizers in emulsion polymerization systems. The fabrication of these Pickering emulsions typically requires strong shear or ultrasonication to force the particles at the liquid/liquid interfaces. The confinement of different physical and chemical properties on either sides of the JPs can generate asymmetric interfaces and open up possibilities for unique ways of interfacial meso-structuration.

Some of the very early accounts on the amphiphilic nature of JPs were reported by both Casagrande [103] and Rossmy [104]. Casagrande *et al.* described the use of 60 μm glass spheres with one hemisphere protected with a varnish and the other hemisphere chemically treated with a silane reagent. The varnish is subsequently dissolved, yielding JPs. Rossmy discussed the absorption of amphiphilic Janus platelets at the oil–water interface of emulsion droplets, resulting in a cauliflower-type appearance of the stabilized droplets. Since then, there has been an explosion in the number of studies conducted in the field of stabilizing emulsions with JPs. The first theoretical explanation of the adsorption capability of spherical JPs versus homogenous spherical JPs was reported by Binks *et al.* [105]. They described an expression for the desorption energy of a spherical amphiphilic particle at the interface of an oil and water dual-phase system based on its surface free energy, $E(\beta)$:

$$E(\beta) = 2\pi R^2 \left[\gamma(JP_\circ 0)(1+\cos\alpha) + \gamma(JP_w 0)(\cos\beta - \cos\alpha) \right.$$
$$\left. + \gamma(JP_w W)(1-\cos\beta) - \frac{1}{2}\gamma(OW)(\sin^2\beta)\right] \quad \text{for } \beta \leq \alpha \qquad (6)$$

$E(\beta)$ is dependent on the immersion depth of the particle at the interface, β, the Janus balance of the JPs, α, and the radius R of the particle. When $\beta = 0°$ and $180°$, the particle is completely immersed into the phases and when $\beta = 90°$, it adsorbs along the equator. For symmetric hemispheres, α becomes 90°. They concluded from their theoretical considerations that desorption energies of homogeneous particles can be enhanced three fold if the amphiphilicity of the JPs can be improved. JPs can also retain their strong adsorption for contact angles of 0° and 180°, but JPs with low or high contact angles can be effective emulsion stabilizers.

Following Bink's theoretical considerations, further contributions to the theory and simulations to this phenomenon appeared, discussing the effect of the Janus balance [14b,106], enhanced adsorption of Janus disks [107] and cylinders [107], orientation of non-spherical JPs at interfaces [108], and differences in the mechanism of droplet coalescence between emulsion droplets stabilized with either homogeneous or JPs [109].

In a thoughtful account by Aveyard, he further calculated that Pickering emulsions stabilized by JPs can lead to a negative free energy change and be considered thermodynamically stable under certain experimental conditions [110]. For systems containing homogeneously wetted particles, the free energy of formation of bare oil–water interface always dominates, and the free energy of forming a particle-coated interface is always positive. For systems with JPs, the magnitude of the adsorption free energy can be increased significantly and the free energy of forming particle-coated interfaces can become negative when the surface coverage by particles is sufficiently high. Furthermore, the adsorbed particles in concentrated monolayers interact laterally. Long-range electrical repulsion through the oil phase can play an important role in affecting the free energy of emulsion formation, while van der Waals attraction appears to be negligible. In almost close-packed particle layers, it was proposed that hydration forces can become essential and, in principle, give rise to a minimum in the free energy of interface (and emulsion) formation as a function of surface coverage by particles.

In an early work by Glaser *et al.*, they verified that dumb bell-shaped Au–Fe_3O_4 JPs exhibit enhanced interfacial activity in comparison to homogeneous particles of the same size [111] (Fig. 22(a)). They monitored the self-assembly of these particles and the evolution of the interfacial tensions of a hexane/water dual-phase system by a pendant drop

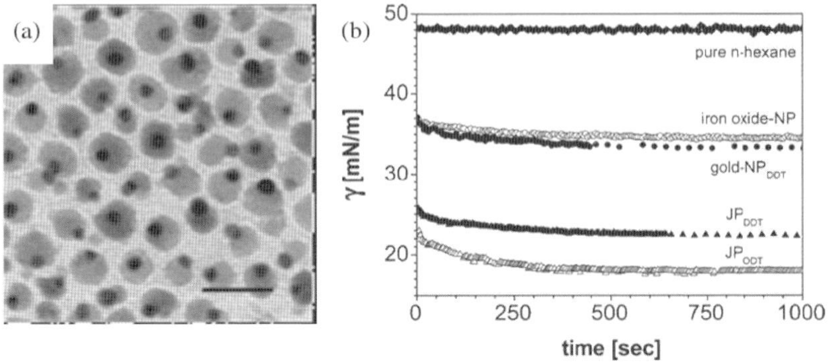

Figure 22. (a) TEM images of the JPs (consisting of gold (darker spheres) and iron oxide (brighter spheres)); The scale bars represent 25 nm. (b) Interfacial tension vs. time. Water was used as the drop phase, and *n*-hexane was used as the ambient phase in which the nanoparticles were diluted. (NP: homogeneous nanoparticles; JP: JPs. The gold moieties were modified using dodecanethiol (DDT) or octadecanethiol (ODT).) Adapted from Ref. [111] with permission. Copyright (2006) American Chemical Society.

tensiometry and found that the quasiequilibrium interfacial tension of the JPs was one-half of that of a homogeneous nanoparticles of similar size. They also further showed that they were able to control the interfacial activity by tuning the particle amphiphilicity via ligand exchange reactions, and that higher particle concentration could force a stronger reduction of the equilibrium interfacial tension. Figure 22(b) shows the decrease of interfacial tension between water and hexane as a function of nanoparticle concentration for different types of particles.

Ou *et al.* created hybrid inorganic multisegmented Au–CNT and Au–Ni–CNT microrods via sequential deposition in AAO membranes and utilized these particles for the stabilization of oil/water emulsion droplets [112]. They immobilized a hydrophobic perfluorinated thiol on the surface of the Au segment and a black droplet was formed when a drop of dichloromethane was introduced into water dispersed with the Au–CNT nanowires. Upon photo-induced cleavage of the thiol, the amphiphilicity was inverted, and the CNT side now favored an orientation to the DCM phase. Hence, the color of the droplet flipped from black to shiny golden. Furthermore, due to the presence of a central magnetic Ni domain in a different triphasic microrod, it was possible to manipulate and trap the emulsified droplet via external magnets (Fig. 23).

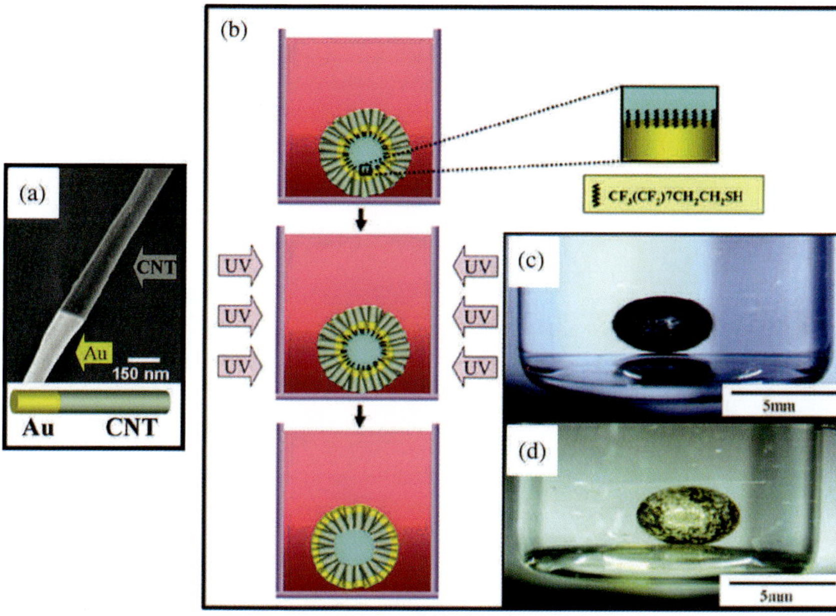

Figure 23. (a) SEM image of a single hybrid nanowire, with (inset) a schematic of a single hybrid CNT–Au nanowire. (b) Schematic showing the photoinduced surface modification of hybrid nanowire assembly using UV irradiation. The nanowire assembly from a black sphere (CNT pointing outward) could be reversed to a golden sphere (Au nanowire facing outward) by selective thiol functionalization of the gold segments, followed by UV irradiation. UV light photo-oxidizes the Au nanowire and removes the thiol group, there by reversing to the original assembly (gold spheres). (c) Optical image of the black sphere formed from the assembly of thiol-functionalized Au–CNT hybrid nanowires dispersed in water, when a drop of dichloromethane is added. (d) Optical image showing the gold sphere formed after a drop of dichloromethane is added after the UV irradiation. Adapted from Ref. [112] with permission. Copyright (2008) American Chemical Society.

Janus nanosheets were produced in large amounts by crushing silica Janus hollow spheres [113]. In the presence of the Janus nanosheets, a stable toluene-in-water emulsion was formed. The droplet size ranged from 0.1 to 0.3 mm and the toluene droplets stabilized with the nanosheets remained stable for at least six months (Fig. 24(i)). The emulsification essentially originated from the amphiphilicity of the Janus nanosheets rather than from the Pickering effect, as shown by the fact that the Janus nanosheets were dispersed in both the water and oil phases. In water, the Janus nanosheets were found to stack into a back-to-back superstructure

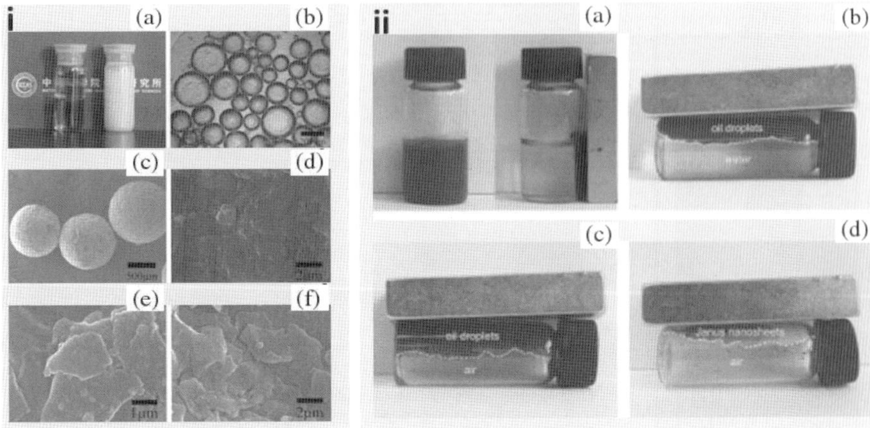

Figure 24. (i) Emulsification of representative immiscible liquid mixtures with the Janus nanosheets: (a) left: immiscible mixture of toluene (top) and water (bottom); right: toluene-in-water emulsion stabilized with the Janus nanosheets; (b) optical microscopy image of the toluene-in-water emulsion; (c) SEM image of the paraffin (T_m: 52–54°C) droplets stabilized with the Janus nanosheets; the weight ratio of Janus nanosheets/paraffin/water is 0.005:1.5:2.5 (Janus nanosheet content 0.12 wt.%); (d),(e) SEM image of the Janus nanosheets on frozen paraffin droplets of the sample as shown in Fig. 2(c) before (d) and after (e) sPS nanoparticles were selectively labeled onto the amine-terminated side; (f) SEM image of the Janus nanosheets on the frozen paraffin droplets; the weight ratio of Janus nanosheets/paraffin/water is 0.025:1.5:2.5 (Janus nanosheet content 0.63 wt.%). (ii) Magnetic manipulation of the emulsion stabilized with the paramagnetic Janus composite nanosheets: (a) the paramagnetic Janus composite nanosheets (as shown in Fig. 1(d) were dispersed in water, and could be driven using a magnet; (b) a bottle was filled with an oil-in-water emulsion stabilized with the paramagnetic Janus nanosheets, and the oil droplets moved toward the magnet on the top of the bottle; (c) water eluted when the bottle was opened, while oil droplets were collected by the magnet; (d) oil eluted after the oil droplets coalesced upon stirring and the collected paramagnetic Janus nanosheets were dried and could be recycled. Adapted from Ref. [113] with permission. Copyright (2011) Wiley.

with the amine group side exposed in the aqueous phase, similar to the hydrophobic association of amphiphilic polymers. This was consistent with amphiphilicity of the Janus nanosheets. The authors conjugated paramagnetic Fe_3O_4 nanoparticles onto the amine-terminated side of the Janus nanosheets to render a magnetic response. The authors demonstrated breaking emulsion when they applied a magnetic field to the toluene-in-water emulsion and the dispersed toluene droplets moves toward the magnet and are thereby collected. As water continuously elutes, the toluene droplets were collected. Upon stirring, the droplets coalesce and oil elutes (Fig. 24(ii)). The Janus composite nanosheets can be collected and recycled for reuse by using the magnet. This technology can find utility in the collection of oil or hazardous chemical spills.

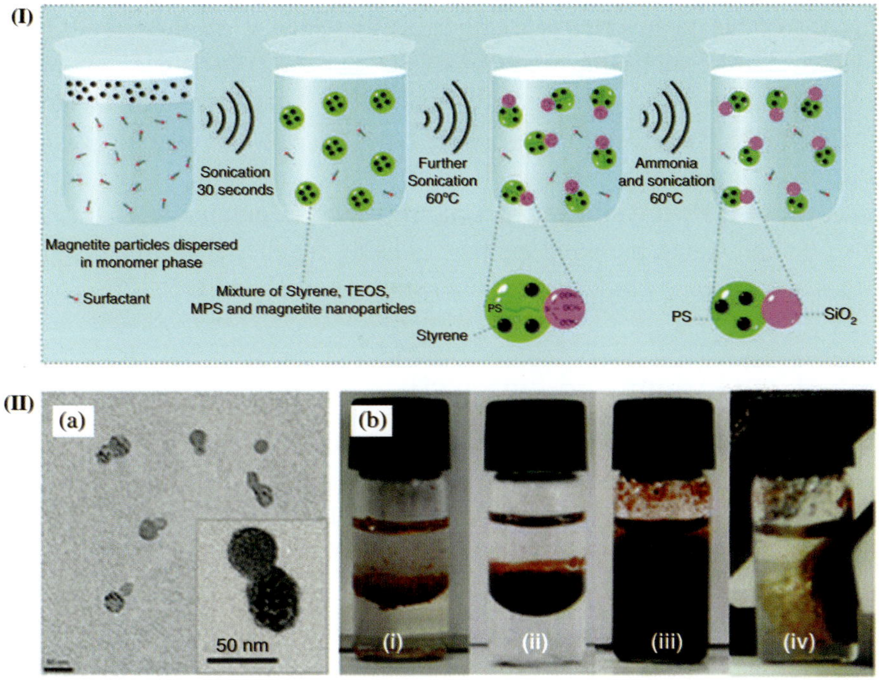

Figure 25. **(I)** Schematic diagram of the procedure for the synthesis of magnetic asymmetric JPs by the sonochemically driven miniemulsion polymerization pathway. **(II)** (a) TEM images of the magnetite JPs. (b) Photographs of various forms of toluene–water–MJP systems: (i) interfacial behavior at low loading with MJPs; (ii) interfacial behavior at high loading with magnetite JPs in the water–toluene dual-phase system; and (iii) an O/W emulsion from (ii) with asymmetric magnetite JPs; the emulsion was stable for at least two days; (iv) manipulation of the MJP system in (iii) by an external magnetic field. The "magnetic" blob could be separated from the continuous phase by the external magnet and the approximate amount of oil + water left in the vial was quantified with a syringe as 0.5 g. The separated oil can be seen between the dark meniscus and the beginning of the yellow emulsion phase.

Very recently, Passas-Lagos and Schüth fabricated iron-based mushroom JPs comprising a poly(sytrene-*co*-divinylbenzene) and a silica moiety both with controllable morphologies [114]. They investigated their JPs as surfactants for Pickering emulsions. Two oil-water dual-phase systems, namely toluene–water and vegetable oil–water, were stabilized with their JPs, producing water-in-oil (W/O) emulsions. By varying several parameters, including JP morphologies and the oil–water ratio, fine-tuning of the emulsion systems was possible; it was even possible to invert the continuous phase to an oil-in-water (O/W) system. Furthermore, the emulsions were stable against coalescence and sedimentation and could be easily separated by centrifugation or a strong magnet. The synthesized mushroom-type JPs are suitable for creating Pickering emulsions and can be used as building blocks for creating nanostructures with tailored properties for specific applications. Teo *et al.* also demonstrated breaking emulsions using magnetic snowman-shaped JPs synthesized via a sonochemical synthesis method [115] (Fig. 25).

4.5. Conclusions and Perspectives

In recent years, research conducted in the field of JPs have flourished. The recent developments of a variety of synthetic approaches have led to a large library of fabrication methods for multifunctional JPs with different chemical and physical properties with tunable geometries. These advances stem from collaborative efforts — from polymer to inorganic nanocrystal synthesis to basic colloidal science research. The understanding of breaking symmetry of homogeneous particles has allowed this technology to blossom and it is now possible to fabricate JPs with tailored functionalities and tunable sizes and shapes in high yields. Potential for the scaling up of the production of JPs could come from emulsion-based polymerization as well as microfluidic-based methods of fabrication. For a cost-efficient production method, it is also critical to move away from methods that require inert atmosphere or vacuum conditions. In consideration of sustainable production methods, fabrication techniques which use organic solvents should be avoided in preference to aqueous solvents. Processes should be kept energy-efficient as well, which means that techniques should focus on being carried out at room temperature.

In regards to the exciting applications of JPs, several proof-of-concept studies have reached an important stage of research with opportunities for their further translation into emerging technologies. For example, in the context of self-propelling particles, researchers have successfully demonstrated the utilization of broken symmetry of anisotropic particles for motion-controlled and directionality by designing highly advanced particles and external fields. Efforts have also been made in the understanding of the relationship between the shape and size with the speed of these functional motors and their ability to pick up and transport cargo. Future research will be focused on fabricating biologically relevant size motors and more biocompatible motors for practical biomedical applications, while taking into account of efficient and controlled nanoscale propulsion under low Re number. In terms of environmental remediation applications, it calls for interdisciplinary collaborative efforts for translating such research activity into practical pilot studies involving different sample sizes.

Bifunctional JPs have also been developed for application in electronic displays switchable by different external fields. This is a significant milestone toward electronic display technologies in terms of enhanced resolution, switching efficiency and frequency, energy consumption, and visibility in day and nighttime.

In conclusion, the field of magnetic anisotropic particles is developing rapidly and we hope that this chapter has provided a comprehensive background on the progress in fabrication methods, physical properties, and applications. We hope that this chapter will encourage interested researchers in taking up this area of research to generate novel ideas and exciting innovations in the field of anisotropic particles.

Acknowledgments

Boon Mian Teo is grateful to the Danish Research Council for a DFF individual postdoc grant and a Sapere Aude level 1 DFF-Ung Eliteforsker Award.

References

[1] de Gennes, P. G. *Rev. Mod. Phys.*, 64(3), 645 (**1992**).

[2] Dawson, C.; Vincent, J. F. V.; Rocca, A.-M. *Nature,* 390(6661), 668 (**1997**).
[3] Howse, J. R.; Jones, R. A. L.; Ryan, A. J.; Gough, T.; Vafabakhsh, R.; Golestanian, R. *Phys. Rev. Lett.,* 99(4), 048102 (**2007**).
[4] Roh, K. H.; Yoshida, M.; Lahann, J. *Materialwissenschaft und Werkstofftechnik,* 38(12), 1008 (**2007**).
[5] Berger, S.; Synytska, A.; Ionov, L.; Eichhorn, K.-J.; Stamm, M. *Macromolecules,* 41(24), 9669 (**2008**).
[6] Wittmeier, A.; Leeth Holterhoff, A.; Johnson, J.; Gibbs, J. G. *Langmui,* 31(38), 10402 (**2015**).
[7] Behrend, C. J.; Anker, J. N.; McNaughton, B. H.; Brasuel, M.; Philbert, M. A.; Kopelman, R. *Phys. Chem., B* 108(29), 10408 (**2004**).
[8] Komazaki, Y.; Hirama, H.; Torii, T.. *J. Appl. Phys.*, 117(15), 154506 (**2015**).
[9] Bärwinkel, S.; Bahrami, R.; Löbling, T. I.; Schmalz, H.; Müller, A. H. E.; Altstädt, V. *Adv. Eng. Mater.*, n/a-n/a (**2015**).
[10] Ling, X. Y.; Phang, I. Y.; Acikgoz, C.; Yilmaz, M. D.; Hempenius, M. A.; Vancso, G. J.; Huskens, J. *Angew. Chem.*, 121(41), 7813 (**2009**).
[11] Chen, C.-H.; Shah, R. K.; Abate, A. R.; Weitz, D. A. *Langmuir*, 25(8), 4320 (**2009**).
[12] Roh, K.-H.; Martin, D. C.; Lahann, J. *Nat. Mater.*, 4(10), 759 (**2005**).
[13] Helgeson, M. E.; Chapin, S. C.; Doyle, P. S. *Curr. Opin. Colloid Interface Sci.*, 16(2), 106 (**2011**).
[14] (a) Wang, Y.; Guo, B.-H.; Wan, X.; Xu, J.; Wang, X.; Zhang, Y.-P. *Polymer*, 50(14), 3361 (**2009**); (b) Jiang, S.; Granick, S. *Langmuir*, 24(6), 2438 (**2008**).
[15] Mock, E. B.; De Bruyn, H.; Hawkett, B. S.; Gilbert, R. G.; Zukoski, C. F. *Langmuir*, 22(9), 4037 (**2006**).
[16] (a) Lattuada, M.; Hatton, T. A. *Nano Today*, 6(3), 286 (**2011**); (b) Walther, A.; Müller, A. H. E. *Chem. Rev.*, 113(7), 5194 (**2013**).
[17] Takahara, Y. K.; Ikeda, S.; Ishino, S.; Tachi, K.; Ikeue, K.; Sakata, T.; Hasegawa, T.; Mori, H.; Matsumura, M.; Ohtani, B. *J. Am. Chem. Soc.*, 127(17), 6271 (**2005**).
[18] Ma, X.; Jannasch, A.; Albrecht, U.-R.; Hahn, K.; Miguel-López, A.; Schäffer, E.; Sánchez, S. *Nano Lett.*, 15(10), 7043 (**2015**).
[19] Kaewsaneha, C.; Tangboriboonrat, P.; Polpanich, D.; Eissa, M.; Elaissari, A. *J. Colloid Interface Sci.*, 409, 66 (**2013**).
[20] Rahman, M. M.; Montagne, F.; Fessi, H.; Elaissari, A. *Soft Matter.*, 7(4), 1483 (**2011**).

[21] Yuet, K. P.; Hwang, D. K.; Haghgooie, R.; Doyle, P. S. *Langmuir*, 26(6), 4281 (**2010**).
[22] Kim, J.; Chung, S. E.; Choi, S.-E.; Lee, H.; Kim, J.; Kwon, S. *Nat. Mater.*, 10(10), 747 (**2011**).
[23] Aketagawa, K.; Hirama, H.; Torii, T. *Magnetic anisotropy for electronic paper using shrinkage-gelation technique*, Solid-State Sensors, Actuators and Microsystems (TRANSDUCERS & EUROSENSORS XXVII), 2013 Transducers & Eurosensors XXVII: The 17th International Conference on, 16–20, pp. 2652–2655 (**2013**).
[24] Jeong, J.; Um, E.; Park, J.-K.; Kim, M. W. *RSC Advances*, 3(29), 11801 (**2013**).
[25] Zhao, N.; Gao, M. *Adv. Mater.*, 21(2), 184 (**2009**).
[26] Wooley, K. L.; Hawker, C. J.; Frechet, J. M. J. *J. Am. Chem. Soc.*, 115(24), 11496 (**1993**).
[27] Lanson, D.; Schappacher, M.; Borsali, R.; Deffieux, A. *Macromolecules*, 40(26), 9503 (**2007**).
[28] Rajaram, S.; Choi, T.-L.; Rolandi, M.; Fréchet, J. M. J. *J. Am. Chem. Soc.*, 129(31), 9619 (**2007**).
[29] (a) Liu; Abetz, V.; Müller, A. H. E. *Macromolecules*, 36(21), 7894 (**2003**); (b) Walther, A.; André, X.; Drechsler, M.; Abetz, V.; Müller, A. H. E. *J. Am. Chem. Soc.*, 129(19), 6187 (**2007**); (c) Erhardt, R.; Zhang, M.; Böker, A.; Zettl, H.; Abetz, C.; Frederik, P.; Krausch, G.; Abetz, V.; Müller, A. H. E. *J. Am. Chem. Soc.*, 125(11), 3260 (**2003**).
[30] (a) Nie, L.; Liu, S.; Shen, W.; Chen, D.; Jiang, M. *Angew. Chem. Int. Ed.*, 46(33), 6321 (**2007**); (b) Zhang, S.; Li, Z.; Samarajeewa, S.; Sun, G.; Yang, C.; Wooley, K. L. *J. Am. Chem. Soc.*, 133(29), 11046 (**2011**).
[31] Kirillova, A.; Schliebe, C.; Stoychev, G.; Jakob, A.; Lang, H.; Synytska, A. *ACS Applied Materials & Interfaces*, 7(38), 21218 (**2015**).
[32] Li, J.; Wang, L.; Benicewicz, B. C. *Langmuir*, 29(37), 11547 (**2013**).
[33] Nie, Z.; Li, W.; Seo, M.; Xu, S.; Kumacheva, E. *J. Am. Chem. Soc.*, 128(29), 9408 (**2006**).
[34] Paxton, W. F.; Sundararajan, S.; Mallouk, T. E.; Sen, A. *Angew. Chem. Int. Ed.*, 45(33), 5420 (**2006**).
[35] (a) Dreyfus, R.; Baudry, J.; Roper, M. L.; Fermigier, M.; Stone, H. A.; Bibette, J. *Nature*, 437(7060), 862 (**2005**); (b) Zhang, L.; Peyer, K. E.; Nelson, B. J. *Lab on a Chip*, 10(17), 2203 (**2010**).
[36] Fischer, P.; Ghosh, A. *Nanoscale*, 3(2), 557 (**2011**).
[37] (a) Paxton, W. F.; Kistler, K. C.; Olmeda, C. C.; Sen, A.; St. Angelo, S. K.; Cao, Y.; Mallouk, T. E.; Lammert, P. E.; Crespi, V. H. *J. Am. Chem. Soc.*,

126(41), 13424 (**2004**); (b) Burdick, J.; Laocharoensuk, R.; Wheat, P. M.; Posner, J. D.; Wang, J. *J. Am. Chem. Soc.*, 130(26), 8164 (**2008**).

[38] (a) Gao, W.; Sattayasamitsathit, S.; Orozco, J.; Wang, J. *J. Am. Chem. Soc.*, 133(31), 11862 (**2011**); (b) Sanchez, S.; Solovev, A. A.; Mei, Y.; Schmidt, O. G. *J. Am. Chem. Soc.*, 132(38), 13144 (**2010**); (c) Dong, B.; Zhou, T.; Zhang, H.; Li, C. Y. *ACS Nano*, 7(6), 5192 (**2013**); (d) Wilson, D. A.; Nolte, R. J. M.; van Hest, J. C. M. *Nat. Chem.*, 4(4), 268 (**2012**).

[39] (a) Baraban, L.; Makarov, D.; Streubel, R.; Mönch, I.; Grimm, D.; Sanchez, S.; Schmidt, O. G. *ACS Nano*, 6(4), 3383 (**2012**); (b) Gao, W.; Pei, A.; Feng, X.; Hennessy, C.; Wang, J. *J. Am. Chem. Soc.*, 135(3), 998 (**2013**).

[40] (a) Palacci, J.; Sacanna, S.; Vatchinsky, A.; Chaikin, P. M.; Pine, D. J. *J. Am. Chem. Soc.*, 135(43), 15978 (**2013**); (b) Ibele, M.; Mallouk, T. E.; Sen, A. *Angew. Chem. Int. Ed.*, 48(18), 3308 (**2009**).

[41] (a) Zhang, L.; Abbott, J. J.; Dong, L.; Peyer, K. E.; Kratochvil, B. E.; Zhang, H.; Bergeles, C.; Nelson, B. J. *Nano Lett.*, 9(10), 3663 (**2009**); (b) Gao, W.; Feng, X.; Pei, A.; Kane, C. R.; Tam, R.; Hennessy, C.; Wang, J. *Nano Lett.*, 14(1), 305 (**2014**); (c) Gao, W.; Sattayasamitsathit, S.; Manesh, K. M.; Weihs, D.; Wang, J. *J. Am. Chem. Soc.*, 132(41), 14403 (**2010**); (d) Ghosh, A.; Fischer, P. *Nano Lett.*, 9(6), 2243 (**2009**).

[42] (a) Wang, W.; Li, S.; Mair, L.; Ahmed, S.; Huang, T. J.; Mallouk, T. E. *Angew. Chem.*, 126(12), 3265 (**2014**); (b) Wang, W.; Castro, L. A.; Hoyos, M.; Mallouk, T. E. *ACS Nano*, 6(7), 6122 (**2012**); (c) Li, J.; Li, T.; Xu, T.; Kiristi, M.; Liu, W.; Wu, Z.; Wang, J. *Nano Lett.*, 15(7), 4814 (**2015**).

[43] (a) Sánchez, S.; Soler, L.; Katuri, J. *Angew. Chem. Int. Ed.*, 54(5), 1414 (**2015**); (b) Gao, W.; Wang, J. *ACS Nano*, 8(4), 3170 (**2014**); (c) Gao, W.; Wang, J. *Nanoscale*, 6(18), 10486 (**2014**); (d) Gao, W.; Sattayasamitsathit, S.; Wang, J. *The Chemical Record*, 12(1), 224 (**2012**); (e) Singh, V. V.; Wang, J. *Nanoscale*, 7(46), 19377 (**2015**).

[44] Nelson, B. J.; Kaliakatsos, I. K.; Abbott, J. J. *Annu. Rev. Biomed. Eng.*, 12(1), 55 (**2010**).

[45] Yesin, K. B.; Vollmers, K.; Nelson, B. J. *Int. J. Robot. Res.*, 25(5–6), 527 (**2006**).

[46] Peyer, K. E.; Zhang, L.; Nelson, B. J. *Nanoscale*, 5(4), 1259 (**2013**).

[47] Leslie-Pelecky, D. L.; Rieke, R. D. *Chem. Mater.*, 8(8), 1770 (**1996**).

[48] Mathieu, J.-B.; Beaudoin, G.; Martel, S. *Biomedical Engineering, IEEE Transactions on*, 53(2), 292 (**2006**).

[49] Martel, S.; Mathieu, J.-B.; Felfoul, O.; Chanu, A.; Aboussouan, E.; Tamaz, S.; Pouponneau, P.; Yahia, L. H.; Beaudoin, G.; Soulez, G.; Mankiewicz, M. *Appl. Phys. Lett.*, 90(11), 114105 (**2007**).

[50] Pawashe, C.; Floyd, S.; Sitti, M. *Appl. Phys. Lett.*, 94(16), 164108 (**2009**).
[51] Ebbens, S. J.; Howse, J. R. *Soft Matter*. 6(4), 726 (**2010**).
[52] Palacci, J.; Sacanna, S.; Steinberg, A. P.; Pine, D. J.; Chaikin, P. M. *Science*, 339(6122), 936 (**2013**).
[53] Kline, T. R.; Paxton, W. F.; Mallouk, T. E.; Sen, A. *Angew. Chem. Int. Ed.*, 44(5), 744 (**2005**).
[54] Sundararajan, S.; Lammert, P. E.; Zudans, A. W.; Crespi, V. H.; Sen, A. *Nano Lett.*, 8(5), 1271 (**2008**).
[55] Garcia-Gradilla, V.; Orozco, J.; Sattayasamitsathit, S.; Soto, F.; Kuralay, F.; Pourazary, A.; Katzenberg, A.; Gao, W.; Shen, Y.; Wang, J. *ACS Nano*, 7(10), 9232 (**2013**).
[56] Orozco, J.; Pan, G.; Sattayasamitsathit, S.; Galarnyk, M.; Wang, J. *Analyst*. 140(**5**), 1421 (**2015**).
[57] Bell, D. J.; Leutenegger, S.; Hammar, K. M.; Dong, L. X.; Nelson, B. J. In *Flagella-like Propulsion for Microrobots Using a Nanocoil and a Rotating Electromagnetic Field*, Robotics and Automation, 2007 IEEE International Conference on, 10–14 April 2007; pp. 1128–1133 (**2007**).
[58] Zhang, L.; Abbott, J. J.; Dong, L.; Kratochvil, B. E.; Bell, D.; Nelson, B. J. *Appl. Phys. Lett.*, 94(6), 064107 (**2009**).
[59] Berg, H. C.; Brown, D. A. *Nature*, 239(5374), 500 (**1972**).
[60] Mhanna, R.; Qiu, F.; Zhang, L.; Ding, Y.; Sugihara, K.; Zenobi-Wong, M.; Nelson, B. J. *Small*, 10(10), 1953 (**2014**).
[61] Qiu, F.; Mhanna, R.; Zhang, L.; Ding, Y.; Fujita, S.; Nelson, B. J. *Sensor. Actuator. B: Chem.*, 196, 676 (**2014**).
[62] Qiu, F.; Fujita, S.; Mhanna, R.; Zhang, L.; Simona, B. R.; Nelson, B. J. *Adv. Funct. Mater.*, 25(11), 1666 (**2015**).
[63] Manesh, K. M.; Campuzano, S.; Gao, W.; Lobo-Castanon, M. J.; Shitanda, I.; Kiantaj, K.; Wang, J. *Nanoscale*, 5(4), 1310 (**2013**).
[64] Gao, W.; Kagan, D.; Pak, O. S.; Clawson, C.; Campuzano, S.; Chuluun-Erdene, E.; Shipton, E.; Fullerton, E. E.; Zhang, L.; Lauga, E.; Wang, J. *Small*, 8(3), 460 (**2012**).
[65] Zhu, W.; Li, J.; Leong, Y. J.; Rozen, I.; Qu, X.; Dong, R.; Wu, Z.; Gao, W.; Chung, P. H.; Wang, J.; Chen, S. *Adv. Mater.*, 27(30), 4411 (**2015**).
[66] Venugopalan, P. L.; Sai, R.; Chandorkar, Y.; Basu, B.; Shivashankar, S.; Ghosh, A. *Nano Lett.*, 14(4), 1968 (**2014**).
[67] Orozco, J.; García-Gradilla, V.; D'Agostino, M.; Gao, W.; Cortés, A.; Wang, J. *ACS Nano*, 7(1), 818 (**2013**).
[68] Guix, M.; Orozco, J.; García, M.; Gao, W.; Sattayasamitsathit, S.; Merkoçi, A.; Escarpa, A.; Wang, J. *ACS Nano*, 6(5), 4445 (**2012**).

[69] Soler, L.; Magdanz, V.; Fomin, V. M.; Sanchez, S.; Schmidt, O. G. *ACS Nano*, **7**(11), 9611 (**2013**).

[70] Orozco, J.; Cheng, G.; Vilela, D.; Sattayasamitsathit, S.; Vazquez-Duhalt, R.; Valdés-Ramírez, G.; Pak, O. S.; Escarpa, A.; Kan, C.; Wang, J. *Angew. Chem. Int. Ed.*, 52(50), 13276 (**2013**).

[71] Gao, W.; Feng, X.; Pei, A.; Gu, Y.; Li, J.; Wang, J. *Nanoscale*, **5**(11), 4696 (**2013**).

[72] (a) Mou, F.; Chen, C.; Ma, H.; Yin, Y.; Wu, Q.; Guan, J. *Angew. Chem. Int. Ed.*, 52(28), 7208 (**2013**); (b) Gao, W.; Uygun, A.; Wang, J. *J. Am. Chem. Soc.*, 134(2), 897 (**2012**).

[73] (a) Anker, J. N.; Behrend, C. J.; Huang, H.; Kopelman, R. *J. Magn. Magn. Mater.*, 293(1), 655 (**2005**); (b) Nguyen, K. V. T.; Anker, J. N. *Sensors and Actuators B: Chem.*, 205, 313 (**2014**); (c) Roberts, T. G.; Anker, J. N.; Kopelman, R. *J. Magn. Magn. Mater.*, 293(1), 715 (**2005**).

[74] Kinnunen, P.; Sinn, I.; McNaughton, B. H.; Newton, D. W.; Burns, M. A.; Kopelman, R. *Biosens. Bioelectron.*, 26(5), 2751 (**2011**).

[75] (a) Mason, T. G.; Weitz, D. A. *Phys. Rev. Lett.*, 74(7), 1250 (**1995**); (b) Squires, T. M.; Mason, T. G. *Ann. Rev. Fluid Mech.*, 42(1), 413 (**2009**).

[76] (a) Mizuno, D.; Head, D. A.; MacKintosh, F. C.; Schmidt, C. F. *Macromolecules*, 41(19), 7194 (**2008**); (b) Wilson, L. G.; Poon, W. C. K. *PCCP*, 13(22), 10617 (**2011**).

[77] Neuman, K. C.; Nagy, A. *Nat. Meth.*, 5(6), 491 (**2008**).

[78] (a) Ghaskadvi, R. S.; Ketterson, J. B.; MacDonald, R. C.; Dutta, P. *Rev. Sci. Instrum.*, 68(4), 1792 (**1997**); (b) Brooks, C. F.; Fuller, G. G.; Frank, C. W.; Robertson, C. R. *Langmuir*. 15(7), 2450 (**1999**); (c) Ding, J.; Warriner, H. E.; Zasadzinski, J. A.; Schwartz, D. K. *Langmuir*, 18(7), 2800 (**2002**); (d) Bantchev, G. B.; Schwartz, D. K. *Langmuir*, 19(7), 2673 (**2003**).

[79] Anguelouch, A.; Leheny, R. L.; Reich, D. H. *Appl. Phys. Lett.*, 89(11), 111914 (**2006**).

[80] Dhar, P.; Cao, Y.; Fischer, T. M.; Zasadzinski, J. A. *Phys. Rev. Lett.*, 104(1), 016001 (**2010**).

[81] Reynaert, S.; Brooks, C. F.; Moldenaers, P.; Vermant, J.; Fuller, G. G. *J. Rheol.*, 52(1), 261 (**2008**).

[82] Choi, S. Q.; Steltenkamp, S.; Zasadzinski, J. A.; Squires, T. M. *Nat. Commun.*, 2, 312 (**2011**)

[83] Tierno, P. *PCCP*, 16(43), 23515 (**2014**).

[84] Wilhelm, C. *Phys. Rev. Lett.*, 101(2), 028101 (**2008**).

[85] Miller, R.; Ferri, J. K.; Javadi, A.; Krägel, J.; Mucic, N.; Wüstneck, R. *Colloid. Polym. Sci.*, 288(9), 937 (**2010**).

[86] Kim, S.-H.; Sim, J. Y.; Lim, J.-M.; Yang, S.-M. *Angew. Chem. Int. Ed.*, 49(22), 3786 (**2010**).

[87] Zhang, L.; Luo, Q.; Zhang, F.; Zhang, D.-M.; Wang, Y.-S.; Sun, Y.-L.; Dong, W.-F.; Liu, J.-Q.; Huo, Q.-S.; Sun, H.-B. *J. Mater. Chem.*, 22(45), 23741 (**2012**).

[88] Hu, S.-H.; Gao, X. *J. Am. Chem. Soc.* 132(21), 7234 (**2010**).

[89] Xu, C.; Wang, B.; Sun, S. *J. Am. Chem. Soc.*, 131(12), 4216 (**2009**).

[90] Senyei, A.; Widder, K.; Czerlinski, G. *J. Appl. Phys.*, 49(6), 3578 (**1978**).

[91] Sun, C.; Lee, J. S. H.; Zhang, M. *Adv. Drug Deliv. Rev.*, 60(11), 1252 (**2008**).

[92] (a) Tromsdorf, U. I.; Bigall, N. C.; Kaul, M. G.; Bruns, O. T.; Nikolic, M. S.; Mollwitz, B.; Sperling, R. A.; Reimer, R.; Hohenberg, H.; Parak, W. J.; Förster, S.; Beisiegel, U.; Adam, G.; Weller, H. *Nano Lett.*, 7(8), 2422 (**2007**); (b) Tromsdorf, U. I.; Bruns, O. T.; Salmen, S. C.; Beisiegel, U.; Weller, H. A *Nano Lett.*, 9(12), 4434 (**2009**); (c) Na, H. B.; Song, I. C.; Hyeon, T. *Adv. Mater.*, 21(21), 2133 (**2009**).

[93] (a) Kim, B. H.; Lee, N.; Kim, H.; An, K.; Park, Y. I.; Choi, Y.; Shin, K.; Lee, Y.; Kwon, S. G.; Na, H. B.; Park, J.-G.; Ahn, T.-Y.; Kim, Y.-W.; Moon, W. K.; Choi, S. H.; Hyeon, T. *J. Am. Chem. Soc.*, 133(32), 12624 (**2011**); (b) Huh, Y. M.; Lee, E. S.; Lee, J. H.; Jun, Y. w.; Kim, P. H.; Yun, C. O.; Kim, J. H.; Suh, J. S.; Cheon, J. *Adv. Mater.*, 19(20), 3109 (**2007**).

[94] Schladt, T. D.; Shukoor, M. I.; Schneider, K.; Tahir, M. N.; Natalio, F.; Ament, I.; Becker, J.; Jochum, F. D.; Weber, S.; Köhler, O.; Theato, P.; Schreiber, L. M.; Sönnichsen, C.; Schröder, H. C.; Müller, W. E. G.; Tremel, W. Au–MnO-"Nanoblumen" — Hybrid-Nanokomposite zur selektiven dualen Funktionalisierung und Bildgebung. *Angew. Chem.*, 122(23), 4068 (**2010**).

[95] Nakhjavan, B.; Tahir, M. N.; Panthofer, M.; Gao, H.; Gasi, T.; Ksenofontov, V.; Branscheid, R.; Weber, S.; Kolb, U.; Schreiber, L. M.; Tremel, W. *Chem. Commun.*, 47(31), 8898 (**2011**).

[96] Hainfeld, J. F.; Slatkin, D. N.; Focella, T. M.; Smilowitz, H. M. *Br. J. Radio.*, 79(939), 248 (**2006**).

[97] (a) Dongkyu, K.; Mi Kyung, Y.; Tae Sup, L.; Jae Jun, P.; Yong Yeon, J.; Sangyong, J. *Nanotechnology*, 22(15), 155101 (**2011**); (b) Narayanan, S.; Sathy, B. N.; Mony, U.; Koyakutty, M.; Nair, S. V.; Menon, D. *ACS Appl. Materi. & Interf.*, 4(1), 251 (**2012**).

[98] Schick, I.; Lorenz, S.; Gehrig, D.; Schilmann, A.-M.; Bauer, H.; Panthöfer, M.; Fischer, K.; Strand, D.; Laquai, F.; Tremel, W. *J. Am. Chem. Soc.*, 136(6), 2473 (**2014**).

[99] (a) Natalio, F.; Kashyap, A.; Lorenz, S.; Kerschbaumer, H.; Dietzsch, M.; Tahir, M. N.; Duschner, H.; Strand, S.; Strand, D.; Tremel, W. *Nanoscale*, 4(15), 4680 (**2012**); (b) Denk, W.; Strickler, J. H.; Webb, W. W. *Science*, 248(4951), 73 (**1990**).

[100] Yin, S.-N.; Wang, C.-F.; Yu, Z.-Y.; Wang, J.; Liu, S.-S.; Chen, S. *Adv. Mater.*, 23(26), 2915 (**2011**).

[101] Yu, Z.; Wang, C.-F.; Ling, L.; Chen, L.; Chen, S. *Angew. Chem. Int. Ed.*, 51(10), 2375 (**2012**).

[102] Petit, T.; Zhang, L.; Peyer, K. E.; Kratochvil, B. E.; Nelson, B. J. *Nano Lett.*, 12(1), 156 (**2012**).

[103] Casagrande, C.; Fabre, P.; Raphaël, E.; Veyssié, M. *Europhys. Lett.*), 9(3), 251 (**1989**).

[104] Rossmy, G. In *Structure, Dynamics and Properties of Disperse Colloidal Systems*, in Rehage, H.; Peschel, G. (Eds). Steinkopff: Darmstadt, pp. 17–26 (**1998**).

[105] Binks, B. P.; Fletcher, P. D. I. *Langmuir*, 17(16), 4708 (**2001**).

[106] Jiang, S.; Granick, S. *J. Chem. Phys.* 127(16), 161102 (**2007**).

[107] Nonomura, Y.; Komura, S.; Tsujii, K. *Langmuir*, 20(26), 11821 (**2004**).

[108] Park, B. J.; Lee, D. *Soft Matter.*, 8(29), 7690 (**2012**).

[109] Fan, H.; Striolo, A. *Soft Matter.*, 8(37), 9533 (**2012**).

[110] Aveyard, R. *Soft Matter.*, 8(19), 5233 (**2012**).

[111] Glaser, N.; Adams, D. J.; Böker, A.; Krausch, G. *Langmuir*, 22(12), 5227 (**2006**).

[112] Ou, F. S.; Shaijumon, M. M.; Ajayan, P. M. *Nano Lett.* 8(7), 1853 (**2008**).

[113] Liang, F.; Shen, K.; Qu, X.; Zhang, C.; Wang, Q.; Li, J.; Liu, J.; Yang, Z. *Angew. Chem. Int. Ed.,* 50(10), 2379 (**2011**).

[114] Passas-Lagos, E.; Schüth, F. *Langmuir*, 31(28), 7749 (**2015**).

[115] Teo, B. M.; Suh, S. K.; Hatton, T. A.; Ashokkumar, M.; Grieser, F. *Langmuir*, 27(1), 30 (**2011**).

Chapter 5

Facile Fabrication of Fe_3O_4–Polymer Anisotropic Magnetic Particles and Applications

Qiuyu Zhang, Baoliang Zhang, Hepeng Zhang, Lei Tian, Nisar Ali, Jiaojun Tan, Shenqiang Wang, Xiangkun Jia, and Mintao Qiao*

Applied Chemistry Department, Northwestern Polytechnical University, 710072 Xian, Shannxi Province, China
*qyzhang@nwpu.edu.cn

Abstract

Magnetic anisotropic particles, including Janus, patchy and multicompartment particles, have been particularly of great interest due to their magnetic response and anisotropy potential application in a wide variety of fields. Some strategies and techniques have been used to fabricate magnetic anisotropic particles. In this chapter, one of the main intension is to impart information about achievements on fabrication strategies and techniques of magnetic anisotropic particles in the last several years. A special emphasis will be made on several facial and scalable techniques. Some applications of these magnetic anisotropic particles are also introduced.

Keywords: Magnetic, Janus, fabrication, application.

5.1. Introduction

For the past 10 years, research hotspots of particulate materials are focused on anisotropic particles [1], which possess asymmetric architecture or composition. So far, various morphologies of anisotropic particles have been fabricated, including asymmetric hollow spheres [2], asymmetric particles (dumbbell-shaped, acorn-shaped particles), asymmetric cages [3], bowls [4], and fibers. They have played an increasing and important role in numerous applications, including solid surfactants [5,6], switchable display devices [7], optical sensors [8,9], antireflection coatings, antimicrobial materials [10], and anisotropic building blocks for complex structure [11] on account of their anisotropy and morphology diversity.

Magnetic anisotropic particles, including Janus, patchy, and multicompartment particles, have been particularly of great interest due to their magnetic response and anisotropy potential application in a wide variety of fields, such as biomedicine and biophysical research, electronic paper [12, 13], and heterogeneous catalysis [14]. The ability to manipulate functionalized magnetic anisotropic particles with external magnetic fields is advantageous for applications such as whole cell and biomolecular separation [15], magnetic resonance imaging contrast [16], and targeted drug delivery [17]. To date, some strategies and techniques have been used to fabricate magnetic anisotropic particles including microfluidic method [18–20], monomer polymerization [21–24], Pickering emulsion method [20,21], solvent evaporation-induced phase separation [25–27], self-assembly or induced assembly [28,29], flame synthesis [30,31], and depositing magnetic materials onto non-magnetic particles [32]. While these magnetic anisotropic particles serve as a ubiquitous and important tool, they often lack uniformity in magnetic responsiveness and perfect biphasic or multiphasic character. Numerous magnetic particles have been reported to have 30–80% variation in magnetophoretic mobility. So, achieving structure control of magnetic anisotropic particles is still problematic and challenging for material scientists. In addition, the fabrication of anisotropic magnetic particles in large quantities is still highly impossible.

In this chapter, one of the main aims is to impart information about achievements in fabrication strategies and techniques of magnetic anisotropic particles in the last several years. A special emphasis will be made on several facial and scalable techniques. Because there is usually some overlap in the definitions of Janus particles (JPs), patchy, and multicompartment particles, in the literature, some anisotropic particles designated patchy were also named as JPs based on the fact that there is complete segregation of the domains to form two opposing faces. Indeed, they belong to the larger family of "dis-symmetrical particles", where the asymmetry is also related their shape and/or the chemical composition of their bulk. We do not discriminate between these concepts in this chapter while we reviewed the relevant literature. Some applications of these magnetic anisotropic particles are also introduced.

5.2. Fabrication Strategies and Techniques of Magnetic Anisotropic Particles

The great and imaginative efforts were devoted to the fabrication of magnetic anisotropic particles with asymmetric morphology recently. More and more bottom-up fabrication approaches have been utilized. In this section, we aim to summarize the fabrication techniques, especially focus on three facile techniques in large quantities: (1) solvent evaporation-induced phase separation of polymer mixtures with magnetic particles by controlling the molecular weights of precursors and surface properties of particles; (2) one-step miniemulsion polymerization to prepare inner asymmetric (IA) PS/Fe_3O_4 composite microspheres; and (3) assembly and *in situ* growing of particles.

5.2.1. Solvent Evaporation-Induced Phase Separation

Solvent evaporation-induced phase separation is a universal method for fabricating magnetic anisotropic particles. This method requires generally two steps: (1) preparation of polymers and (2) phase separation of polymer blends and/or inorganic particles dissolved in volatile organic solvents such as dichloromethane (DCM), trichloromethane, and tetrahydrofuran as the solvent evaporation.

Wang *et al.* [25] produced different morphologies of poly (methylmethacrylate)/polystyrene (PMMA/PS) JPs by solvent evaporation-induced phase separation. By introducing hexadecane as the template, hollow, bowl-like particles, and hemispheres were prepared with different types and dosages of the emulsifier. Moreover, the morphology can be further regulated by the ratio of hexadecane/polymer and the two polymers. When Fe_3O_4 nanoparticles with incompatible coating were

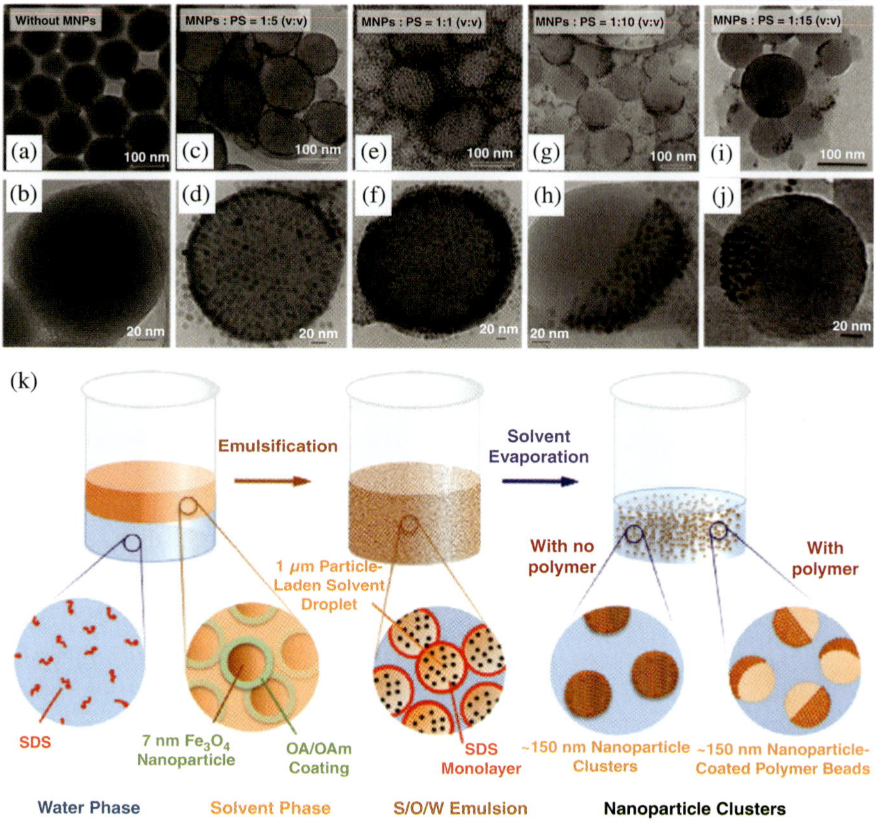

Figure 1. The effect of magnetic particle concentration on the morphologies of MJPs and schematic of the synthesis procedure.

introduced into the emulsion droplets, the magnetic particles would be excluded from the polymer-rich domain, and magnetic Janus particles (MJPs) are formed. Hatton *et al.* [26] investigated the effect of volatilization temperature, which decided the rate of solvent evaporation and magnetic particle concentration, etc., on the morphologies of magnetic clusters and JPs. When the volume ratio of magnetic nanoparticles (MNPs) to PS was 1:5, a well-ordered Fe_3O_4 nanoparticle monolayer was coated on the surface of PS particles. With the increasing of the ratio, MJPs were formed (1:15) as shown in Fig. 1. In an external magnetic field, those MJPs self-assembled into magnetic nanochains with the alignment of non-uniform structure.

Yabu *et al.* [27] developed a solvent evaporation method named self-organized precipitation (SORP) to prepare γ-Fe_2O_3@PS/PI MJPs. The mechanism of this method is shown in Fig. 2. γ-Fe_2O_3 nanoparticles were stabilized by amphiphilic block-copolymer PS-*b*-PAA. By solvent evaporation of the tetrahydrofuran solution containing polymer-stabilized γ-Fe_2O_3 nanoparticles, PS, and polyisoprene, aqueous dispersed magnetic Janus microparticles were successfully prepared, and the γ-Fe_2O_3 nanoparticles were selectively introduced.

A facile solvent evaporation-induced phase separation procedure was adopted by our group [33] to fabricate magnetic anisotropic particles. A typical recipe was shown in Table 1 (sample 5). The oil phase in this reaction system consists of the same ratios of PS/poly(glycidyl methacrylate) (PS/PGMA). With the evaporation of DCM, ternary phase separation of PS, PGMA, and Fe_3O_4 NPs could be achieved under vigorous stirring. The scanning electron microscopy (SEM) image and particle size distribution were shown in Fig. 3. In Fig. 3(a), particles which one side was relatively smooth and the other side had small protuberances, presenting obvious double-hemisphere morphology. Particle size measurement results showed that particles size of Fe_3O_4@PS/PGMA MJPs was mainly concentrated in 18–30 μm range (>86%), and the average particle size was 23 μm.

The effects of the amount of Fe_3O_4 nanoparticles, the ratio of PS and PGMA, the surfactant concentration, and emulsification speed on the formation of Fe_3O_4@PS/PGMA magnetic anisotropic particles have been investigated. Table 2 provides the magnetic contents of Fe_3O_4@PS/PGMA MJPs of samples with different mass ratios of Fe_3O_4 nanoparticles and polymer precursors, 3:20, 1.5:20, and 1:20, respectively.

The magnetic contents were respectively 8.9%, 5.4%, and 3.6%. It turned out that with the increase of the dosage of Fe_3O_4 nanoparticles, the magnetic content of magnetic anisotropic particles increased. However, when the added amount of Fe_3O_4 nanoparticles was more than 3:20, preparation of Fe_3O_4@PS/PGMA MJPs were not a success. Possible reason might be that the stability of magnetic emulsion deteriorated in the presence of an excess of Fe_3O_4 nanoparticles.

Morphology of magnetic anisotropic particles was largely dependent on the ratio of two polymer precursors, which could cause different degrees of phase separation. Therefore, the influence of the ratio of

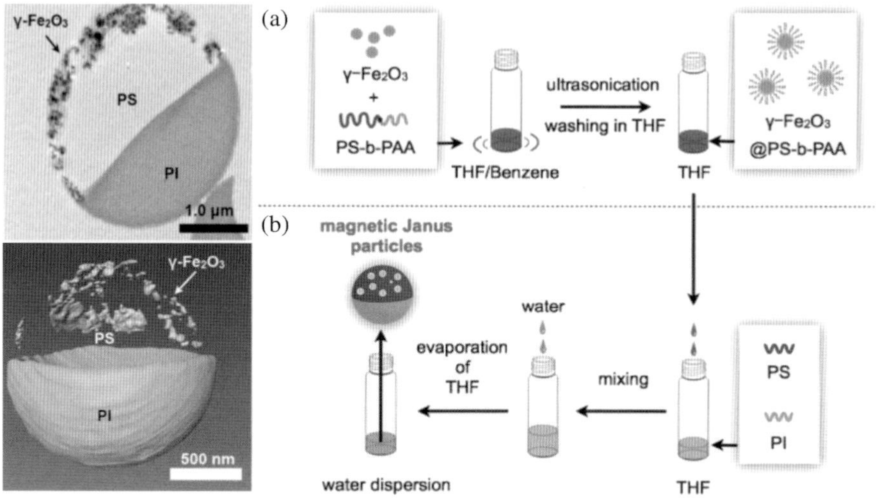

Figure 2. Schematic illustrations of preparation of MJPs and cross-sectional transmission electron microscope (TEM) image and STEMT image of MJP on the surface of PS phase. After coloring with pigments or dyes, the particles can be applied as pixels for electronic papers.

Table 1. Recipes for the production of MJPs.

Sample	SDBS/(g)	PS/(g)	PGMA/(g)	Fe_3O_4/(g)	Stirring rate (rpm)
1	0.5	1.0	1.0	0.3	500
2	0.5	1.0	1.0	0.15	500
3	0.5	1.0	1.0	0.1	500
4	0.5	1.2	0.8	0.2	500
5	0.5	1.0	1.0	0.2	500
6	0.5	0.8	1.2	0.2	500
7	0.3	1.0	1.0	0.2	500
8	0.4	1.0	1.0	0.2	500
9	0.6	1.0	1.0	0.2	500
10	0.5	1.0	1.0	0.2	600
11	0.5	1.0	1.0	0.2	700
12	0.5	1.0	1.0	0.2	800

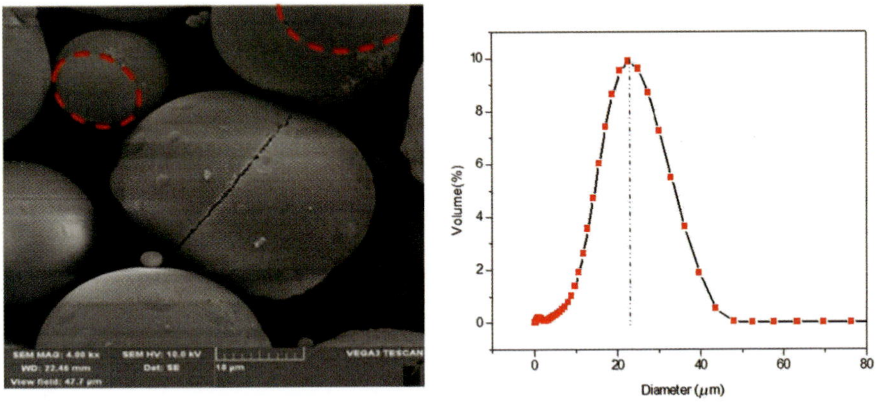

(a) (b)

Figure 3. SEM image of Fe_3O_4–PS/PGMA MJPs (a) and particle size distribution (b).

Table 2. The amount of Fe_3O_4 nanoparticle in Fe_3O_4@PS/PGMA MJPs.

Sample	3:20[a]	1.5:20	1:20
Amount	4.47%	1.98%	1.89%

[a]Mass ratios of Fe_3O_4 nanoparticles and polymer precursors.

polymer precursors was investigated, as shown in Fig. 4. The sizes of two hemispheres, which could be changed by various ratios of polymer precursors, were basically the same when the ratio of PS and PGMA was 1:1 in Fig. 4(b). To confirm the components of two hemispheres, we decreased the proportion of PGMA (Fig. 4(a)) and increased the amount of PS (Figs. 4(c) and 4(d)). The results showed that with the increasing of the dosage of PGMA, magnetic anisotropic particles experienced a change of shape from snowman-shape to dumbbell-shape and then finally to snowman-shape. The surface of the rough hemisphere continued to enlarge at the same time. In addition, it was found that the component of the bigger rough hemisphere was mainly PGMA, and the smaller smooth hemisphere was mainly PS.

To determine the distribution of Fe_3O_4 nanoparticles in Fe_3O_4@PS/PGMA magnetic anisotropic particles, two types of MJPs, in which the ratio of PS and PGMA were 3:2 and 2:3, respectively, were analyzed by

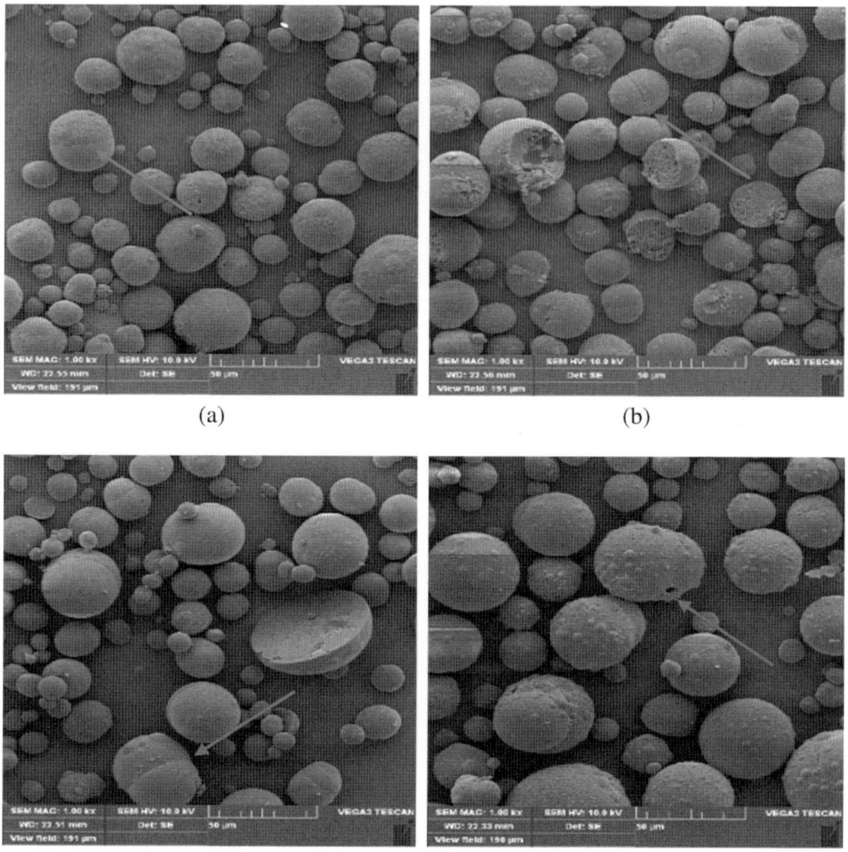

(a) (b) (c) (d)

Figure 4. SEM images of Fe_3O_4–PS/PGMA MJPs, the ratios of PS and PGMA were 3:2 (a); 1:1 (b); 2:3 (c); 1:2 (d).

optical microscope as shown in Fig. 5. It could be seen that Fe_3O_4 nanoparticles distribution is concentrated on one side. When PS:PGMA was 3:2, Fe_3O_4 nanoparticles distributed on the bigger side. And when PS:PGMA was 2:3, Fe_3O_4 nanoparticles distributed on the smaller side. Therefore, Fe_3O_4 nanoparticles were observed to distribute on the PS side. The primary cause was that the compatibility of oleic acid-modified Fe_3O_4 nanoparticles with hydrophobic PS was relatively stronger than with that with PGMA.

The morphology of magnetic anisotropic particles prepared by phase separation was a state of minimum interfacial free energy of surface

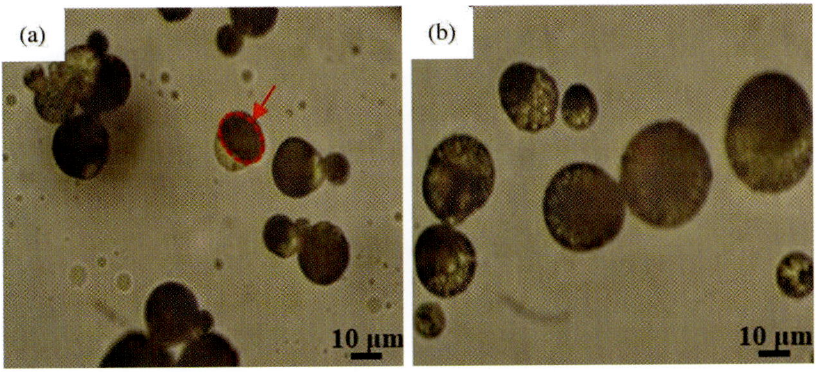

Figure 5. Optical micrographs of Fe_3O_4–PS/PGMA MJPs, the ratios of PS and PGMA were 3:2 (a) and 2:3 (b).

composed of different components [26]. So, the concentration of surfactant indirectly influenced the morphology of magnetic anisotropic particles. It was found that the proportion of sodium dodecyl benzene sulfonate (SDBS) was low at 0.15%, and magnetic anisotropic particles were difficult to be prepared. This was due to the fact that the emulsifying agent was not enough to stabilize the oil–water interface. With the increase in the dosage of SDBS, the morphology of particles tended to be snowman-shaped. When 0.25% SDBS was used, it achieved the optimal situation, which was a dumbbell-shape with the same size of hemispheres. As the amount of SDBS continued to increase, magnetic anisotropic particles were gradually divided into two single particles.

Based on these results, a dual-phase separation mechanism was expected as shown in Scheme 1. Two kinds of phase separation conducted synchronously, included: (1) separation of PS and PGMA [30]; (2) separation of PGMA and modified Fe_3O_4 nanoparticles. The dual phase separation occurred at the same time, and ended up with the formation of Fe_3O_4@PS/PGMA magnetic anisotropic particles. This mechanism was supported by optical images of particles at each reaction time point, as shown in Fig. 6.

The formation of magnetic anisotropic particles experienced emulsification (Fig. 6(a)), solvent evaporation induced internal phase separation (Figs. 6(b) and 6(c)), and two phases separated completely to form magnetic anisotropic particles (Fig. 6(d)). This means that oil phase was

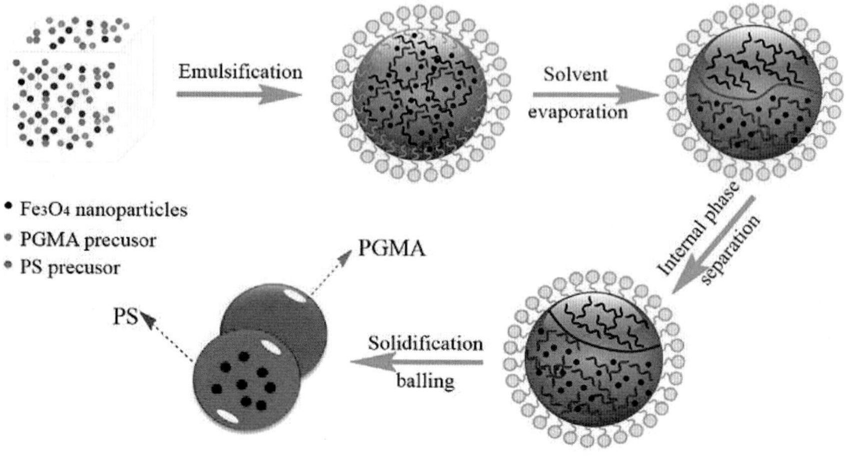

Scheme 1. Schematic illustration of the formation of Fe_3O_4–PS/PGMA MJPs.

emulsified once the monomer was poured into the water phase. Hydrophobic Fe_3O_4 NPs were clearly embedded in the emulsion droplets as shown in Fig. 6(a). At 36 min, with the evaporation of the solvent, emulsion droplet started to solidify, and internal phase separation started to take place (as shown in Fig. 6(b)). Black moieties represent droplets located with Fe_3O_4 NPs and white areas of the droplets were relatively clear to be polymer blends, mainly the PGMA. We can draw a conclusion that PS and Fe_3O_4 NPs separated from PGMA simultaneously. Internal phase separation was basically completed within 1 h (Fig. 6(c)). Nevertheless, it was an intermediate state with different particle morphologies, like three-hemispheres and core–shell structure. Fe_3O_4 NPs were mainly distributed on the interface of the two phases due to the system-increased viscosity. Subsequently, the system completed phase separation at 2 h (Fig. 6(d)), and formed magnetic anisotropic particles. Composite particles, which had a three-hemisphere structure, provided a possibility for part of the formation of three-hemisphere structure under 30 mL DCM.

Beyond doubt, solvent evaporation-induced phase separation is a universal method for fabricating magnetic anisotropic particles. Although solvent evaporation-induced phase separation has its inherent shortcomings such as the particles size is distributed mainly in micrometers,

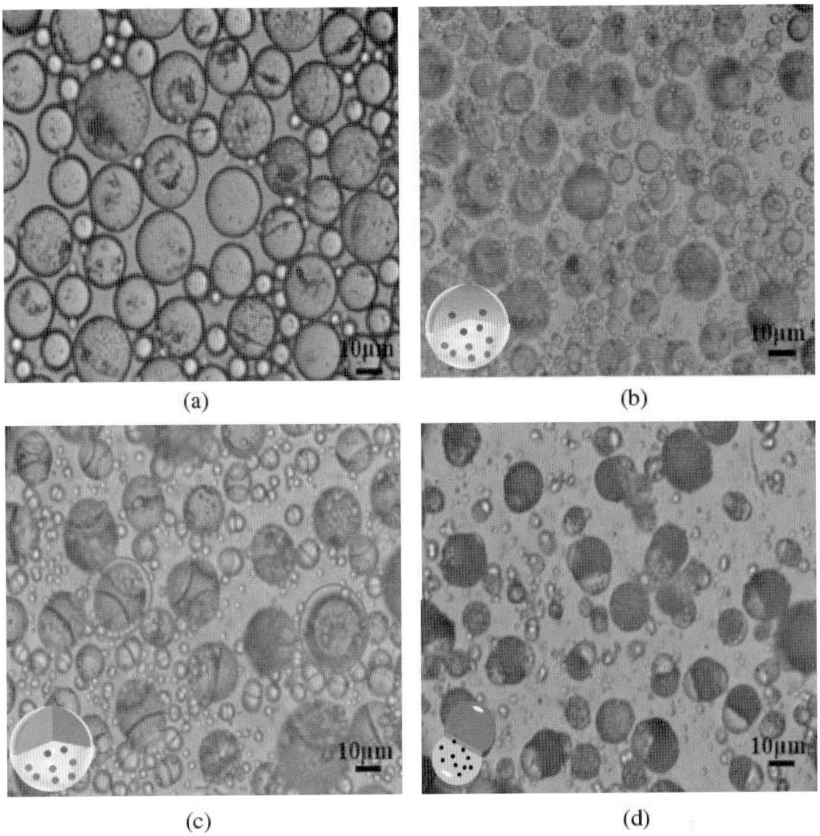

Figure 6. Optical microscope images of Fe_3O_4–PS/PGMA MJPs at different phase separation times: 7 min (a); 36 min (b); 1 h (c); 2 h (d).

although the size of particles prepared by solvent evaporation-induced phase separation can be controlled by the concentration of the solution and the mixing ratio of the good solvent to the poor solvent [3]. In the future, more work should be done to achieve a precise control of the distribution of MNPs in the particles.

5.2.2. Miniemulsion Polymerization

Due to time-consuming multiple steps needed for most of template-assisted fabrication methods, a convenient one-pot process can be used to

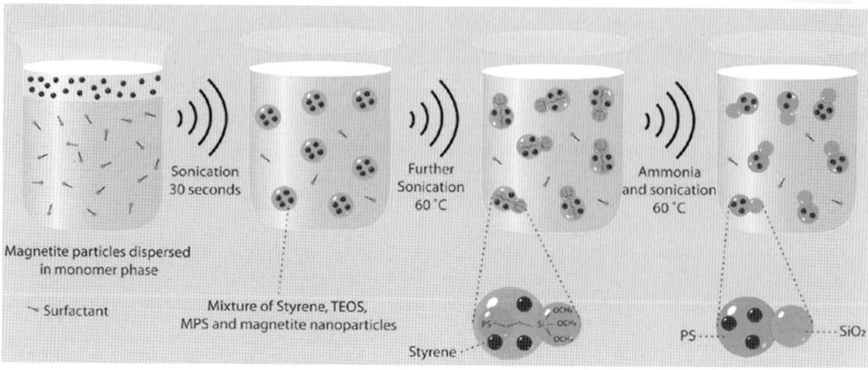

Figure 7. Schematic diagram of the procedure for the synthesis of magnetic asymmetric JPs by the sonochemically driven miniemulsion polymerization pathway.

fabricate magnetic anisotropic particles. In principle, miniemulsion polymerization is an ideal way to fabricate magnetic anisotropic particles. Teo *et al.* [34] reported a convenient one-pot process to prepare magnetic anisotropic particles by the sonochemically driven miniemulsion polymerization pathway (Fig. 7). Magnetic anisotropic particles were composed of silica and PS, with the PS portion loaded with nanosized magnetite particles. Upon application of a strong magnetic field, the particles can be separated from the oil droplets, leading to an almost completely phase-separated system. It can be anticipated that such anisotropic JP materials will find many applications in areas such as magnetically controlled optical switches, the fabrication of microengines, the assembly of colloid crystal structures with novel symmetries and emulsion stabilization–destabilization processes.

Wang *et al.* [35] have synthesized asymmetric Fe_3O_4/silica/PS composite particles with high anisotropy via a one-pot process combining miniemulsion polymerization with sol–gel technology (Fig. 8). The asymmetricity was established by having two portions of the particle composed of both PS and silica, among which the latter was conjugated with magnetite nanoparticles. The asymmetric magnetic composite particles (AMCPs) show not only the asymmetric geometry with associated chemical compositions but also the amphiphilic surface property. The mechanism of phase separation has been identified. In this process, the morphological control of the magnetic composite particles can be achieved

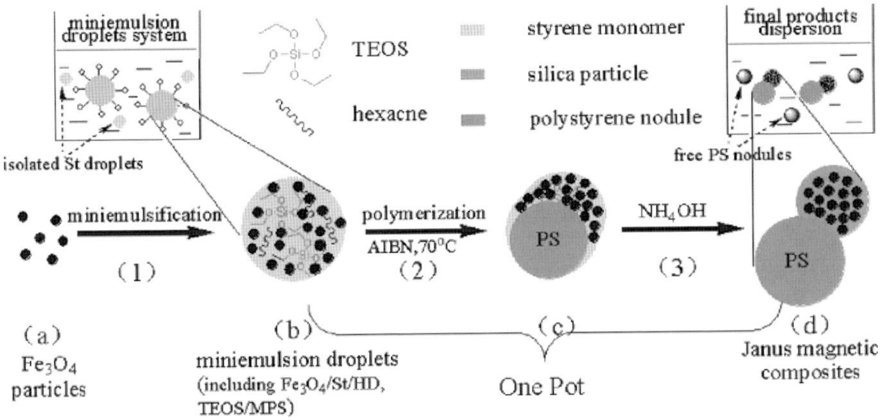

Figure 8. Schematic diagram showing the formation of the asymmetric Fe_3O_4–silica/PS composite particles by the one-pot process combining the *in situ* miniemulsion polymerization of the Styrene monomer and the sol–gel reaction of tetraethyl ortho silicate (TEOS) in the same miniemulsion droplet containing an iron oxide core.

by adjusting the processing parameters, including 3-(trimethoxysilyl) propyl methacrylate (MPS) and divinylbenzene (DVB) content, and the time at which the suspension pH was changed. The as-synthesized AMCPs were shown to be superparamagnetic and amphiphilic, which can serve as an ideal candidate for biomedical applications such as nucleic acid extraction, protein purification, cancer diagnosis and treatment, and biosensors.

Zhang [36] reports a novel approach for the fabrication of Janus magnetic microcapsules via miniemulsion polymerization (Fig. 9). By taking advantage of phase separation between TEOS and the growing polymers, the TEOS phase is compressed and restricted as a liquid core and then converted into silica, leading to the formation of a hollow structure. By adding MPS-modified MNPs in oil phase, magnetic microcapsules were obtained with MNPs embedded in the polymeric shell. Furthermore, magnetic microcapsules could be fabricated by the encapsulation of hydrophobic Fe_3O_4 nanoparticles in the polymeric shell and Janus magnetic microcapsules could be obtained with the help of an external magnetic field.

They have succeeded in restricting the oily inorganic precursor (TEOS) as a liquid core, pushing the compressed TEOS phase out of the miniemulsion droplets, dividing the compressed TEOS phase into

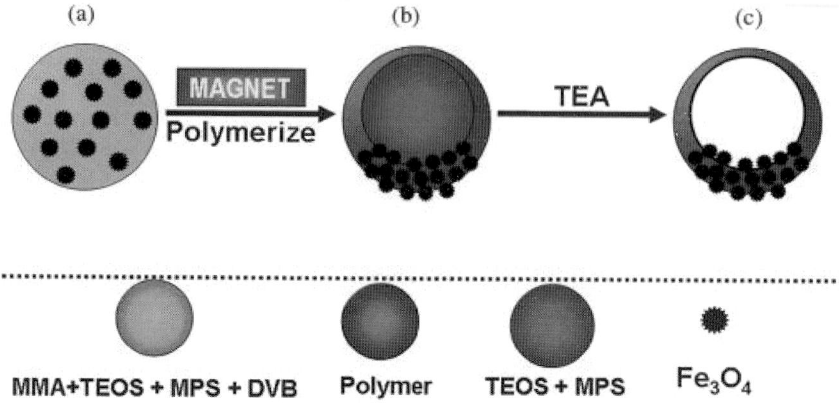

Figure 9. Illustration of the formation of Janus magnetic microcapsules by miniemulsion polymerization.

nanodroplets, or pushing the compressed TEOS phase to one side within the monomer droplets when the organic monomers have polymerized. As a result, hybrid microspheres with various morphologies have been fabricated due to the different compression directions of the TEOS phase. These magnetic microcapsules could find applications in controlled release systems, energy storage and conversion, and catalysis.

Recently, a novel inner magnetic asymmetric composite microspheres [37] and microcapsules [38], in which most magnetic particles encapsulated into the polymer matrix aggregated in one side of the microsphere, while only a few particles randomly dispersed in the remaining part, has been prepared by our group via miniemulsion polymerization.

TEM of these composite microspheres is shown in Figs. 10(a) and 10(b). The magnetic particles without modification that are uniformly dispersed in the PS microspheres are shown in the TEM image in Fig. 10(a), while the surface-modified magnetic particles distributed asymmetrically in the PS microspheres (Fig. 10(b)). Most of the encapsulated surface-modified magnetic particles are assembled into clusters on one side of the microspheres, while in the other part of the microspheres, there were quite a few particles. So, one important conclusion is that the inner structure of composite microspheres was related to the surface property of magnetic particles. The IA composite microspheres could be prepared by simply modifying the surface of the magnetic particles with oleic acid (OA).

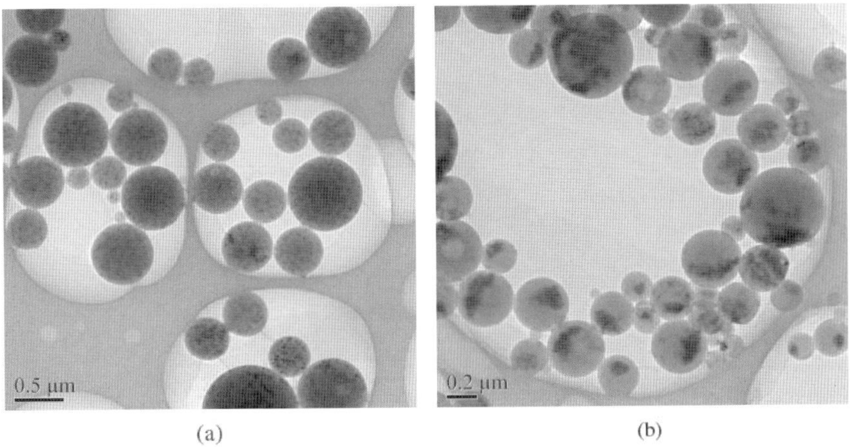

Figure 10. TEM of composite microspheres: (a) prepared with magnetic particles, (b) prepared with surface modified magnetic particles.

The mechanism of this process was not very clear yet, and one possible reason was shown in Scheme 2. When the surfaces of the magnetic particles were modified with oleic acid, they could disperse well in oil (e.g., styrene) and aggregate in water. By using these particles to prepare magnetic composite microspheres via miniemulsion, the mini-emulsified droplets of styrene were stabilized by surfactants (SDS) and a costabilizer (CA), and the surface-modified magnetic particles were evenly dispersed in droplets. At beginning of the polymerization, there were many short PS chains dissolved in styrene monomer and the magnetic particles might be also evenly dispersed in droplets (step 1). As the polymerization proceeded, the polymer chains became longer and PS precipitated out from the monomer. The chains began to entangle mutually, and there were three phases (PS(S), magnetic particles(S), and styrene(L)) existing in the droplet at the time, because the modified magnetic particles had higher affinity with styrene than the ordinary magnetic particles. Thus, it was difficult for the particles to entangle in PS but could be able to move into styrene-like extraction. The lubrication of oleic acid on the surface of the modified magnetic particles promoted the movement of the surface-modified magnetic particles among the polymers. Most of styrene polymerized as the reaction proceeded, and due to mutual

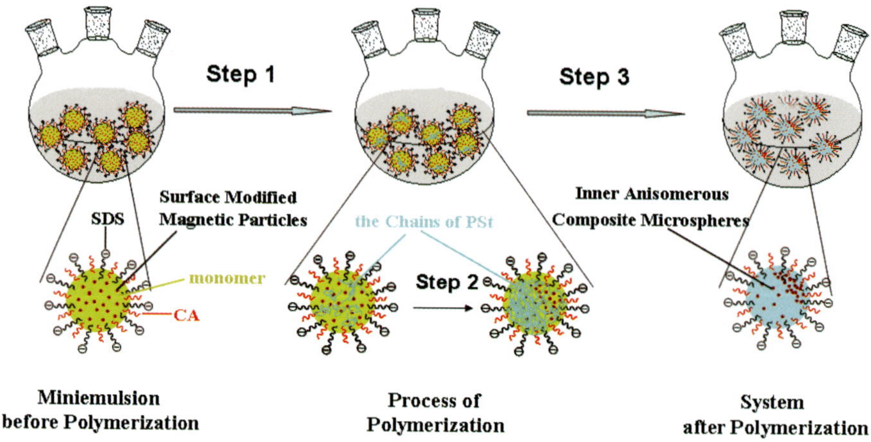

Scheme 2. Schematic of the formation of IA composite microspheres.

entanglement of PS chains, the volume of microsphere shrunk. The residual styrene in microsphere was gradually extruded to one side of the microsphere and most of surface-modified magnetic particles moved to the same side with styrene. Thus, only a few magnetic particles were encapsulated by PS (step 2). During the final period of polymerization, the conversion of styrene was further improved and the little residual styrene on one side of the microsphere continuously polymerized. Also, the modified magnetic particles were hydrophobic and they could not migrate into water. In order to decrease the system energy, they gathered into clusters and were gradually encapsulated by PS. Finally, the asymmetric composite microspheres were formed (step 3).

When magnetic particles, which are not modified by oleic acid, are used to prepare PS/iron oxides composite microspheres by miniemulsion, the affinity between the magnetic particles and the styrene is poor, and the interactions between iron oxide particles with styrene and PS are different from those indicated in the last paragraph. The magnetic particles can't freely move along with styrene in microspheres, so the composite microspheres, in which the magnetic particles are evenly dispersed, are obtained.

The magnetic hysteresis loop and thermogravimetric curve of the novel IA composite microspheres are shown in Fig. 11(a). The saturation

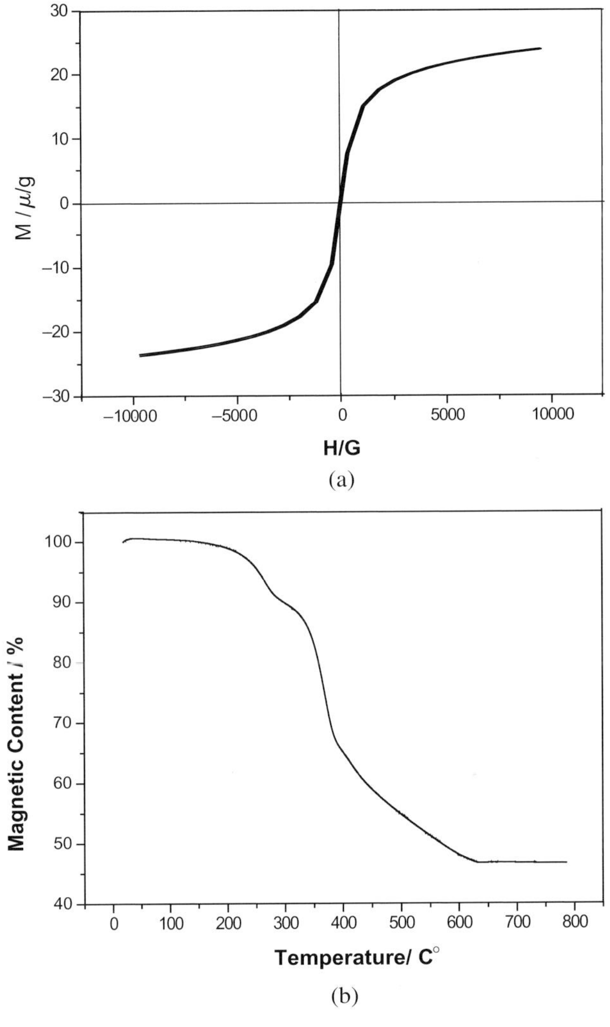

Figure 11. The magnetic hysteresis loop (a) and thermogravimetric curve (b) of the novel IA composite microspheres.

magnetization of the novel microspheres is 23.8 μg^{-1}. The zero residual magnetism and zero coercive force of magnetic microspheres suggest that the composite microspheres prepared in this work are superparamagnetic. Figure 11(b) indicates that the degradation process can be divided into two stages: the first weight loss stage of 150–300°C is mainly the volatilization and decomposition of oleic acid, while the second weight loss stage of 320–620°C is mainly the decomposition of PS. The magnetic content of this asymmetric composite microsphere is about 46.7%.

Similarly, IA microcapsules were prepared via miniemulsion. TEM images and particle size distributions of OA-Fe_3O_4 nanoparticles and IA-PS/Fe_3O_4/*n*-heptane (HP) IA magnetic microcapsule are shown in Fig. 12. In Fig. 12(a), OA-Fe_3O_4 nanoparticles showed favorable dispersion property in HP. OA cannot only adjust the hydrophilic and hydrophobic properties, but also reduce the particle surface energy and act as release agent for improving the dispersion of nanoparticles. The particle size distribution of OA-Fe_3O_4 nanoparticles is shown in Fig. 12(e), and the particle size was mainly concentrated at about 8–10 nm and the average particle size was 7.23 nm. Figures 12(b) and 12(c) are the TEM images of IA-PS/Fe_3O_4/HP IA magnetic microcapsule. From the higher magnification TEM image, OA-Fe_3O_4 was found embedded and dispersed inside the capsule wall. The presence of cystic structure inside the microcapsule can be clearly seen. The capsule-core was not in the center of the microcapsule. The capsule–core was close to one side of outer wall of the microcapsule and showed IA structure. OA-Fe_3O_4 nanoparticles were uniformly dispersed in the capsule–wall of IA-PS/Fe_3O_4/HP. Meanwhile, it can be seen that the volume ratio of capsule–core to the whole microcapsule in each IA-PS/Fe_3O_4/HP was nearly the same. This can contribute to the same component of each oil droplet before polymerization and there was a 1:1 correspondence between initial monomer droplets and final obtained particles in miniemulsion polymerization process, which was experimentally demonstrated for the first time by Ugelstad *et al.* [39]. Figure 12(d) is the SEM image of IA-PS/Fe_3O_4 /HP. The capsules with a smooth surface were spherical. One side of the surface of some spheres shown a cupped structure. The cupped side was the cavity side. The particle size distribution of IA-PS/Fe_3O_4/HP was determined and shown in Fig. 12(f), from which we can clearly see that the range of particle size was from 0.5 to 1.3 μm, and the average particle size was 0.8 μm.

Several IA-PS/Fe_3O_4/HP microcapsules with different cavity volumes were prepared by adjusting an additional amount of HP. TEM images of these IA microcapsules were shown in Figs. 13(a)–13(d) indicate that the

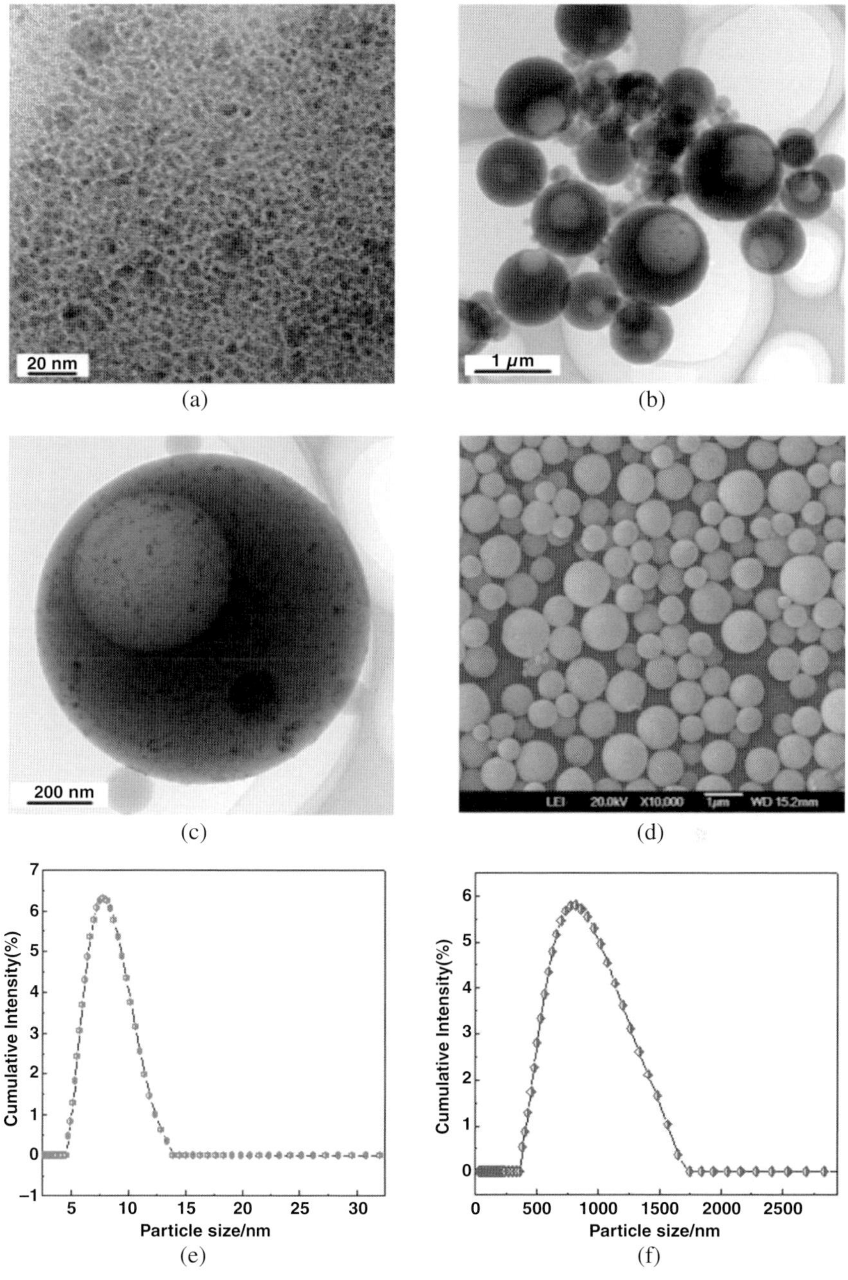

Figure 12. TEM images of Fe_3O_4 MNPs modified by OA (a) and IA-PS/Fe_3O_4 IA magnetic composite microcapsules (b) and (c); SEM image of IA-PS/Fe_3O_4/HP (d); the particle size distribution curves of Fe_3O_4 MNPs (e) and IA-PS/Fe_3O_4/HP (f).

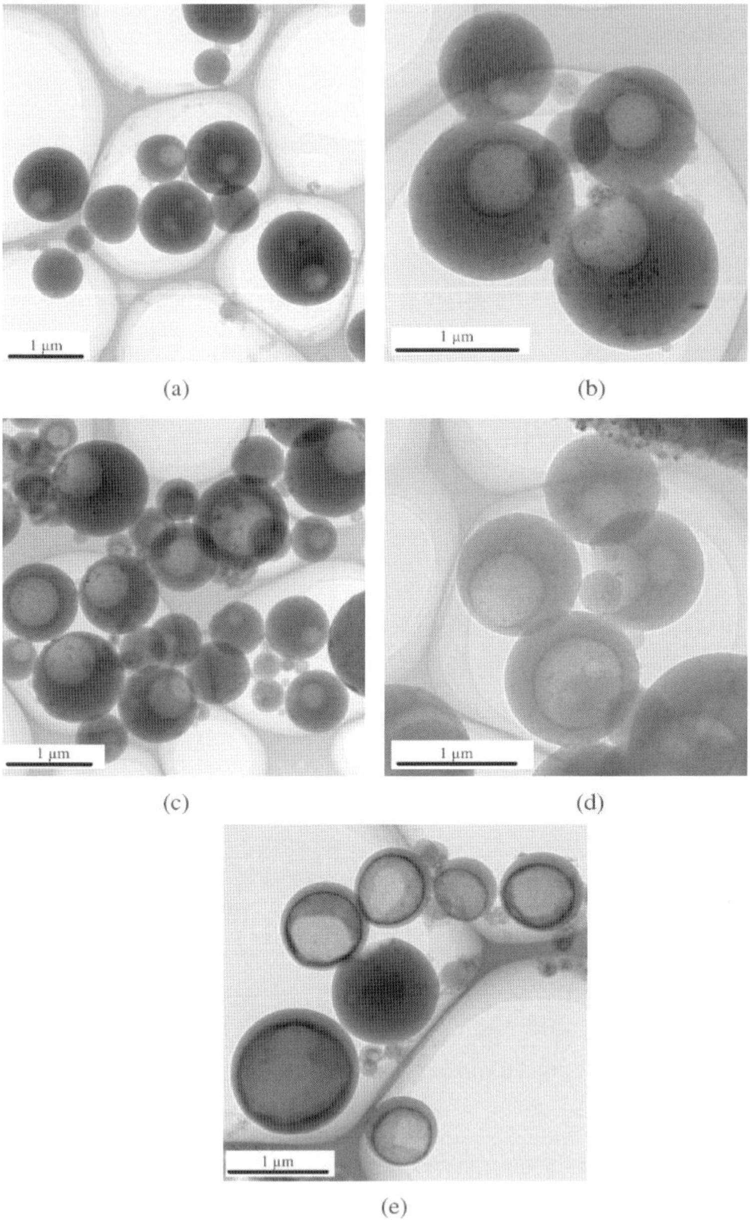

Figure 13. TEM images of IA magnetic composite microcapsules with different additional amounts of *n*-heptane: IAM-1 (a); IAM-2 (b); IAM-3 (c); IAM-4 (d); IAM-5 (e).

cavity volume of IA-PS/Fe_3O_4/HP microcapsules increased with increasing the additional amount of HP, but the thickness of the thinner-side capsule-wall of the microcapsules decreased. The capsule ruptured when the additional amount of HP was more than 2.0 g. The ruptured capsule was bowl-shaped and most of MNPs distributed at the interface of concave, as shown in Fig. 13(e). The reason why nanoparticles moved to the interface may be explained as follows. The rupture was because of the big cavity. So, during the polymerization, the thin capsule wall may be permeated by the external water phase. The thinner the capsule wall is, the more obvious permeation becomes. This will change the hydrophilic and hydrophobic properties of the capsule core. This change may take effect in the migration of MNPs. The migration may be explained by the theory of Pickering emulsion. OA-Fe_3O_4 nanoparticles moving to the interface in the presence of a large amount of HP could reduce the interfacial energy.

Till now, a wide range of magnetic anisotropic particles has been fabricated in recent years. Miniemulsion polymerization has been shown to provide efficient routes for the fabrication of magnetic anisotropic particles. The limitation of miniemulsion polymerization is that it is difficult to control uniformity in magnetic responsiveness and perfect biphasic or multiphasic charter of the particles. So, we have to find new strategies to overcome this limitation.

5.2.3. Assembly and *In Situ* Growing of Particles

Some work has been done to prepare magnetic clusters of defined size by self-assembly, while most are not reversible [40,41]. The preparation of reversible clusters in response to stimuli remains challenging. However, JP might meet these needs, because they possess a higher level of nanoparticle functionalization than the common particles. Hatton *et al.* have prepared Janus MNPs (~20 nm) by grafting either sodium PS sulfonate or poly(*N,N*-dimethylamino-2-ethyl methacrylate) to the surface of the poly(acrylic acid) (PAA)-coated MNPs preabsorbed on positively charged silica beads [28] (Fig. 14). Controlled aggregation of Janus MNPs was formed in a low pH condition, while this was reversible in that individual particles could recover by restoring the pH to high values. Besides, the effects of the molecular weight of the grafted polymer, the MNP surface

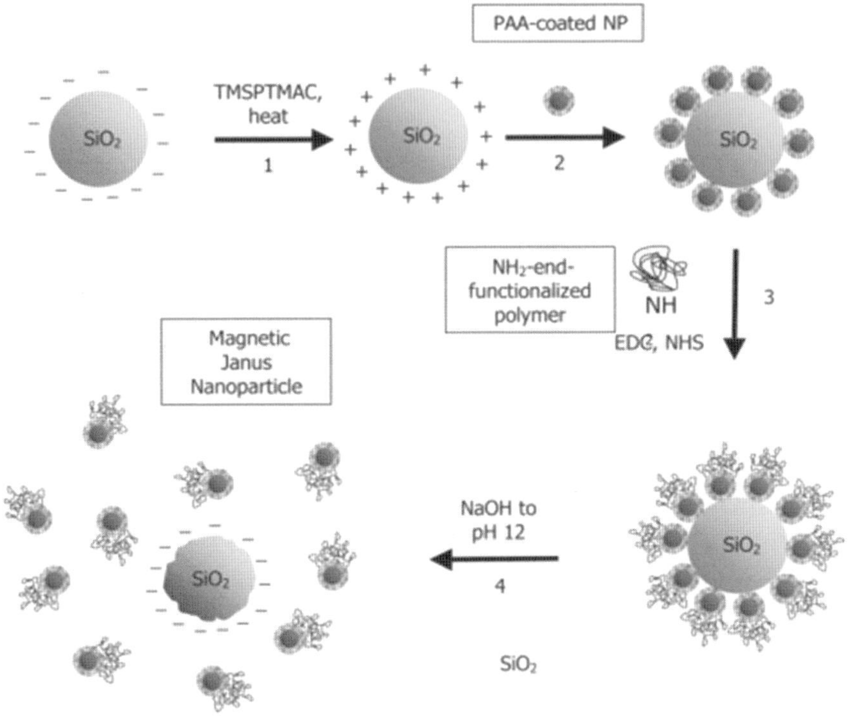

Figure 14. Schematic representation of masking procedure for the asymmetric functionalization of PAA-coated MNPs.

availability for grafting, and the addition of small amounts of PAA-coated MNPs on the size of the stable clusters were also investigated.

Based on their previous work, Hatton *et al.* have synthesized Janus nanoparticles consisting of ~5 nm MNPs with a pH-dependent and temperature-independent polymer (PAA) on one side and a pH-independent polymer (sodium PS sulfonate, PSSNa) or a temperature-dependent polymer (poly(*N*-isopropyl acrylamide), PNIPAM) on the other side [29]. At high pH values and low temperatures, these Janus nanoparticles are individually dispersed stably, but at low pH values (PSSNa) or at high temperatures (>31°C) (PNIPAM), they can self-assemble to stable dispersions of clusters, which are analyzed by the dynamic light scattering and zeta potential measurements (Fig. 15).

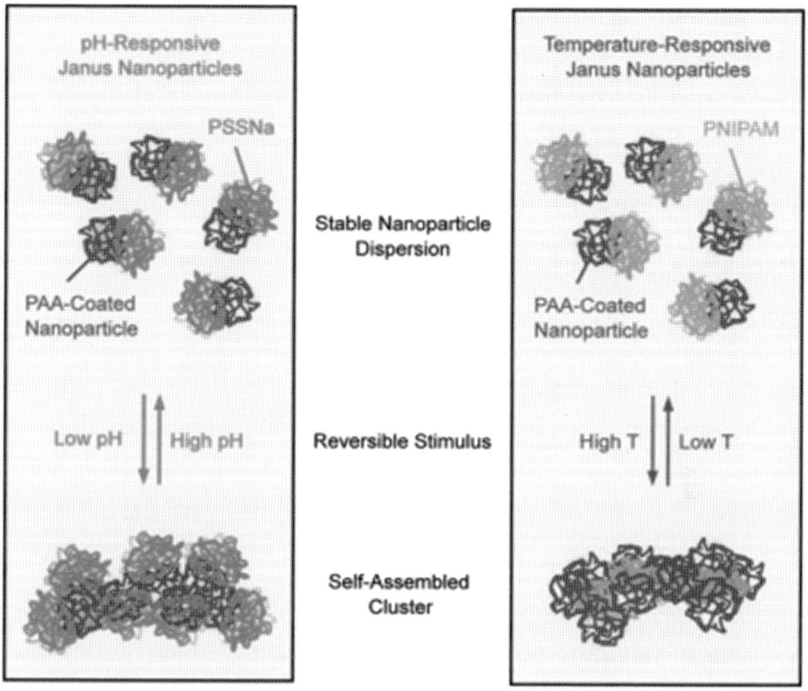

Figure 15. Schematic representation of the Janus nanoparticles and their self-assembled structures on application of an appropriate stimulus. Note that the PSSNa polymers constitute the stabilizing shell of the clusters, while the PNIPAM chains collapse and form the hydrophobic core of the clusters.

Dyab *et al.* [42] reported a novel technique for preparing anisotropic magnetic microparticles based on "arresting" the alignment of oleic acid-coated magnetite nanoparticles (OCMNs) dispersed in polymerizable oil drops. This method included the following steps: (i) preparing a polymerizable magnetic O/W emulsion, which was stabilized by a polymerizable anionic surfactant; (ii) gelling the aqueous phase in the cooling process of the mixture of emulsion and hot agarose solution; (iii) heating the gelled magnetic emulsion triggered the polymerization of oil drops and applying an external magnetic field simultaneously; (iv) releasing and washing the particles and redispersing them in an aqueous phase. These authors demonstrated that anisotropic microparticles, with the diameter ranging from tens to hundreds of micrometers, could be prepared by adjusting the stirring speed. And a surfactant-free method was developed by altering the pH of the aqueous continuous phase of the emulsion. The resulting anisotropic microparticles had high coercivity (maximum value of H_c is 707 Oe) and possessed anisotropic optical and magnetic properties. And it is indicated that these anisotropic microparticles find potential application in optical switches and valves, microengines, and assembling of novel structures.

Sacanna [43] has introduced a new class of spherical colloids that reversibly self-assemble into well-defined nonlinear structures by virtue of "magnetic patches". When the magnetic patches are in close proximity to one another, but are not sufficient to cause magnetic flocculation, the resulting magnetic forces could be strong enough to bind particles together. The magnetic bonds that are established among interacting particles are directional, reversible, and form regardless of the chemical environment surrounding it.

The synthesis protocol, which is shown in Fig. 16, starts with the preparation of hematite microcubes. The resulting cubes are uniform in size and shape, and possess a permanent magnetic moment. Next, the magnetic cubes are encapsulated into polymerizable silicon oil droplets. In the final step of this synthesis, a radical polymerization hardens the oil phase, yielding patchy colloids. The nucleation and growth mechanism is as follows: Each cube, initially suspended in the water phase, acts as a nucleation site for the formation of an oil droplet, thus producing droplets with a single magnetic inclusion. A key feature of this encapsulation mechanism is that, during the phase transfer from water to oil, the cubes

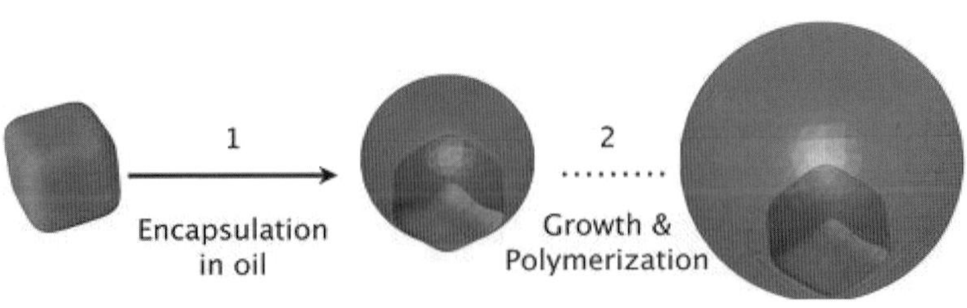

Figure 16. The synthetic steps involved in the preparation of particles with a single magnetic patch.

are trapped at the interface by surface tension, thus leaving one face of the cube exposed to the water phase. A further radical polymerization hardens the oil phase, yielding patchy colloids. They could control the droplet's size with accuracy and produce monodisperse droplets (polydispersity <5%), which cannot be prepared with common homogenization methods and, in the absence of external fields, the particles can self-organize into well-defined clusters or bind at specific magnetic targets on a substrate.

Yang *et al.* [44] have prepared the anisotropic JPs with strong magnetic iron oxide domain and polymer microspheres domain with large scale in one batch by *in situ* growing of particles in a solvothermal situation in the presence of poly(methylenebisacrylmide-*co*-methacrylic acid) (P(MBA-*co*-MAA)) microspheres, as illustrated in Fig. 17. The coordination interaction between the carboxyl acid groups and inorganic species played an essential role during the formation of uniform P(MBA-*co*-MAA)/iron oxide core–shell microspheres via the *in situ* controlled hydrolysis of ferric chloride in the initial stage. The formation of P(MBA-*co*-MAA)/Fe_3O_4 JPs may originate from a thermodynamic incompatibility between the P(MBA-*co*-MAA) domain and the Fe_3O_4 domain during the partial reduction

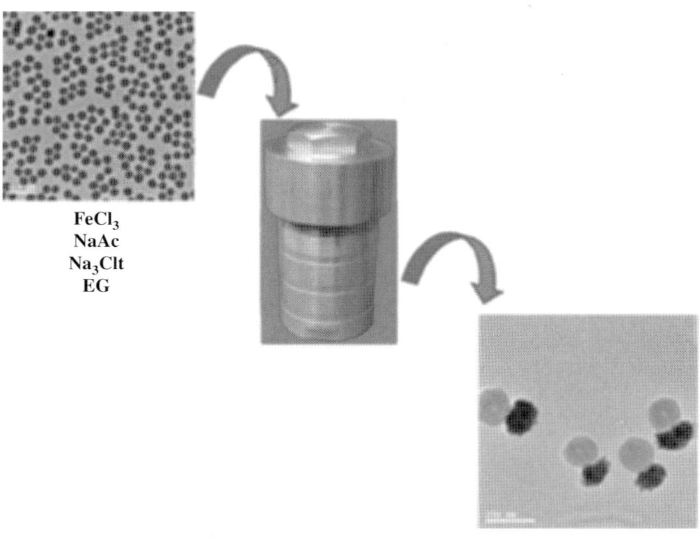

Figure 17. The process for preparation of anisotropic JPs.

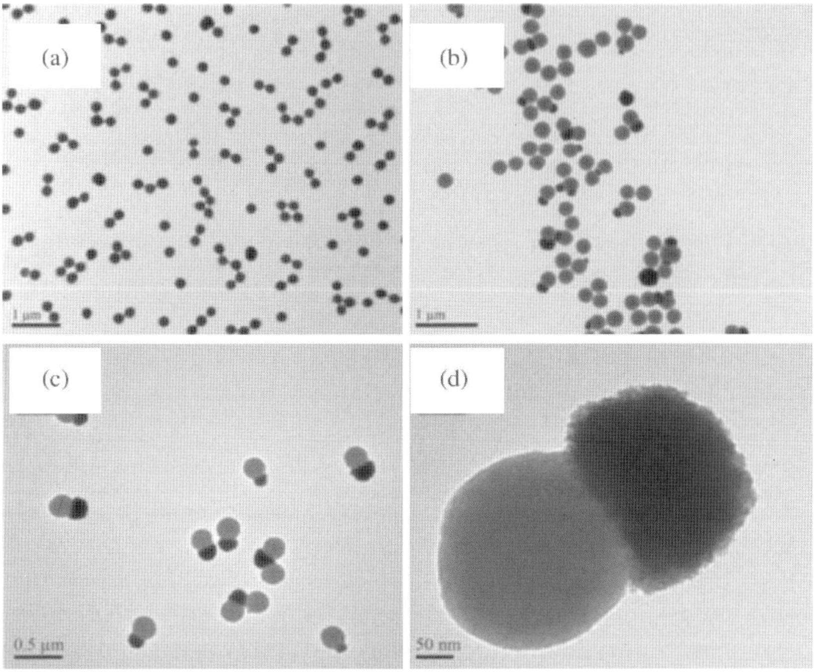

Figure 18. The process for preparation of anisotropic JPs. TEM image: (a) P(MBA-MAA) microspheres; anisotropic JPs of (b) before magnetic separation; (c) after magnetic separation; (d) with a higher magnification.

and crystallization of inorganic component to lower the interfacial energy change by reducing the interfacial area between the polymeric and inorganic magnetic domain.

The corresponding TEM image in Fig. 18(a) showed that the P(MBA-*co*-MAA) microspheres had a well-defined spherical shape and smooth surface with a diameter of 180 nm. The TEM image of P(MBA-*co*-MAA)/Fe_3O_4 anisotropic JPs is shown in Fig. 18(b). In this image, many particles are of anisotropic Janus structure together with some neat polymer spheres and few neat magnetite particles. However, after the purification of the sample by magnetic separation, nearly all of the neat polymer microspheres can be efficiently removed as TEM observation in Fig. 18(c). In Fig. 18(d), the resultant anisotropic JPs were stable as an integrated structure via a clearer observation with a higher magnification. Further,

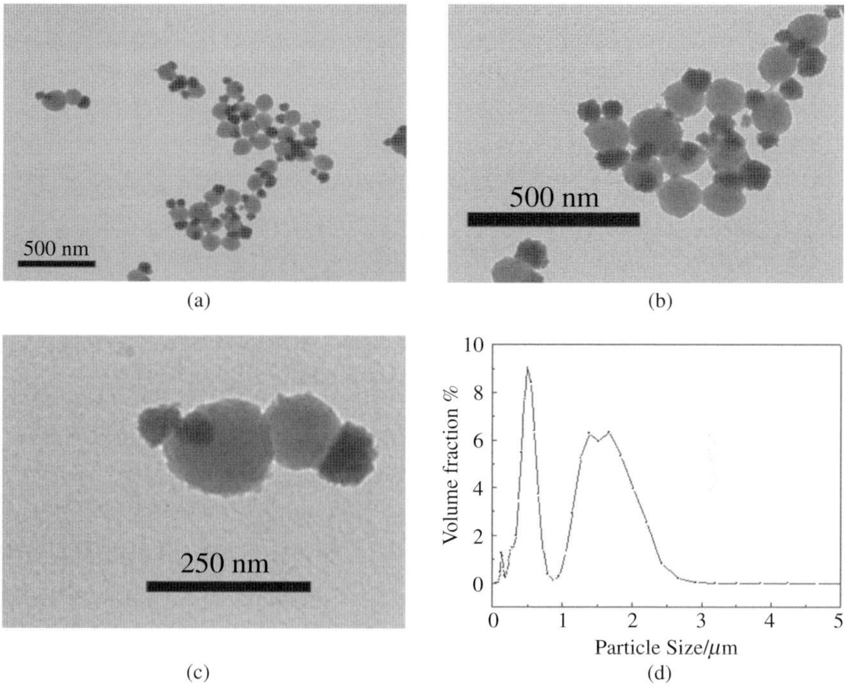

Figure 19. TEM images of P(MMA-AA-DVB)/Fe_3O_4 MJPs showing (a)–(c) with different magnification and (d) particle size distribution of P(MMA-AA-DVB)/Fe_3O_4 MJPs.

the sizes of the inorganic magnetite domains could be well controlled via altering the ratio of P(MBA-*co*-MAA) microspheres to ferric chloride inorganic precursors.

Similar magnetic anisotropic particles, magnetic P(MMA-AA-DVB) microspheres, had also been fabricated via this method [45]. TEM of P(MMA-AA-DVB) microspheres is shown in Fig. 19. The asymmetrical hybrid structure of P(MMA-AA-DVB)/Fe_3O_4 comprises two parts: inorganic iron oxide at one side and organic polymer nanoparticles at the other side. Janus MNPs are dually functionalized and possess stable asymmetric morphology. The synthesized P(MMA-AA-DVB)/Fe_3O_4 magnetic Janus nanoparticles possess anisotropic structure but there are few neat polymer spheres and magnetite particles. The anisotropic particles were retained in the resultant nanoparticles after purification by an external magnetic field with ultrasonic irradiation and vigorous

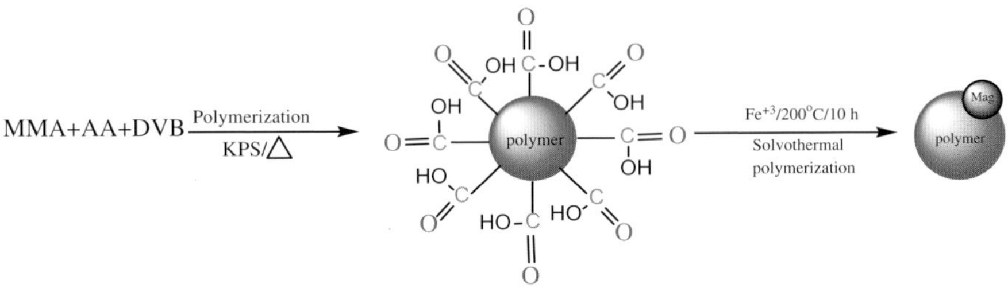

Scheme 3. Synthetic procedures of P(MMA-AA-DVB)/Fe_3O_4 Janus magnetic composite microspheres.

magnetic stirring. During the formation of magnetic anisotropic particles, the chelation of citrate with iron prevents the process of nucleation and the adsorption of citrate on the iron oxide nuclei, resulting in hydrolysis and the growth of the nuclei [46]. The size distribution of the sample is determined as shown in Fig. 19(d). It shows that the average particle size is about 200 nm.

The formation mechanism of P(MMA-AA-DVB)/Fe_3O_4 Janus magnetic composite microspheres is shown in Scheme 3. In the first stage of this process, Fe^{3+} was coordinated with the carboxyl groups on the P(MMA-AA-DVB) microspheres and then hydrolyzed as iron(III) oxide to form an integrated Janus structure. The formation of Fe_3O_4 from Fe_2O_3 results from heating the Fe_2O_3 inorganic component, which was partially reduced to Fe_3O_4 [47]. The thermodynamic factor also demonstrates the equilibrium morphology of anisotropic JPs formation, which have minimum interfacial free energy. The difference in thermodynamic properties between P(MMA-AA-DVB) and the Fe_3O_4 domains may serve as the driving force for the formation of the anisotropic morphology of the resultant P(MMA-AA-DVB)/Fe_3O_4 MJPs, similar to the synthesis of anisotropic magnetic Fe_3O_4/PS hybrid microparticles via emulsion polymerization [45,48]. During heating, the phase separation between P(MMA-AA-DVB) polymeric and the inorganic Fe_3O_4 domain lowers the interfacial energy change by reducing the interfacial area between the polymeric and inorganic magnetic domains, which leads to the formation of one-directional growth of Fe_3O_4 component and occurring on one side of the P(MMA-AA-DVB) microspheres to result in a Janus structure.

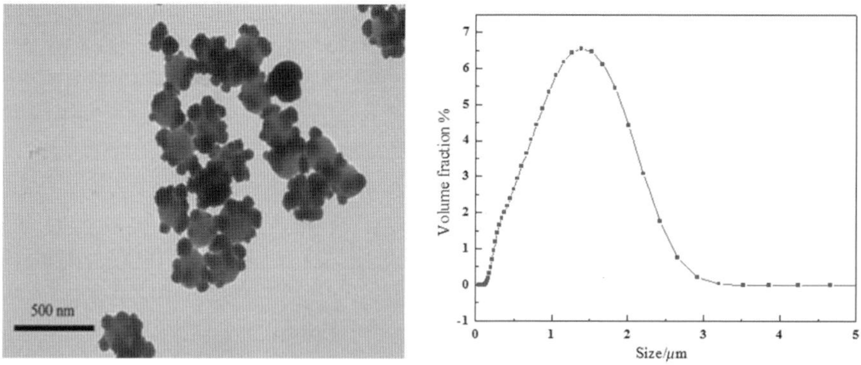

(a) (b)

Figure 20. TEM image of raspberry-like P(MMA-AA-DVB)/Fe_3O_4 magnetic composite microspheres with raspberry-like structure (a) and the particle size distribution (b).

By adjusting the experimental factors, raspberry-like P(MMA-AA-DVB)/Fe_3O_4 magnetic anisotropic particles [49] were obtained and the morphology of such kind of particles was shown in Fig. 20(a). These magnetic anisotropic particles are uniform and are nearly monodispersed spherical particles with an average diameter of about 140 nm, and there are some black Fe_3O_4 nanoparticles properly grown on the surface of the gray color P(MMA-AA-DVB) polymeric precursor microspheres. It is more like a raspberry in shape. The calculated average particle size of the magnetic anisotropic particles is about ~230 nm. The size distribution of the sample was determined as displayed in Fig. 20(b). From the curves, it is clear that the average size is about 200 nm. The formation mechanism of raspberry-like P(MMA-AA-DVB)/Fe_3O_4 magnetic anisotropic particles was suggested as in Scheme 4.

This is a very simple fabrication method for magnetic anisotropic particles. The resultant particles were in well uniform structure with small size in nanoscale. The size of the magnetic anisotropic particles was controlled to 200 nm by combining polymer particles by solvothermal process with partial reduction of iron oxide. The synthesized microspheres exhibit higher effective content of the magnetite in the final anisotropic magnetic microspheres with 25 μg^{-1}. Indeed, the synthesized

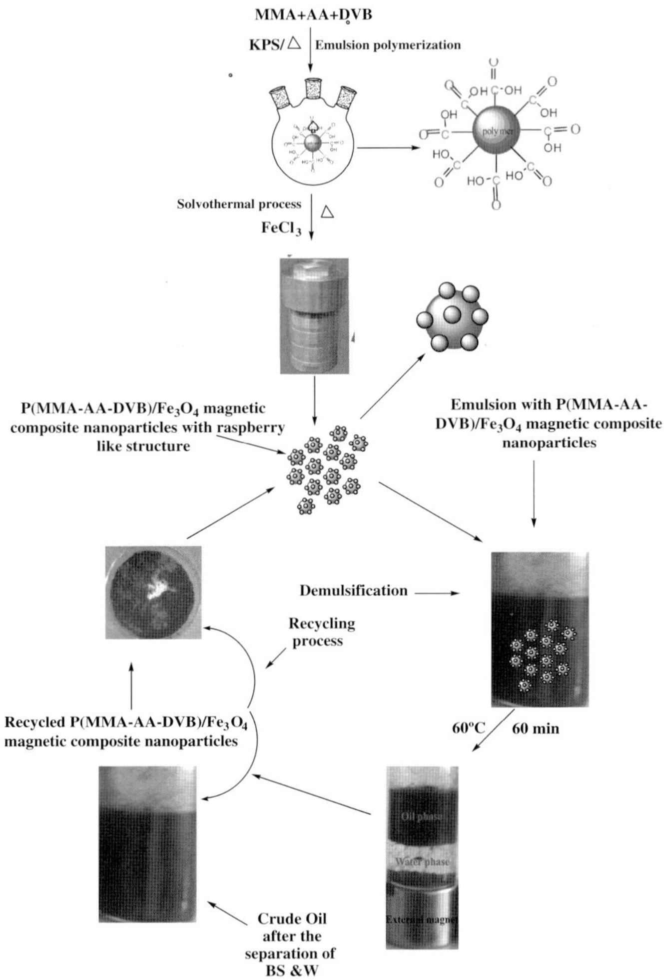

Scheme 4. Synthetic procedures of raspberry-like P(MMA-AA-DVB)/Fe_3O_4 magnetic composite microspheres.

magnetic anisotropic particles possess tunable surface properties and higher effective content of magnetite.

5.2.4. Pickering Emulsion

Pickering emulsion is also an effective way to fabricate magnetic anisotropic particles.

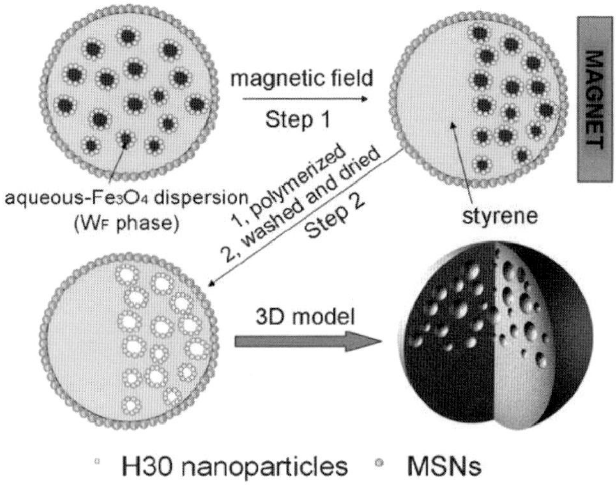

Scheme 5. Schematic representation of a W_F/O/W emulsion template.

Ning *et al.* [20] presented a facile approach to the preparation of Janus microspheres with a multihollow structure (possessing MNPs) located within one side of the solid microsphere. Scheme 5 shows the process that has been used for the fabrication of Janus microparticles. First, a stable aqueous Fe_3O_4 dispersion in oil-in-water (W_F/O/W) double Pickering emulsion was successfully generated by using hydrophobic amorphous fumed silica and recently synthesized hydrophilic mesoporous silica nanoparticles (MSNs) to stabilize the primary inner W_F/O droplets and outer O/W droplets, respectively. The aqueous Fe_3O_4 dispersion is intensively chosen as the inner water phase (W_F), which adds the magnetic property to the inner emulsions.

After that, the polymerization of the middle oil phase of the double emulsions was started while applying an external magnetic field on the desired position of the sample. The Janus microspheres with the multihollow structures concentrating on one side of the sphere were obtained after completing the polymerization and washed as well as redispersed in an aqueous phase.

Moreover, they have systematically investigated the influence of Fe_3O_4 particle concentration, hydrophobic silica particle content, and volume ratio of the inner water phase to middle oil phase (W_F/O) on the double emulsion formation and consequently on the structure of the resulting Janus

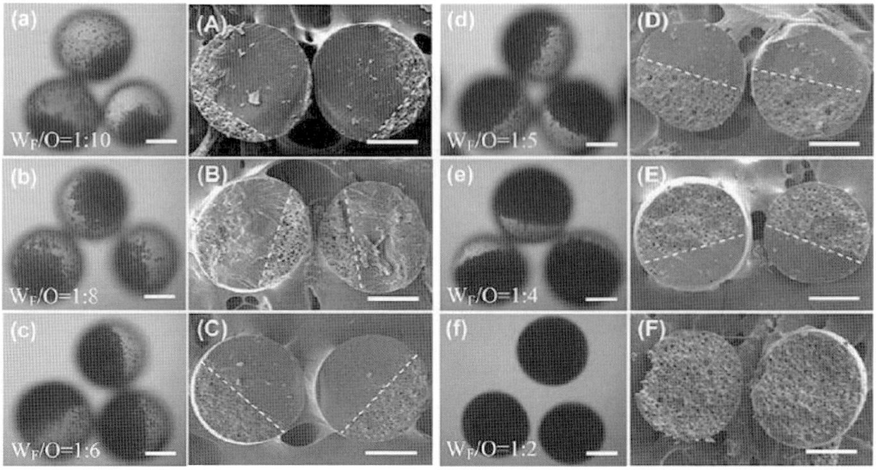

Figure 21. Optical micrographs (left column) and the corresponding SEM images (right column) of microspheres produced under various W_F/O volume ratios. All scale bars represent 150 μm.

microspheres [21]. Figure 21 shows the as-prepared anisotropic microspheres with different distributions of the multihollow inside the microspheres. It can be seen that the volume of multihollow part increases gradually as the W_F/O volume ratio increases from 1:10 to 1:2. Meanwhile, the multihollow part became a microsphere after polymerization of the middle oil phase of the double emulsions under different conditions. The results show that the distribution of the multihollow structures within the prepared microspheres can be accurately tailored by adjusting the ratio of W_F/O.

The as-prepared Janus microspheres can be magnetically orientated due to the special internal structure and magnetic asymmetric distribution. Figure 22 shows the process of how to orientate the as-prepared Janus microparticles. Before taking Fig. 22(f), all these Janus microspheres were orientated under an external magnetic field (Fig. 22(c)), making their multihollow part closely attached to the magnet since multihollow part possesses magnetite particles that preferentially stick with the magnet. Then, these orientated Janus spheres were transferred to a transparent adhesive tape (Fig. 22(d)), and finally the photographs were taken. From Fig. 21(f), it appears that the Janus microspheres are opaque since they are photographed with the solid part down while multihollow part on the up, as shown in Fig. 21(e).

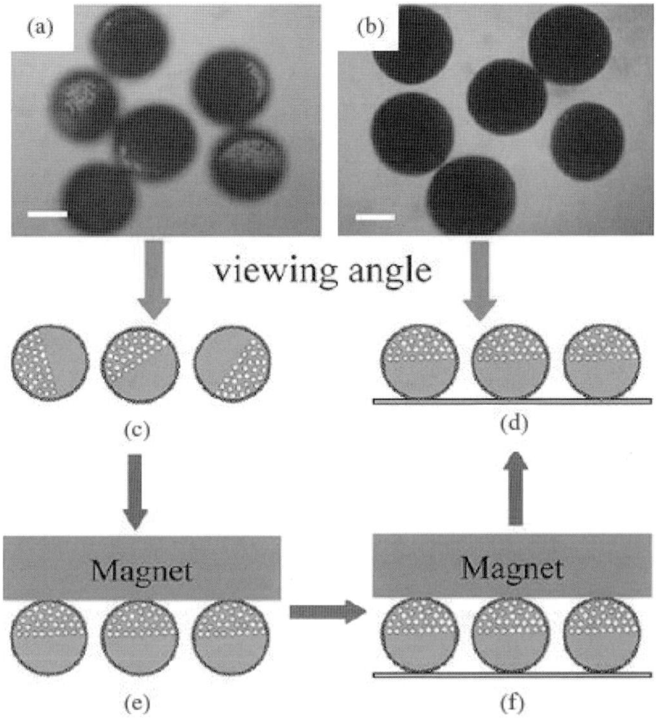

Figure 22. (a) Optical micrograph of Janus microspheres before orientation; (b) random pattern; (c) Janus microspheres being orientated under a magnet; (d) transporting onto a transparent adhesive tape; (e) orientated Janus microspheres being attached on a transparent adhesive tape; (f) optical micrograph of the orientated Janus microspheres. The scale bar represents 150 μm.

Recently, the mechanism of Pickering emulsion to fabricate magnetic anisotropic particles was understood. It is also becoming an effective method. Of particular importance is its magnetic orientation property that makes it a good candidate for synthesizing Janus membranes.

5.2.5. Seed Emulsion Polymerization

Rahman [50] reported a seed emulsion polymerization approach for the preparation of Janus magnetic microparticles containing inorganic iron oxide nanoparticles and organic polymer starting from oil-in-water magnetic emulsion (Fig. 23). The approach included two-step controlled processes: (a) the swelling of preformed magnetic oil in water ferrofluid

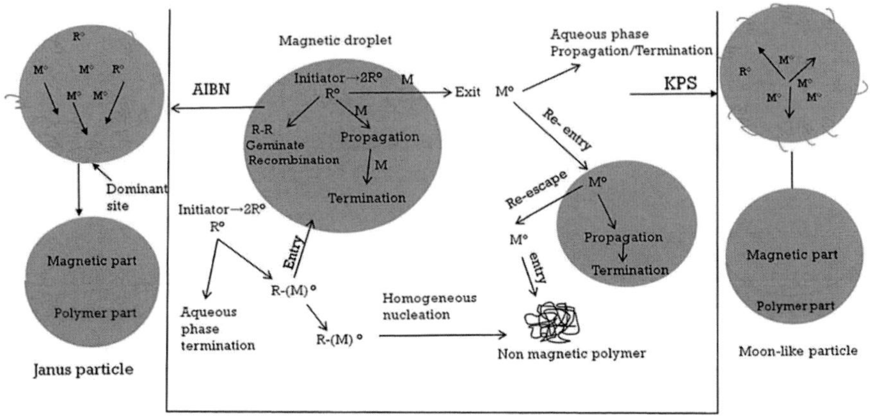

Figure 23. The polymerization mechanism proposed for anisotropic magnetic polymer formation (Rc is a primary free radical, *M* is a styrene unit, Mc is a propagating radical).

droplets with styrene and (b) a seed emulsion polymerization of styrene in the presence of monomer-swollen ferrofluid as seed, were accomplished. They produced a series of Janus magnetic microparticles by emulsion polymerization of styrene in the presence of (O/W) magnetic emulsion as a seed and found out that the polymer shell varies with the concentration of styrene, initiator, and initiator types. They found out that the incompatibility between the PS phase and the phase of the ferrofluid droplets led to particles exhibiting a phase-separated morphology of a Janus type.

Wang group [51] has developed a surfactant-free seeded emulsion polymerization with Fe_3O_4 colloidal nanoclusters as the seed to prepare asymmetric magnetic composite microspheres. Fe_3O_4 colloidal nanoclusters mediated by MPS were selected as the seed. The morphology of the composite microspheres can be regulated from raspberry-shaped, flower-like shapes to eccentric structures by adjusting the feeding weight ratio of the seed to the monomer and the amount of cross-linker DVB (Fig. 24). Moreover, these composite microspheres can be functionalized with carboxylic, hydroxyl, and epoxy groups by copolymerization with AA, GMA, and HEMA, respectively.

In fact, to distinguish uniform magnetic anisotropic particles prepared by seed emulsion polymerization, this way has attracted more interest, especially to fabricate magnetic anisotropic particles with patchy and multicompartment structure.

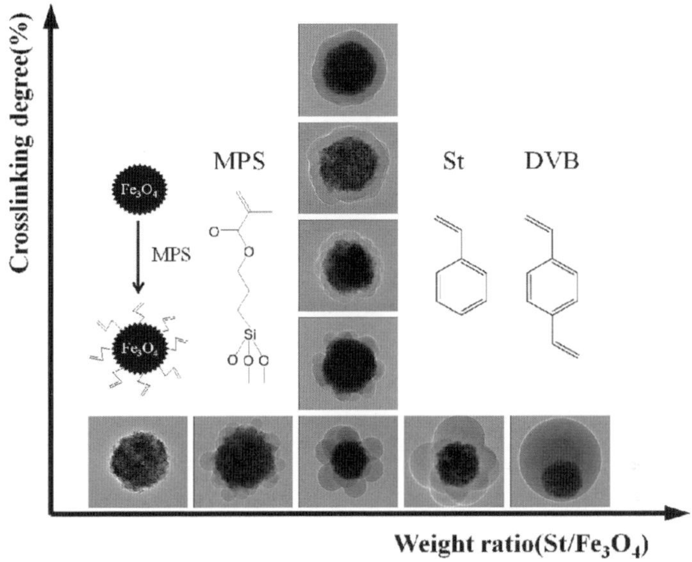

Figure 24. Morphology evolution of Fe_3O_4@PS composite microspheres on the feeding ratio of St to Fe_3O_4 and the concentration of cross-linker DVB.

5.2.6. Preparation of Versatile Janus Particles via Microfluidics

The preparation of uniform particles with multifunction via microfluidics has recently become a popular approach as it provides an excellent route to achieve precise and ultimate control over size distribution [52]. This technique is an advanced version of suspension polymerization, in which droplets are created at an orifice by injecting disperse phase into continuous phase [53]. Uniform droplets with CV below 2% can be easily produced by microfluidic setups. Meanwhile, microfluidics is insensitive to monomer components, so various monodisperse particles can be expediently prepared via changing ingredients [54,55], which makes it become an ideal candidate to prepare anisotropic magnetic particles with unconventional shapes and morphologies.

Doyle introduced Fe_3O_4 nanoparticles and photocurable monomers to microfluidics, and prepared multifunctional Janus hydrogel particles with anisotropic superparamagnetic properties and chemical composition for assembly of hydrogel superstructures [56]. The obtained Janus magnetic particles assembled into chain-like or mesh-like superstructure forms when an external uniform magnetic field was applied, which could be potentially used as building blocks to construct targeted superstructures for tissue engineering. This method demonstrated the feasibility of generate multifunctional JPs with two advantages: (i) direct capsule and precise special distribution of biological substance and (ii) selective surface functionalization.

Chen [19] prepared monodisperse single-phase and Janus supraballs via microfluidic device, respectively (Fig. 25). As shown in Fig. 25(b), a microfluidic device with a two-needle microchannel was able to make two different oil solutions to flow simultaneously, and this device was adopted to prepare bifunctional Janus supraballs to realize fluorescent and magnetic-responsive performances within one anisotropic structure.

In many cases, applications of these QD-polymer nanocomposites were limited because of nanocrystal (NC) aggregation and quenching of the QDs. Surprisingly, the bifunctional particles exhibited long-term stability of the photoluminescence (PL), since the polymer of solution offered better passivation on QDs to heal surface defects. Meanwhile, the

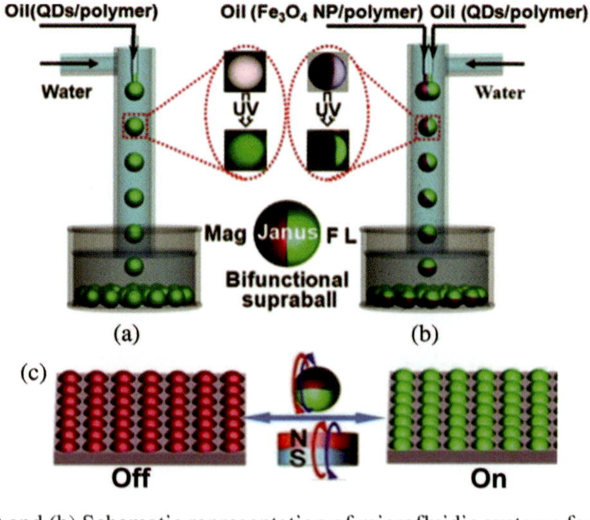

Figure 25. (a) and (b) Schematic representation of microfluidic systems for the synthesis of monodisperse single-phase and Janus supraballs, respectively. (c) Schematic representation of a fluorescent switch of JPs controlled by varying the direction of an external magnetic field.

micromanipulation of the Janus supraballs by a small magnet was achieved due to their superparamagnetic property. Thus, this kind of JPs could be designed for magneto-driven fluorescent switch and bead display (Fig. 25(c)). That is to say, the orientation of JPs, representing the state of ON and OFF, can be altered by varying the direction of the external magnetic field. This work may provide a new method to fabricate novel flexible bead displays with Janus supraballs.

5.2.7. Spin-Coating and Thermal Evaporation Method

Some physical fabrication methods were also developed to fabricate magnetic anisotropic particles during the last several years. Two examples are given here.

Guan developed a simple electrospray procedure via a subsequent non-equilibrium calcination process to fabricate asymmetric magnetic γ-Fe_2O_3/TiO_2 Janus hollow bowls [57] (Fig. 26). Initially, two kinds of precursor solution (γ-Fe_2O_3 and TiO_2) were constructed, and both

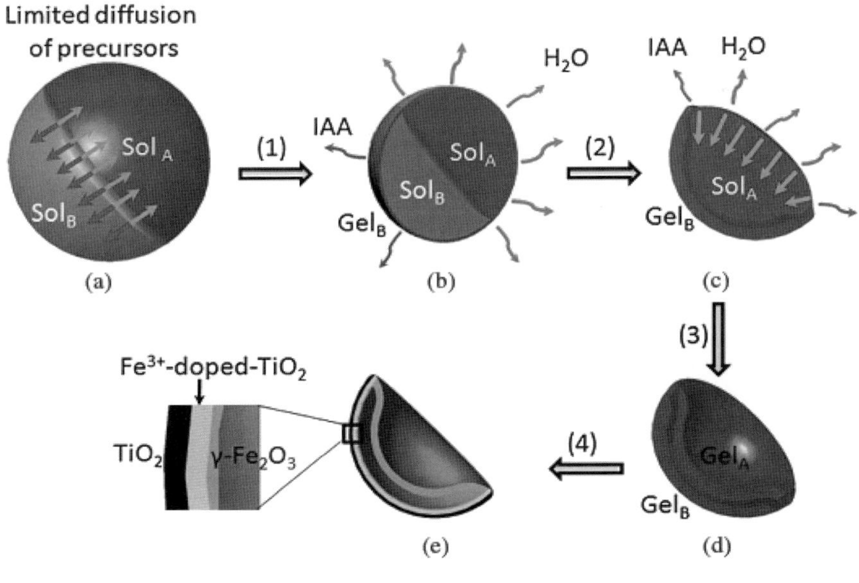

Figure 26. The illustration for the formation process of the γ-Fe_2O_3/TiO_2 JHBs from a Sol A/Sol B Janus droplet jetted by a side-by-side electrospray method.

solutions were transferred into the separate syringe connected with metallic nozzles. Then, a dual component fluid jet was generated when a high voltage of 15 K was applied, and the electrospraying gel particles were collected and then annealed. Since an asymmetric shrinkage occurred during the gelation process, magnetic γ-Fe_2O_3/TiO_2 Janus hollow bowls were obtained. The as-prepared Janus hollow bowls exhibited efficient visible-light photocatalytic activity for RhB degradation and magnetic separation behavior, indicating that the Janus hollow bowls could serve as an ideal photocatalyst for environmental purification owing to their efficient photocatalytic performance and convenience of separation and recycling.

Kopelman *et al.* [58] fabricated magnetically uniform JPs with tunable magnetic properties using a spin-coating and thermal evaporation method. The schematic illustration of the fabrication and the resulting samples are exhibited in Fig. 27.

The ferromagnetic particles were fabricated by evaporating 30 nm Ni on 1 μm particles, 60 nm Ni on 2 μm particles, 300 nm Ni on 1 μm particles, and 300 nm Ni on 100 μm particles. The superparamagnetic particles were fabricated by sequentially evaporating 5 nm Ni and 20 nm Au onto both the 2 and 10 μm diameter particles. The Au coating was aided in the visualization of the particles. After fabrication, the particles were magnetized in a 60 mT magnetic field. Up to a calculated value of 10^9, MJPs can be fabricated on a single 4″ wafer, assuming 50% area

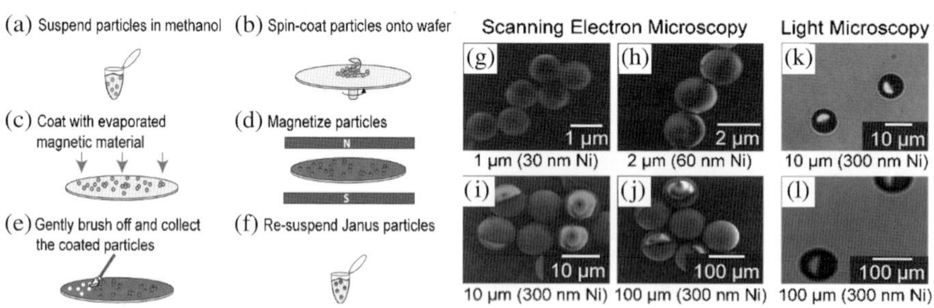

Figure 27. Fabrication schematic for MJPs. (a)–(d) SEM images and (e) and (f) light microscopy images of MJPs. The particle size and corresponding Ni coating thickness are stated below each figure.

coverage. The ferromagnetic particles were imaged using SEM and light microscopy, as shown in Fig. 27.

5.3. Applications

As mentioned before, magnetic anisotropic particles have many potential applications due to their unique properties. However, most of these applications still are in the principle phase only. In-depth researches are required in order to use these novel materials in practical applications.

5.3.1. Assembly

Velev [59] reported on the reconfigurable assembly of structures composed of JPs–polystyrene spheres with a thin magnetic metal shell evaporated on one hemisphere. The anisotropic particles assemble in low-symmetry structures, which were well defined and stable in the absence of applied fields or supplied energy. After applying magnetic fields, the particles assembled quickly into "staggered chain" or "double chain" structure. The metal parts of the particles thus formed a metal lane in the middle of the chain, in which two adjacent particles were trapped facing on the same side. The chains not only oriented on the direction of the applied field, but were also strong enough to withstand drag forces. Additionally, this self-assembly of novel magnetic structures could rapidly disassemble on-demand by removing demagnetization. This made it become possible to achieve reversible structures, and reuse and recycle particle building blocks.

5.3.2. Biomedical Application

Superfine [60] reported a template-based technique for fabricating magnetic Janus nanorods and showed that the magnetic rotation of these nanorods near a floor results in predictable translational motion. Using this method, they demonstrated considerable rod steerability. They used Janus rods labeled on one side (the exposed Au surface) with 1,6-hexanedithiol to capture and translate a 1 μm metal-coated PS bead. When contact was made between the thiol-coated face of the rod and the bead, a covalent bond was formed between the AuPd surface layer of the PS bead and the exposed thiol groups on the Au segment of the Janus rod, whereas manipulation of chemically unmodified Janus rods in contact with AuPd-coated PS beads resulted in no observable binding interaction. Later, they coupled nanorods to HBE-16 cells via non-specific binding to membrane proteins. The process was motivated by the field-aligning behavior of the Janus nanorods, leading to controllable single-cell translational manipulation near the surface of the glass slide substrate.

Dong [36] synthesized novel magnetic-mesoporous JPs by a simple sol–gel method. The aspect ratio of JPs could be precisely controlled by varying the feed weight ratio. Under an external magnetic field, the particles were aligned into 1D necklace-like structures, in which dipole–dipole interactions between adjacent magnetic particles couple them together and force the formation of anisotropic structures. It can be concluded that the novel JPs exhibit high magnetic sensitivity and manipulability under an external magnetic field. In addition, when the particles were covalently coupled with an amine-reactive dye fluorescein isothiocyanate (FITC), they would be able to be used in bioseparation and related applications, such as *in situ* detection and monitoring of the movement of the doped cells under an external magnetic field.

Sacanna [50] introduced a new class of organosilica polymer sphere with a single hematite micromagnet embedded. The spheres magnetically "click", forming directional bonds by means of their magnetic patches. Each cube acts as a nucleation site for the formation of an oil droplet to produce droplets with a single magnetic inclusion. During the process of transferring from water to oil, the cubes are trapped at the interface by surface tension. The particles self-assemble into clusters driven by magnetostatic interactions between their patches without an applied field. The patchy particles self-assemble into dimers or trimers, which is determined by the steric hindrance parameter. Besides, the magnetostatic interactions allow the particles to reversibly unbind the assemblies or to reconfigure them into different structural arrangements, such as linear chains and dense layered sediments. While imposing an external magnetic field enables the clusters to unbind or change their geometry, the creation of materials with a reconfigurable structural arrangement can be achieved. This may provide a new road for drug delivery application where a particle with its cargo of active material can be delivered to a specific location.

Gao [61] produced a magneto-optical combination by incorporating fluorophores and MNPs in separate compartments. These new multifunctional nanocomposites are not only compact in size but are also sensitive to spinning magnetic fields. Besides, the nanocomposites were applied in an innovative application field, magnetolytic therapy, which is achieved through magnetic field modulated cell membrane damage. Fluorescence imaging confirms that the nanocomposites can be quickly attached to the cell surface under a magnetic field. During the therapy process, cells were placed in a spinning magnetic field (50 rpm), which creates a mechanical force on the cell membrane, leading to the compromised integrity of the cell membrane and promoting apoptosis, as shown in Fig. 28. There's no

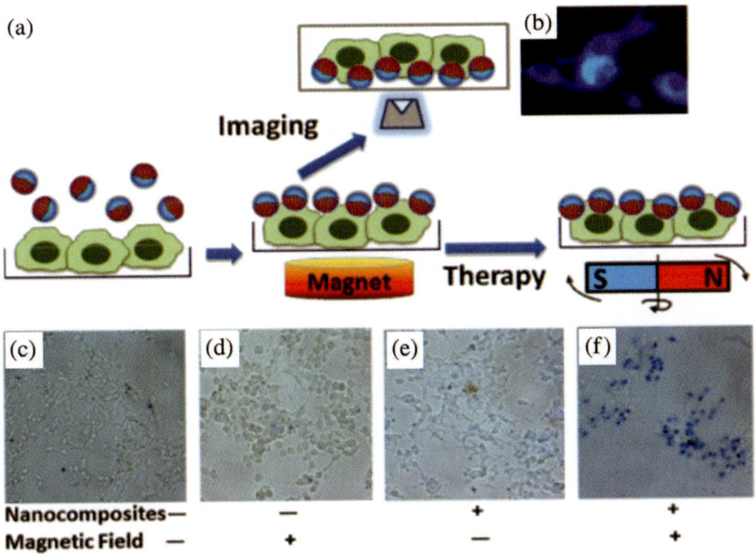

Figure 28. Simultaneous imaging and treatment of cancer cells using biphasic nanocomposites. (a) Schematic illustration of the experimental conditions and magnetolytic therapy. Nanocomposites are first brought down to cell surface by magnetic attraction. Cells can then be examined by fluorescence microscopy or treated with oscillating magnetic fields, which result in mechanical forces to break down cell membranes. (b) Fluorescence imaging of cells coated with nanocomposites. (c)–(f) Magnetolytic therapy on tumor cells. In comparison with the controls, missing either magnetic fields or nanocomposites, the viability of cells exposed to both is significantly reduced (dead cells appear blue due to Trypan blue staining). Note the low cell density in the treatment group (f) resulting from detachment of dead cells from cell culture dishes. The number of live cells in (f) compared with that in (c) is reduced by 77%, whereas those in (d) and (e) show greater than 99.5% cell survival.

Table 3. Saturated adsorption and encapsulation efficiency of IA magnetic composite microcapsules.

Sample	IAM-1	IAM-2	IAM-3	IAM-4
Saturated adsorption (mg/mg)	5.3/50	18.5/50	29.2/50	33.7/50
Encapsulation efficiency (%)	10.6	37.0	58.4	67.4

doubt that this enabling technology will open up exciting opportunities in nanomedicine, electronics, spintronics, and catalysis.

Drug loading properties of IA-PS/Fe_3O_4 microcapsule synthesized in our group had been evaluated [40]. Phenolphthalein, commonly used as a laxative, was chosen as the model drug. Saturated adsorption and encapsulation efficiency of IAM-1, IAM-2, IAM-3, and IAM-4 were listed in Table 3. It was clear to see that saturated adsorption and encapsulation efficiency increased gradually from IAM-1 to IAM-4. The reason was that cavity volume increased from IAM-1 to IAM-4. The maximum encapsulation efficiency belonged to IAM-4, and the value was 67.4%. The drug-loading capacity of prepared novel IA microcapsule was higher than parts of stimulus-responsive magnetic composite microspheres designed for drug release.

Figure 29 features the relationship curves between loading temperature and saturated adsorption time of microcapsules with different morphologies. The results showed that saturated adsorption time decreased with increasing loading temperature for the same IA-PS/Fe_3O_4 microcapsule. It could be explained by the increase of drug diffusion speed under higher temperature. Under the same temperature, the saturated adsorption time of IAM-1, IAM-2, IAM-3, and IAM-4 was shortened in sequence. This could be attributed to the reason that one side of the capsule-wall became thinner from IAM-1 to IAM-4 (Fig. 13). The thinner the capsule wall, the easier for the drug to pass through. The above results suggested that the main channel for drug loading was from the thinner side of the capsule wall. Moreover, as displayed in Fig. 29, it was found that the effect of loading temperature on saturated adsorption time was significant when the capsule wall was thick. The main reason was that the diffusion speed was also limited to the diffusion length.

The release behaviors of IA-PS/Fe_3O_4 microcapsule under different temperatures were investigated and are shown in Fig. 30. It was found that

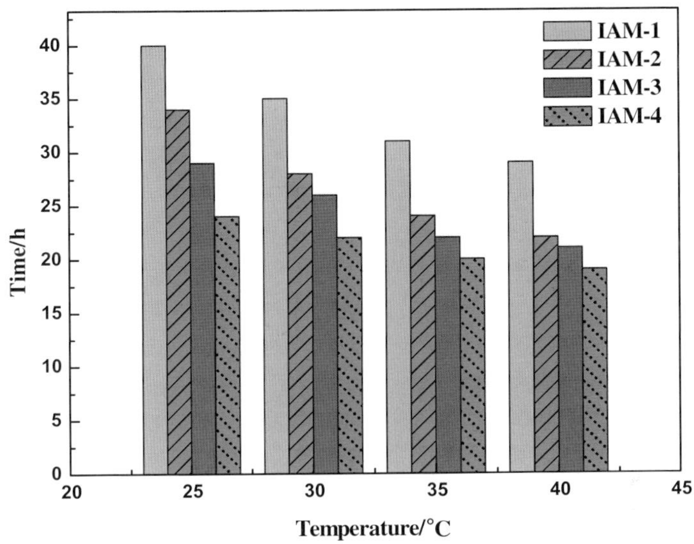

Figure 29. Saturated adsorption time of IA magnetic composite microcapsules under different temperatures.

cumulative release amount of IAM-1, IAM-2, IAM-3, and IAM-4 was 56.6%, 91.4%, 95.2%, and 98.9% at 96 h, respectively. The total releasing time of the above four microcapsules in turn were 96 h, 84 h, 64 h, and 36 h. In the first 2 h, the cumulative release amount of the microcapsules was nearly the same, which were not affected by the difference in the microstructure of the microcapsules. We hypothesized that this part of drug was loaded into the surface layer of the capsule wall. The drug loading of different microcapsules was almost the same, which was attributed to the same particle size. The release rate and cumulative release amount of IAM-1 was obviously lower than others as time progressed. The reason was that the thick capsule wall was difficult to pass through for loaded drugs. Total releasing time was decreased with the decreasing of the thickness of the thinner-side capsule wall of the microcapsule. On the contrary, the cumulative release amount was increased at the same time. It suggested that the main channel for drug release was from the thinner side of capsule wall of IA-PS/Fe_3O_4 microcapsule, as shown in Scheme 6. Therefore, this kind of novel microcapsules showed potential application value in unilateral or contact point controlled and concentrated release.

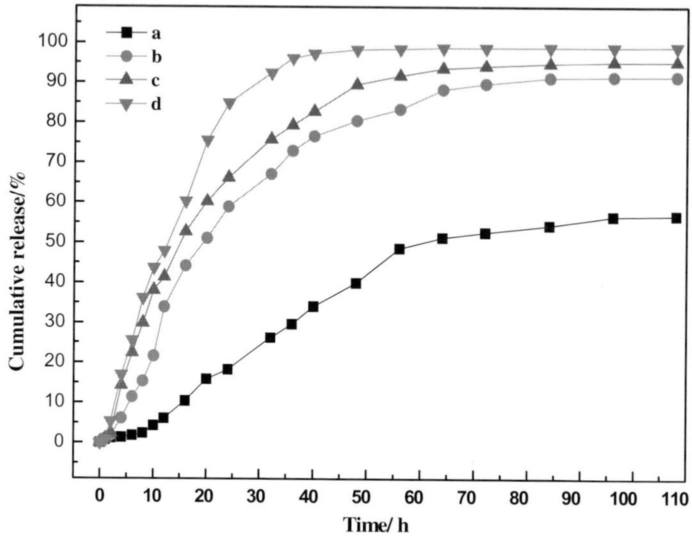

Figure 30. The curves of cumulative drug release amount of IA magnetic composite microcapsules: IAM-1 (a); IAM-2 (b); IAM-3 (c); IAM-4 (d).

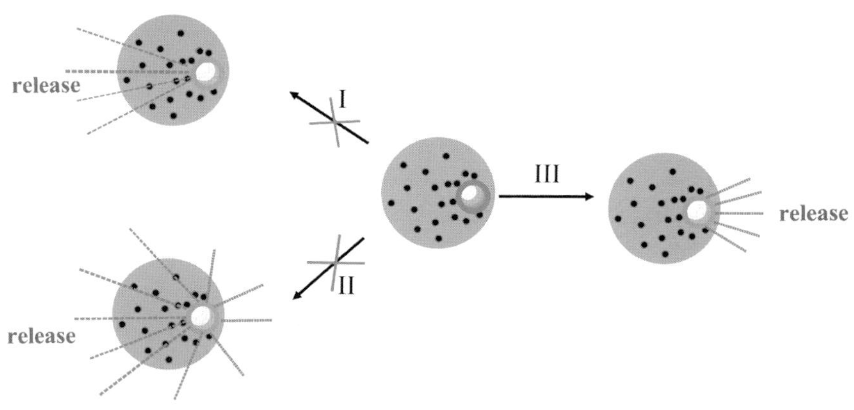

Scheme 6. Schematic of drug release route of IA magnetic composite microcapsules.

5.3.3. Demulsification

Due to interfacial activity and magnetic responsibility of magnetic anisotropic particles, several kinds of magnetic anisotropic particles with Janus, raspberry morphologies were used as affective demulsifiers by us. The effects of settling time, temperature, mixing conditions, and the concentration

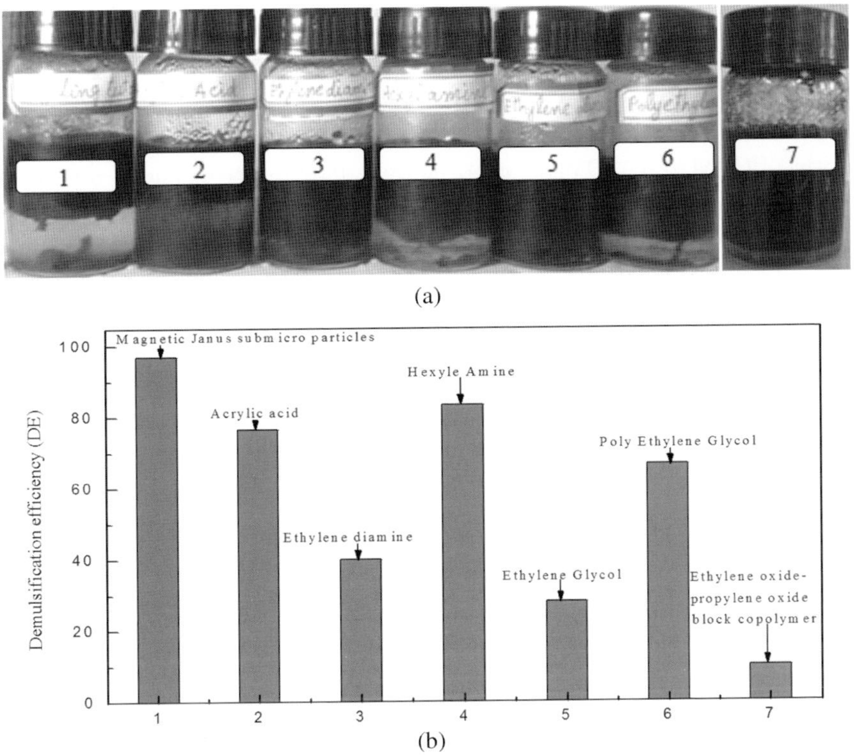

Figure 31. The comparison of P(MMA-AA-DVB)/Fe_3O_4 with chemical demulsifiers. Images (a) after settling on a magnet for 2 h at 60°C. (b) is the graphical representation.

of the magnetic anisotropic particles on the separation were experimentally investigated. After optimizing the coalescence temperature and time, the demulsification efficiency of Janus magnetic microspheres at 600 ppm addition was compared with different commercial demulsifiers in the heavy crude oil emulsion having 40% initial water content. The results are shown in Figs. 31(a) and 31(b). Commercial demulsifiers are constituted by an emulsion breaker, a coalescence promoter, and clarifier components. An amount of 130 ppm of acrylic acid, ethylene glycol, polyethylene glycol, ethylene diamine, hexyl amine, and ethylene oxide-propylene oxide block copolymer was used as commercial chemical demulsifiers. The water removal efficiency for acrylic acid, ethylene glycol, polyethylene glycol, ethylene diamine, hexyl amine, and ethylene oxide–propylene oxide block copolymer is found to be 76%, 28%, 68%, 40%, 83.3%, and 10%, respectively at 60°C and 2 h time interval. The demulsification efficiency of hexyl amine (83.33%) is much better than other commercial demulsifiers, but on the other hand, the demulsification efficiency of our synthesized magnetic composite Janus microspheres is 98%, which is around 17.5% higher than that of hexyl amine. P(MMA-AA-DVB)/Fe_3O_4 shows better performance in water removal than other demulsifiers, possibly because of its high interfacial properties and enhanced coalescence and settling of magnetically tagged water droplets by an external magnetic field. These results indicate that P(MMA-AA-DVB)/Fe_3O_4 Janus magnetic composite microspheres can be considered the most effective demulsifier for heavy crude oil and water emulsion.

Unlike other commercial and chemical demulsifiers, P(MMA-AA-DVB)/Fe_3O_4 Janus magnetic composite microspheres have the ability to be recycled and reused after demulsification. Recycling reduces its cost in practical applications in the oil and gas industry.

A series of demulsification tests were conducted at 60°C keeping the demulsification time at 2 h to study the recycling capability of P(MMA-AA-DVB)/Fe_3O_4 as a demulsifier. After each cycle, the top oil phase was separated by decanting while holding the P(MMA-AA-DVB)/Fe_3O_4 Janus magnetic composite microspheres tagged at the bottom of the test bottle with a hand magnet. The collected (spent) P(MMA-AA-DVB)/Fe_3O_4 Janus magnetic composite microspheres were washed three times with ethanol absolute and deionized water to remove residues and used for the subsequent demulsification test. The demulsification efficiency in each cycle is shown in Fig. 32. P(MMA-AA-DVB)/Fe_3O_4 Janus magnetic composite microspheres showed excellent and constant demulsification efficiency in the first five cycles tested. All the five tests gave nearly constant results of ~95% water removal. It is clear from these results that P(MMA-AA-DVB)/Fe_3O_4 Janus magnetic composite microspheres exhibit excellent stability and high demulsification efficiency. Our study demonstrates that P(MMA-AA-DVB)/Fe_3O_4 Janus magnetic composite microspheres is a very good magnetic demulsifier for practical applications in the petroleum industry.

The mechanism of demulsification of crude oil and water emulsions by P(MMA-AA-DVB)/Fe_3O_4 Janus magnetic composite microspheres is explained in Scheme 7.

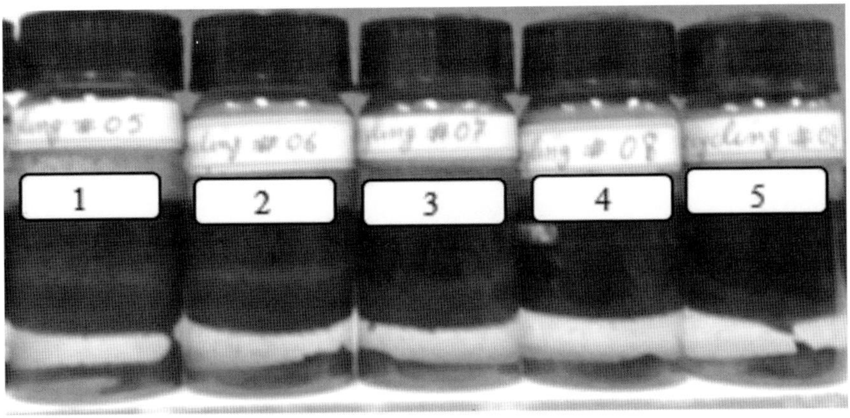

(a)

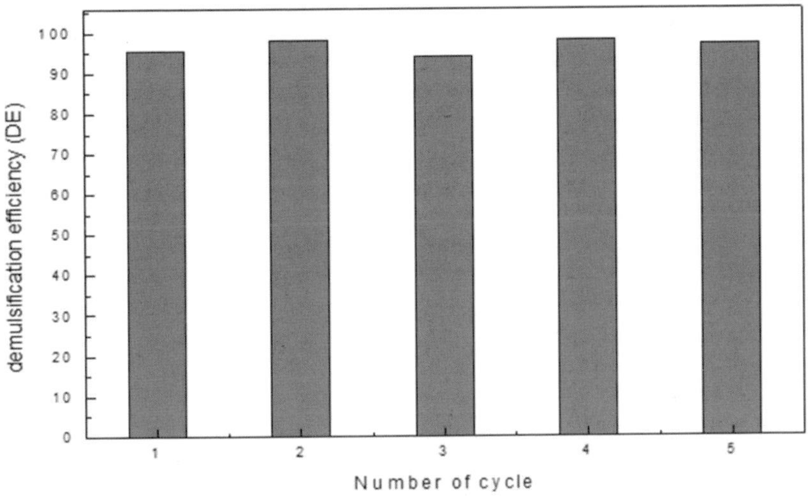

(b)

Figure 32. Images (a)1–(a)5 showing five time recycling each of P(MMA-AA-DVB)/Fe_3O_4 Janus magnetic composite microspheres for demulsification of heavy crude oil and water emulsion at 60°C and 2 h and (b) is the graphical representation.

(i) P(MMA-AA-DVB)/Fe_3O_4 Janus magnetic composite microspheres enter the oil–water interface and break up the interfacial film formed by surface-active components in the crude oil–water interface due to more surface active nature of P(MMA-AA-DVB)/Fe_3O_4.

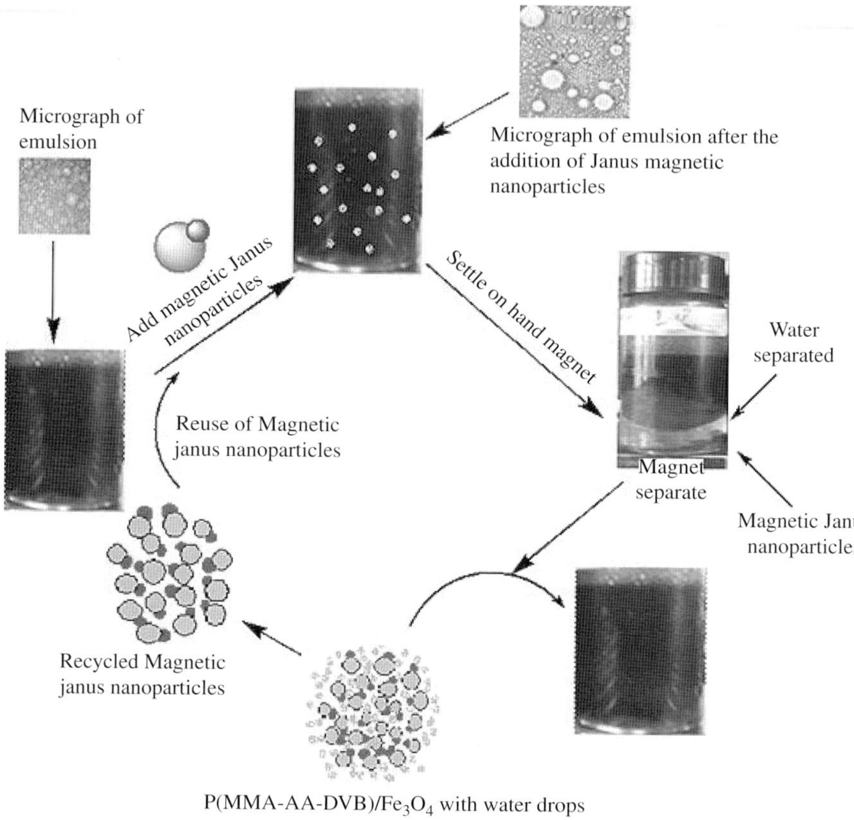

Scheme 7. Procedures explaining the process of demulsification and recycling by using of P(MMA-AA-DVB)/Fe_3O_4 Janus magnetic composite microspheres.

(ii) Janus magnetic composite microspheres flocculates water droplets through bridging flocculation, leading to their coalescence. P(MMA-AA-DVB)/Fe_3O_4 Janus magnetic composite microspheres may lead to increase in the rate of partition coefficient and increase in the adsorption of demulsifier molecules at the oil–water interface. The P(MMA-AA-DVB)/Fe_3O_4 Janus magnetic composite microspheres tagged water droplets are then easily removed by an external magnetic field. Figure 33 explains different images of heavy crude oil and water emulsion with and without Janus magnetic microspheres.

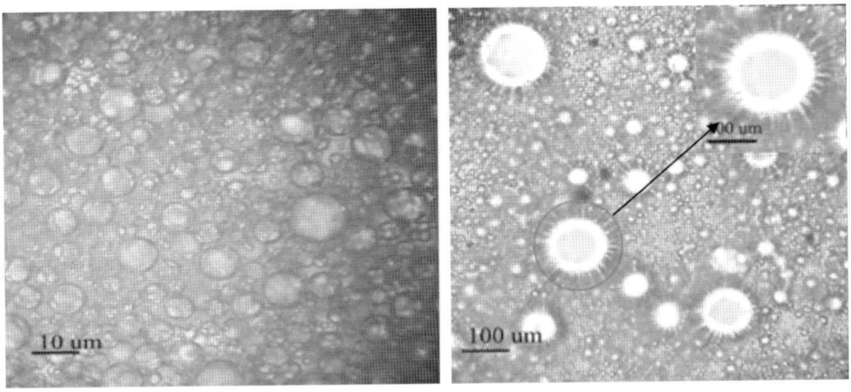

(a) (b)

Figure 33. Image (a) showing micrograph of heavy crude oil and water emulsion and (b) showing micrograph of emulsion with P(MMA-AA-DVB)/Fe_3O_4 Janus magnetic composite microspheres.

5.3.4. Other Applications

Sheridon [62] reported a new kind of JP, known as magnetoelectric coupling, by combining permanent electric and magnetic moments in particles. It consisted of electric dipolar molecules, and then application of an electric field would exert a torque and align the molecules with the field. The molecules also possessed a magnetic moment parallel to the axis of the electric dipole moment, and so orienting one moment should fix the other in space, provided the electric and magnetic dipoles were "tied together". Once the electric-dipolar ferromagnetic particle was magnetized, each particle carried its own magnetic field, which rendered it anti-symmetric under time reversal symmetry and thus allowed for magnetoelectric effects.

Jenness [63] developed a new method for the controlled synthesis of dot JPs that have sufficiently small surface coverage of cobalt film to allow compatibility with an optical trap. Since optical trapping is more easily accomplished with dot JPs than with half-coated JPs, the dot JPs behave more like normal dielectric particles in an optical trap, while also responding to magnetic forces and torques produced by an external magnetic field.

In addition, MJPs have already shown a lot of potential applications in a wide variety of fields, such as biomedicine and biophysical research, electronic paper [12,13], and heterogeneous catalysis [14]. Undoubtedly, it is a kind of functional material with a great prospect.

5.4. Conclusions

This chapter reviewed the recent advances in the preparation and application of magnetic anisotropic particles. In particular, magnetic anisotropic particles are now available in a wide sizes and diverse morphologies. Some of the suspected properties of magnetic anisotropic particles have been characterized. Their abilities in drug delivery and as demulsifiers were discussed. The results are promising. Magnetic anisotropic particles also show good deemulsifying properties in emulsion of oil and water. This complementary finding has enriched application fields of anisotropic particles that normally were used as a stabilizer in mixture of oil and water. Further efforts shall be directed at developing robust and reliable fabrication ways that can be extrapolated to large-scale productions. Solvent evaporation-induced phase separation, miniemulsion polymerization, and seed polymerization will very probably be more appropriate from their yield and cost points of view.

References

[1] Gennes, P. *Rev. Mod. Phys.*, **64**, 645 (**1992**).
[2] Liang, F. X.; Liu, J.; Zhang, C.; Qu. X.; Li, J.; Yang, Z. *Chem. Commun.*, **47**, 1231 (**2011**).
[3] Walther, A.; Hoffmann, M.; Müller, A. *Angew. Chem.*, **120**, 723 (**2008**).
[4] Mou, F.; Xu, L.; Ma, H.; Guan, J.; Chen, D.; Wang, S. *Nanoscale*, **4**, 4650 (**2012**).
[5] Aveyard, R. *Soft Matter.*, **8**, 5233 (**2012**).
[6] Meilikhov, M.; Furukawa, S.; Hirai, K.; Fischer, R.; Kitagawa, S. *Angew. Chem. Int. Ed.*, **52**, 341 (**2013**).
[7] Howse, J.; Jones, R.; Ryan, A.; Gough, T.; Vafabakhsh, R.; Golestanian, R. *Phys. Rev. Lett.*, **99**, 048102 (**2007**).
[8] Alexeev, A.; Uspal, W.; Balazs, A. *ACS Nano*, **2**, 1117 (**2008**).
[9] McConnell, M.; Kraeutler, M.; Yang, S.; Composto, R. *Nano Lett.*, **10**, 603 (**2010**).

[10] Yuet, K.; Hwang, D.; Haghgooie, R.; Doyle, P. *Langmuir* **26**, 4281 (**2010**).
[11] Glotzer, S.; Solomon, M. *Nat. Mater.*, **6**, 557 (**2007**).
[12] Arita, T.; Kanahara, M.; Motoyoshi, K.; Koike, K.; Higuchi, T.; Yabu, H. *Mater. Chem. C.*, **1**, 207 (**2013**).
[13] Fu, X.; Song, L.; Liu, J.; Li, X.; Zhang, X.; Jia, Y. *Macromol. Chem. Phys.*, **213**, 1663 (**2012**).
[14] Wang, Y.; Zhang, C.; Tang, C.; Li, J.; Shen, K.; Liu, J.; Yang, Z. *Macromolecules*, **44**, 3787 (**2011**).
[15] Saiyed, Z.; Telang, S.; Ramchand, C. *Biomagn. Res. Technol.* **1**, 2 (**2003**).
[16] Chambon, C.; Clement, O.; Le Blanche, A.; Schouman-Claeys, E.; Frija, G. *Reson. Imaging*, **11**, 509 (**1993**).
[17] Dobson, J. *Drug Dev. Res.*, **67**, 55 (**2006**).
[18] Prasad, N.; Perumal, J.; Choi, C.; Lee, C.; Kim, D. *Adv. Funct. Mater.*, **19**, 1656 (**2009**).
[19] Yin, S.; Wang, C.; Yu, Z.; Wang, J.; Liu, S.; Chen, S. *Adv. Mater.*, **23**, 2915 (**2011**).
[20] Ning, Y.; Wang, C.; Ngai, T.; Yang, Y.; Tong, Z. *RSC Adv.*, **2**, 5510 (**2012**).
[21] Ning, Y.; Wang, C.; Ngai, T.; Tong, Z. *Langmuir*, **29**, 5138 (**2013**).
[22] Ning, Y.; Wang, C.; Ngai, T.; Tong, Z. *Langmuir*, **2**, 7207 (**2011**).
[23] Zhang, J.; Qiu, S.; Zhu, Y.; Huang, Z.; Yang, B.; Yang, W.; Yang, J. *Polym. Chem.*, **4**, 1459 (**2013**).
[24] Yu, H.; Chen, M.; Rice, P.; Wang, S.; White, R.; Sun, S. *Nano Lett.*, **5**, 379 (**2005**).
[25] Wang, Y.; Guo, B.; Wan, X.; Xu, J.; Wang, X.; Zhang, Y. *Polymer*, **50**, 3361 (**2009**).
[26] Isojima, T.; Suh, S. K.; Vander Sande, J. B.; Hatton, T. *Langmuir*, **25**, 8292 (**2009**).
[27] Yabu, H.; Kanahara, M.; Shimomura, M.; Arita, T.; Harano, K.; Nakamura, E.; Jinnai, H. *ACS Appl. Mater. Interfaces*, **5**, 3262 (**2013**).
[28] Lattuada, M.; Hatton, T. *J. Am. Chem. Soc.*, **129**, 12878 (**2007**).
[29] Isojima, T.; Lattuada, M.; Vander Sande, J.; Hatton, T. *ACS Nano*, **2**, 1799 (**2008**).
[30] Sotiriou, G.; Hirt, A.; Lozach, P.; Teleki, A.; Krumeich, F.; Pratsinis, S. *Chem. Mater.*, **23**, 1985 (**2011**).
[31] Ren, B.; Ruditskiy, A.; Song, J.; Kretzschmar, I. *Langmuir*, **28**, 1149 (**2012**).
[32] Jian, L.; Zhang, Q.; Zhang, B. *Fine Chemicals*, **28**, 18 (**2011**).
[33] Tian, L.; Zhang, B.; Li, W.; Li, X.; Fan, X.; Jia, X.; Zhang, Q. *RSC Adv.*, **4**, 27152 (**2014**).

[34] Teo, B.; Suh, S.; Hatton, T.; Ashokkumar, M.; Grieser, F. *Langmuir*, **27**, 30 (**2011**).
[35] Wang, Y.; Xu, H.; Ma, Y.; Guo, F.; Wang, F.; Shi, D. *Langmuir*, **27**, 7207 (**2011**).
[36] Zhang, L.; Zhang, F.; Dong, W.; Song, J.; Huo, Q.; Sun, H. *Chem. Commun.*, **47**, 1225 (**2011**).
[37] Zhang, H.; Zhang, Q.; Zhang, B.; Xu, H.; Fan, X. *Polym Int.*, **61**, 990 (**2012**).
[38] Zhang, B.; Zhang, H.; Tian, L.; Li, X.; Li, W.; Fan, X.; Zhang, Q. *Chem. Eng. J.*, **275**, 235 (**2015**).
[39] Azad, A.; Fith, R.; Ugelstad, J. Abstracts of Papers of the American Chemical Society. 77 (**1974**).
[40] Ditsch, A.; Laibinis, P.; Wang, D.; Hatton, T. *Langmuir* **21**, 6006 (**2005**).
[41] Berret, J.; Schonbeck, N.; Gazeau, F.; El Kharrat, D.; Sandre, O.; Vacher, A.; Airiau, M. *J. Am. Chem. Soc.*, **128**, 1755 (**2006**).
[42] Dyab, A.; Ozmen, M.; Ersoz, M.; Paunov, V. *J. Mater. Chem.*, **19**, 3475 (**2009**).
[43] Sacanna, S.; Rossi, L.; Pine, D. *J. Am. Chem. Soc.*, **134**, 6112 (**2012**).
[44] Liu, B.; Zhang, W.; Zhang, D.; Yang, X. J. *Colloid. Interf. Sci.*, **385**, 34 (**2012**).
[45] Ali, N.; Zhang, B.; Zhang, H.; Li, W.; Zaman, W.; Tian, L.; Zhang, Q. *Fuel*, **141**, 258 (**2015**).
[46] Cheng, F.; Su, C.; Yang, Y.; Yeh, C.; Tsai, C.; Wu, C.; Shieh, D. *Biomaterials*, **7**, 729 (**2005**).
[47] Ali, N.; Zhang, B.; Zhang, H.; Zaman, W.; Li, W.; Zhang, Q. *Particuology*, **17**, 59 (**2014**).
[48] Ali, N.; Zhang, Q.; Zhang, B.; Zaman, W.; Ali, S. *Adv. Mater. Res.*, **1105**, 264 (**2015**).
[49] Ali, N.; Zhang, B.; Zhang, H.; Zaman, W.; Li, X.; Li, W.; Zhang, Q. *Colloid. Surface A.*, **472**, 38 (**2015**).
[50] Rahman, M.; Montagne, F.; Fessi, H.; Elaissari, A. *Soft Matter*, **7**, 1483 (**2011**).
[51] Xu, S.; Ma, W.; You, L.; Li, J.; Guo, J.; Hu, J.; Wang, C. *Langmuir*, **28**, 3271 (**2012**).
[52] Gokmen, M.; Du Prez, F. *Prog. Polym. Sci.*, **37**, 365 (**2012**).
[53] Abate, A.; Lee, D.; Do, T.; Holtze, C.; Weitz, D. *Lab Chip.*, **8**, 516 (**2008**).
[54] Zhang, J.; Coulston, R.; Jones, S.; Geng, J.; Scherman, O.; Abell, C. *Science*, **335**, 690 (**2012**).

[55] Trindade, A.; Canejo, J.; Patrício, P.; Brogueira, P.; Teixeira, P.; Godinho, M. *J. Mater. Chem.*, **22**, 22044 (**2012**).

[56] Kawasaki, H.; Yamamoto, H.; Fujimori, H.; Arakawa, R.; Iwasaki, Y.; Inada, M. *Langmuir*, **26**, 4281 (**2009**).

[57] Mou, F.; Xu, L.; Ma, H.; Guan, J.; Chen, D.; Wang, S. *Nanoscale*, **4**, 4650 (**2012**).

[58] I. Sinn, P. Kinnunen, S. N. Pei, *et al*, *Appl. Phys. Lett.*, **98**, 024101 (**2011**).

[59] Sinn, I.; Kinnunen, P.; Pei, S.; Clarke, R.; McNaughton, B.; Kopelman, R. *Soft Matter*, **5**, 1285 (**2009**).

[60] Mair, L.; Evans, B.; Hall, A.; Carpenter, J.; Shields, A.; Ford, K.; Superfine, R. *J. Phys. D: Appl. Phys.*, **44**, 125001 (**2011**).

[61] Hu, S.; Gao, X. *J. Am. Chem. Soc.* **132**, 7234 (**2010**).

[62] Ghosh, A.; Sheridon, N.; Fischer, P. *arXiv preprint arXiv*, **0708**, 1126 (**2007**).

[63] Erb, R.; Jenness, N.; Clark, R.; Yellen, B. *Adv. Mater.*, **21**, 4825 (**2009**).

Chapter 6

Synthesis, Self-Assembly, and Applications of Amphiphilic Janus and Triblock Janus Nanoparticle Analogs

Yang Yang[*,†], *Kyle Webb*[*], *Yijing Liu*[*], *Kun Liu*[†], and *Zhihong Nie*[*,‡]

[*]Department of Chemistry and Biochemistry,
University of Maryland, College Park, Maryland 20742, USA

[†]State Key Laboratory of Supramolecular Structure and Materials,
College of Chemistry, Jilin University, Changchun 130012, China

[‡]znie@umd.edu

Abstract

This chapter reviews recent progress in the emerging field of synthesis, self-assembly, and applications of amphiphilic Janus nanoparticles (JNPs) and triblock JNP analogs formed by the phase separation of ligands on the inorganic nanoparticle (NP) surface. In particular, JNPs, composed of NPs tethered with phase-separated polymer brushes, have recently emerged as a novel class of building blocks for fabricating new functional hybrid materials. This type of JNP shows great promise in the nanoscale design of new hybrid materials with increasing complexity and functionality. On the one hand, the JNPs can be structurally considered as organic molecular analogs that are capable of self-assembly at multiple hierarchical levels. On the other hand, the assembly of JNPs is significantly different from molecular assembly owing to their unique dimensions, hybrid composition, complex geometry, and multiscale interactions throughout the course of assembly. In this chapter, we will summarize the recent progresses on the self-assembly of JNPs into functional materials and their corresponding biological applications.

Keywords: Janus particles, triblock Janus particles, hybrid nanostructure, self-assembly, amphiphilic.

6.1. Introduction

Throughout the past few decades in the fields of nanoscience and nanotechnology, there has been a drastic shift from the synthesis of individual nanoparticles (NPs) to the construction of functional composite architectures using NPs as building blocks [1]. This transition is stimulated by the emerging collective properties and tailored electronic, magnetic, and optical characteristics of NP assemblies that may be radically different from those of the constituent NP building blocks and the corresponding bulk materials. These interesting properties are mainly induced by coupling interactions (e.g., plasmon–plasmon, plasmon–exciton, and magnetic–magnetic interactions) between the subunits within NP ensembles. The utilization of these unique properties has since led to many groundbreaking findings in various fields, for example, energy, sensing, metamaterials, biomedicine, and optoelectronics [2].

Incidentally, great progress has been achieved in the controlled self-assembly of NPs into complex, hierarchical nanostructures. Major strategies for the self-assembly of NPs include the utilization of templates (e.g., carbon nanotubes and polymer scaffolds) [3], biomolecular recognition (e.g., DNAs) [4], directional physical and chemical binding [5], and external fields/stimuli (e.g., electric and magnetic fields) [6]. Inspired by the well-established self-assembly of amphiphilic molecular building blocks, such as phospholipids, surfactants, and block copolymers (BCPs) [7], the use of so-called colloidal amphiphilic Janus nanoparticles (JNPs), which contain both hydrophobic and hydrophilic regions, has been recently explored for the assembly of inorganic NPs into ensembles with desired properties [8]. The synthesis of these amphiphilic JNP building blocks is achieved through modification of the NP surface with one or more species of small molecules or polymers. The amphiphilic distribution or phase separation of the ligands results in the formation of both JNPs, where one half of the NP is hydrophilic and the other is hydrophobic, and triblock JNPs (TJNPs) [9] where the hydrophobic region forms between two hydrophilic regions or *vice versa* [10]. Driven by the directional interactions derived from Janus features, the JNP amphiphiles can then spontaneously self-assemble into the desired nanostructures.

6.2. Fabrication Strategies of JNPs and TJNPs

6.2.1. Selective Modification of Gold Nanorods

Gold nanorods (Au NRs) can be selectively modified with thiol/disulfide-terminated tethers at both ends, owing to the preferential binding of cetyl trimethylammonium bromide (CTAB) along the {100} facet of the longitudinal side of the NRs. This leaves the {111} facet deprived of CTAB and allows for the binding of thiol/disulfide-terminated tethers to the ends of the NRs [11]. When the hydrophilic NRs are tethered with hydrophobic ligands at both ends, they behave as amphiphilic ABA TJNPs. As an exemplary system, Kumacheva *et al.* selectively grafted hydrophobic polystyrene (PS) homopolymers to both ends of the hydrophilic Au NRs [12]. The Au NRs, covered with a double layer of CTAB, constitute the central rigid hydrophilic block, while the thiol-terminated PS molecules, strongly anchored to the {111} facets at the ends of the NRs, form the flexible hydrophobic end blocks. These TJNPs can then reversibly assemble into a broad range of nanostructures with different degrees of organization and collective behavior complexity through selective solvent modifications. This is allowed due to both the physical binding of the organic tethers and phase separation of the constituent blocks, which results in a multitude of structures including rings, nanochains, bundled nanochains, bundles, nanospheres, and vesicles (Fig. 1). In particular, the assembly of TJNPs into nanochains linked by physical bonds is intriguing for it follows the theory of step-growth polymerization, which enables superior selectivity of 1D nanostructures by assembling NP chains with predetermined lengths [13]. However, the selective modification method is only limited to anisotropic NPs with different facets.

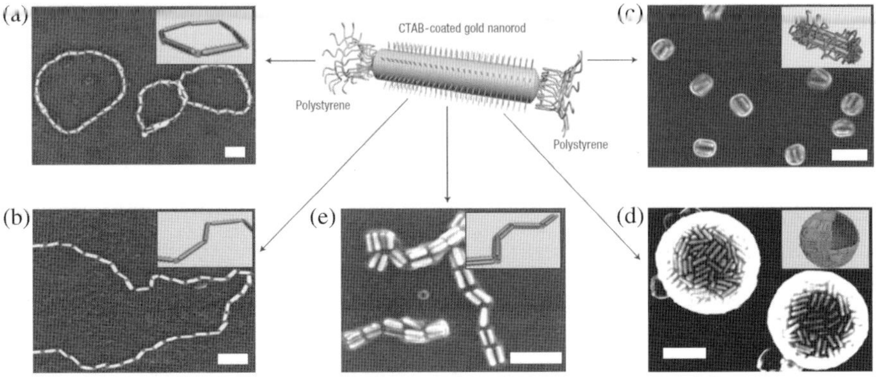

Figure 1. Self-assembly of polymer-tethered Au NRs in selective solvents. An amphiphilic Au NR carrying a double layer of CTAB along the longitudinal side and PS molecules grafted to both ends. (a)–(e) SEM images of the self-assembled NR structures: rings (a) and chains (b) self-assembled in the dimethyl formamide/water (DMF/H_2O) mixture at water contents of 6 and 20 wt.%, respectively, side-to-side aggregated bundles of NRs (c) and nanospheres (d) self-organized in the tetrahydrofuran/water (THF/H_2O) mixture at water contents of 6 wt.% and 20 wt.%, respectively, and bundled NR chains obtained in the ternary DMF/THF/H_2O mixture at a weight ratio of liquids 42.5:42.5:15 (e). The scale bars are 100 nm. The insets show corresponding schematic diagrams of the NR assemblies. Reproduced with permission from Ref. [12a]. Copyright (2007) Nature Publishing Group.

6.2.2. Selective Modification with Flat Substrate Masking

The flat substrate masking technique can also be considered as a flexible method for the preparation of JNPs [14]. In general, NPs are first adsorbed onto the substrate, followed by modification of the exposed NP surface. These NPs are then released from the substrate, exposing the now unprotected surface, which can then be modified to produce JNPs. The strength of this technique is that the masking can produce JNPs modified with a wide variety of functional groups. For instance, Li *et al.* demonstrated the synthesis of JNPs, functionalized with two distinct polymer brushes on opposite sides of the Au NPs, using lamellar single crystals of thiol-terminated polymers as the solid substrate mask [15]. "Solid-state grafting to" and "grafting-from" methods are combined in this synthesis strategy (Scheme 1). Initially, polymer single crystals are crystallized from solution, with thiol groups excluded on the surface, followed by the immobilization of Au NPs on the crystal surface by mixing the Au NP solution

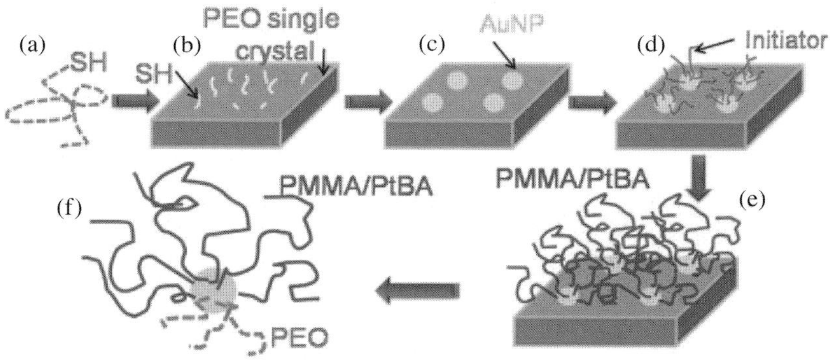

Scheme 1. Schematic illustration of polymer-functionalized Janus Au NPs by flat substrate masking method. Reproduced with permission from Ref. [15a]. Copyright (2008) American Chemical Society.

and thiol-terminated poly(ethylene oxide) (HS-PEO) single-crystal suspension. The thiol-terminated initiator is then applied to replace the ammonium ligands on the "free" surface of Au NPs (top part in Scheme 1(c)), forming JNPs with segregated polymer and initiator sides. The other polymer, (poly (methyl methacrylate) (PMMA), poly(*tert*-butyl acrylate) (P*t*BA), or poly(acrylic acid) (PAA)), is grown by atom transfer radical polymerization from the now anchored initiator. Janus Au NPs, including amphiphilic PEO-Au-PMMA, PEO-Au-P*t*BA, and hydrophilic PEO-Au-PAA, are finally obtained after the single-crystal substrates are dissolved. Thanks to the Janus feature, the JNPs are able to self-assemble into clusters and wormlike aggregates by adjusting the solvent conditions [16]. Due to the versatility of this method, homo- and hetero-NP dimers and chains [17], as well as thermo-responsive amphiphilic Janus silica NPs [18], have been successfully fabricated. Particularly, the thermo-responsive JNPs showed a lower thermal transition temperature and a much narrower transition range when compared with uniform poly (*N*-isopropylacrylamide) (PNIPAM) grafted NPs.

6.2.3. Interfacial Engineering Method

Liquid–liquid and air–liquid interfaces have also been proven capable of providing the asymmetric environment, in terms of surface modification, ideal for the synthesis of JNPs [19]. For example, the Langmuir technique has been exploited by Chen *et al.* to prepare nanosized JNPs with segregated hydrophobic and hydrophilic halves at the air–water interface [20]. Briefly, NPs stabled with hydrophobic alkanethiolate ligands are dispersed at the air–water interface at low surface pressures. Upon mechanical compression, the NPs form a closely packed monolayer on the water surface at high surface pressures, in which the interfacial mobility of NPs is dramatically confined with ligand intercalation between adjacent NPs. Thus, injection of hydrophilic ligands into the water subphase will result in ligand exchange only with the NP face submerged in the aqueous phase, leading to the formation of JNPs.

As another example, Bishop *et al.* demonstrated a general method to prepare JNPs with asymmetrical hydrophobic and hydrophilic ligand domains on their respective hemispheres [21]. Driven by the reduction of interfacial energy, the NPs and two competing ligands, one hydrophobic dodecanethiol (DDT) and the other hydrophilic mercaptoundecanoic acid (MUA), spontaneously organize at the interface of water and toluene, which produces JNPs with amphiphilic surfaces (Scheme 2). The asymmetric surface chemistry of the JNPs was further confirmed by contact angle measurements. This approach has also been extended for the preparation of amphiphilic iron oxide NPs with two competing diol-terminated ligands.

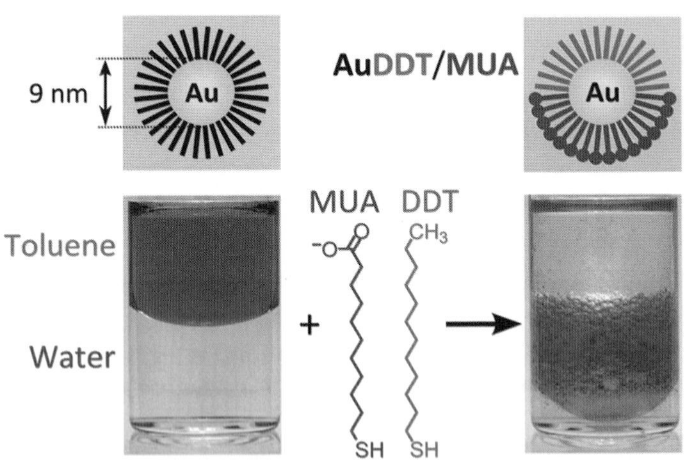

Scheme 2. Schematic illustration of the surface functionalization procedure of Au NPs for the preparation of JNPs. Reproduced with permission Ref. [21]. Copyright 2012, American Chemical Society.

6.2.4. Block Copolymer Template

Recently, Liu *et al.* proposed a new approach to synthesize ultrasmall anisotropic Au NPs (<4 nm) monofunctionalized with a single amphiphilic triblock copolymer chain per NP, PEO-AuNP-PS, by using a triblock copolymer as the template [22]. The outer blocks of triblock copolymers are composed of PEO and PS, while the middle block is poly(lipoic acid 2-hydroxy-3-(methacryloyloxy)-propyl ester-*co*-glycidyl methacrylate) [P(LAMP-*co*-GMA)], which is responsible for the functionalization of Au NPs through multidentate NP-binding (Fig. 2(a)). By taking advantage of the Janus feature, the as-synthesized amphiphilic NPs are capable of exhibiting rich self-assembly morphologies in aqueous solutions, including nanoclusters, spherical micelles, vesicles, rods, and large compound micelles (Figs. 2(b)–2(e)).

6.2.5. Phase Separation of Incompatible Ligand Mixture

Recently, the most common method for the preparation of JNPs is based on the phase separation of incompatible ligand mixtures. JNP amphiphiles, with conformational flexibility of all ligands, are generated by attaching a mixture of hydrophilic and hydrophobic ligands onto the surface of the NPs, which is driven by the phase separation of the incompatible ligands [23]. The ligands on the NP surface can dynamically rearrange in response to changes in the local environment, rather than remain anchored. As a consequence, the asymmetric distribution of ligands on the NP surface, in response to the organization of NPs in solution, leads to the formation of JNPs. An example system was developed by Meskers and Schenning *et al.* where amphiphilic JNPs were synthesized by varying the ratio of the hydrophobic and hydrophilic ligands [24]. Different molecular ratios of dithiol mono-capped hydrophobic alkyl and hydrophilic ethylene oxide ligands were introduced for ligand exchange with the tetraoctylammonium bromide capping on the Au NPs, which yielded amphiphilic JNPs with a wide array of ratios between the hydrophobic and hydrophilic ligands. In water, phase separation of the ligands on NP surface takes place to form JNPs, similar to those formed at interfaces. These JNPs, with a well-defined ratio of hydrophobic and hydrophilic ligands, can

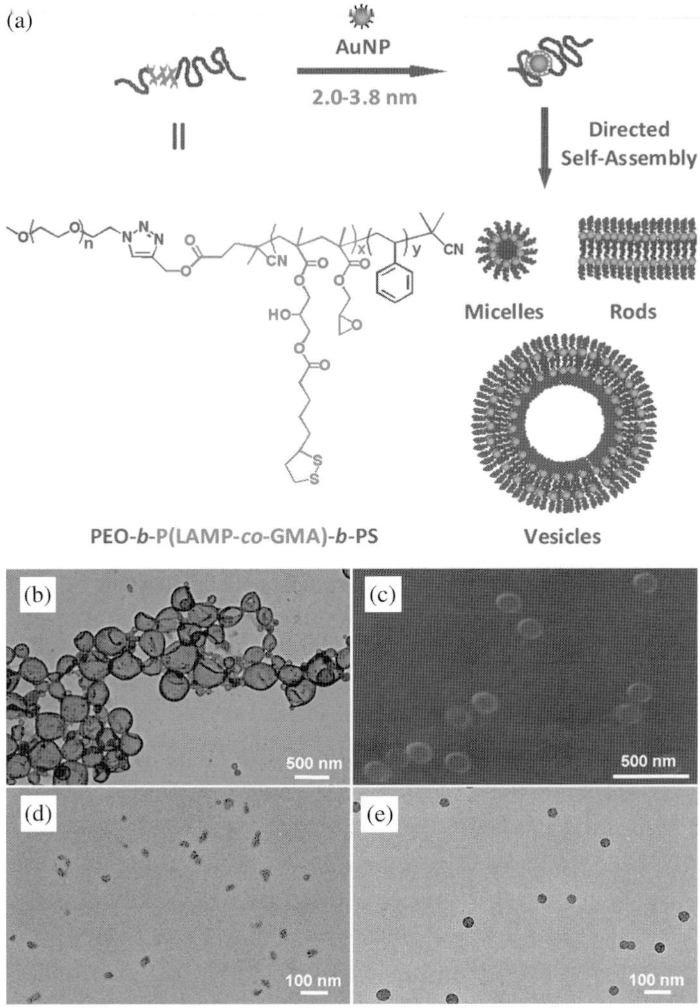

Figure 2. (a) Schematic illustration of the template-free, size-selective fabrication of ultrasmall Au NPs (2.0–3.8 nm) anchored with a single triblock copolymer chain and the hierarchical self-assembly of the resulting Au NP hybrid amphiphilic triblock copolymers PEO–AuNP–PS in aqueous media. (b)–(e) NP assemblies in aqueous solution. (b) High-resolution TEM and (c) atomic force microscopy (AFM) images of PEO_{45}–$AuNP_{2.0}$–PS_{260}. HRTEM images of (d) PEO_{113}–$AuNP_{2.0}$–PS_{60} and (e) PEO_{113}–$AuNP_{2.0}$–PS_{300}. Reproduced with permission from Ref. [22]. Copyright (2012) American Chemical Society.

further self-assemble into disk-shaped micelles. However, the lack of polymer-like flexibility of these building blocks may cause constraints in the NP packing during self-assembly.

Subsequently, Weller and Förster partially modified hydrophobic CdSe/CdS core–shell NPs with hydrophilic PEO brushes through a ligand exchange reaction. As the polymer/NP ratio decreases, a structural evolution takes place, where, in selective solvents, spheres will form and then transform into cylinders, then networks, and finally into vesicles along with a decrease of the grafting density (δ) of the PEO brushes [25]. However, unlike molecular amphiphiles, one of the amphiphilic components (i.e., hydrophilic portion and hydrophobic portion) for the above-mentioned NP amphiphiles is composed of rigid NPs. In this case, the amphiphilic Janus feature arises from a combination of small molecular ligands, which stabilize the NPs during their synthesis, and polymer brushes post-modified on the NP surface.

Additionally, Moffitt *et al.* were able to synthesize CdS NPs decorated with ionic triblock copolymers with hydrophobic and partially charged hydrophilic chains (Figs. 3(a) and 3(b)). Driven by phase separation, the NPs rearrange into JNPs and self-assemble in polar solvent mixtures to form complex hierarchical nanostructures, including supermicelles, segmented wormlike assemblies with internal NP lamellae, and bilayer vesicles with concentric layers of close-packed NPs (Figs. 3(c)–3(f)) [26]. These NP self-assembly pathways can then be chemically tuned by adjusting the strength of electrostatic interactions between hydrophilic chains through addition of salt or acid.

Another example of amphiphilic Janus Au NPs was demonstrated by Duan *et al.* where they designed Au NPs and NRs modified with mixed hydrophilic poly(ethylene glycol) (PEG) and hydrophobic PMMA polymer brush coatings. A strong indication of the Janus morphology was the spontaneous formation of vesicular nanostructures with strong interparticle plasmon coupling properties (Fig. 4) [27]. The vesicular structures combine the inherent physical and chemical properties of both the polymers and inorganic cores, which facilitates the disruption of the plasmonic vesicles through a change in pH or light irradiation. The group was able to utilize these properties by substituting the PMMA polymer brush with polylactide (PLA), which made the plasmonic vesicles biodegradable. This allows for the vesicles to be broken down by both enzymatic degradation and near

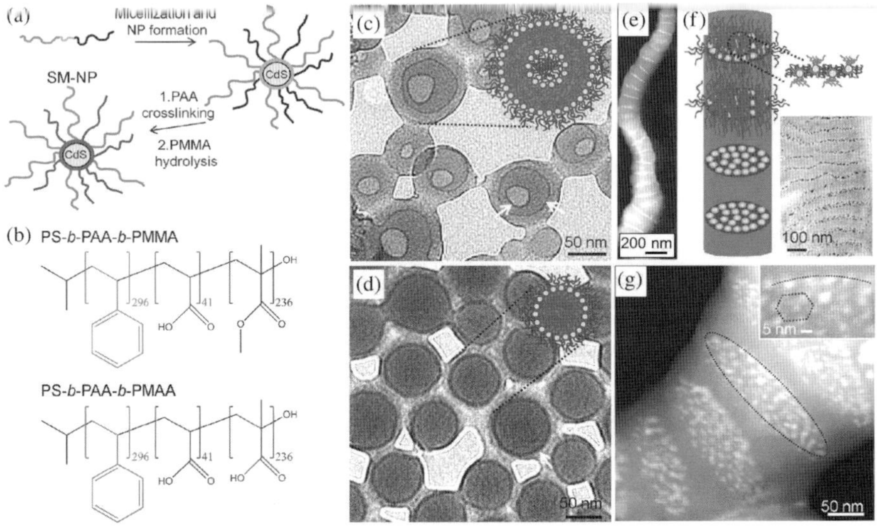

Figure 3. (a) Preparation of JNP building blocks and (b) molecular structures of the constituent PS-*b*-PAA-*b*-PMMA triblock copolymer and its hydrolyzed form, PS-*b*-PAA-*b*-PMAA. (c) TEM images of vesicles consisting of a bilayer arrangement of JNPs (inset) forming two concentric shells of densely packed CdS NPs (indicated by arrows). (d) TEM image of equilibrium supermicelles consisting of a spherical arrangement of JNPs (inset) forming a shell of densely packed CdS NPs on the PS core surface (indicated by an arrow). (e) High-angle annular dark field (HAADF) image showing the segmented structure of the assemblies. (f) Schematic of lamellar ordering of JNPs; TEM image of a microtomed section of a wormlike assembly indicates internal alternating layers of PS (gray) and CdS NPs (dark dots) dispersed within PMAA (light). (g) HAADF image highlighting regularly spaced disks containing 2D arrays of CdS NPs (bright dots) with preferred hexagonal ordering. Reproduced with permission from Ref. [26]. Copyright (2011) American Chemical Society.

infrared (NIR) photothermal heating for theranostic applications [28]. However, it is difficult to quantitatively control the relative δ of polymer mixtures on the NP or NR surface due to the different bonding strengths and absorption kinetics of different polymers onto the NPs [21, 29].

6.2.6. Phase Separation of Block Copolymers

More recently, functionalization of NPs with V-shaped or linear amphiphilic BCPs has emerged as a simple yet effective route for the construction

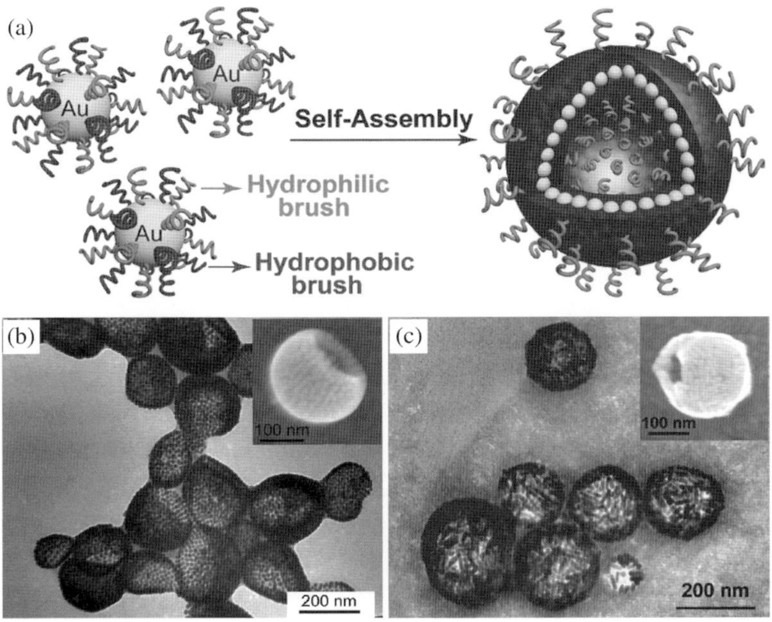

Figure 4. (a) Schematic illustration of self-assembly of amphiphilic nanocrystals with mixed polymer brushes into vesicular structures. Representative TEM and (inset) SEM images of the plasmonic vesicles assembled from (b) 14 nm Au NPs and (c) Au NRs (~13 nm × 51 nm) with mixed PEG and PMMA brushes. Reproduced with permission from Ref. [27] Copyright (2011) American Chemical Society.

of novel assembly building blocks (Scheme 3) [30]. This approach is based on the spontaneous phase separation of BCPs on NP surfaces via molecularly controlled interactions within ligand shells, which offers a more delicate control over the surface chemistry (e.g., the lengths, chemical compositions, and δ of polymer tethers), overall architectural complexity, and diversity of the building blocks. The unique geometry and dimensions of such NP amphiphiles may diversify the nanostructures that can be produced during self-assembly. In addition, TJNPs with carefully designed constituent block length ratios appear to be amphiphilic, as reflected by their preferential isolation at the water/chloroform interface. This implies that BCPs on a NP surface can thermodynamically self-adapt to changes in their local environment via contortion of their chain conformations rather than behaving as static sandwich layers. Driven by

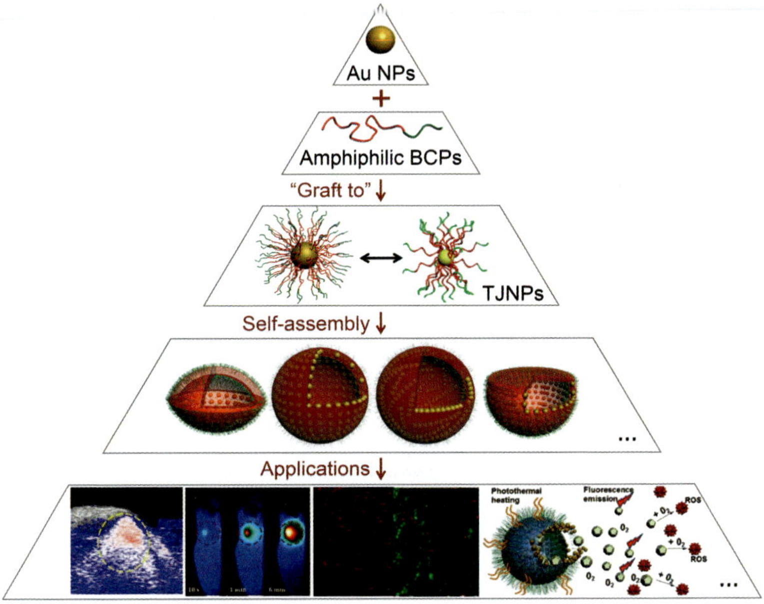

Scheme 3. Schematic illustration of the preparation and application of functional hybrid nanostructures via TJNP self-assembly.

conformational changes of tethered BCP chains, JNPs are capable of assembling into a variety of inorganic/polymer hybrid nanostructures with unique properties. The morphologies and geometries of these assembled nanostructures can in turn be controlled by the size of NPs (*d*), molecular weight (M_w), δ, composition, architecture of BCPs, and solvent composition [25, 31]. The manipulation of long flexible BCPs in the vicinity of the NP surface offers us an effective tool for the engineering of interactions between NPs and hence the complexity of NP assembly. In the following, we will focus on summarizing the recent progress on the self-assembly of amphiphilic BCP-tethered JNPs into functional materials along with their corresponding biological applications. First, we will highlight strategies for the assembly of amphiphilic JNPs, including film rehydration, selective solvents by solvent mixing or dialysis, and microfluidic approaches. Second, we will discuss the utilization of JNP vesicles in the field of biomedical imaging and delivery. Finally, we will outline current challenges and perspectives on this frontier.

6.3. Strategies for Assembling JNP/TJNP Amphiphilies

6.3.1. Film Rehydration Method

The film rehydration method has been widely used for the preparation of vesicular nanostructures through self-assembly of molecular amphiphiles (e.g., lipids, BCPs). This method relies on the rehydration of a dried thin film of amphiphiles in water with additional heating or sonication. It is simple and straightforward, not involving organic solvents, which facilitates the encapsulation of bioactive species (e.g., drugs, DNAs, RNA, and peptides) in the hollow vesicular cavity.

Recently, this technique has also been proven to be effective for the preparation of assembled TJNP nanostructures (Fig. 5(a)) [31a, 32]. Briefly, a NP amphiphile solution in THF is first dried to form a thin film on a glass substrate (e.g., the wall of glass vials) under a N_2 stream. Then the film is rehydrated in water (or an aqueous solution of species to be encapsulated) with sonication or heating, thus triggering the assembly process.

For instance, when Au NPs are tethered with BCPs of PS-*b*-PMEO$_2$MA or PEO-*b*-PS with thiol end-functionalized PS, vesicular or tubular assemblies are formed (Figs. 5(b) and 5(c)). Both assembled nanostructures are hollow, composed of a monolayer shell of hexagonally packed Au NPs, as confirmed by high magnification SEM images and their corresponding fast Fourier transform (FFT) pattern. The assembly behavior is predominantly dependent on the hydrophobic/hydrophilic balance, that is, the relative size of hydrophobic/hydrophilic constituents. Thus, by fixing the PEO block length, the morphologies of assembled structures can be reproducibly designed, where structures are largely determined by the ratio between the length of the hydrophobic block in BCPs (R) and the relative d of Au NPs (Fig. 5(d)). When $R/d < 0.5$, TJNPs favorably assemble into vesicles, while when $R/d \approx 0.5$, tubules are preferred; however, when $R/d > 0.5$, TJNPs cannot be rehydrated in water. These results arise from the surface curvature-dependent hydrophilicity of NP amphiphiles, which is determined by the density (D) of hydrophilic terminals (PEO blocks). This can then be described by the equation, $D = \delta/[4(0.5 + R/d)^2]$, where the density of the hydrophilic block is shown to exponentially decrease with increasing R/d. For

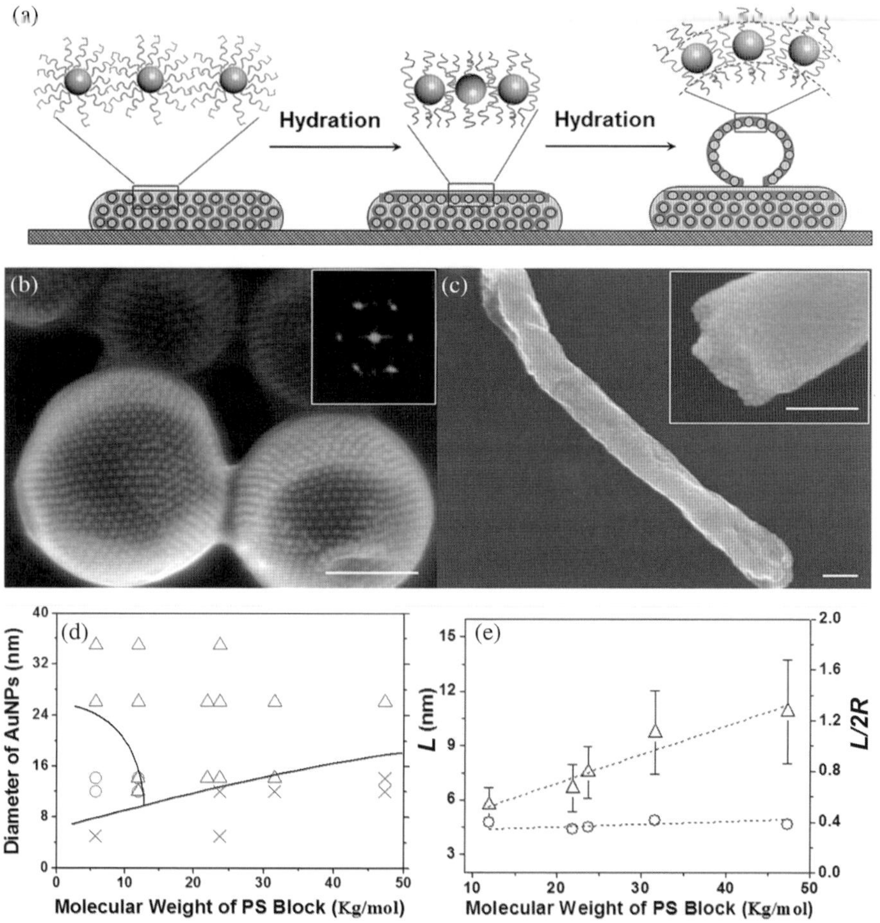

Figure 5. (a) Schematic illustration of self-assembly of TJNPs by film rehydration. (b) and (c) Representative SEM images of vesicles and tubules obtained from self-assembly of TJNPs. Inset in (b) is the FFT pattern of SEM images. (d) Self-assembly "product diagram" with varying diameter of Au NPs and M_w of PS block. The dashes lines indicate the given morphology transition boundary; △, vesicles; ○, tubules; and ×, precipitates. (e) The plot of L (Δ) and $L/2R$ (O) as a function of the M_w of the PS block. Reproduced with permission from Ref. [31a]. Copyright (2012) American Chemical Society.

$R/d < 0.5$, the D value is high, so there is sufficient hydrophilicity for TJNPs to rehydrate and assemble in water. However, when $R/d > 0.5$, the density of hydrophilic chain terminals is insufficient to trigger the rehydration and assembly of TJNPs in water.

It is appealing that even though the TJNPs are isotropic in terms of their geometries and chemical functionalities, they can still spontaneously induce the anisotropic interactions between TJNPs necessary to drive the assembly. The spontaneous formation of BCP anisotropy is believed to be caused by the conformational change of the long and flexible BCP chains, although the lateral diffusion of polymers on the surface of NPs may also play an important role. The mechanism is supported by the fact that inter-particle spacing (L), between NPs in the membrane of vesicular assemblies, is much smaller than that in the thin film before rehydration. The decrease in L indicates the relocation of BCP segments from the gap between pairs of NPs to the sides of NPs due to polymer conformation change. In this way, the exposure of hydrophilic blocks in water is maximized, while hydrophobic blocks are shielded, enabling the minimization of interfacial energy. With further rehydration, the TJNP monolayers detach, roll over, and enclose to form sealed BCP vesicles with cavities.

Additionally, the localized surface plasmon resonance (LSPR) of NP ensembles is known to predominately rely on the d, L, and the NP arrangement. For near-field plasmon coupling of Au NPs, the shift in the LSPR peaks of Au NP pairs is proportional to $e^{-(L/d)}$ [33]. Compared with individual Au NPs, the reduction L between adjacent Au NPs within assembled nanostructures leads to a redshift of the LSPR peak due to enhanced plasmon coupling. Typically, for the tubular assemblies, the LSPR splits into two modes: a maximum and a weak absorption peak corresponding to the longitudinal and transverse plasmon modes, respectively [34]. As the M_w of BCPs is gradually increased, L approximately linearly increases (Fig. 5(e)), leading to a blue-shift of the LSPR peak due to the weaker plasmon coupling.

Besides d, the shape and morphology of the core NPs also have a significant influence on the LSPR of NP assemblies. For example, owing to the rough and irregular surface, Au nanoflowers (NFs) show LSPR absorption in the range of 500–700 nm depending on the d and surface roughness (Fig. 6(a)) [35]. Both the NP surface area and plasmon field around NPs can be significantly increased by the sharp protrusions of Au NFs, thus expanding their applicability to use in surface-enhanced Raman scattering (SERS), catalysts, and biosensors. These irregular shaped Au NFs tethered with BCPs (Fig. 6(b)) can then be assembled into plasmonic vesicles via the film rehydration method (Figs. 6(c) and

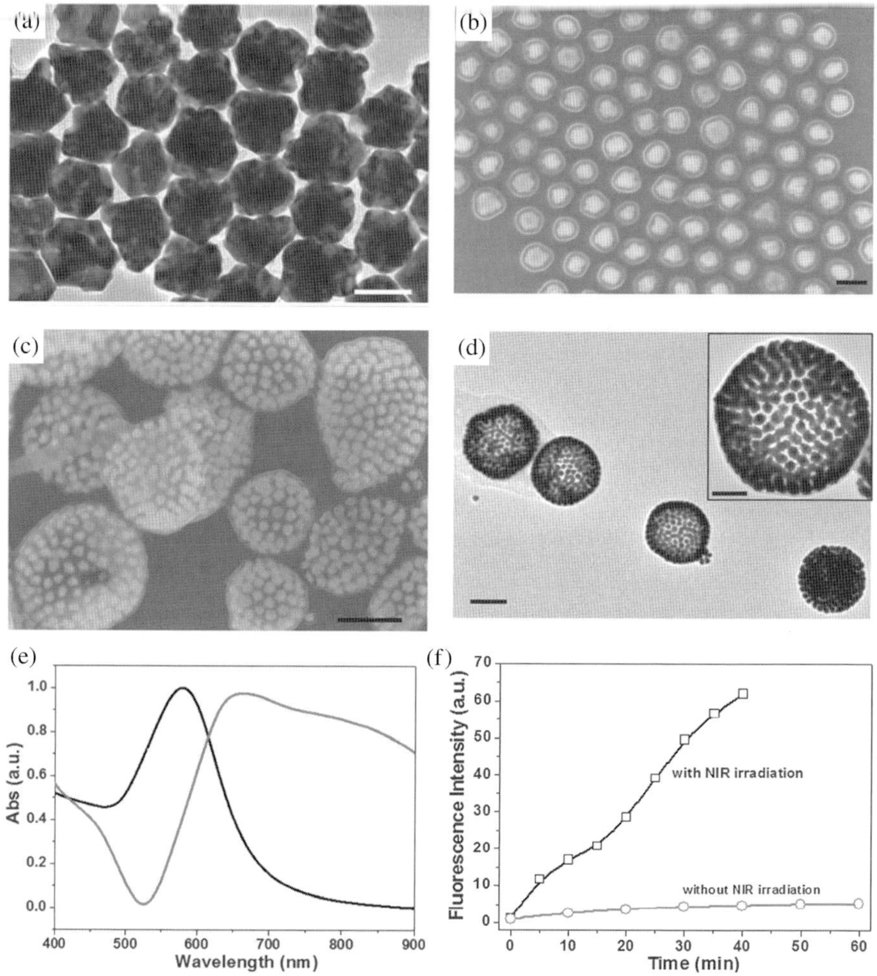

Figure 6. (a) Representative TEM image of 63 nm Au NFs synthesized using a seed-mediated growth method. (b) SEM image of Au NFs modified with amphiphilic BCPs of PEO_{45}-*b*-PS_{455}. (c) and (d) SEM and TEM images of vesicles assembled from amphiphilic Au NFs. (e) UV–Vis spectra of Au NFs before (black curve) and after (gray curve) their assembly into vesicles. (f) Normalized fluorescence intensity of RB as a function of time with or without laser irradiation. Scale bars are 50 nm in (a), 100 nm in (b), 300 nm in (c), 500 nm in (d), and 200 nm in the inset of (d). Reproduced with permission from Ref. [36]. Copyright (2013) Royal Society of Chemistry.

6(d)) [36]. It is worth noting that the strong plasmon coupling between adjacent Au NFs within vesicular membranes leads to a significant increase in the absorption of the vesicular structures in the NIR (Fig. 6(e)). The strong NIR absorption enables the light-triggered release of compounds encapsulated in the vesicles (Fig. 6(f)), which can be potentially used for various applications, such as controlled drug delivery and release, biosensors, and chemical reactors.

6.3.2. Selective Solvent Method

Amphiphilic BCPs can spontaneously organize into a variety of solvent-dependent nanostructures [37]. In desired solvents, the insoluble blocks of amphiphilic BCPs segregate together to minimize interfacial free energy, while the soluble blocks are further exposed to external solvent media. Through careful manipulation of these solvent conditions, various assembly structures can then be prepared. This strategy has also recently been employed for the effective self-assembly of JNPs into similar functional nanostructures. More importantly, this assembly mechanism allows us to fully explore the critical role of kinetic and thermodynamic factors during the assembly process, which tends to be missing in film rehydration process.

When BCPs are grafted on NPs, with the junction point of two blocks serving as the anchor, the NPs serve as the junctions of hydrophobic and hydrophilic blocks. Recently, Zubarev *et al.* exploited this strategy by covalently modifying the surface of Au and Ag NPs with amphiphilic V-shaped PEO-*b*-PS (Fig. 7(a)) [38]. The amphiphilicity drives the self-assembly of the resulting JNPs into 1D tubular structures in selective solvents (Figs. 7(b)–7(d)). In such structures, inorganic NPs are located at the interface between the PS core and PEO corona. The formation of defined nanostructures is explained by the local spatial separation of the hydrophobic and hydrophilic arms of V-shaped BCPs around the NP core upon water addition, resulting in the Janus-like building blocks. Moreover, the assembling size and morphology are largely dependent on the dialysis solvents, concentration of the JNP solution, the d of NPs, and the length of polymer.

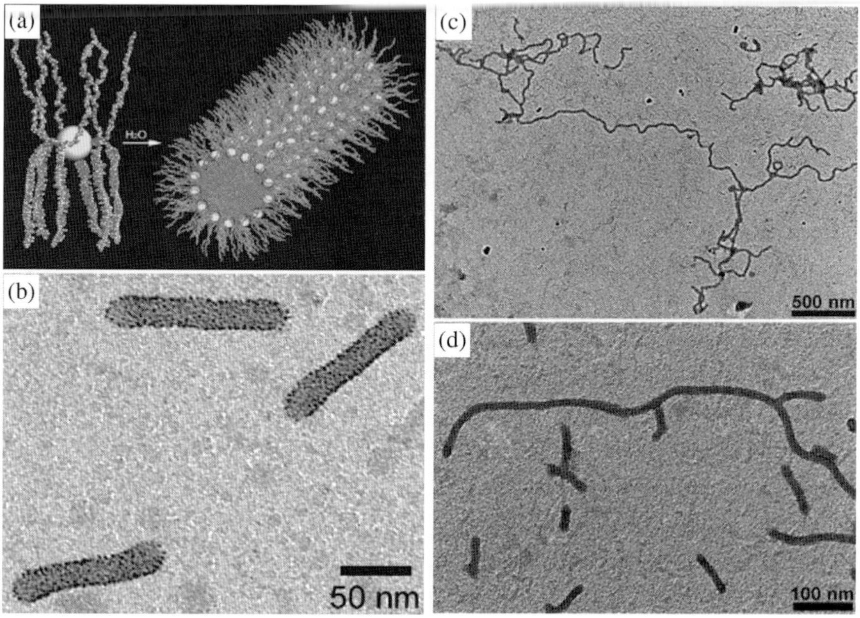

Figure 7. (a) Schematic representation of the amphiphilicity-driven self-assembly of Au–(PS-PEO)*n* NPs. (b) TEM image of a sample prepared from an aqueous solution of Au–(PS-PEO)*n* NPs after dialysis of a THF/H_2O (1:3 vol.) solution against DI water. (c) Wormlike assemblies of amphiphilic Au–(PS_{40}-PEO_{50})*n* NPs from an aqueous solution after dialysis from a DMF/H_2O (1:3 vol.) mixture. (d) 1D assemblies of Ag-(PS-PEO)*n* NPs (dialysis from a THF/H_2O mixture). Reproduced with permission from Ref. [38]. Copyright (2006) American Chemical Society.

When amphiphilic BCPs with thiol end-terminated hydrophobic blocks are used to assist the self-assembly of NPs, precise control over both the *L* and type of assembled nanostructure is achieved, thus fine-tuning the plasmon properties of assembled NPs. As a typical example, our group utilizes this strategy to assemble TJNPs, with Au NPs tethered with linear amphiphilic PEO-*b*-PS BCPs, into various superstructures, such as unimolecular micelles, clusters, and hollow vesicles (Fig. 8(a)) [31b]. The assembly process is triggered via slow addition of a poor solvent (e.g., water) for the hydrophobic blocks, into a solution of TJNPs dispersed in an organic solvent (e.g., THF). The organic solvent is subsequently removed by dialysis of the solution against water. During the course of self-assembly, the

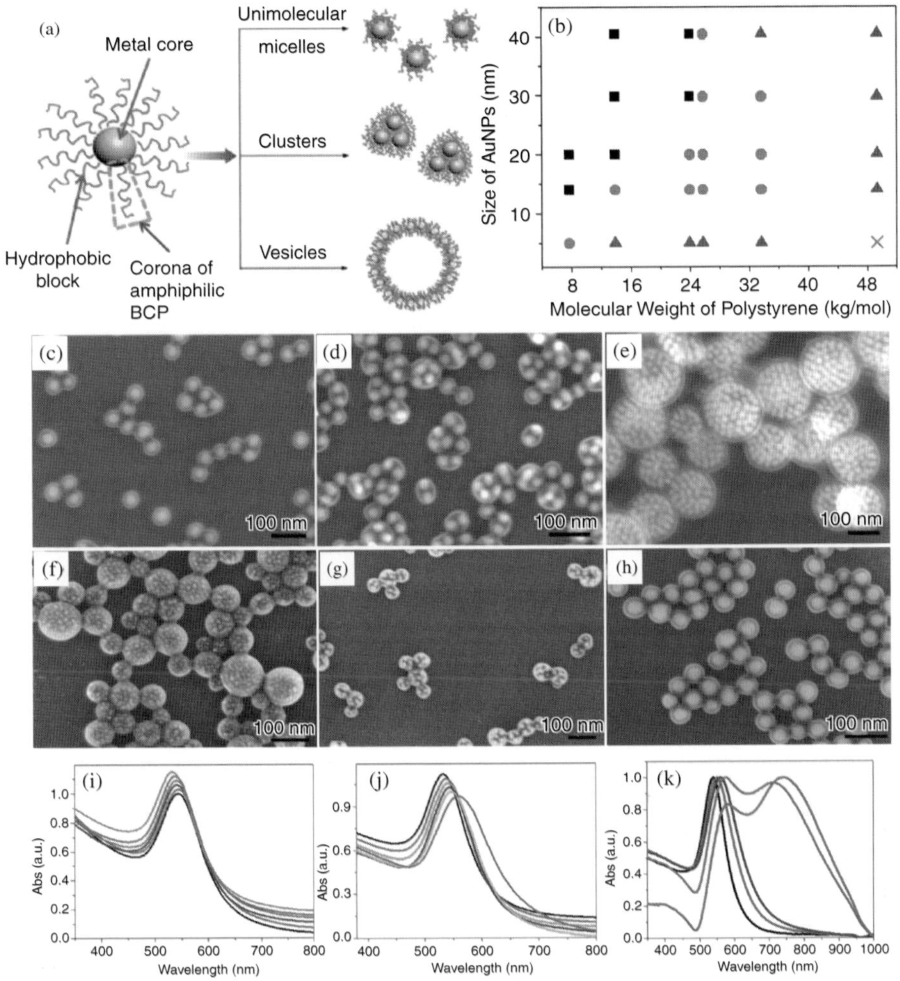

Figure 8. (a) Schematic illustration of the self-assembly of TJNPs via selective solvent method. (b) Phase diagram of the self-assembly of TJNPs with different M_w of BCPs and *d* of Au NPs: unimolecular micelles (■); clusters (●); vesicles (▲); and precipitates (×). (c)–(e) SEM images of TJNP assemblies using various M_w of PS in the tethered BCPs: (c) 14 k, (d) 24 k, and (e) 49 k. (f)–(h) SEM images of TJNP assemblies using various *d* of Au cores: (f) 5 nm; (g) 15 nm; and (h) 30 nm. (i)–(k) UV–Vis spectra of TJNP assemblies using various M_w of PS in the tethered BCPs. The M_w of BCPs monotonically increases from left to right, along with the redshift of plasmon peaks. The *d* of Au cores: (i) 14 nm, (j) 20 nm, and (k) 40 nm. Reproduced with permission from Ref. [31b]. Copyright (2013) American Chemical Society.

addition of poor solvent causes the collapse of the PS blocks of tethered BCP chains in order to minimize the overall free energy of the system, leading to the formation of various nanostructures. This implies a strong dependence on the M_w of the PS block and the d of Au NPs, thus allowing TJNPs, assembled in a mixture solvent of H_2O/THF, to adopt various superstructures, including unimolecular micelles, clusters with controlled number of TJNPs, and vesicles (Fig. 8(b)). Experimentally, it was found that as the M_w of hydrophobic PS blocks increases, the assembled nanostructures transition from unimolecular micelles, to clusters, and eventually to vesicles (Figs. 8(c)–8(e)). This observation is in good agreement with the morphological transition dependent on the classic geometrical packing parameter (p) for molecular amphiphiles [39]. p can be defined as $p = v/a_oR$, where v is the volume of the hydrophobic compositions, a_o is the cross-sectional area of the hydrophilic segment, and R is the length of the hydrophobic portions. In general, as p increases within $0 < p < 1$, the assembled structures shift from spherical micelles, to cylindrical micelles, and eventually to vesicles or planar bilayers. However, the assembled structures of TJNPs are dependent not only on the relative fraction of hydrophilic and hydrophobic blocks, but also on the d of Au NPs. This can be proven based on the morphological transition dependent on the d of Au cores, where a change in structure from vesicles to clusters and unimolecular micelles is triggered by increasing the value of d (Figs. 8(f)–8(h)). Conversely, this transition is opposite to the trend observed in molecular assembly, if the NP cores are considered as hydrophobic portions. This would suggest that restriction of the metal cores has a significant influence on the conformation of BCP chains, for example, the stretching and compacting of polymers.

The delicate interplay of the enthalpy/entropy contribution to the overall free energy along with the spherical architecture of the TJNPs plays a crucial role in the self-assembly behavior. Individual BCPs grafted on Au NPs provide much less entropic and enthalpic contributions than free linear BCPs, in terms of the overall free energy associated with the segregation of individual polymer chains during the micellization process. In order to elaborate, dissipative particle dynamics (DPD) simulations with bottom-up coarse-grained (CG) potentials can be used to achieve a more comprehensive understanding of the assembly kinetics and mechanism. The simulation results of the chain length effect show that TJNP assembly is strongly dependent on the redistribution of polymer segments on the NP surface and hence the deformability of flexible polymer layers. When TJNPs are tethered with hydrophobic PS blocks of insufficient length, the redistribution of hydrophilic PEO segments on the NP surface cannot be induced. Therefore, the TJNPs preferentially aggregate to form small clusters. Once the hydrophobic PS block is of sufficient length, the TJNPs can effectively deform. Thus, when several NPs approach one another, the hydrophobic blocks can accommodate new configurations, which allows for the redistribution of hydrophilic blocks on both sides of the NP during aggregation, enabling the TJNPs to assemble into vesicles.

The plasmon coupling of TJNP assemblies predominately depends on the *d* of NPs, the aggregation number of NPs, the interparticle *L*, and the assembly morphologies. The aggregation number increases with respect to the M_w of PS blocks, leading to a large redshift, despite a slight increase of *L* (Figs. 8(i) and 8(j)). When the plasmon coupling between Au NPs is weak, the redshift of the LSPR peak, with the increase of BCP M_w, is due to both the increase of the NP aggregation number within the assemblies and the morphological transition from unimolecular micelles to vesicles (Fig. 8(i)). When the plasmon coupling is strong ($L/d \leq 0.23$), with the increase of BCP M_w, there is not only a significant plasmon shift, but also a splitting of the plasmon peak into two well-defined peaks (Fig. 8(k)). The splitting of the plasmon peak is caused by the occurrence of "plasmon hybridization", which only occurs due to strong plasmon coupling between Au NPs and has been rarely observed in previous vesicular assemblies of discrete NPs.

Very recently, we have carried out the detailed study to determine how δ effects the assembled structure of TJNPs [31d]. By tuning the δ of TJNPs, the formation of hollow chain vesicles with membranes composed of NP strings is achieved, rather than vesicles with a uniform NP distribution (referred to as non-chain vesicles) (Figs. 9(a)–9(c)). δ is controlled by tuning the weight ratio of BCPs and NPs during surface modification. At low δ values (~0.03 chain/nm^2), the assembly process produces chain vesicles with a monolayer of NP strings in the vesicular membranes. Compared with non-chain vesicles, chain vesicles exhibit two distinct peaks located in the visible and NIR ranges, respectively (Figs. 9(d) and 9(e)), even with the small *d* of NPs (<~15 nm), which is more favorable

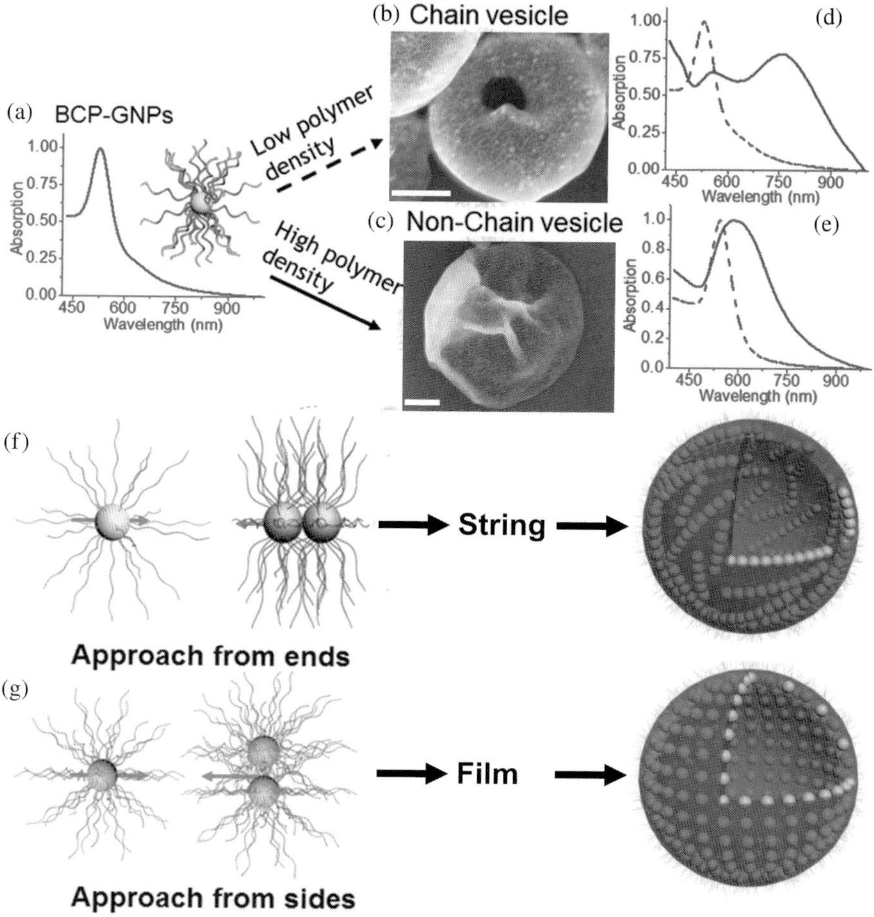

Figure 9. (a)–(c) Schematic illustration of the self-assembly of TJNPs into chain vesicles and non-chain vesicles. Representative SEM images of chain vesicle (b) and non-chain vesicle (c) made from TJNPs. UV–Vis spectra of individual TJNPs (a), chain vesicles (d), and non-chain vesicles (e). Mechanisms of the formation of chain (f) and non-chain vesicles (g). Scale bars: 250 nm. Reproduced with permission from Ref. [31d]. Copyright (2015) Wiley-VCH.

with respect to clearance from animal body. These hollow chain vesicles, composed of NP strings in the membrane, are formed by two critical steps: the organization of individual NPs into 1D strings with a strong plasmon coupling and the wrapping of the strings into hollow vesicles (Fig. 9(f)).

Au NPs are in such close contact with one another along the string that, in some cases, the Au NPs fuse together; however, this close proximity is what gives rise to the strong plasmon coupling and thus the NIR absorption. When Au NPs grafted with a higher δ (≥ 0.05 chain/nm^2) are used, non-chain vesicles are formed, with NPs that are relatively uniform in their distribution in the vesicular membranes (Fig. 9(g)).

The competition between the formation of chain or non-chain vesicles is controlled by the δ as illustrated in Figs. 9(f) and 9(g). When the δ is low, the van der Waals force between neighboring Au NPs is stronger. This results in the squeezing out of BCPs from the gaps between NP pairs, leading to a higher polymer density at the interface around the center than at either pole of the NP pair. In the early stage of chain vesicle formation, upon addition of water, 1D NP strings are produced due to repulsive forces at the center and attractive van der Waals interactions at both poles. This can be confirmed by the fact that the interchain distance between strings is larger than the L between NPs in these strings. Through further addition of water, the NP strings become unstable and group together to form chain vesicles in order to minimize the interfacial tension. When the δ is high, the compression of the polymer brushes between neighboring NPs is depressed, resulting in a smaller repulsive force at the interface. Thus, the van der Waals attractions become too weak to drive the NPs close, and steric repulsions are mainly balanced by attractive hydrophobic interactions. To maximize hydrophobic interactions, individual TJNPs approach from the sides of clusters to form 2D monolayer films and eventually non-chain vesicles.

As previously stated, TJNPs have emerged as a remarkable class of amphiphilic building blocks, which can self-assemble into various functional nanostructures. The concurrent assembly of TJNPs with BCPs, which combines the inherent properties of both building blocks, can offer a new platform for a higher-level complex hierarchical self-assembly. In order to design nanostructures with increasing complexity and new functionality, multiple types of amphiphiles are used for simultaneous assembly into discrete architectures. Eisenberg *et al.* developed an approach in which TJNPs are isolated in the central portion of BCP rods, micelles, or vesicular walls through coassembly of BCPs and TJNPs (Fig. 10) [40]. There are two sequential steps in the strategy. First, the NPs are stabilized with BCPs that are the same as or similar to the BCPs that are used for the preparation of the BCP rods, micelles, or vesicles. Next, the BCP assembly into rods, micelles, or vesicles, in the presence of TJNPs, is conducted. The specific placement of NPs in the central portion of BCP rods, micelles, or vesicular walls is largely controlled due to both entropic and steric effects.

As another example, which we discovered, various hybrid vesicles with well-defined shapes, morphologies, and surface patterns can be achieved via the concurrent assembly of TJNPs with free amphiphilic linear BCPs by mediating the entropy-driven lateral phase separation of binary mixtures within membranes (Fig. 11(a)) [31c]. Typical hybrid vesicles include spherical, hemispherical, and disk-like Janus-like vesicles (JVs) with distinguished BCP and TJNP halves (Figs. 11(b)–11(e)) along with patchy vesicles (PVs) with multiple small TJNP domains surrounded

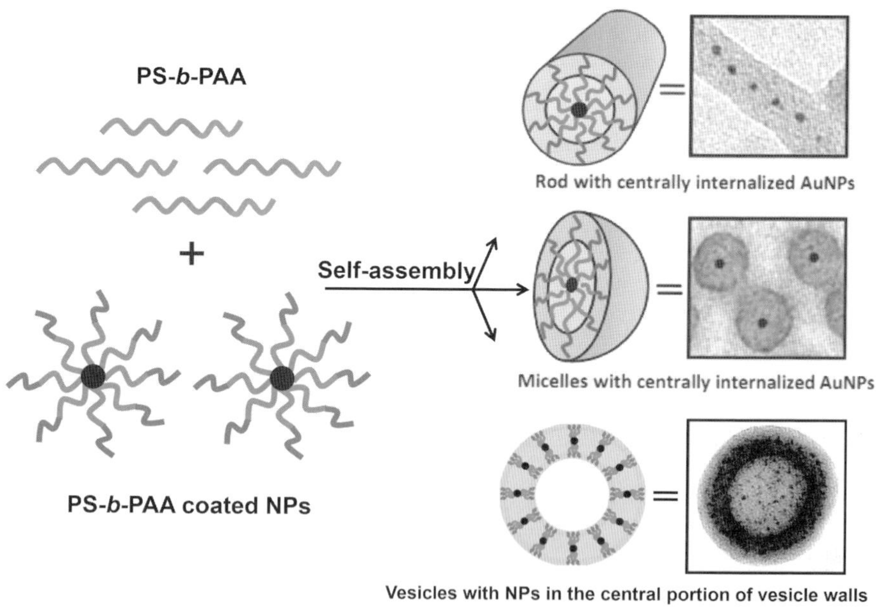

Figure 10. Schematic illustration of the incorporation of BCP-coated NPs into the central portion of BCP rods, micelles or vesicular walls, and corresponding TEM images. Reproduced with permission from Ref. [40]. Copyright (2010) and (2011), American Chemical Society.

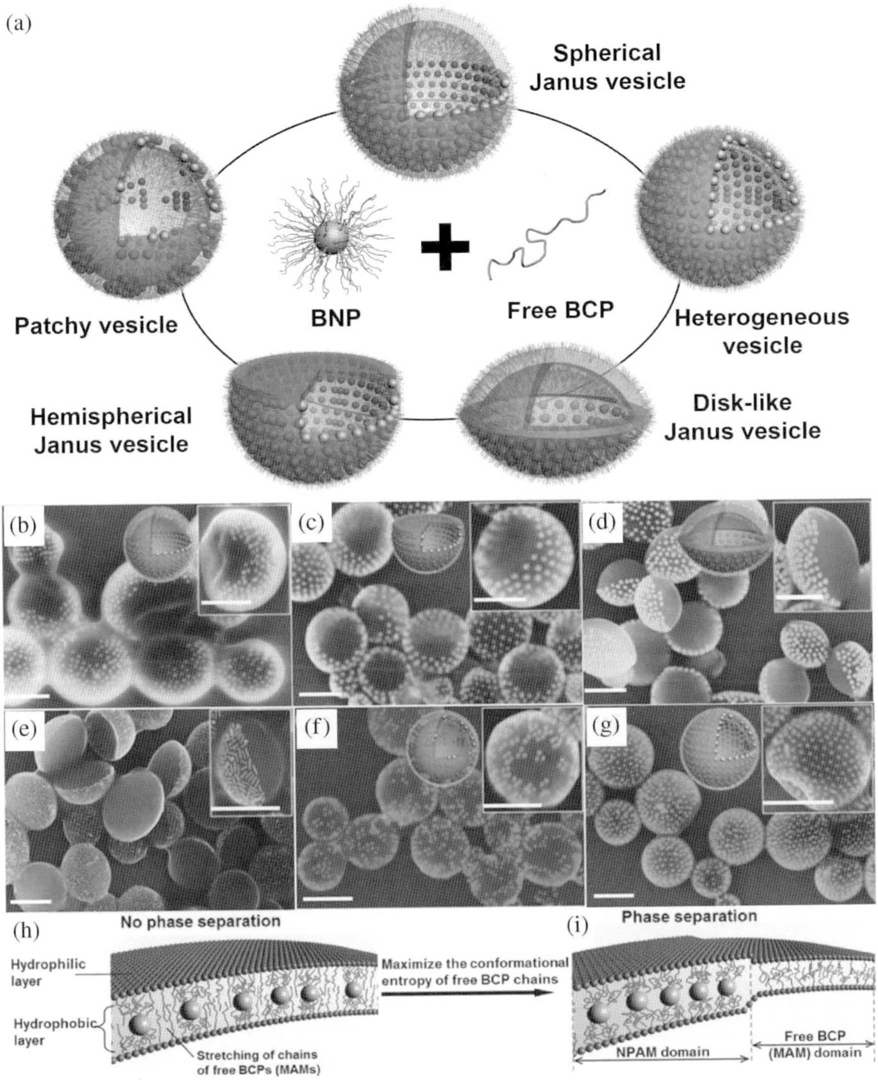

Figure 11. (a) Schematic illustration of the concurrent self-assembly of TJNPs and free BCPs into hybrid vesicles. (b)–(g) Representative SEM images of (b) spherical JVs, (c) hemispherical JVs, (d) disk-like JVs, (e) disk-like JVs with Au NRs, (f) PVs, and (g) HVs. The scale bars: 400 nm in (b), 150 nm in (c), 200 nm in (d), (f), (g), and 600 nm in (e). (h) In membrane without phase separation, the TJNPs are stretched in the direction normal to the membrane due to the compression from neighboring TJNPs. (i) The effective attraction between TJNPs is induced by the need for maximizing conformational entropy of TJNPs. Reproduced with permission from Ref. [31c]. Copyright (2014) American Chemical Society.

by a continuous BCP phase (Fig. 11(f)), and heterogeneous vesicles (HVs) with uniformly distributed TJNPs (Fig. 11(g)). These are all largely dependent on the size of TJNPs and the constituent block lengths of both the tethered and free BCPs. The increase in the length of the BCP hydrophobic block gives rise to a morphological transition, of the assemblies, from PVs to JVs and lastly to HVs. The presence of both the rigid inorganic core and the soft polymer shell significantly increases the range of Young's modules that can be tuned by controlling the relative dimensions of both the organic and inorganic components. The dimensional mismatch of TJNPs and BCPs, the entanglement of BCP chains, and the mobility of TJNPs synergistically determine the formation of different nanostructures. Moreover, it was found that the entropy arisen from the maximization of conformational entropy of the BCP chains plays a dominant role in driving the lateral phase-separation of TJNPs and BCPs in the vesicular membrane.

Since the tethered and free BCPs are both PEO-*b*-PS, the free BCPs serve not only as building blocks in the course of self-assembly, but also as depleting agents to induce depletion interactions between TJNPs. When the membrane thickness is mismatched for TJNPs and BCPs, the depletion effect excludes the BCPs from adjacent TJNPs and hence increases the conformational entropy of BCPs in assemblies (Figs. 11(h) and 11(i)). This results in the reduction of the overall free energy of the system, and the lateral phase separation of BCPs and TJNPs within the vesicular membranes. When the TJNPs and BCPs strongly segregate, JVs with a distinct boundary of BCP and TJNP domains form, while weaker segregation of TJNPs and BCPs results in the formation of PVs or HVs. Thanks to the unique characteristics of TJNPs, such as their dimensions, bending modulus, and rigidity, the morphologies of vesicular nanostructures (i.e., spherical or non-spherical shape) can be finely tuned. Remarkably, the intriguing non-spherical hybrid JVs, which can be hemispherical and disk-like shaped, can be formed due to the interplay of bending modules and line tension of multiple domains of TJNPs and BCPs [41]. By tuning the ratio of TJNPs to BCPs, the JV surface area ratio between the TJNP and BCP side can be readily controlled. When the membrane thickness is equal, TJNPs and BCPs prefer to distribute uniformly in the vesicle membranes. This is interesting for these JVs mimic the shape deformation and budding of multicomponent lipid vesicles found in nature.

6.3.3. Microfluidic Method

Both the film rehydration and selective solvent methods have been proven to be efficient strategies for the self-assembly of amphiphilic JNPs; however, these methods exhibit poor reproducibility and limited control with respect to the scope of assemblies (e.g., the vesicles are usually 50–500 nm in diameter). The dimension of JNPs is about one or two orders of magnitude larger than conventional BCPs and other small molecular amphiphiles. The relatively large size of JNP building blocks leads to a much slower relaxation throughout the course of assembly, thus allowing kinetics to start playing a critical role in the self-assembly. Although this poses a challenge in obtaining pristine equilibrium structures void of defects, the capability of controlling assembly kinetics enables us the ability to fabricate functional structures with increasing complexity, for example, non-equilibrium structures. Microfluidics has emerged as a powerful tool not only for the manipulation of fluids at the nano or picoliter scale but also for tuning the flow hydrodynamics at the microscale, while demonstrating superior control over the self-assembly of small molecular amphiphiles and amphiphilic BCPs [42]. This technique offers us a new tool to control the kinetic aspects of JNP self-assembly by tuning the hydrodynamics of fluids to fabricate functional materials [43]. As shown in Fig. 12(a), TJNP self-assembly is typically triggered by focusing a solution of TJNPs in THF between two water streams at a junction through a common outlet channel in microfluidic flow-focusing devices (MFFDs) [44]. Consequently, the low Reynolds number (ranging from 0.1 to 5.0) in microchannels causes the streams of two miscible fluids to form a laminar flow, where the mixing between fluids is controlled by diffusion. Additionally, the width of the central stream as well as the diffusive mixing can be precisely controlled by varying the volume flow rates (Q) of each fluid. Thus, easy manipulation of the Q directly influences the diffusive mixing along the transverse direction, changing the quality of solvents at the biphasic boundary, which triggers simultaneous self-assembly.

As a typical example, Au NRs tethered with BCPs are chosen as the building blocks, owing to their tunable LSPR absorption in the NIR region. The microfluidic platform is then applied for the continuous self-assembly of these building blocks, where precise manipulation of the hydrodynamic flow enables reproducible generation of various

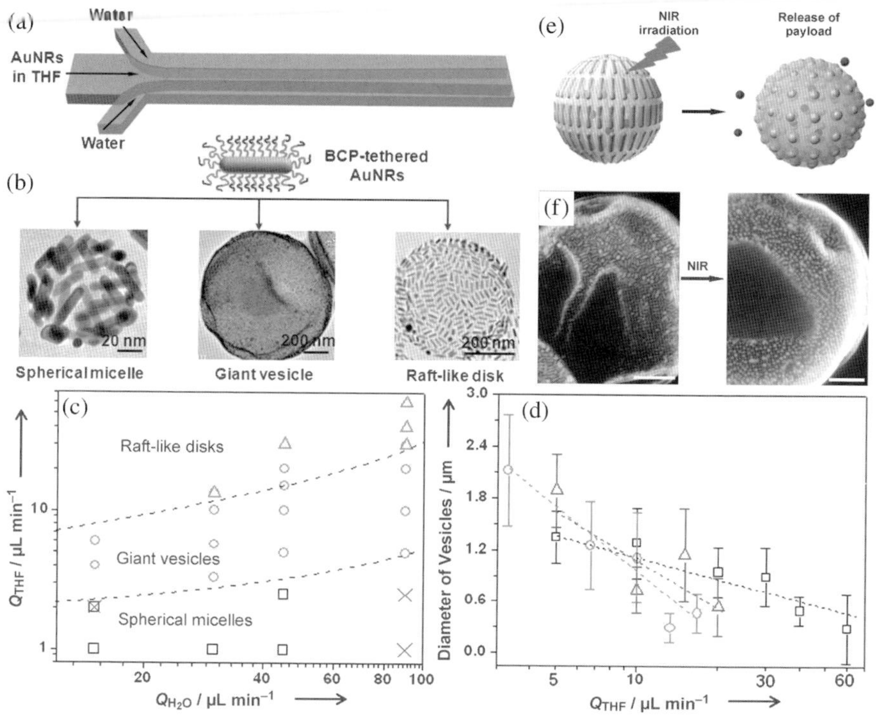

Figure 12. (a) Hydrodynamic self-assembly of amphiphilic NPs tethered with BCPs using MFFDs. A THF solution of polymer-tethered NPs is focused by two water streams on the sides. (b) TEM images of assemblies of PEO45-b-PS211 tethered to Au NRs and obtained at different hydrodynamic conditions in MFFDs. (c) The product diagram of the self-assembly of polymer tethered Au NRs in the [log Q_{THF}, log Q_{H_2O}] space: micelles (□), disks (Δ), giant vesicles (○), and no assemblies (×). (d) The size of giant vesicles as a function of the flow rate of central THF phase at a fixed water flow rate: 90 μLmin^{-1} (□), 45 μLmin^{-1} (Δ), and 30 μLmin^{-1} (○). (e) NIR-triggered release using giant vesicles. The photothermally induced shape deformation of Au NRs to spherical Au NPs creates extra spacing between Au NPs for the release of encapsulated molecules. (f) SEM images of giant vesicles before and after exposed to NIR laser. The Au NRs in the vesicles deform to spherical Au NPs upon irradiation. Scale bars are 200 nm. Reproduced with permission from Ref. [45]. Copyright (2013) Wiley-VCH.

non-equilibrium structures such as micelles, giant vesicles, and disks with a high degree of dimensional control (Fig. 12(b)) [45]. In particular, the monodispersed giant vesicles, comprising a monolayer of Au NRs in the membranes, can be tuned to various sizes by slightly adjusting the flow rate of each fluid. As stated earlier, assembly morphologies are mainly determined by the hydrodynamic conditions, such as the flow-rate ratio (FRR) of water to THF (Q_{H_2O}/Q_{THF}) (Fig. 12(c)), rather than the solvent compositions. When the FRR > 7.5, small micelles with diameters less than 100 nm are obtained due to the strong focusing effect imposed on TJNPs by the two side water streams. As a result, the fast depletion of the NP phase induces a quick quenching of the assembly. When the FRR is reduced to 2.5 < FRR < 7.5, the assembled structures sharply transition from micelles to giant vesicles with diameters larger than 500 nm. In this case, the solely diffusive mixing between neighboring phases offers a long-duration intermediate environment for the continuous recruitment of Au NRs at a temporal spatial concentration gradient region of the boundary. As the FRR is further decreased below 2.5, the assembled structures transition into disk-like micelles, composed of a single layer of Au NRs, with diameters in the range of 300–500 nm. In addition to morphology control, the size of the resulting structures can also be manipulated by adjusting the FRR. For example, the size of the giant vesicles begins at 500 nm and can be increased up to 2.0 μm by decreasing Q_{THF} when Q_{H_2O} is held constant due to the approximate linear dependence on Fig. 12(d). This is opposite to the microfluidic self-assembly of molecular amphiphiles due to the significant difference of diffusion coefficients. Additionally, the TJNP size also plays an important role in determining the assembly dimensions, as a result of the competitive diffusion of solvent/TJNPs along the transverse direction. Larger TJNPs with small diffusion coefficient lead to a longer mixing time and hence the formation of large assemblies, and *vice versa*. Our experimental observations are in good agreement with the COMSOL simulation results.

Additionally, the strong photothermal effects of giant vesicles enable the remote-controlled release of payloads (Figs. 12(e) and 12(f)). Upon NIR light irradiation, giant vesicles are disrupted and ruptured due to localized intense heating, resulting in the delivery of drug payloads. After laser exposure, Au NRs are deformed into spherical Au NPs due to the photothermal melting of Au NRs. The shape transformation is ascribed to the electron–photon relaxation process, during which the absorbed Au NR photons transfer energy, resulting in the heating of free electrons.

More recently, microfluidic approaches have been developed for the continuous synthesis of nanomotors (Figs. 13(a) and 13(b)) [46]. JVs

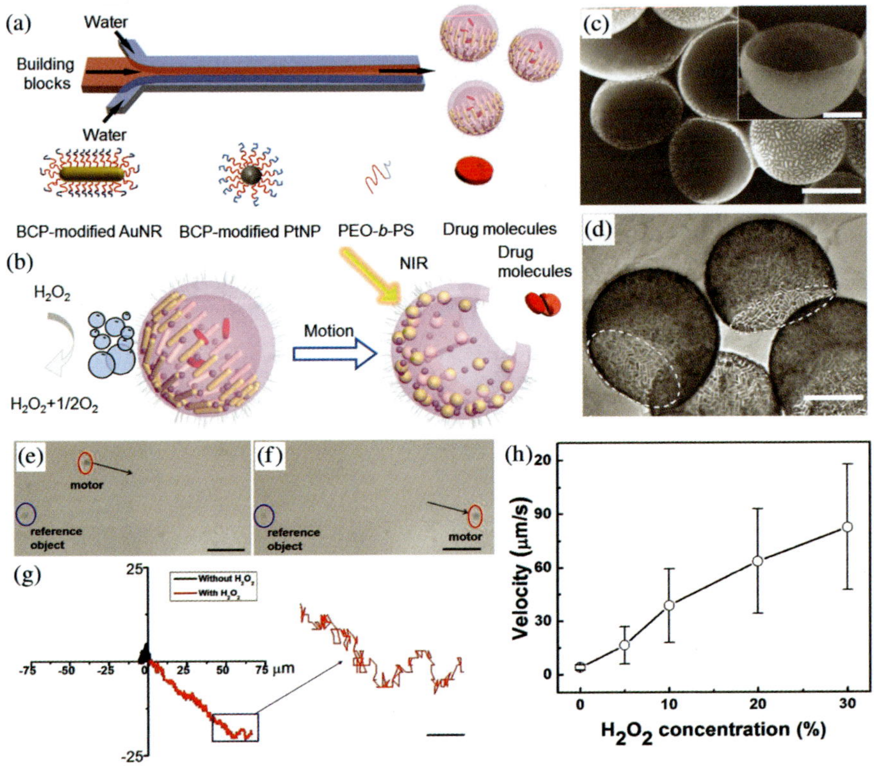

Figure 13. (a) Schematic illustration of the self-assembly of a mixture of building blocks (i.e., BCP modified AuNRs/PtNPs and free BCPs) into JVs in MFFDs. (b) Scheme of the self-propelling of JVs in the presence of H_2O_2 as fuel and the payload release from JVs upon the irradiation of NIR light. (c) SEM and (d) TEM of JVs assembled from a mixture building blocks of free PEO45-*b*-PS455 and Au NRs tethered with PEO45-*b*-PS455-SH. Scale bars: 500 nm (c), 200 nm for inset in (c), and 500 nm (d). The motion of JVs asymmetrically integrated with PtNPs as catalytic motors. (e, f) The snapshots of the movement of vesicular motor (red circle) in the presence of 10% H_2O_2 with polymer microbeads as reference objects (blue circle). The scale bars are 20 μm. (g) The trajectory of vesicular motor with (black) and without (red) the addition of H_2O_2. Scale bar is 4 μm. (h) The average velocity of the JV motors as a function of the concentration of H_2O_2. Reproduced with permission from Ref. [46]. Copyright (2015) Wiley-VCH.

with two distinct halves of NP-rich domains and polymer-rich domains (Figs. 13(c) and 13(d)) are prepared using MFFDs through the assembly of a mixture of amphiphilic BCPs, BCP-tethered Pt NPs, and BCP-tethered Au NRs. This strategy enables fine control over both the relative portion of the NP domain in the vesicular membranes and the overall diameter of the JVs. When the free BCP concentration is increased from 0.1 to 0.3 mg mL^{-1}, the coverage percentage of the polymer-rich domain increases from 15.4% to 40.4%. Using the principle of microfluidic flow rate control, the size of the JVs can be tuned from 400 nm to 2.6 μm by decreasing the FRR. The asymmetrical integration of Pt NPs and Au NRs in the JV membrane imparts the system with both the ability to undergo autonomous propulsion as well as remote-controlled drug release.

When Pt NPs are integrated into one half of JVs, the JVs can undergo self-propulsion with hydrogen peroxide (H_2O_2) serving as the fuel (Figs. 13(e)–13(g)). The movement of JV motors is driven by the catalytic decomposition of H_2O_2 at the Pt hemisphere. The velocity achieved can then be directly controlled by the concentration of H_2O_2 (Fig. 13(h)). Additionally, the controlled release of payload was achieved due to the inclusion of Au NRs within the JV membrane.

6.4. Biological Application of JNP/TJNP Assembly

Individual inorganic NPs have been extensively explored for their application in biomedical imaging and delivery due to their unique size and intrinsic properties [47]. Intriguingly, when assembled into complex architectures, these NP building blocks develop collective properties that can be radically different from their individuals. For example, the assembly of NPs leads a significant reduction of *L*, resulting in a redshift of the extinction spectra due to enhanced plasmon coupling between adjacent NPs. This creates a new paradigm and way to control the desired absorption bands by means other than size and shape manipulation. The ability to control and develop new parameters might lead to new applications of NPs or new solutions to problems that cannot be resolved by individual NPs.

The block copolymer-based strategy for assembling NPs offers us the ability to fine-tune the assembled structures (e.g., morphologies of assemblies, *d*, and *L* between neighboring NPs) and thus their resulting collective properties. As mentioned above, the LSPR peak of the ensembles can be engineered to fall within the NIR range through vesicle formation, due to the strong plasmon coupling between NP subunits in the assemblies. Particularly, NIR light is favorable for biomedical applications due to its ability to penetrate deep into tissues (up to several centimeters), such as skin, tissues, and hemoglobin, which are relatively transparent to NIR light [48]. The strong and tunable NIR absorption band facilitates the application of vesicular assemblies in biomedical imaging, delivery, and photothermal therapy [49]. Particularly, our group and others have recently explored the use of plasmonic vesicles for effective cancer imaging. The strong plasmon coupling induced between NPs within vesicular membranes has been shown to significantly enhance the imaging capability of assembled structures in various imaging modalities, such as multiphoton absorption-induced luminescence (MAIL) imaging, photothermal (PT) imaging, and photoacoustic (PA) imaging, making it an attractive option for further study.

For instance, the enhanced MAIL imaging and photothermal therapy of cancer cells, using TJNP assemblies with strong coupling between NP subunits, have been demonstrated by our group [31b]. In our experiment, 4T1 breast cancer cells are internalized with various assemblies, including unimolecular micelles, monodisperse clusters with controlled numbers of TJNPs, and vesicles. MAIL imaging of the cells was then collected upon excitation with 800 nm light. As expected, the MAIL signals gradually increased with the coupling of electromagnetic field enhancement, along with the increase of the TJNP aggregation number (Fig. 14(a)). Compared with individual TJNPs, the MAIL signal is enhanced by a factor of ~7 for vesicular structures (Fig. 14(b)). In addition to MAIL signal enhancement, Au-based nanostructures can also absorb light and transfer it into heat with high efficiency by the Landau damping effect. Therefore, the vesicular assemblies, with efficient heating upon irradiation with NIR light, are viable candidates for enhanced photothermal therapy (Fig. 14(c)). Upon NIR light irradiation, the cell viability of 4T1 breast cancer cells internalized with vesicular assemblies was only $7.2 \pm 3.4\%$. The *in vivo* therapeutic studies showed that with laser irradiation, tumors treated with vesicular assemblies were effectively ablated without reoccurrence (Fig. 14(d)).

The vesicular assemblies have also been used as a multifunctional theranostic platform for effective combination therapies, such as synergistic PT ablation and photoactivated chemotherapy. Resembling polymersomes or liposomes, the plasmonic vesicles can be used to encapsulate either hydrophilic drugs within the aqueous lumen or hydrophobic drugs within the hydrophobic membrane. Compared to molecular vesicles, the Au vesicles are more stable and show improved payload retention within the cavity.

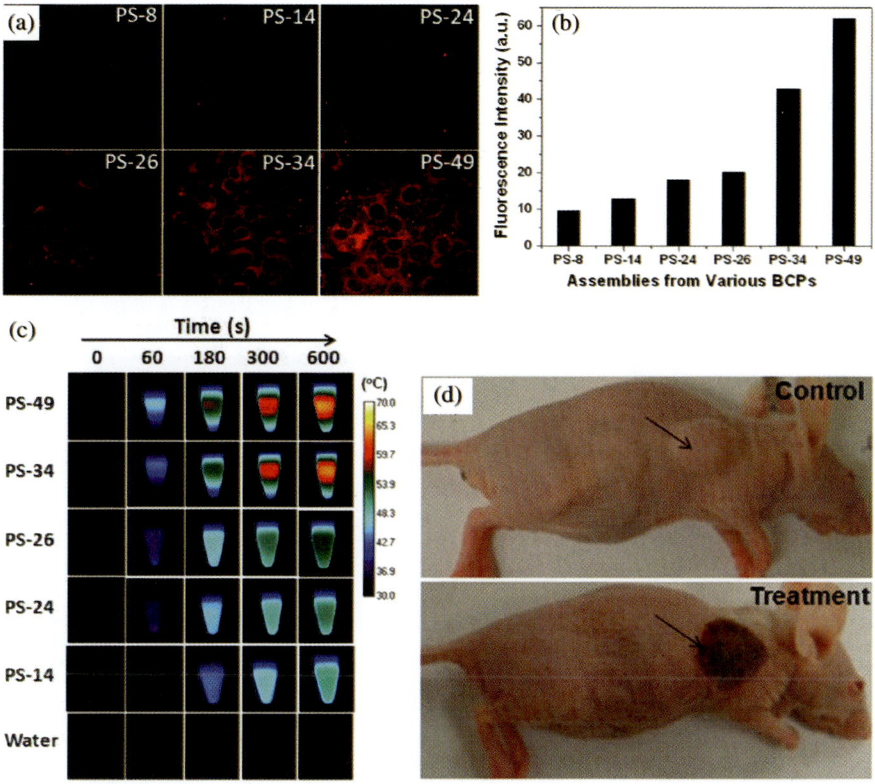

Figure 14. (a) MAIL images of 4T1 cancer cells loaded with assemblies from various TJNPs for 3 h at a concentration of 0.05 mg/mL. (b) The average fluorescence intensity of MAIL images shown in (a). (c) Temperature mapping during the 808 nm laser irradiation of various assemblies in distilled water by real-time thermal imaging using a thermal camera. (d) Mice treated with and without vesicles under laser irradiation [31b]. Copyright (2013) American Chemical Society.

The significantly reduced leakage is possibly due to the kinetic trapping of NPs in the vesicular membranes. Additionally, the release of payloads can be externally controlled by NIR light irradiation, which can potentially reduce chemotherapy side effects that arise from non-specific delivery of therapeutic agents. For example, our group demonstrated a synergistic combination therapy involving PT therapy and photodynamic (PD) therapy for *in vitro* (Fig. 15(a)) and *in vivo* (Fig. 15(b)) tumor treatment based on photosensitizer-loaded plasmonic vesicles [32]. In this study, the plasmonic

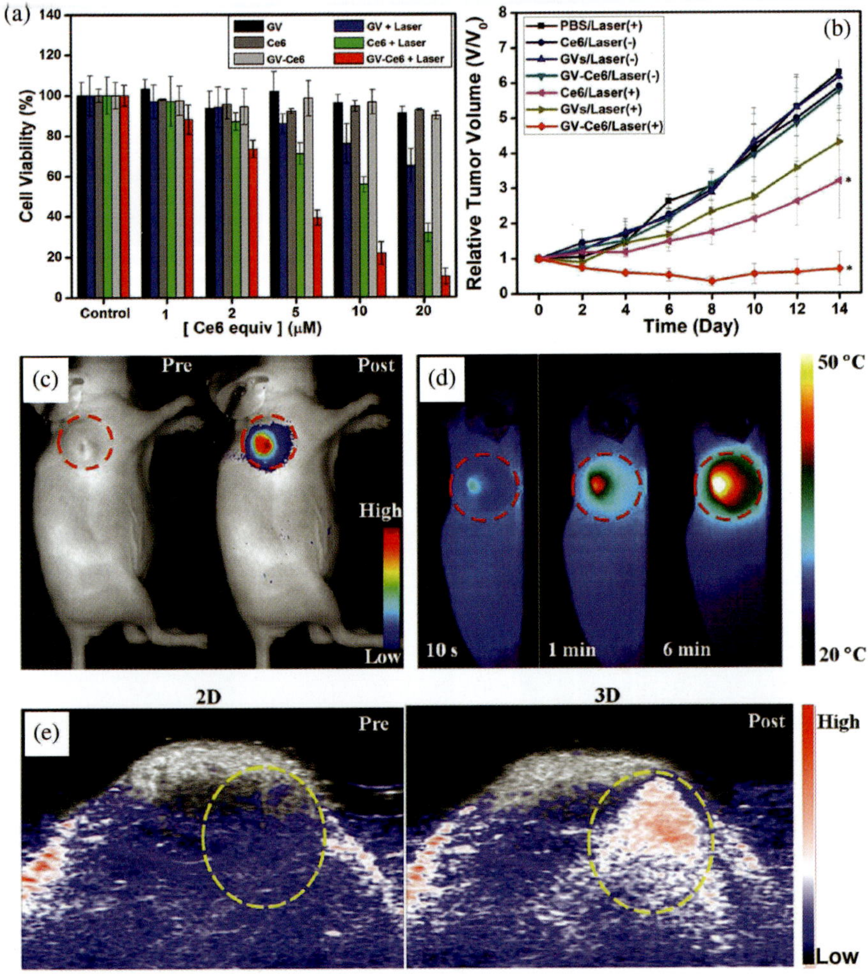

Figure 15. (a) Cell viability at different concentrations of plasmonic vesicles, Ce6, and Ce6-loaded vesicles for 12 h at 37°C with or without irradiation for 3 min with a 671 nm laser (2 W/cm^2). (b) Tumor growth curves of different groups of tumor-bearing mice after treatment. (c) *In vivo* NIR fluorescence image of tumor-bearing mice at preinjection and postinjection of Ce6-loaded vesicles. (d) Thermal images of tumor-bearing mice exposed to 671 nm laser for 6 min at postinjection of Ce6-loaded vesicles. Red circles indicate the location of tumors. (e) *In vivo* PA images of tumor tissues at preinjection and postinjection of Ce6-loaded vesicles. Yellow circles indicate the injected location. Reproduced with permission from Ref. [32]. Copyright (2013) American Chemical Society.

vesicles were loaded with photosensitizers (Ce6) during assembly, which were effectively internalized into cells via energy-dependent endocytosis. The strong NIR absorption coupled with the high loading efficiency of Ce6 enabled the implementation of a trimodal NIR fluorescence/PT/PA imaging-guided synergistic PT/PD therapy with enhanced anticancer therapeutic efficacy. Based on the strong absorption band of Ce6 at 671 nm, the vesicle absorption band was tuned to allow for the simultaneous excitement of both species to produce the heat and singlet oxygen needed to kill cancer cells. Upon laser irradiation, the Ce6-loaded vesicles are disrupted to release the payload, due to localized intense heating. The released Ce6 additionally reacts with oxygen present in tissues, generating singlet oxygen that can kill cancer cells. According to the fluorescence (Fig. 15(c)), PT (Fig. 15(d)), and PA (Fig. 15(e)) signals, upon irradiation with NIR light, the tumor tissues can be selectively destroyed in a non-invasive approach. Both *in vitro* and *in vivo* studies indicated the enhanced therapeutic efficacy of the Ce6-loaded vesicles via the synergetic PT and PD treatment of tumors, which is highly selective and localized. In addition, the multifunctional vesicles have shown to be incredibly stable and biocompatible, thus facilitating their use in biomedical applications.

As previously demonstrated, the chain vesicles exhibit strong NIR absorption due to the strong coupling between Au NPs. Much like with the work that was just discussed, this property makes them an ideal platform for remote NIR light-released drug delivery. Moreover, Au NPs in chain vesicles can absorb light, leading to the emission of acoustic waves. This property enables their detection via ultrasound, thus facilitating PA imaging of biological tissues upon NIR irradiation (Fig. 16(a)). Upon *in vivo* testing of both chain vesicles and non-chain vesicles, the PA signal of chain vesicles was found to have a remarkable enhancement around eightfold times greater than that of the non-chain vesicles (Figs. 16(b)–16(e)). Furthermore, the chain vesicles showed heightened stability when exposed to physiological conditions. These results demonstrate the potential of chain vesicles as more efficient contrast agents for use in PA imaging.

Additionally, recent advances have demonstrated the ability to introduce biodegradation into the assembly and drug delivery process. When the BCPs are transposed with biodegradable PEG-*b*-PCL BCPs, with a

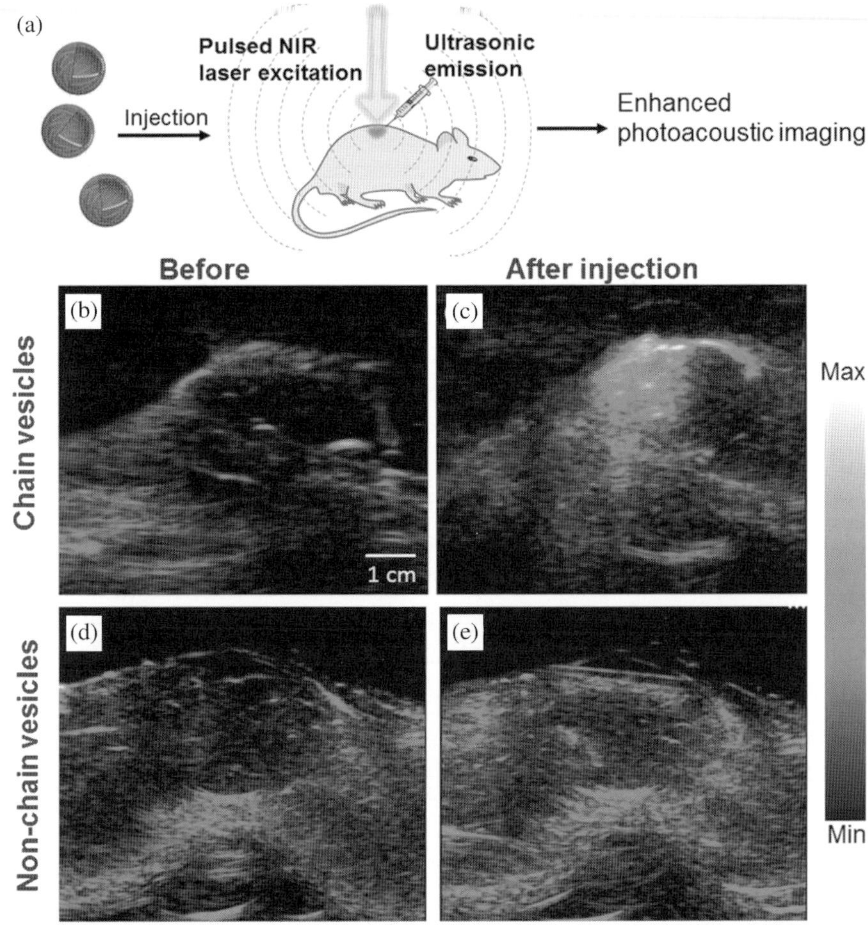

Figure 16. (a) Schematic illustration of enhanced PA imaging with chain vesicles. *In vivo* 2D PA imaging of mouse tissue before and after the injection of chain vesicles (b), (c) or non-chain vesicles (d), (e). Reproduced with permission from Ref. [31d]. Copyright (2015) Wiley-VCH.

disulfide bond at its terminus, the TJNP assembly can lead to the formation of vesicular ensembles with broadened LSPR peaks in NIR region (~800–1000 nm) [50]. The dense packing of TJNPs during the assembly process induces ultrastrong NIR absorption, due to the higher degree of flexibility of the PEG-*b*-PCL tethers when compared to PEG-*b*-PS, which contains rigid benzene rings. The strong NIR absorption coupled with a high photothermal conversion efficiency (37%) of these vesicles facilitates both their effective multimodal PT and PA (Fig. 17(a)) *in vivo* imaging of cancers and enhanced PT therapy (Fig. 17(b)). The assembled biodegradable vesicles demonstrated significantly enhanced PT and PA signals when compared with individual Au NPs, NRs, and nanostars, due to a much higher PT conversion efficiency, which is largely controlled by *L*. Particularly, the *in vivo* PA signal of the biodegradable vesicles was found to be approximately twice as strong as the system assembled with PEG-*b*-PS, due to the significantly reduced *L* between NPs and hence increased PT conversion efficiency. In addition to a higher PT conversion, the greatest advantage of this system can be attributed to thermal

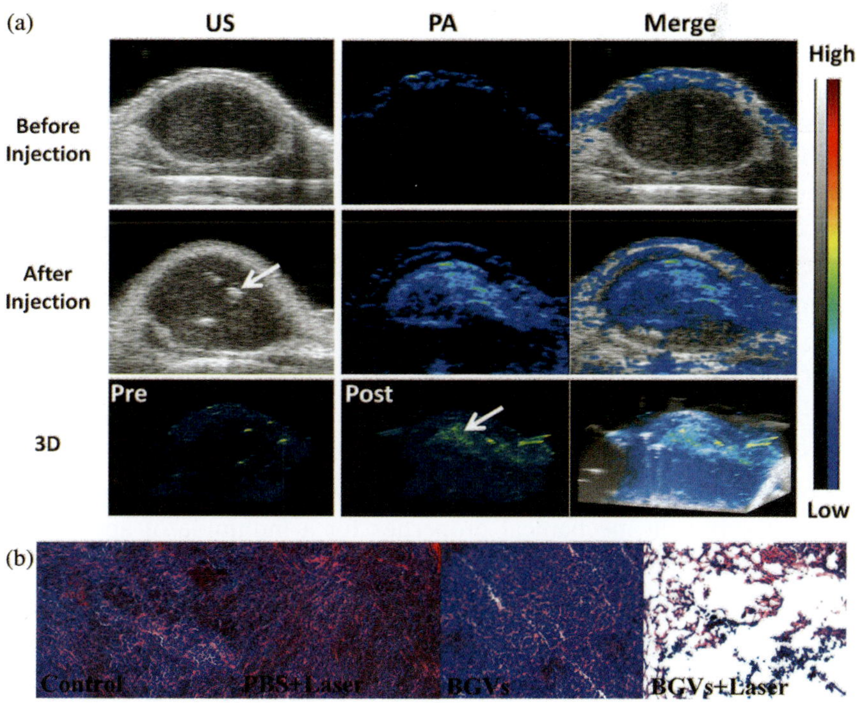

Figure 17. (a) *In vivo* 2D ultrasonic (US) and PA images and 3D PA images of tumor tissues before and after the injection of biodegradable vesicles. Arrows indicate the location of vesicles. (b) Hematoxylin and eosin staining of tumor sections collected from different groups of mice after laser irradiation. Reproduced with permission from Ref. [50]. Copyright (2013) Wiley-VCH.

degradation of the vesicles into constituent building blocks via elevated temperature or longer exposure. After the completion of PT treatment, the tumor area is intentionally irradiated with NIR light to elevate the degradation rate of biodegradable polymers and hence the dissociation rate of plasmonic vesicles. Preliminary results suggest that the disassembling of NP ensembles may help expedite the removal of theranostic NPs from the body when the NP surface is properly designed.

6.5. Conclusions and Outlook

This chapter reviews recent progress in the synthesis, self-assembly, and applications of amphiphilic JNP and TJNP analogs, which have demonstrated great promise in the nanoscale design of novel hybrid materials with increasing complexity and functionality. Structurally, JNPs can be considered as organic molecular analogs that are capable of varying degrees of hierarchical self-assembly. Conversely, JNP assemblies are significantly different from molecular assemblies, owing to their unique dimensions, hybrid composition, complex geometry, and multiscale interactions throughout the course of assembly. The unique combination of BCPs and inorganic NPs shows at least three advantages over existing assembly systems [30a]:

1. These JNPs naturally combine the intrinsic properties of both the inorganic cores (i.e., optical, electronic, and magnetic properties) and the polymers (i.e., flexibility, processability, mechanical, and dielectric properties). Particularly, the polymers will endow the NP assemblies with sufficient mechanical properties for a multitude of applications such as in optoelectronic devices.
2. The (directional dependent) anisotropic interactions between NPs arise spontaneously to drive the hierarchical self-assembly of NPs without the need for complex functionalization of the NPs. The isotropic functionalization of NPs is extremely simple, which makes this methodology particularly intriguing, from the perspective of cost, scaling-up, and processing.
3. The length of a BCP chain is generally in the range of several to tens of nanometers, which features the critical dimension of inorganic

NPs. The compatible length scale of BCPs and NPs broadens the range of tunable forces for assembling NPs and tuning the physical properties of assembled structures, therefore enhancing the capabilities of assembling NPs.

In light of the success in the field of molecular self-assembly, we expect that the compositionally and structurally complex JNPs will lead to a wealth of functional materials with exquisite control over the assembly process.

Despite encouraging potential in the assembly and application of JNPs, challenges still remain at this frontier. First, a better fundamental understanding of JNP assembly at multiple dimensions is urgently needed to achieve a more predictive framework for the design and assembly of JNP building blocks into targeted structures. This requires the collaboration of various tools, including synthetic/assembly techniques, theoretical modeling, and advanced characterization, to quantitatively correlate the building blocks into the final assembly structures. Current studies are still focused on the manipulation of fairly simple NP interactions (e.g., hydrophobic–hydrophobic); thus, in order to achieve more precise and complex assemblies, it is crucial to integrate more intricate interactions such as electrostatic interaction, dipole–dipole interaction, and external fields (e.g., electric and magnetic fields) into the assembly conditions. Second, a systematic structure–property correlation is of great value in order to promote the development of future applications. Particularly, current experimental studies of coupling interactions between NPs are centered on structures that are fabricated on a planar substrate. In contrast, the coupling effect of three-dimensionally arranged NP ensembles is not well understood, largely due to the deficiency of model systems in which parameters can be precisely controlled to predict behavior. However, recent advances in DNA-based NP assemblies may provide model systems for revealing the relationship between structures and properties [51]. Finally, more efforts need to be explored in the development of interesting assembly applications along with critical evaluations of their potential. For example, in order to facilitate the design and application of NP assemblies in the field of biomedicine, it is crucial to systematically assess the risks associated with organ distribution, accumulation, clearance, immunogenicity, and inflammation. The fundamental interactions between NP assemblies and their biological environment need further evaluation before their use in clinical daily practice.

References

[1] (a) Li, M.; Schnablegger, H.; Mann, S. *Nature*, **402**, 393 (**1999**). (b) Lu, Y.; Fan, H.; Stump, A.; Ward, T. L.; Rieker, T.; Brinker, C. J. *Nature*, **398**, 223 (**1999**). (c) Boal, A. K.; Ilhan, F.; DeRouchey, J. E.; Thurn-Albrecht, T.; Russell, T. P.; Rotello, V. M. *Nature*, **404**, 746 (**2000**). (d) Ofir, Y.; Samanta, B.; Rotello, V. M. *Chem. Soc. Rev.*, **37**, 1814 (**2008**). (e) Nie, Z.; Petukhova, A.; Kumacheva, E. *Nat Nano* **5**, 15 (**2010**). (f) Jones, M. R.; Osberg, K. D.; Macfarlane, R. J.; Langille, M. R.; Mirkin, C. A. *Chem. Rev.*, **111**, 3736 (**2011**). (g) Rycenga, M.; Cobley, C. M.; Zeng, J.; Li, W.; Moran, C. H.; Zhang, Q.; Qin, D.; Xia, Y. *Chem. Rev.*, **111**, 3669 (**2011**). (h) Wang, L.; Xu, L.; Kuang, H.; Xu, C.; Kotov, N. A. *Acc. Chem., Res.* **45**, 1916 (**2012**).

[2] (a) Daniel, M.-C.; Astruc, D. *Chem. Rev.*, **104**, 293 (**2004**). (b) Talapin, D. V.; Lee, J.-S.; Kovalenko, M. V.; Shevchenko, E. V. *Chem. Rev.*, **110**, 389 (**2010**).

[3] (a) Lopes, W. A.; Jaeger, H. M. *Nature*, **414**, 735 (**2001**). (b) Xia, Y.; Yin, Y.; Lu, Y.; McLellan, J. *Adv. Funct. Mater.*, **13**, 907 (**2003**). (c) Correa-Duarte, M. A.; Pérez-Juste, J.; Sánchez-Iglesias, A.; Giersig, M.; Liz-Marzán, L. M. *Angew. Chem., Int. Ed.*, **44**, 4375 (**2005**).

[4] Liu, J.; Lu, Y. *J. Am. Chem. Soc.*, **125**, 6642 (**2003**).

[5] (a) Zhang; Glotzer, S. C. *Nano Lett.*, **4**, 1407 (**2004**). (b) DeVries, G. A.; Brunnbauer, M.; Hu, Y.; Jackson, A. M.; Long, B.; Neltner, B. T.; Uzun, O.; Wunsch, B. H.; Stellacci, F. *Science*, **315**, 358 (**2007**).

[6] (a) Sahoo, Y.; Cheon, M.; Wang, S.; Luo, H.; Furlani, E. P.; Prasad, P. N. *J. Phys. Chem., B* **108**, 3380 (**2004**). (b) Leunissen, M. E.; Vutukuri, H. R.; van Blaaderen, A. *Adv. Mater.*, **21**, 3116 (**2009**).

[7] (a) Discher, B. M.; Won, Y.-Y.; Ege, D. S.; Lee, J. C.-M.; Bates, F. S.; Discher, D. E.; Hammer, D. A. *Science*, **284**, 1143 (**1999**). (b) Mai, Y.; Eisenberg, A. *Chem. Soc. Rev.*, **41**, 5969 (**2012**).

[8] (a) Shan, J.; Tenhu, H. *Chem. Commun.*, 4580 (**2007**). (b) Lattuada, M.; Hatton, T. A. *Nano Today*, **6**, 286 (**2011**). (c) Walther, A.; Müller, A. H. E. *Chem. Rev.*, **113**, 5194 (**2013**). (d) Moffitt, M. G. *J. Phys. Chem. Lett.*, **4**, 3654 (**2013**).

[9] Chen, Q.; Bae, S. C.; Granick, S. *Nature*, **469**, 381 (**2011**).

[10] Gentilini, C.; Pasquato, L. *J. Mater. Chem.*, **20**, 1403 (**2010**).
[11] (a) Caswell, K. K.; Wilson, J. N.; Bunz, U. H. F.; Murphy, C. J. *J. Am. Chem. Soc.*, **125**, 13914 (**2003**). (b) Wang, L.; Zhu, Y.; Xu, L.; Chen, W.; Kuang, H.; Liu, L.; Agarwal, A.; Xu, C.; Kotov, N. A. *Angew. Chem. Int. Ed.*, **49**, 5472 (**2010**).
[12] (a) Nie, Z.; Fava, D.; Kumacheva, E.; Zou, S.; Walker, G. C.; Rubinstein, M. *Nat. Mater.*, **6**, 609 (**2007**). (b) Nie; Fava, D.; Rubinstein, M.; Kumacheva, E. *J. Am. Chem. Soc.*, **130**, 3683 (**2008**).
[13] Liu, K.; Nie, Z.; Zhao, N.; Li, W.; Rubinstein, M.; Kumacheva, E. *Science*, **329**, 197 (**2010**).
[14] (a) Wu, S.; Tan, S. Y.; Ang, C. Y.; Nguyen, K. T.; Li, M.; Zhao, Y. *Chem. Commun.*, **51**, 11622 (**2015**). (b) Lattuada, M.; Hatton, T. A. *J. Am. Chem. Soc.*, **129**, 12878 (**2007**). (c) Sardar, R.; Heap, T. B.; Shumaker-Parry, J. S. *J. Am. Chem. Soc.*, **129**, 5356 (**2007**).
[15] (a) Wang, B.; Li, B.; Zhao, B.; Li, C. Y. *J. Am. Chem. Soc.*, **130**, 11594 (**2008**). (b) Wang, B.; Dong, B.; Li, B.; Zhao, B.; Li, C. Y. *Polymer*, **51**, 4814 (**2010**).
[16] Wang, B.; Li, B.; Dong, B.; Zhao, B.; Li, C. Y. *Macromolecules*, **43**, 9234 (**2010**).
[17] Dong, B.; Li, B.; Li, C. Y. *J. Mater. Chem.*, **21**, 13155 (**2011**).
[18] Zhou, T.; Wang, B.; Dong, B.; Li, C. Y. *Macromolecules*, **45**, 8780 (**2012**).
[19] Liu, G.; Tian, J.; Zhang, X.; Zhao, H. *Chemistry — An Asian Journal*, **9**, 2597 (**2014**).
[20] Pradhan, S.; Xu, L.; Chen, S. *Adv. Funct. Mater.*, **17**, 2385 (**2007**).
[21] Andala, D. M.; Shin, S. H. R.; Lee, H.-Y.; Bishop, K. J. M. *ACS Nano*, **6**, 1044 (**2012**).
[22] Hu, J.; Wu, T.; Zhang, G.; Liu, S. *J. Am. Chem. Soc.*, **134**, 7624 (**2012**).
[23] (a) Zhao, B.; Zhu, L. *Macromolecules*, **42**, 9369 (**2009**). (b) Lee, H.-Y.; Shin, S. H. R.; Drews, A. M.; Chirsan, A. M.; Lewis, S. A.; Bishop, K. J. M. *ACS Nano*, **8**, 9979 (**2014**). (c) Iida, R.; Kawamura, H.; Niikura, K.; Kimura, T.; Sekiguchi, S.; Joti, Y.; Bessho, Y.; Mitomo, H.; Nishino, Y.; Ijiro, K. *Langmuir*, **31**, 4054 (**2015**). (d) Bourone, S. D. M.; Kaulen, C.; Homberger, M.; Simon, U. *Langmuir*, **32**, 954 (**2016**). (e) Harkness, K. M.; Balinski, A.; McLean, J. A.; Cliffel, D. E. *Angew. Chem., Int. Ed.*, **50**, 10554 (**2011**). (f) Kim, H.; Carney, R. P.; Reguera, J.; Ong, Q. K.; Liu, X.; Stellacci, F. *Adv. Mater.*, **24**, 3857 (**2012**). (g) Chen, T.; Yang, M.; Wang, X.; Tan, L. H.; Chen, H. *J. Am. Chem. Soc.*, **130**, 11858 (**2008**). (h) Chen, T.; Chen, G.; Xing, S.; Wu, T.; Chen, H. *Chem. Mater.*, **22**, 3826 (**2010**). (i) Choueiri, R. M.; Klinkova, A.; Thérien-Aubin, H. s.; Rubinstein, M.;

Kumacheva, E. *J. Am. Chem. Soc.*, **135**, 10262 (**2013**). (j) Singh, C.; Ghorai, P. K.; Horsch, M. A.; Jackson, A. M.; Larson, R. G.; Stellacci, F.; Glotzer, S. C. *Phys. Rev. Lett.*, **99**, 226106 (**2007**).

[24] (a) van Herrikhuyzen, J.; Portale, G.; Gielen, J. C.; Christianen, P. C. M.; Sommerdijk, N. A. J. M.; Meskers, S. C. J.; Schenning, A. P. H. J. *Chem. Commun.*, 697 (**2008**). (b) Jakobs, R. T. M.; van Herrikhuyzen, J.; Gielen, J. C.; Christianen, P. C. M.; Meskers, S. C. J.; Schenning, A. P. H. J. *J. Mater. Chem.*, **18**, 3438 (**2008**).

[25] Nikolic, M. S.; Olsson, C.; Salcher, A.; Kornowski, A.; Rank, A.; Schubert, R.; Frömsdorf, A.; Weller, H.; Förster, S. *Angew. Chem., Int. Ed.*, **48**, 2752 (**2009**).

[26] Guo, Y.; Harirchian-Saei, S.; Izumi, C. M. S.; Moffitt, M. G. *ACS Nano*, **5**, 3309 (**2011**).

[27] Song, J.; Cheng, L.; Liu, A.; Yin, J.; Kuang, M.; Duan, H. *J. Am. Chem. Soc.*, **133**, 10760 (**2011**).

[28] Song, J.; Pu, L.; Zhou, J.; Duan, B.; Duan, H. *ACS Nano*, **7**, 9947 (**2013**).

[29] Zhao, B.; Haasch, R. T.; MacLaren, S. *J. Am. Chem. Soc.*, **126**, 6124 (**2004**).

[30] (a) Liu, Y.; Liu, Y.; Yin, J.-J.; Nie, Z. *Macromol. Rapid Commun.*, **36**, 711 (**2015**). (b) Spaeth, J. R.; Kevrekidis, I. G.; Panagiotopoulos, A. Z. *J. Chemi. Phys.*, **135**, 184903 (**2011**). (c) Dong, J.; Li, J.; Zhou, J. *Langmuir*, **30**, 5599 (**2014**).

[31] (a) He, J.; Liu, Y.; Babu, T.; Wei, Z.; Nie, Z. *J. Am. Chem. Soc.*, **134**, 11342 (**2012**). (b) He, J.; Huang, X.; Li, Y.-C.; Liu, Y.; Babu, T.; Aronova, M. A.; Wang, S.; Lu, Z.; Chen, X.; Nie, Z. *J. Am. Chem. Soc.*, **135**, 7974 (**2013**). (c) Liu, Y.; Li, Y.; He, J.; Duelge, K. J.; Lu, Z.; Nie, Z. *J. Am. Chem. Soc.*, **136**, 2602 (**2014**). (d) Liu, Y.; He, J.; Yang, K.; Yi, C.; Liu, Y.; Nie, L.; Khashab, N. M.; Chen, X.; Nie, Z. *Angew. Chem., Int. Ed.*, **54**, 15809 (**2015**).

[32] Lin, J.; Wang, S.; Huang, P.; Wang, Z.; Chen, S.; Niu, G.; Li, W.; He, J.; Cui, D.; Lu, G.; Chen, X.; Nie, Z. *ACS Nano*, **7**, 5320 (**2013**).

[33] (a) Reinhard, B. M.; Siu, M.; Agarwal, H.; Alivisatos, A. P.; Liphardt, J. *Nano Lett.*, **5**, 2246 (**2005**). (b) Jain, P. K.; Huang, W.; El-Sayed, M. A. *Nano Lett.*, **7**, 2080 (**2007**).

[34] Zhong, Z.; Patskovskyy, S.; Bouvrette, P.; Luong, J. H. T.; Gedanken, A. *J. Phys. Chem. B.*, **108**, 4046 (**2004**).

[35] Xie, J.; Zhang, Q.; Lee, J. Y.; Wang, D. I. C. *ACS Nano*, **2**, 2473 (**2008**).

[36] He, J.; Zhang, P.; Babu, T.; Liu, Y.; Gong, J.; Nie, Z. *Chem. Commun.*, **49**, 576 (**2013**).

[37] Discher, D. E.; Eisenberg, A. *Science*, **297**, 967 (**2002**).

[38] Zubarev, E. R.; Xu, J.; Sayyad, A.; Gibson, J. D. *J. Am. Chem. Soc.*, **128**, 15098 (**2006**).

[39] Motornov, M.; Roiter, Y.; Tokarev, I.; Minko, S. *Prog. Polym. Sci.*, **35**, 174 (**2010**).

[40] (a) Mai, Y.; Eisenberg, A. *J. Am. Chem. Soc.*, **132**, 10078 (**2010**). (b) Mai, Y.; Eisenberg, A. *Macromolecules*, **44**, 3179 (**2011**).

[41] Hu, J.; Weikl, T.; Lipowsky, R. *Soft Matter*, **7**, 6092 (**2011**).

[42] (a) McDonald, J. C.; Whitesides, G. M. *Acc. Chem. Res.*, **35**, 491 (**2002**). (b) Stroock, A. D.; Whitesides, G. M. *Acc. Chem. Res.*, **36**, 597 (**2003**).

[43] Tumarkin, E.; Kumacheva, E. *Chem. Soc. Rev.*, **38**, 2161 (**2009**).

[44] He, J.; Wang, L.; Wei, Z.; Yang, Y.; Wang, C.; Han, X.; Nie, Z. *ACS Appl. Mater. Interfaces*, **5**, 9746 (**2013**).

[45] He, J.; Wei, Z.; Wang, L.; Tomova, Z.; Babu, T.; Wang, C.; Han, X.; Fourkas, J. T.; Nie, Z. *Angew. Chem., Int. Ed.*, **52**, 2463 (**2013**).

[46] Wang, L.; Liu, Y.; He, J.; Hourwitz, M. J.; Yang, Y.; Fourkas, J. T.; Han, X.; Nie, Z. *Small*, **11**, 3762 (**2015**).

[47] (a) Kennedy, L. C.; Bickford, L. R.; Lewinski, N. A.; Coughlin, A. J.; Hu, Y.; Day, E. S.; West, J. L.; Drezek, R. A. *Small*, **7**, 169 (**2011**). (b) Dreaden, E. C.; Alkilany, A. M.; Huang, X.; Murphy, C. J.; El-Sayed, M. A. *Chem. Soc. Rev.*, **41**, 2740 (**2012**).

[48] Wang, Y.; Black, K. C. L.; Luehmann, H.; Li, W.; Zhang, Y.; Cai, X.; Wan, D.; Liu, S.-Y.; Li, M.; Kim, P.; Li, Z.-Y.; Wang, L. V.; Liu, Y.; Xia, Y. *ACS Nano*, **7**, 2068 (**2013**).

[49] Song, J.; Zhou, J.; Duan, H. *J. Am. Chem. Soc.*, **134**, 13458 (**2012**).

[50] Huang, P.; Lin, J.; Li, W.; Rong, P.; Wang, Z.; Wang, S.; Wang, X.; Sun, X.; Aronova, M.; Niu, G.; Leapman, R. D.; Nie, Z.; Chen, X. *Angew. Chem., Int. Ed.*, **52**, 13958 (**2013**).

[51] Sharma, J.; Chhabra, R.; Cheng, A.; Brownell, J.; Liu, Y.; Yan, H. *Science*, **323**, 112 (**2009**).

Chapter 7

Janus Ellipsoids: Self-Assembly and Applications

Wei Li, Jeffrey M. Rickman*[†,‡]*, Ya Liu*[§]*, and James D. Gunton*[†,¶]

*Materials Research Laboratory,
University of California, Santa Barbara, CA 93106, USA
[†]Department of Physics, Lehigh University,
Bethlehem, PA 18015, USA
[‡]Department of Materials Science and Engineering,
Lehigh University, Bethlehem, PA 18015, USA
[§]Department of Chemical and Petroleum Engineering,
University of Pittsburgh, Pittsburgh, PA 15261, USA
[¶]jdg4@lehigh.edu

Abstract

We review the results of recent numerical and experimental studies of the self-assembly of Janus ellipsoids to examine the role of particle aspect ratio and patch orientation in dictating associated cluster morphologies and properties. Our main emphasis is on the use of simple models in computer simulation to describe the process and structural and thermodynamic consequences of self-assembly. More specifically, we consider both oblate and prolate spheroidal particles and examine the implications of particle geometry on cluster formation. The discussion is then expanded to include the case of patchy ellipsoids in which the size of spheroidal surface patches can be systematically varied. The overarching aim of this review is to highlight the broad range of structures and rich thermodynamic behavior that characterize Janus ellipsoidal systems.

Keywords: Janus ellipsoid, patchy particle, self-assembly, phase diagram, selective encapsulation.

7.1. Introduction

Self-assembly describes the process of autonomous organization of pre-existing components from a disordered system into ordered structures or patterns [1,2]. It is well established that the self-assembly of isotropic colloidal particles results in both amorphous and crystalline structures [2]. However, given the inherent simplicity of building block shape and interactions, traditional colloidal crystals are limited to the phases of simple symmetry, for example, face-centered cubic, hexagonal close-packed, and body-centered cubic, etc. [3]. In recent years, owing to advances in synthetic technology, colloidal particles exhibit a range of anisotropies [3]. As a consequence, building blocks with more complex shapes and directional interactions can be constructed, significantly extending the spectrum of accessible phases for self-assembled structures and the potential applications of the corresponding bulk materials [3–7].

Many experimental and simulational studies have focused on the impact of "anisotropy dimensions" [3], such as interaction patchiness (via surface coverage or surface patterning), aspect ratio, and faceting, on particle self-assembly [8–21]. Colloidal spheres with varying degrees of patchiness (e.g., different number, size, and arrangement of patches) have been fabricated experimentally to study their associated aggregates [15,22,23]. In particular, the Janus sphere, a particle with two hemi-surfaces coded so as to be chemically distinct (e.g., hydrophobic/hydrophilic, charged/uncharged, metallic/polymeric, etc.) has received increasing attention [10,11,15,24–31].

Examples of Janus particles (JPs) exist not only in colloidal science but also in protein science. Most proteins are asymmetric in shape and/or have different types of patches. The patchy particle models of the Kern–Frenkel [8] type have been widely used in studying protein crystallization.

For instance, they have been used in models of phase diagrams for proteins which undergo undesirable aggregation, leading to various types of diseases, including cataracts, sickle cell anemia, and amyloid diseases such as Alzheimer's disease. A recent review of this rapidly growing field is presented in Ref. [32]. Furthermore, there is a class of proteins that is closely related to the JPs discussed in this book. These are hydrophobins, a class of fungi proteins, which have in common a cysteine residue structure that gives rise to a particular hydrophobic patch and an overall amphiphilic nature. Consequently, hydrophobins can undergo self-organization and additional fibrillization. A comprehensive review of this type of protein is given in Ref. [30].

A variety of stable cluster structures has been found for the Janus system under different environmental conditions. Given the range of parameter space, computer simulation has been employed to examine the phase behavior of this system. For numerical convenience, the dissimilarity of the two Janus surfaces is often depicted by a primitive Kern–Frenkel model [10,11]. Monte Carlo simulations using this model have revealed a number of interesting features of the Janus sphere system [11,13]. Moreover, the influence of additional anisotropy dimensions on JP self-assembly has also been of considerable interest, and several recent studies have highlighted methods for the synthesis of non-spherical JPs [33–36]. In particular, by varying the aspect ratio, one can create Janus ellipsoids that have different packing characteristics *vis-à-vis* Janus spheres, thereby expanding the range of assembled cluster morphologies [37–42].

In this chapter, we summarize the findings of recent numerical and experimental studies of Janus ellipsoids, paying particular attention to simulational work based on a pair potential description. Our primary interest is to highlight the role of particle aspect ratio in dictating cluster morphologies and properties associated with self-assembly. In the next section, we outline our computational model and simulation methodology. We consider two types of Janus systems, namely those comprising oblate and prolate spheroids, and we examine the morphology and distribution of clusters in these systems and the implications of clustering on observed properties. We then extend the discussion to the case of patchy ellipsoids in which the patch size can be varied on the particle surface. Finally, in the last section, we present a brief summary.

7.2. Model and Simulation Method

We begin with a description of the model for Janus ellipsoids and an outline of the simulation methodology. In particular, we present a simplified, computationally tractable model of ellipsoidal interactions comprising both hard-core and square-well contributions. The Monte Carlo simulation method in this context is then briefly reviewed.

7.2.1. Spheroidal Geometry

An ellipsoid has semiprincipal axes of length *a, b,* and *c*. If this object is centered at $\mathbf{r}_0 = (x_0, y_0, z_0)$ and has axial orientations $\mathbf{u}^{\mathrm{T}} = (u_1, u_2, u_3)$, $\mathbf{v}^{\mathrm{T}} = (v_1, v_2, v_3)$, $\mathbf{w}^{\mathrm{T}} = (w_1, w_2, w_3)$, respectively, then the corresponding equation for its surface is given by

$$\frac{[\mathbf{u}\cdot(\mathbf{r}-\mathbf{r}_0)]^2}{a^2}+\frac{[\mathbf{v}\cdot(\mathbf{r}-\mathbf{r}_0)]^2}{b^2}+\frac{[\mathbf{w}\cdot(\mathbf{r}-\mathbf{r}_0)]^2}{c^2}=1. \tag{1}$$

Equation (1) may be conveniently rewritten in terms of a matrix representation as $\mathbf{X}^{\mathrm{T}}\mathbf{A}\mathbf{X}=0$, where $\mathbf{X}^{\mathrm{T}}=(\mathbf{x}, \mathbf{y}, \mathbf{z}, 1)$ and $\mathbf{A}$ is a 4×4 symmetric matrix. The 10 independent elements $\mathbf{A}$ are summarized as follows:

$$\begin{aligned}
\mathrm{A}_{11} &= \frac{u_1^2}{a^2}+\frac{v_1^2}{b^2}+\frac{w_1^2}{c^2}\\
\mathrm{A}_{12} &= \frac{u_1u_2}{a^2}+\frac{v_1v_2}{b^2}+\frac{w_1w_2}{c^2}\\
\mathrm{A}_{13} &= \frac{u_1u_3}{a^2}+\frac{v_1v_3}{b^2}+\frac{w_1w_3}{c^2}\\
\mathrm{A}_{14} &= -x_0\mathrm{A}_{11}-y_0\mathrm{A}_{12}-z_0\mathrm{A}_{13}
\end{aligned}$$

$$\begin{aligned}
\mathrm{A}_{22} &= \frac{u_2^2}{a^2}+\frac{v_2^2}{b^2}+\frac{w_2^2}{c^2}\\
\mathrm{A}_{23} &= \frac{u_2u_3}{a^2}+\frac{v_2v_3}{b^2}+\frac{w_2w_3}{c^2}\\
\mathrm{A}_{24} &= -x_0\mathrm{A}_{21}-y_0\mathrm{A}_{22}-z_0\mathrm{A}_{23}
\end{aligned}$$

$$A_{33} = \frac{u_3^2}{a^2} + \frac{v_3^2}{b^2} + \frac{w_3^2}{c^2},$$
$$A_{34} = -x_0 A_{13} - y_0 A_{23} - z_0 A_{33},$$
$$A_{44} = -1 - x_0 A_{14} - y_0 A_{24} - z_0 A_{34}.$$

Here, **A** is normalized such that a point in the interior of the ellipsoid, $\mathbf{X}_0$, satisfies $\mathbf{X}_0^{\mathrm{T}}\,\mathbf{A}\mathbf{X}_0 < 0$, implying that det ($\mathbf{A}$) < 0. This geometric formulation is useful in determining spatial relations between ellipsoids in a multiparticle system.

A Janus ellipsoid is defined as a hard-core particle covered with a hemisurface patch whose normal is aligned along one of its principal axes. For simplicity, we focus on a specific case, namely that of a spheroid for which $a = b \neq c$ (i.e., $a = b = \sigma_\perp$, $c = \sigma_\parallel$). Given the aspect ratio, $\varepsilon = \sigma_\parallel/\sigma_\perp$, the spheroid is characterized as oblate (prolate) if $\varepsilon < 1.0$ ($\varepsilon > 1.0$). Thus, taking the patch orientation into account, there are four varieties of Janus spheroids.

7.2.2. Interaction Potential

We model the interaction between Janus spheroids in terms of a pair potential that depends on particle separation and orientation (see Fig. 1). In particular, we employ the Kern–Frenkel patch model [8], in which the interaction potential for particles i and j having a separation vector $\vec{r}_{ij} \equiv \vec{r}_i - \vec{r}_j$ and patch orientations $\hat{n}_i$ and $\hat{n}_j$, respectively, is given by

$$U_{ij} = U(r_{ij}) f(\hat{r}_{ij}, \hat{n}_i, \hat{n}_j), \tag{2}$$

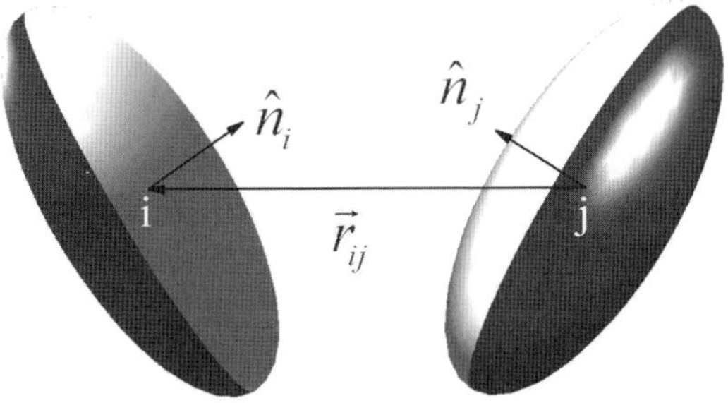

Figure 1. Illustration of two interacting Janus ellipsoids. Light gray hemisurfaces attract each other; black hemisurface interacts with other surfaces through a hard-core repulsion.

with the angular-dependent part,

$$f(\hat{r}_{ij},\hat{n}_i,\hat{n}_j)=\begin{cases}1 & \text{if } \hat{n}_i\cdot\hat{r}_{ij}\le 0,\ \hat{n}_j\cdot\hat{r}_{ij}\ge 0\\ 0 & \text{otherwise,}\end{cases} \tag{3}$$

where the hat denotes a unit vector.

Analogous to the standard square-well potential model, the separation-dependent potential $U(r_{ij})$ is given by [37]

$$U(\hat{r}_{ij})=\begin{cases}\infty & \text{if particles overlap}\\ -u^* H(\sigma_{ij}+w\sigma_0-r_{ij}) & \text{otherwise,}\end{cases} \tag{4}$$

where u^* is the well depth, w is the well width, σ_{ij} is an orientation-dependent range, and σ_0 is the unit of length. The Heaviside step function, $H(\sigma_{ij} + w\sigma_0 - r_{ij})$, truncates the interaction for a center-to-center distance, r_{ij}, which is greater than $\sigma_{ij} + w\sigma_0 - \sigma_{ij}$, is introduced here as an approximation to the orientation-dependent distance between two spheroidal objects, which is generated from the Gaussian overlap model of Berne and Pechukas [43]. In their model, one defines a spheroidal Gaussian by $G(\vec{r})=\exp(-\vec{r}\cdot\boldsymbol{\gamma}^{-1}\cdot\vec{r})$, where $\vec{r}$ is the 3D position vector and the range matrix is written in dyadic form as $\boldsymbol{\gamma}=(\sigma_{\parallel}^2-\sigma_{\perp}^2)\hat{n}\hat{n}+\sigma_{\perp}^2\boldsymbol{I}$. In this expression, $\hat{n}$ is the unit vector along the principal axis of $\sigma_{\parallel}$ and $\boldsymbol{I}$ is the unit matrix. One can then write an interaction potential between two spheroids as

$$\begin{aligned}V(\hat{n}_i,\hat{n}_j,\vec{r}_{ij}) &\sim \left|\boldsymbol{\gamma}_i\right|^{-1/2}\left|\boldsymbol{\gamma}_j\right|^{-1/2}\int d\vec{r}\exp[-\vec{r}\cdot\boldsymbol{\gamma}_i^{-1}\cdot\vec{r}-(\vec{r}-\vec{r}_{ij})\cdot\boldsymbol{\gamma}_j^{-1}\cdot(\vec{r}-\vec{r}_{ij})]\\ &\sim \left|\boldsymbol{\gamma}_i+\boldsymbol{\gamma}_j\right|^{-1/2}\exp[-\vec{r}_{ij}\cdot(\boldsymbol{\gamma}_i+\boldsymbol{\gamma}_j)^{-1}\cdot\vec{r}_{ij}],\end{aligned} \tag{5}$$

which is more conveniently expressed in the simple form

$$V(\hat{n}_i,\hat{n}_j,\vec{r}_{ij})=\varepsilon(\hat{n}_i,\hat{n}_j)\exp[-r_{ij}^2/\sigma^2(\hat{n}_i,\hat{n}_j,\hat{r}_{ij})], \tag{6}$$

where $\varepsilon(\hat{n}_i,\hat{n}_j)$ is an orientation-dependent strength parameter [43].

By extending the mathematical development in the original paper (Ref. [43]), one derives an explicit form for $\sigma(\hat{n}_i, \hat{n}_j, \hat{r}_{ij})$, namely [44]

$$\sigma(\hat{n}_i, \hat{n}_j, \hat{r}_{ij}) = (\sigma_{i\perp}^2 + \sigma_{j\perp}^2)^{\frac{1}{2}} \left[1 - \left(\frac{\lambda_i(\hat{r}_{ij} \cdot n_i)^2 - 2\lambda_i\lambda_j(\hat{r}_{ij} \cdot \hat{n}_i)(\hat{n}_i \cdot \hat{n}_j)(\hat{r}_{ij} \cdot \hat{n}_j) + \lambda_j(\hat{r}_{ij} \cdot \hat{n}_j)^2}{1 - \lambda_i\lambda_j(\hat{n}_i \cdot \hat{n}_j)^2} \right) \right]^{-\frac{1}{2}}, \quad (7)$$

where

$$\lambda_i = (\sigma_{i\|}^2 - \sigma_{i\perp}^2) / (\sigma_{i\|}^2 + \sigma_{j\perp}^2), \quad \lambda_j = (\sigma_{j\|}^2 - \sigma_{j\perp}^2) / (\sigma_{j\|}^2 + \sigma_{i\perp}^2). \quad (8)$$

Specifically, in the case of two identical spheres, each with radius r_0, $\sigma(\hat{n}_i, \hat{n}_j, \hat{r}_{ij}) = \sqrt{2}r_0$. Hence, to define a hard-core interaction potential, a simple choice is $\sigma_{ij} = \sqrt{2}\sigma(\hat{n}_i, \hat{n}_j, \hat{r}_{ij})$, such that there is no overlapping interaction if $r_{ij} \geq \sigma_{ij}$.

We note that this approximation for σ_{ij} has been employed to model other anisotropic particles, such as those found in liquid crystals and granular material [20,21,45], and therefore $H(\sigma_{ij} + w\sigma_0 - r_{ij})$ prescribes an orientation-dependent interaction potential that can be regarded as a "quasi-square-well" potential [37]. In Fig. 2, we present several specific

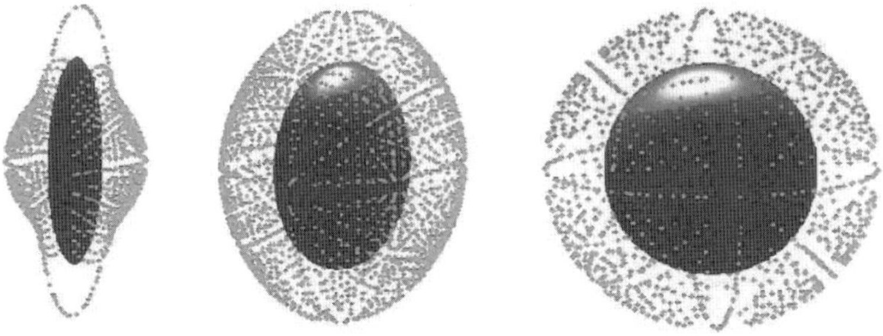

Figure 2. Illustration of interaction regime for an individual ellipsoid under the condition $\hat{n}_i = \hat{n}_j$, $\varepsilon = 0.1$, 0.5, and 0.9 from left to right. The gray surfaces represent the attraction range of quasi-square-well potential and the black bodies for hard cores associated with the individual ellipsoids. Reprinted with permission from Ref. [37]. Copyright (2011) American Chemical Society.

cases corresponding to aspect ratios $\varepsilon = 0.1, 0.5$, and 0.9 (oblate spheroids) that highlight different interaction regimes. In each case, for particles i and j, $\hat{n}_i = \hat{n}_j$, and two particles begin to interact once their gray shells touch each other.

The hard-core nature of $U(r_{ij})$ is enforced in practice as follows [46,47]. Given two ellipsoids, denoted by A and B, with corresponding surface equations $\mathbf{X}^T\mathbf{A}\mathbf{X} = 0$ and $\mathbf{X}^T\mathbf{B}\mathbf{X} = 0$, respectively, a fourth-degree characteristic polynomial $F(\lambda) = \det(\lambda\mathbf{A} - \mathbf{B})$ is constructed, ($\mathbf{A}$ and $\mathbf{B}$ are normalized so that the interiors of A and B satisfy $\mathbf{X}^T\mathbf{A}\mathbf{X} < 0$ and $\mathbf{X}^T\mathbf{B}\mathbf{X} < 0$). The characteristic equation $F(\lambda) = 0$ always has two positive real roots, and the remaining two roots and their geometric implications can be described as follows:

(1) A and B are separate, corresponding to two distinct negative roots.
(2) A and B touch each other externally, corresponding to two equal (negative) roots.
(3) Otherwise, A and B overlap.

For numerical implementation, Sturm sequence methods are employed to determine the nature of the roots and, hence, the geometrical relationship between the two spheroids [47].

In summary, a Janus spheroid comprises two dissimilar hemi-surfaces that interact differently with other particles. For this model, as shown in Fig. 1, it is evident that (a) the interactions between two gray (i.e., attractive) hemisurfaces of Janus spheroids are described by a quasi-square-well potential, and (b) the interaction between a black region and other regions is described by a hard-core repulsion. Finally, we note that this model corresponds to the standard square-well potential model as used in Janus sphere studies when $\varepsilon = 1.0$ [11,13].

7.2.3. Simulation Method

Monte Carlo methods are powerful tools for studying the thermodynamic properties and the cluster morphologies of self-assembling systems. For a more detailed discussion of the methodology itself, the reader is referred to the literature (for example, Ref. [48]). Here, we discuss the application of Monte Carlo simulations to the Janus spheroid model described above. The Monte Carlo simulations are performed in the canonical ensemble. Specifically, the simulation starts from a spatially random initial configuration of monomers that is confined in a 3D cell under periodic boundary conditions. In each Monte Carlo step, trial moves include particle translations and rotations, and these moves are accepted or rejected based on the usual Metropolis criterion [48]. The equilibration of the system is monitored by tracking the temporal evolution of the potential energy of the system, and statistically significant information is obtained from an ensemble average using multiple runs.

We note that the systems examined here are primarily micelle-rich gases, and so the main clustering mechanism is via monomer addition rather than cluster–cluster aggregation [11,13,40,49]. Hence, the conventional Metropolis Monte Carlo scheme employing local particle moves is satisfactory in this context. For systems having a much higher particle density such that hierarchical self-assembly becomes important, a more sophisticated geometric cluster algorithm should be employed to enhance the efficiency of phase-space sampling [50].

7.3. Janus Oblate Spheroids

We begin by reviewing some numerical studies of the aggregation of Janus oblate spheroids. Specifically, the patch is oriented along the principal axis of oblate spheroid with semi-diameter $\sigma_{\parallel}$ (as shown in Fig. 1). For this system, the aspect ratio $\varepsilon < 1.0$ and the length of the longer semi-diameter is $\sigma_{\perp} = \sigma_0/2$ (σ_0 is the unit of length). The well-depth of the interaction potential is $u^* = u_0$, with u_0 the unit of energy and the temperature T given in units of u_0. As is customary, all quantities are reported in reduced units.

7.3.1. The Special Case of Janus Spheres ($\boldsymbol{\varepsilon} = 1.0$)

Before considering the effect of anisotropy, we first discuss the special case for which $\varepsilon = 1.0$, i.e., Janus spheres. Sciortino *et al.* [11] presented a numerical study of the phase behavior of Janus spheres. In their model, Janus spheres interact via a square-well potential with interaction range

$w = 0.5$. The phase diagram is shown in Fig. 3(a). (The phase diagram is updated in a later report, and the estimated gas–liquid critical point is $T_c \approx 0.302$. [13]) By contrast with the normal gas–liquid coexistence curve (exhibiting an upper-critical solution temperature), the size of the coexistence region in this system shrinks as the temperature, T, decreases. Two branches of the coexistence curve shift toward higher density, ρ, and approach each other on cooling. Such behavior creates a wide window for a stable gas-like phase in the low-temperature, low-density regime.

The gas phase comprises large aggregates of micelles and vesicles, each having a well-defined number of particles [11,13,49]. For example, a typical single-layer micelle (in which the attractive patches are oriented inward) consists of approximately 10 particles (Fig. 3(b)), while a typical double-layer vesicle has approximately 40 particles (Fig. 3(c)). For both types of the aggregates, the hard-core parts of the constituent outer-shell particles are oriented outward, thereby inhibiting cluster–cluster interactions. Moreover, the ordered organization of the particles inside the aggregates lowers the average potential energy per particle in the gas phase, in

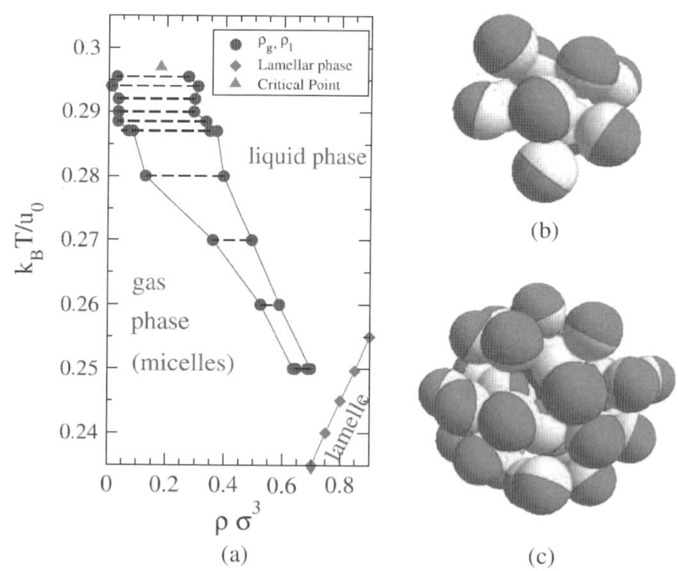

Figure 3. (a) Phase diagram of the Janus spheres in the T–ρ plane, (b) Typical cluster of size 10 (micelle), (c) Typical cluster of size 40 (vesicle). Reproduced from Ref. [11]. Copyright (2009) American Physical Society.

comparison with that in the liquid phase. Nevertheless, the liquid phase possesses orientational disorder and therefore a large entropy. This interplay between the enthalpic and entropic contributions to the free energy leads to the unconventional gas–liquid phase behavior that is characteristic of this system.

7.3.2. Gas Phase Behavior

Monte Carlo simulations have also been applied to study the self-assembly of Janus oblate spheroids [37,40,44]. The aim of these investigations is to elucidate the effect of geometric anisotropy on cluster morphologies. In particular, the systems examined here are located in the gas-phase regime of the phase diagram. For concreteness, we consider explicitly the role of aspect ratio and interaction range on the aggregation of Janus oblate spheroids.

7.3.2.1. *Effect of aspect ratio*

The quasi-square-well potential used for Janus spheroids reduces to the standard square-well model for $\varepsilon = 1.0$, as in this limit, $\sigma_{ij} = \sigma_0$. Thus, in order to compare the results with those obtained in previous studies of Janus sphere aggregation, we first consider the case $w = 0.5$ for the oblate spheroidal system. To examine the impact of aspect ratio on phase behavior, Fig. 4 displays a series of snapshots of the system at temperature

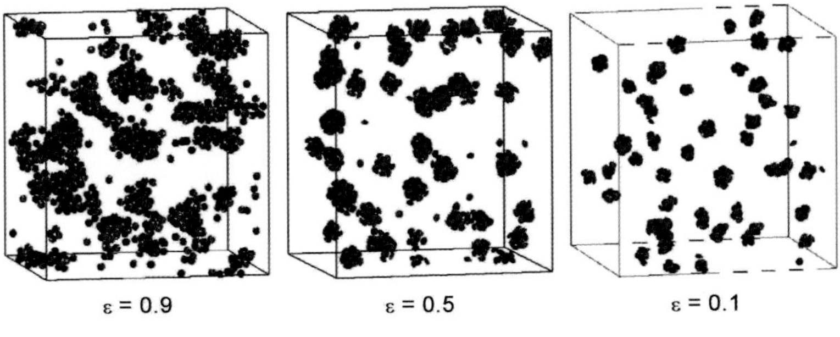

Figure 4. Snapshots of the oblate spheroidal system ($w = 0.5$, $T = 0.33$, $\rho = 0.037$) for different vales of ε. Reproduced with permission from Ref. [37]. Copyright (2011) American Chemical Society.

$T = 0.33$ and particle number density $\rho = 0.037$ for $\varepsilon = 0.9$, 0.5, and 0.1, respectively. As is evident from the figure, phase separation occurs in these three systems. For $\varepsilon = 0.9$, a range of aggregates, such as monomers, oligomers, and giant clusters, are observed. For $\varepsilon = 0.5$ and 0.1, fewer oligomers are formed and the clusters become more uniform in shape. These results suggest that the phase behavior of Janus oblate spheroids is quite similar to that of Janus spheres, for which the gas phase comprises large aggregates. Hence, one would expect that the gas–liquid coexistence curve for the Janus oblate spheroidal system is similar to that found for Janus spheres, as shown in Fig. 3(a).

The distribution of cluster size for different aspect ratios is shown in Fig. 5. As ε increases, the dispersion in cluster size increases. Large clusters are formed through cluster–cluster aggregation, which indicates that the smaller clusters may have open structures with residual attractive patches on the surface. By contrast, for small aspect ratios, typical peaks occur at distinct cluster sizes, implying the prevalence of structurally well-defined clusters in this limit. One should also note that the monomer density decreases as ε decreases as monomers are absorbed by clusters. These results imply that, at $T = 0.33$, a system with larger ε is closer to the

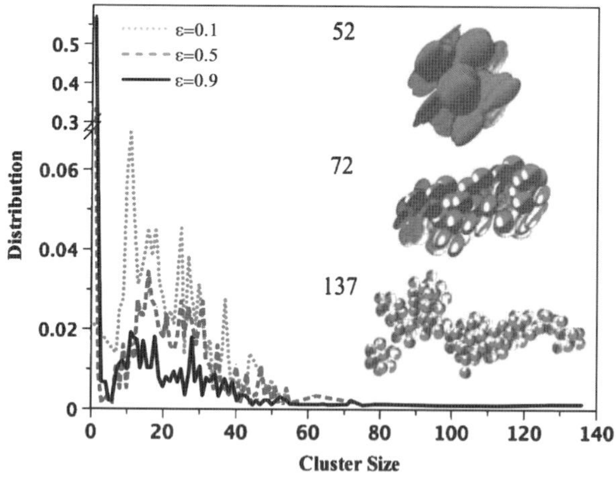

Figure 5. Distribution of cluster size for Janus oblate spheroids for $\varepsilon = 0.1$, 0.5, and 0.9. Insets are the morphologies of the associated largest cluster from the simulations. The number beside each cluster is the cluster size. Reprinted with permission from Ref. [37]. Copyright (2011) American Chemical Society.

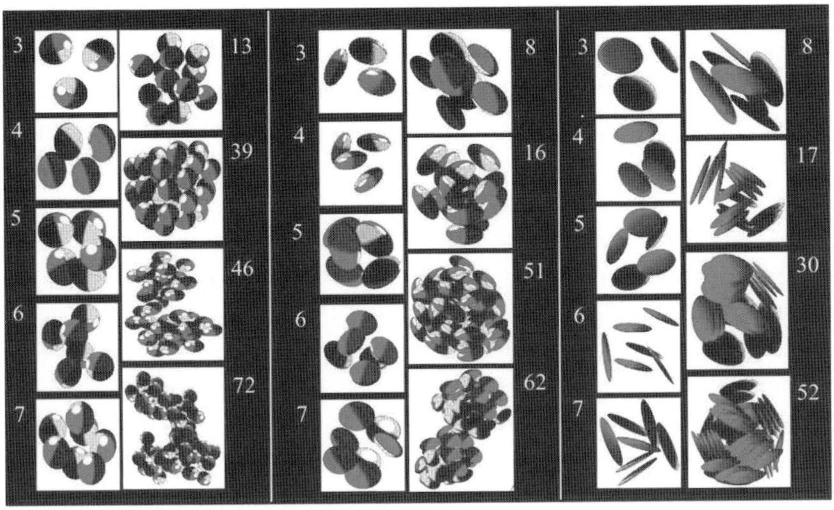

Figure 6. Plot of typical cluster morphologies for $\varepsilon = 0.9$, 0.5, and 0.1 from left to right. The number next to each panel is the cluster size. Reprinted with permission from Ref. [37]. Copyright (2011) American Chemical Society.

gas–liquid critical point. Therefore, the gas–liquid coexistence curve is shifted toward higher T and larger ρ when ε is reduced from 1.0.

Some examples of cluster morphology for different size clusters are shown in Fig. 6. Small oligomers, such as trimer, tetramer, pentamer, hexamer, and heptamer, are similar in shape and monomers form normal polyhedra, except in the case of $\varepsilon = 0.1$. As ε decreases, the ratio between the volume of the interaction range and the particle size increases for Janus spheroids. The system gains more entropy through the variation in cluster morphology while maintaining the same potential energy. Moreover, for $\varepsilon < 0.5$, the minor axis ($2\sigma_{\|}$) of the spheroid is less than the well-width of the interaction potential. From enthalpy considerations, multilayer structures are favored to lower the system's potential energy. Thus, particle anisotropy affects both the enthalpy and entropy of the Janus system and, hence, dictates the self-assembled morphologies.

As the system gets close to the critical point, thermal fluctuation effects become increasingly important, with the consequence that aggregates formed at $T = 0.33$ have more open structures for $\varepsilon = 0.9$. Moreover, for large ε, attractive patches on the aggregate surface cause further interactions among clusters, leading to larger clusters with irregular shapes. For smaller ε the fluctuation effects are relatively less important as the associated system is getting away from the critical point. Hence, the clusters are more organized and essentially spherical. The non-interacting patches of the constituent particles dominate the surface of clusters, and the hard-core repulsion inhibits further aggregation among clusters, thereby resulting in a relatively small thermal correlation length.

To further explore cluster structure and its relationship with the aspect ratio, we study the conformation of bonded pair particles. For this purpose, a scalar product $\hat{n}_1 \cdot \hat{n}_2$ (where $\hat{n}_1$ and $\hat{n}_2$ are patch orientation vectors) is computed for all the bonded pairs in the system and its distribution function $P(\hat{n}_1 \cdot \hat{n}_2)$ is plotted for different values of ε in Fig. 7. From the figure, characteristic peaks are found for $P(\hat{n}_1 \cdot \hat{n}_2)$ curves. In particular, for $\varepsilon = 0.9$, there are two distinct peaks, a result similar to that found for Janus spheres [11]. Specifically, the peak near –1.0 is associated with two face-to-face interacting particles, while the peak near 1.0 indicates a pair of interacting particles that are nearly parallel to each other. Two

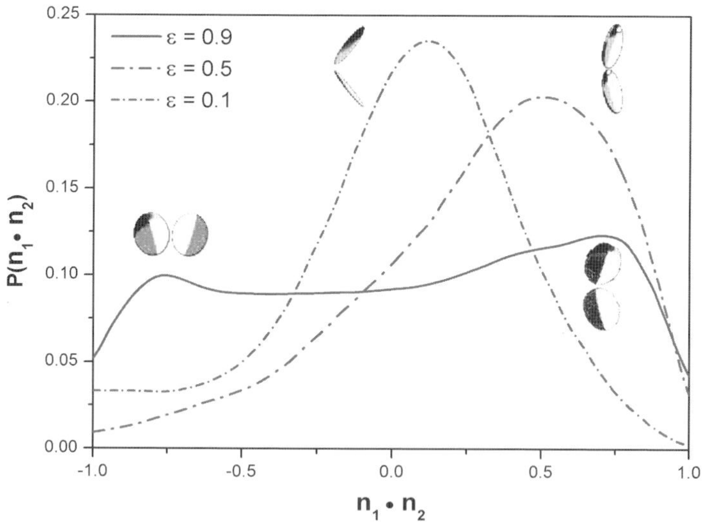

Figure 7. Distribution of $P(\hat{n}_1 \cdot \hat{n}_2)$ over all pairs of bonded Janus oblate spheroids for different vales of ε. Typical configurations corresponding to the characteristic peaks are shown beside the curves. Reproduced with permission from Ref. [37]. Copyright (2011) American Chemical Society.

dissimilar patterns of bonded pairs provide more possible configurations for the self-assembly of large complex clusters.

As ε decreases, the peak near +1.0 shifts toward smaller values of $\hat{n}_1 \cdot \hat{n}_2$. This suggests that a smaller angle between two patches is preferred in the system as the Janus spheroid gets thinner. Hence, considering the packing of particle pairs, clusters associated with smaller internal angles are more compact in structure, as shown in Fig. 6. In contrast, the peak near –1.0 decreases in height as ε varies from 0.9 to 0.5, indicating that the face-to-face particle configuration is less favored. This trend is reversed from $\varepsilon = 0.5$ to 0.1, since complex multilayer clusters are formed at $\varepsilon = 0.1$. Moreover, as T decreases, thermal fluctuation effects are suppressed and the ordering of particle organization inside clusters is enhanced. Therefore, one would expect that the characteristic peaks become more prominent in the distribution curves of $P(\hat{n}_1 \cdot \hat{n}_2)$ (results not shown).

7.3.2.2. *Effect of interaction range*

The study of aggregation in the system with $w = 0.5$ suggests that the gas phase behavior of Janus oblate spheroids is, in general, quite similar to that of Janus spheres. In addition to micelles and vesicles, clusters with multilayer structures are formed upon reducing ε, as dictated by enthalpy considerations. Next, we examine the effect of interaction range on self-assembly morphology. The interaction range can, in practice, be tuned by modifying the environmental conditions of the self-assembling system, e.g., the ionic strength in solution. In Fig. 8(a), the distribution of $P(\hat{n}_1 \cdot \hat{n}_2)$ as a function of w is shown for the system with $\varepsilon = 0.6$ and $\rho = 0.08$ at $T = 0.25$.

We find that there are essentially two types of distribution curves. For $w = 0.5$, the peak near $\hat{n}_1 \cdot \hat{n}_2 = -1.0$ vanishes, consistent with the findings in Fig. 7 (but for a lower T). As w decreases, a peak starts to develop for smaller values of $\hat{n}_1 \cdot \hat{n}_2$ and the distribution curve becomes bimodal, which signifies a change in cluster structure with a change in interaction range. Two typical clusters from systems with $w = 0.2$ and $w = 0.5$ are shown in Figs. 8(b) and 8(c), respectively. These two clusters have a similar size and are both well rounded. To quantitatively characterize the cluster structure, for each particle i the distance, $r(i)$, from the cluster's center of mass is computed for each cluster and summarized in Fig. 8(d).

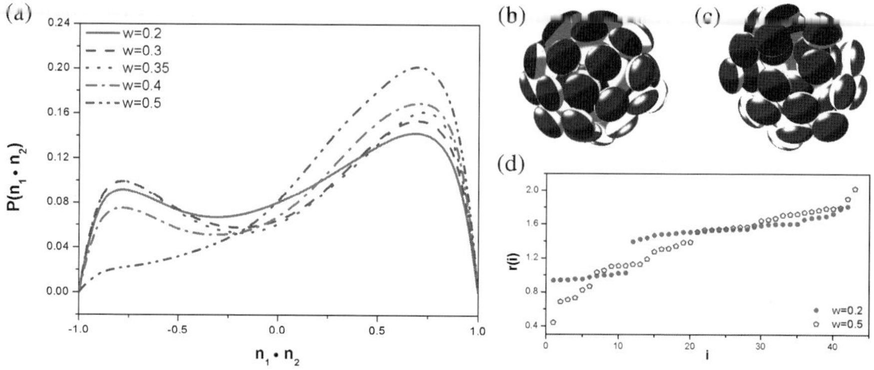

Figure 8. (a) Distribution of $\hat{n}_1 \cdot \hat{n}_2$ over all pairs of bonded Janus oblate spheroids with $\varepsilon = 0.6$ for different w at $\rho = 0.08$ and $T = 0.25$. (b) A cluster of size 42 in a system of $w = 0.2$. (c) A cluster of size 43 in a system of $w = 0.5$. (d) Distribution of particle distances from cluster's center of mass for constituent particle i, for the two clusters shown in (b) and (c). Reprinted with permission from Ref. [44]. Copyright (2015) AIP Publishing LLC.

From the plot, we find that the cluster for the $w = 0.2$ system is approximately a double-layer vesicle. Specifically, the normals of the attractive patches of its inner-shell particles point outward to face the corresponding outer-shell patches (hence the inner shell's surface is mostly gray in color from Fig. 8(b)), similar to the structure of Janus sphere vesicles (see Fig. 3(c)). The average distances for two shells to the center of mass are 0.98 ± 0.03 and 1.57 ± 0.10, respectively, which indicates that the two shells are barely touching, as their separation is about $2\sigma_{\|}$ ($\sigma_{\|} = 0.3$ for $\varepsilon = 0.6$).

By contrast, the cluster for the $w = 0.5$ case has a relatively loose structure. With such a long-range interaction, the particles need not remain in proximity inside the cluster. Hence, there is space for cluster reorganization. As a result, particle pairs in an open configuration are preferred such that the interaction regime for the attractive patches is larger. Consequently, face-to-face pairing is suppressed, leading to a reduction of the peak height close to $\hat{n}_1 \cdot \hat{n}_2 = -1.0$ in the distribution curve. Thus, one can broadly classify the interaction range into two categories for Janus spheroids, namely, short-ranged and long-ranged. Under the same conditions, clusters are more well-organized for short-ranged interactions and are more "liquid-like" for long-ranged interactions.

7.3.2.3. *Critical micelle concentration*

As noted by Sciortino *et al.* [13], the onset of micellization can be captured by the relationship between the monomer density, ρ_1, and the system particle density, ρ. An example is shown in Fig. 9 for a system having $\varepsilon = 0.6$ and $w = 0.2$ at various temperatures. The flatness of this curve in certain regions indicates that ρ_1 is rather insensitive to changes in ρ; in other words, the number of monomers is about the same upon adding more particles to the aggregating system. Hence, the critical micelle concentration (cmc) can be located by tracing the deviation from ideal gas behavior (i.e., $\rho_1 = \rho$) for a given temperature. For instance, the cmc is about $\rho = 4 \times 10^{-3}$ for $T = 0.25$ from Fig. 9. It is also clear that the cmc is less than $\rho = 10^{-3}$ for a temperature below $T = 0.21$.

Figures 10(a) and 10(b) show the cluster size distribution for the system at fixed $T = 0.19$ for various ρ and at fixed $\rho = 0.001$ for various T, respectively. At $T = 0.19$, the cmc is below $\rho = 10^{-3}$. Thus, micellization occurs for the values of ρ investigated in Fig. 10(a). Moreover, the vesicle

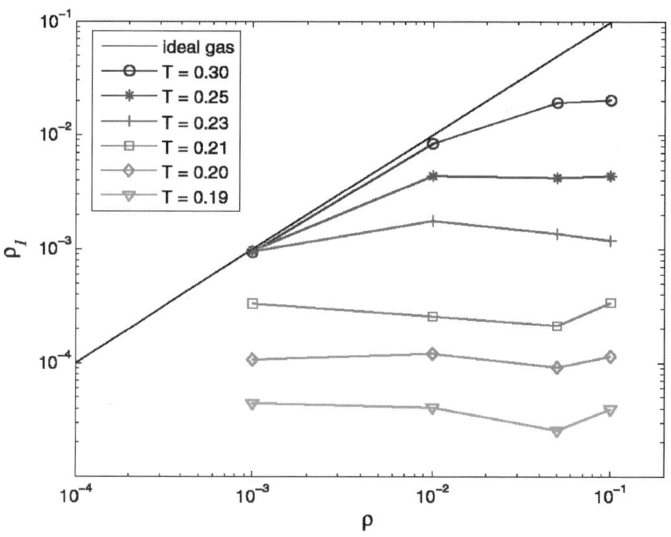

Figure 9. The relationship between the monomer density, ρ_1, and the system particle density, ρ, at different temperatures, T, for Janus oblate spheroid systems having $\varepsilon = 0.6$ and $w = 0.2$. Reprinted with permission from Ref. [40]. Copyright (2014) AIP Publishing LLC.

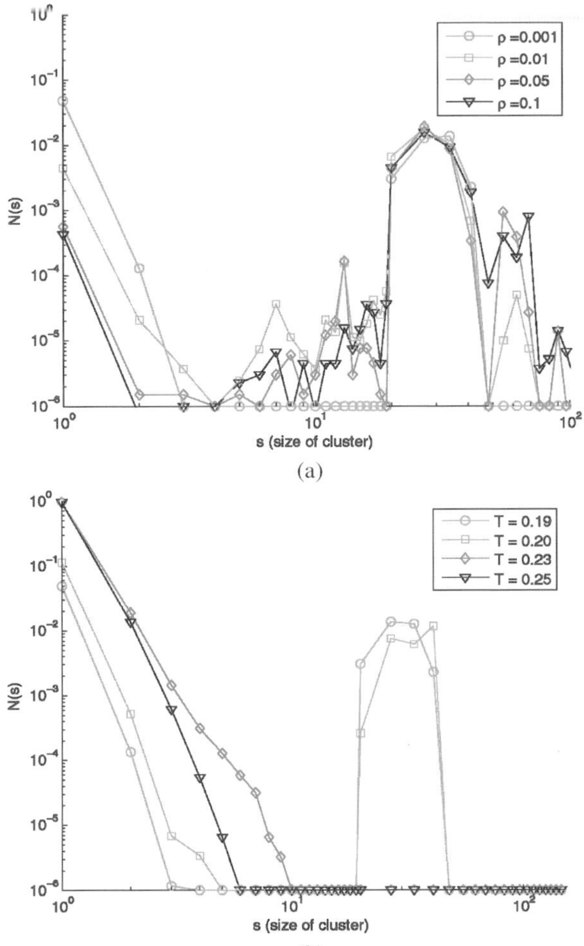

Figure 10. Cluster size distribution, $N(s)$, for a system of spheroids having $\varepsilon = 0.6$ and $w = 0.2$ at (a) $T = 0.19$ for different ρ, and (b) $\rho = 0.001$ for different T. Reprinted with permission from Ref. [40]. Copyright (2014) AIP Publishing LLC.

peak around the size 40 is dominant, which suggests that vesicles are the lower-energy structures in comparison with micelles. This behavior is also found in Janus sphere systems [13]. For the system with $\rho = 0.001$, it is shown from Fig. 10(b) that the cmc is above $\rho = 10^{-3}$ for $T = 0.23$ and 0.25, which is consistent with the results presented in Fig. 9.

7.3.3. Potential Application: Selective Encapsulation

Microencapsulation is an intriguing and important subject given the abundant applications of associated technologies [51–53]. In general, encapsulation can protect a guest object from the outside environment, thereby increasing the stability and lifetime of the guest. In addition, encapsulation can be applied to control the release rate of the guest substance. Therefore, capsules produced at the micro/nanometer scale have broad applications in various fields, such as pharmaceuticals [51,54], materials science [55,56], the chemical [52,57], and food industries [53,58,59], etc. For instance, micron capsules acting as carriers in drug delivery systems are employed not only to mitigate drug transport problems (e.g., limited water solubility, high systemic toxicity, etc.) [54,60], but also to achieve targeted drug delivery [60–62].

A number of microencapsulation strategies have been proposed and implemented in recent years [51,63–67]. Among these strategies, those that promote self-assembly into capsules under specific conditions are especially important. These self-assembling systems consist primarily of amphiphilic molecules and macro molecules, such as lipids and amphiphilic copolymers [61,64,65]. The liposomes and micelles (or vesicles) formed span from tens of nanometers to a few microns in size, providing a versatile platform for encapsulation applications. As discussed above, the JP, as an amphiphile, also forms closed structures via self-assembly and hence may provide another route for encapsulation.

In a preliminary study [68], an encapsulation system consisting of Janus oblate spheroids and isotropic spheres is investigated by means of Monte Carlo simulations. In the system, the spheroids are taken as the encapsulating agents and the spheres are encapsulated guests. The sphere interacts with the attractive patch part of the Janus oblate spheroid but not the hardcore part. As in the case of other self-assembling systems (e.g., lipids and copolymers), the ability to form shells by self-assembly suggests that Janus spheroids are good encapsulating agents, which are over length scales of the order of microns. In addition, the associated voids created in the Janus shells may result in porous capsules, a desirable feature for materials having specific applications, such as catalysis and controlled drug delivery systems [69]. The study [68] suggests that for guest spheres whose diameter equals the length of the major axis of the spheroids, encapsulation by the Janus spheroids is highly efficient, especially for an aspect ratio equal to 0.6 [68].

In a later report [44], a systematic investigation has been conducted to further expose the encapsulation behavior of the Janus oblate spheroids in the same encapsulation system. Specifically, the system is located in the low-temperature and low-density regime, having an excess of spheroids compared with guest spheres to ensure encapsulation sufficiency. The impact of variations in temperature, particle size, interparticle interaction range, and strength on encapsulation process is examined. From the study [44], some interesting aspects of the encapsulation system have been revealed. First, the system is found to be quite adaptive to the external environment factors, such as temperature and solution condition, etc. More importantly, the selectivity of the encapsulation with respect to guest particle size and agent–guest interaction strength is highlighted. In other words, the agent spheroids act differently upon guest spheres having different sizes and different affinities with the agent.

The collective effect of the selectivity in encapsulation is summarized in the form of an encapsulation map for Janus spheroids of $\varepsilon = 0.6$, as shown in Fig. 11 [44]. To construct such a map, a simplified system, which has only one guest sphere, is considered. Different number densities of the Janus oblate spheroids are examined to check the generality of the results. Moreover, the number densities studied are beyond the cmc in the system, such that micellization of the spheroids occurs. Also, the number of spheroids is large enough to ensure that the sphere can be fully covered from geometric considerations. In the map, the radius of the guest sphere, r, and the agent–guest interaction strength, u_{es}, are varied, while the agent–agent interaction strength is set to be 1.0. At each sampling point, whether the sphere is fully encapsulated or not is then determined by the visual observation of the assembly configuration with the lowest potential energy from control groups.

From the figure, we find that the micellization of spheroids is more prominent for $u_{es} \leq 0.5$ while encapsulation dominates for $u_{es} \leq 1.5$, for a single guest sphere with radius from 0.1 to 1.0. In the intermediate regime, the role of the guest size effect is manifest. It is easier for Janus spheroids to encapsulate relatively small guests rather than large ones. Moreover,

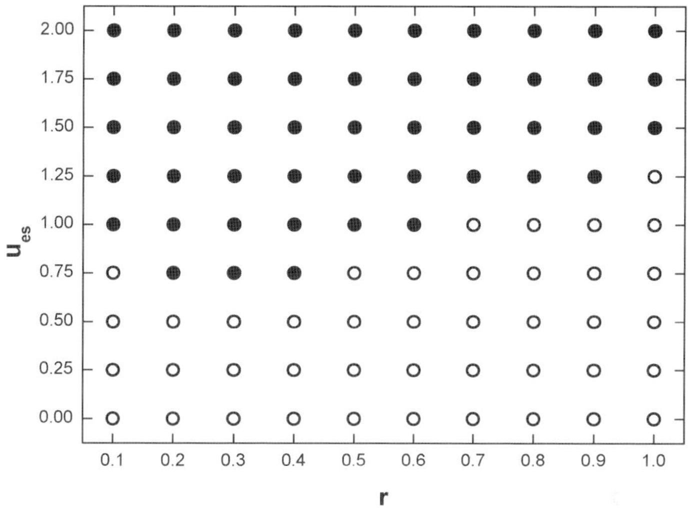

Figure 11. An encapsulation map summarizing the encapsulation behavior for the systems having different agent–guest interaction strengths, u_{es}, and guest sphere size, r, at temperature $T = 0.25$. The solid dots denotes complete encapsulation of the guest sphere while the open dots indicate that the guest sphere cannot be fully encapsulated. Reprinted with permission from Ref. [44]. Copyright (2015) AIP Publishing LLC.

very small guest particles can also not be fully encapsulated, due to entropy considerations that particles can lower the free energy of the system via the increase in entropy associated with sampling more of configuration space. It is to be noted that while the guest cannot be fully encapsulated in this context, encapsulation may still take place in a system having many guest particles. Simply, the guests can be trapped inside completely by complex aggregates formed by small clusters of agents and guests. Nevertheless, the open dots in the encapsulation map may suggest that the encapsulation efficiency is low for a system at such points in parameter space.

7.4. Janus Prolate Spheroids

We next consider systems of Janus prolate spheroids for which $\varepsilon > 1.0$. Based on the patch orientation, there are two kinds of Janus prolate spheroid, denoted as type I and type II, respectively. The type I Janus prolate spheroid has the attractive patch along its major axis, while for type II spheroids, the patch is along a minor axis of the particle.

7.4.1. Type I Janus Prolate Spheroid

We review a Monte Carlo study [38] (in a low-temperature, low-density regime) of the equilibrium morphology of type I Janus prolate spheroids. This work is complementary to the earlier discussion of oblate spheroids and considers the case in which the aspect ratio $\varepsilon > 1.0$ (the length of the longer semi-diameter is $\sigma_{\|} = \sigma_0/2$). Prolate spheroids provide a wider variety of morphological structures than those found for oblate spheroids. In order to investigate the dependence of the morphology on temperature and aspect ratio, the second virial coefficient, B_2, is considered. This often has been used as a convenient quantity to characterize the effective (temperature dependent) interaction between particles and is given by [8]

$$B_2 = \frac{1}{2V} \int \mathrm{d}r_i \int \mathrm{d}r_j \int \mathrm{d}\hat{n}_i \int \mathrm{d}\hat{n}_j [1 - \mathrm{e}^{-\beta U_{ij}}], \tag{9}$$

where the notation for the patch vectors is as before. (In particular, the quantity discussed here is a scaled B_2, namely, $B_2^* = B_2/B_2^{\mathrm{hs}}$ where $B_2^{\mathrm{hs}} = 2\pi\sigma^3/3$ stands for the second viral coefficient of a hard sphere whose diameter is σ. For simplicity, the '*' in the notation is omitted for the B_2's in the following discussion.)

The numerical calculation of B_2 for Janus ellipsoids shows that its magnitude decreases as either the temperature or the aspect ratio increases. One thus expects that, in this system, monomers combine more readily at lower temperature and for smaller aspect ratios. In addition, the value of B_2 approaches that for a pure hard-core repulsion as the aspect ratio increases. A more detailed explanation of the behavior of B_2 is given in the paper [38]. To study the impact of aspect ratio on self-assembly morphology of the prolate spheroids, two somewhat arbitrary regimes of the aspect ratio are introduced: regime 1, $\varepsilon = 2.0$–10.0, in which the structure is relatively uninteresting and regime 2, $\varepsilon = 1.0$–2.0, in which the structure behavior is richer. Two sampling step sizes ($\Delta\varepsilon$) are employed to sweep the value of ε, namely, $\Delta\varepsilon = 1.0$ in regime 1 and $\Delta\varepsilon = 0.1$ in regime 2.

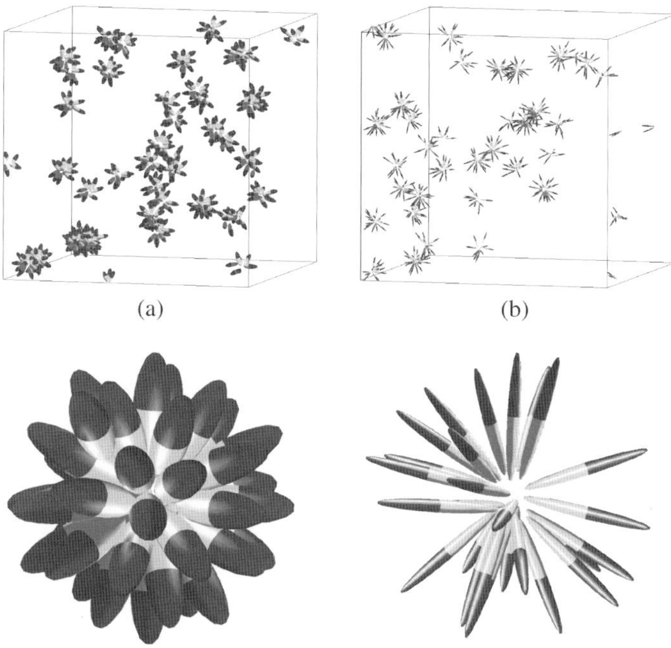

Figure 12. Configuration of the system ($T = 0.25$, $\rho = 0.02$) for ellipsoid aspect ratios (a) $\varepsilon = 3.0$ and (b) $\varepsilon = 10.0$. (c) and (d) show the largest clusters in those two configurations with cluster sizes $N_p = 42$ and 26, respectively. Reproduced with permission from Ref. [38]. Copyright (2013) American Chemical Society.

The simulations are carried out at a fixed system temperature, $T = 0.25$, which is expected to be lower than the critical temperatures for the systems examined. Again, one would expect that the results shown for this temperature are, in general, typical for the entire gas phase regime, as is the case for the Janus sphere system discussed earlier. The snapshots and associated cluster structures (with cluster size N_p) are shown in Fig. 12 for the systems in regime 1 with $\varepsilon = 3.0$ and 10.0, respectively. As is evident from the figure, in both cases shown, the clusters are micelles (i.e., single-layer structures, either complete or incomplete).

In Fig. 13, typical configurations for systems having different aspect in regime 2 are presented. The morphologies are essentially the same for aspect ratios $\varepsilon = 1.0$, 1.1, and 1.2 and the systems consist of small oligomers, micelles and vesicles. However, as ε approaches 1.3, the

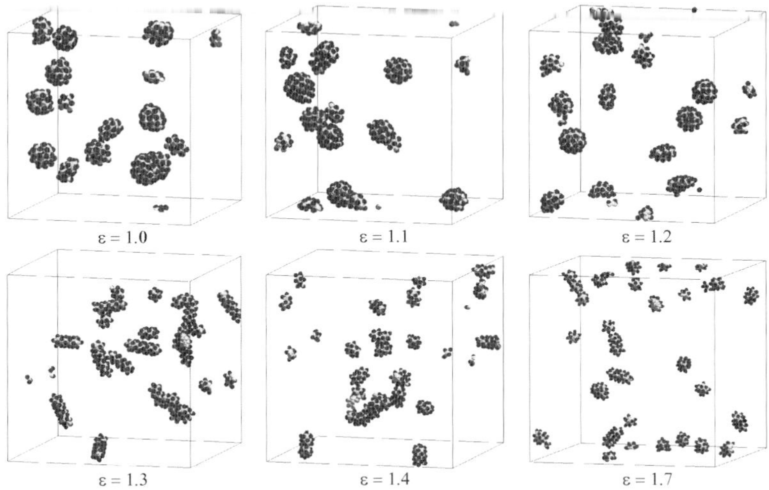

Figure 13. Configuration of the system (T = 0.25, ρ = 0.02) for different values of ε Reproduced with permission from Ref. [38]. Copyright (2013) American Chemical Society.

vesicle structure is no longer stable and a tubular micelle structure dominates. This structure survives for even larger aspect ratios, e.g., ε = 1.4 and 1.7, but the cylindrical length decreases as the aspect ratio increases. The morphology mainly changes to micellelike beyond ε = 1.3. Thus, there is a rather narrow stability window around 1.3 for the tubular micelle phase of the Janus prolate spheroids.

Some typical cluster structures for ε = 1.2–1.4 are shown in Fig. 14, It is manifest to see the transition from vesicle to tubular micelle and micelle-like structures, including the order found in the tube at ε = 1.3. Furthermore, Fig. 14(b) shows how such a tubular structure forms via the attachment of building blocks one by one, with a reorganization of the particles into ordered geometrical arrangements. For ε = 1.4, however, the change in length of semiaxes $a = b = c/\varepsilon$ from 0.385 to 0.357 (compared with ε = 1.3) reduces the formation of the higher hierarchical order through such an attachment process, leading to a reduction in the total length.

The cluster probability distribution function is plotted in Fig. 15 (using a bin size $\Delta N_p = 5$). An approximate association of these peaks with cluster morphologies can be made. For instance, the dominant peak at $N_p \cong 40$ for ε = 1.0 corresponds to a vesicle. Such behavior has been seen before for Janus spheres and Janus oblate spheroids [11,13,37,40]. The

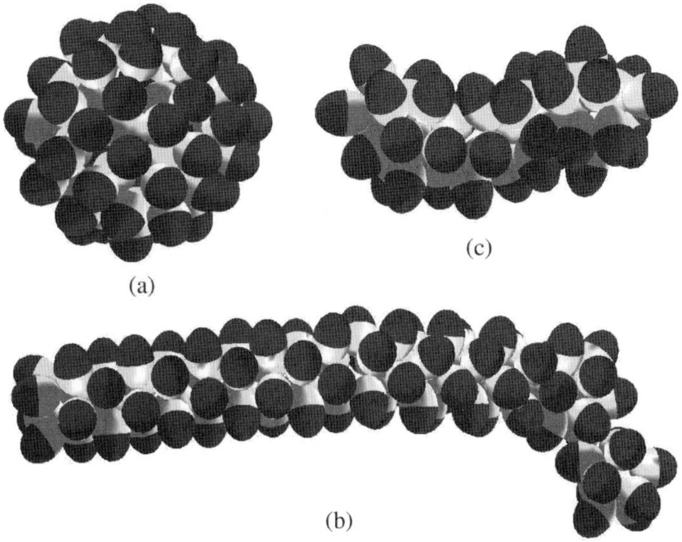

Figure 14. Typical clusters for (a) $\varepsilon = 1.2$, $N_p = 69$, (b) $\varepsilon = 1.3$, $N_p = 88$, and (c) $\varepsilon = 1.4$, $N_p = 39$. Reproduced with permission from Ref. [38]. Copyright (2013) American Chemical Society.

secondary peak at $N_p = 20$ corresponds to micelle clusters. An interesting case is the distribution curve of $\varepsilon = 1.3$, which exhibits a more oscillatory type of behavior, with peaks at approximately $N_p = 5$, 15, 25, and 35. One interpretation of these peaks is that the tubular structure is formed from building blocks of size $N_p = 5$–9, rather than from simple monomers. A further discussion of the formation of such structures is given in the paper [38]. It is to be noted that, due to the late-stage cluster–cluster aggregation, small changes in the cluster distribution function may occur with increasing simulation time, especially for $\varepsilon = 1.3$. The tubular micelle structure may continue to grow until it exhausts the basic building blocks in its neighborhood.

To study the orientational order inside the clusters, the probability distribution function for the scalar product $\hat{n}_1 \cdot \hat{n}_2$ is plotted. As described in the previous section, this function is calculated for all pairs of bonded spheroidal particles in the system. The results for different values of ε are shown in Fig. 16. The similarity of the distribution functions for

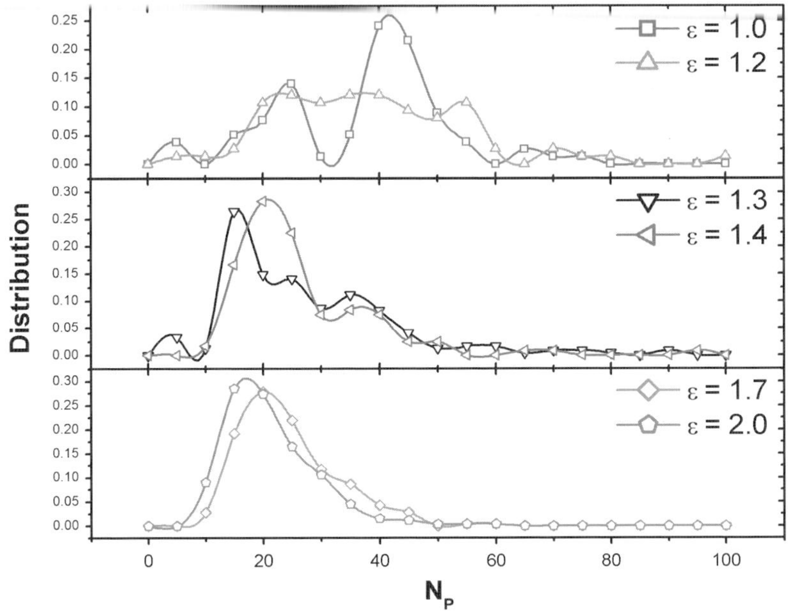

Figure 15. Distribution of cluster size (N_p) inside the system (after 10^7 Monte Carlo steps) for the different vales of ε. Reprinted with permission from Ref. [38]. Copyright (2013) American Chemical Society.

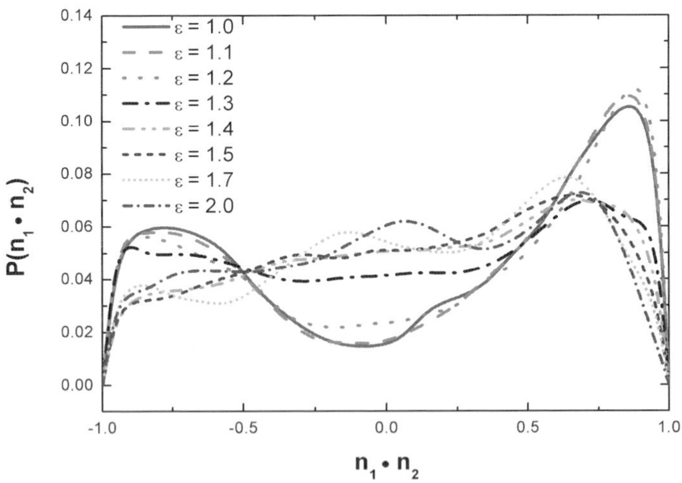

Figure 16. Distribution of $\hat{n}_1 \cdot \hat{n}_2$ over all pairs of bonded spheroidal particles for different values of ε. Reprinted with permission from Ref. [38]. Copyright (2013) American Chemical Society.

$\varepsilon = 1.0–1.2$ is consistent with the similarity of the cluster morphologies in these three cases. Additional order develops within the intermediate region due to the typical close-packing structures resulting from the corresponding geometrical shapes of spheroidal particles. This effect becomes more obvious for larger aspect ratios, as the major cluster structures are micelles.

By contrast with the other two groups of curves ($\varepsilon = 1.0–1.2$; $\varepsilon = 1.4–2.0$), the behavior of the system for $\varepsilon = 1.3$ is quite different. Namely, the different types of particle pairs have an equal probability of occurrence, corresponding to the plateau region of the distribution function. This is due to the cylindrical symmetry of the ordered tubes, in contrast to the spherical symmetry of micelle structures. Combining with the case of Janus oblate spheroids [37], we note that this type of Janus spheroidal particle (with the patch along the c-axis) preferentially forms bilayer or multilayer structures for $\varepsilon = 0.1–1.2$, tubular micelle structures for $\varepsilon \cong 1.3$, and primarily micelle structures for $\varepsilon > 1.3$ in the low-density regime.

7.4.2. Type II Janus Prolate Spheroid

We next study the morphologies formed by the type II Janus prolate spheroid, for which the attractive patch orientation is along a minor axis. Solomon *et al.* [39,41] experimentally synthesized this type of particle and investigated its self-assembly behavior in aqueous solutions. Specifically, the Janus prolate spheroids prepared are polystyrene spheroids with the hemi-surfaces sequentially coated with chrome and gold. The preference of gold–gold binding makes the gold faces the strong attractive patches.

Through tuning the concentration of sodium chloride (NaCl) in solution, the Debye layer thickness and hence the electrostatic repulsion are adjusted between the particles. In consequence, different self-assembled phases can be achieved through varying the salt concentration in the system. The experimental results [41] are summarized in Fig. 17. Four phases have been observed, including free particle (FP) phase, close-packed cluster (C1C) phase, ordered fiber (OrF) phase, and disordered fiber (DiF) phase. The confocal laser scanning microscopy images of the typical system configurations are shown in Fig. 17(a). In particular, the clusters

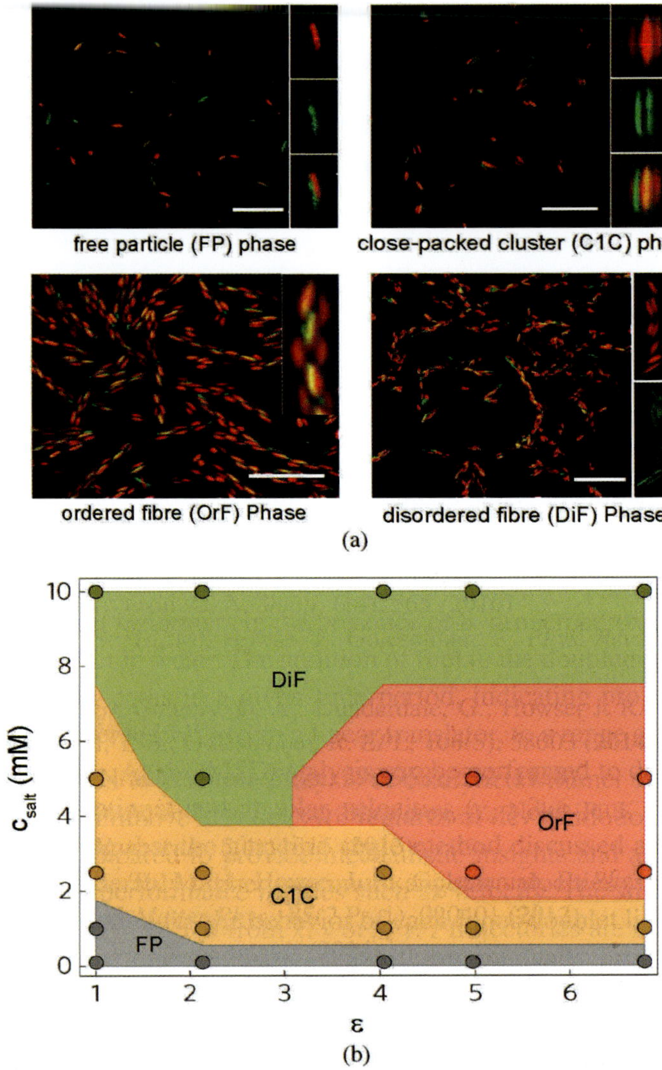

Figure 17. (a) Experimental observations of assembled phases for type II Janus prolate spheroids at different salt concentrations, showing FP, C1C, OrF, and DiF phases. (b) Phase diagram of assembled phases from the experiment as a function of the aspect ratio ε and salt concentration. Adapted from Ref. [41]. Copyright (2014) Nature Publishing Group.

in the OrF phase are positionally and orientationally ordered. On the contrary, there is no internal order in the DiF phase. The strong gold–gold binding hinders the reorientation of particles inside the clusters, which leads to numerous defects. Hence, the DiF phase is more likely a kinetically trapped phase, but not an equilibrium one [41].

A schematic phase diagram (Fig. 17(b)) is shown to describe the self-assembly behavior of the type II Janus prolate spheroid as a function of the aspect ratio ε and salt concentration. Interestingly, although the OrF phase is observed at different salt concentration, it occurs only for $\varepsilon > 3.5$. This suggests that the OrF phase requires certain geometric anisotropies, as is the case for the tubular micelle phase of the type I Janus prolate spheroid.

The molecular dynamics simulations have been carried out to study the type II Janus prolate spheroid system [41]. Specifically, the spheroid is modeled with 10 smoothly overlapping beads packed inside a spheroidal shell. Three potentials are included for the pair interaction between the beads: (1) a shifted Weeks–Chandler–Andersen potential for volume exclusion; (2) an angularly modulated, shifted Lennard–Jones potential for the van der Waals attraction between gold patches; and (3) a Yukawa potential for the screened electrostatic repulsion due to surface charges. A detailed recipe of the potential forms and simulation method can be found in the paper by Shah *et al.* [41]. A phase diagram for the simulation study is presented in Fig. 18 for type II Janus prolate spheroids having $\varepsilon = 5.0$ as a function of temperature and Debye screening length. Besides the four phases found in the experiment, two more phases, namely fluid fiber (F1F) phase and horizontal fiber (HF) phase, are suggested from the simulations.

The ordering of the OrF phase is further investigated by the simulations. As shown in Fig. 19, a fiber in the OrF phase is found to consist of four strands of aligned spheroids ($\varepsilon = 5.0$), with the strands lying on the vertices of a rhomb. The spheroids in adjacent strands are shifted by one half-particle length and their patch orientations are perpendicular to each other. In this way, the positional and orientational ordering inside the fiber is constructed and the gold–gold contact is maximized. Hence, a

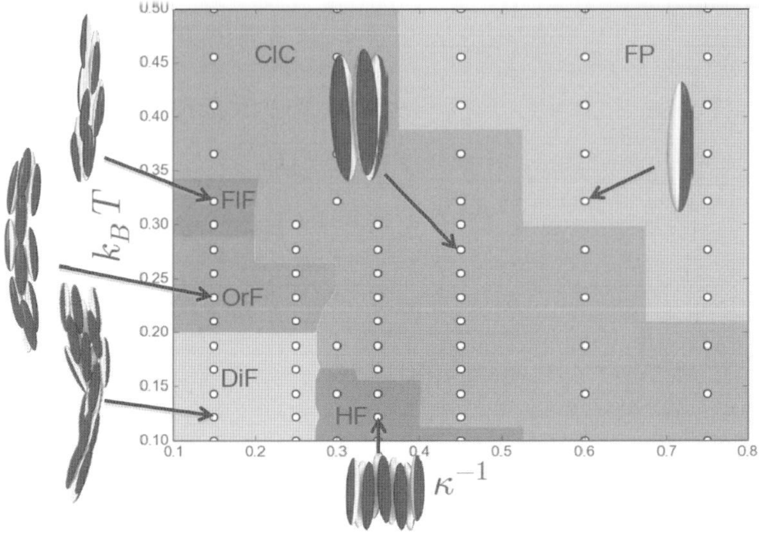

Figure 18. Phase diagram of assembled phases from molecular dynamics simulations for type II Janus prolate spheroids of $\varepsilon = 5.0$ as a function of temperature and Debye screening length. The phases observed are FP, C1C, F1F, OrF, DiF, and HF. Figure reproduced from Ref. [41]. Copyright (2014) Nature Publishing Group.

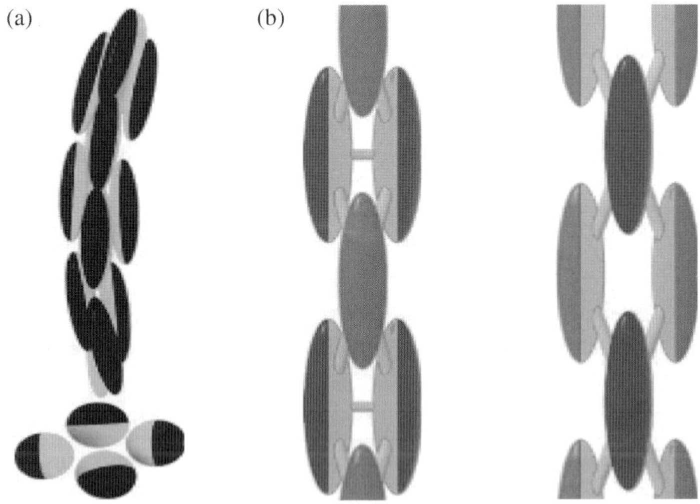

Figure 19. (a) Closeup view of an OrF and its cross-section; (b) perpendicular views of the contact network in the 1D fiber structure. Figure reproduced from Ref. [41]. Copyright (2014) Nature Publishing Group.

self-limiting, 1D growth mechanism for the OrF phase is revealed by the simulations [41], which is comparable to a step-growth polymerization of the four-particle unit-cell components along with the internal rearrangement of the unit cell. Such a scenario resembles the formation of the tubular micelle phase discussed earlier (see Fig. 14(b)), which again highlights the associated effect of particle shape anisotropy and Janus functionality in self-assembly.

The stability of the OrF phase is examined by applying an external alternating current (a.c.) electric field in the experiment [41]. Interestingly, the OrF structures are rapidly extended under the a.c. electric field, resulting in a new chain-link (CL) phase that has the same symmetry as common jewelry chains. Such an elongation is measured as $36 \pm 7\%$ in length from the experiment (while the predicted strain is 33.1% from geometric considerations). Once the electric field is removed, the fibers are reconfigured from the CL phase to the OrF phase. This process is reversible upon controlling the application of a.c. electric fields, which suggests that the OrF structure is a 1D shape-memory material. Furthermore, the force required to extend from the OrF to the CL conformation is found to be comparable to the stall force of molecular motors, such that the ordered fiber formed by the type II Janus prolate spheroids may apply as colloidal actuators with muscle-like function for soft robotics [41].

7.5. Patchy Spheroidal Particles

So far, we have discussed the studies of JPs, for which the patch covers its half surface. A natural extension to these studies is to consider the patch size effect. He *et al.* [42] carried out a simulation study on spheroidal particles with different aspect ratio and "patchiness." A coarse-grained model of patchy spheroids is presented. Specifically, a hollow patchy spheroid (as shown in Fig. 20) consists of two types of coarse-grained beads, denoted as A (gray) and B (black), which are almost homogeneously distributed on the spheroid surface. The beads of two spheroids interact via Lennard–Jones potential. The interaction strength varies for different bead–bead interactions. In particular, the interaction between A and A is much stronger than the other two types (i.e., A–B interaction and B–B interaction). Hence, the A beads in a spheroid form the attractive patch. The component ratio (describing the patch size) is defined as

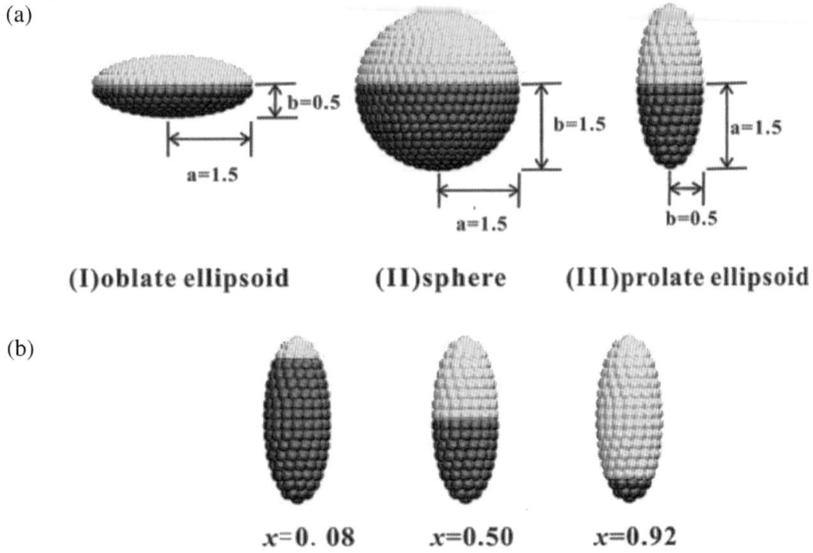

Figure 20. (a) Representatives of Janus spheroids. Particle I is a Janus oblate spheroid ($\varepsilon = 0.33$), particle II is a Janus sphere ($\varepsilon = 1.0$), and particle III is a Janus prolate spheroid ($\varepsilon = 3.0$). (b) Patchy prolate spheroids ($\varepsilon = 3.0$) with different component ratios. Figure reproduced with permission from Ref. [42]. Copyright (2015) The Royal Society of Chemistry.

$x = N_A/(N_A + N_B)$, where N_A and N_B are the number of A beads and B beads, respectively.

The interaction between two patchy spheroids is thus given by the sum of the interactions between the coarse-grained beads of each spheroid. To reduce the computational cost and improve the simulation efficiency, a coarse-grained potential form is proposed to depict the pair interaction between two spheroids. With this potential, the rigid body Brownian dynamics simulation is employed, which not only explores the assembled morphology, but also gives a dynamical description of the self-assembly process. The detailed interpretation of the coarse-graining technique and simulation method is given in the paper [42]. Here, we focus on the self-assembled structures of the patchy spheroids found from their study.

The system ($\rho = 0.0086$) is quenched from a high-temperature (dispersed) state to the aggregate temperature. The simulations are run

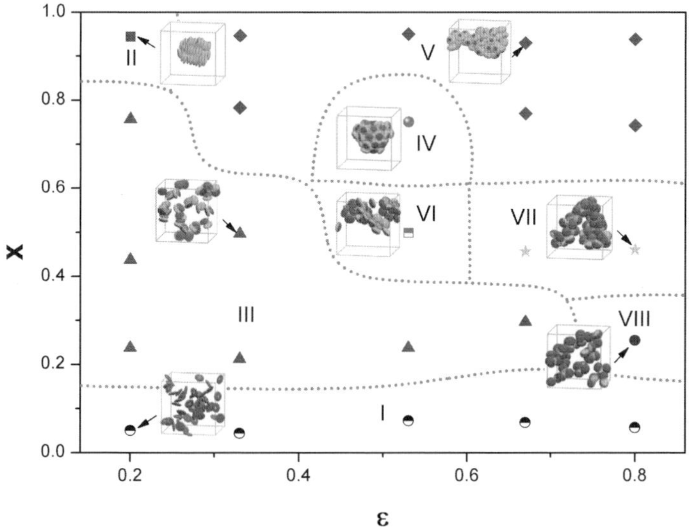

Figure 21. Phase diagram of patchy oblate spheroids as a function of the attractive patch component ratio x and the aspect ratio ε. The regions of equilibrium self-assembled structures are: (I) monomers, (II) columnar aggregates, (III) sandwich-like dimers, (IV) vesicles, (V) random aggregation structure, (VI) lamellae, (VII) wormlike structure, and (VIII) spherical micelles. Typical morphologies of each region are shown as inset. Figure adapted with permission from Ref. [42]. Copyright (2015) The Royal Society of Chemistry.

sufficiently long to reach final equilibrium structures. The aspect ratio ε and the component ratio x are varied in the study [42]. The phase diagrams of morphologic behavior for both patchy oblate and prolate spheroids are obtained, as given in Figs. 21 and 22, respectively. The effect of system temperature on the stability of the final aggregates is examined. The phase diagram is found to remain the same for a reasonable range of temperatures [42].

As summarized in the paper [42], there are eight regions in the phase diagram of the patchy oblate spheroids, including (I) monomers, (II) columnar aggregates, (III) sandwich-like dimers, (IV) vesicles, (V) random aggregation structure, (VI) lamellae, (VII) wormlike structure, and (VIII) spherical micelles. For the patchy prolate spheroids, there are six regions: (I) monomers, (II) liquid crystals, (III) spherical micelles, (IV) wormlike micelles, (V) random aggregation structure, and (VI) dimers. Some typical cluster structures are also shown in the figures. As is evident from the phase diagrams, a number of different morphological structures

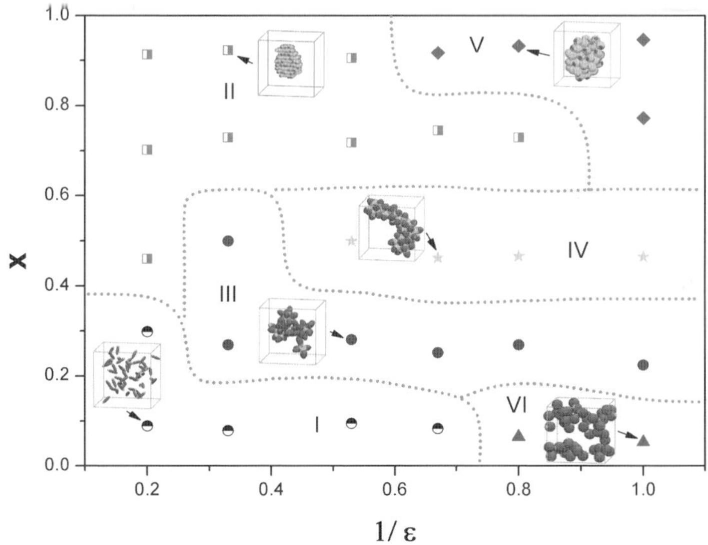

Figure 22. Phase diagram of patchy prolate spheroids as a function of the patch component ratio x and $1/\varepsilon$. The regions of equilibrium self-assembled structures are (I) monomers, (II) liquid crystals, (III) spherical micelles, (IV) wormlike micelles, (V) random aggregation structure, and (VI) dimers. Typical morphologies of each region are shown as inset. Figure adapted with permission from Ref. [42]. Copyright (2015) The Royal Society of Chemistry.

emerge due to the combined effect of the aspect ratio and component ratio, as the system prefers to maximize the contact of strong attractive patches. Hence, the patch size provides another degree of freedom to consider when designing the self-assembly building blocks for practical applications. Associated with the patchiness anisotropy, the spectrum of accessible assembled morphologies for spheroidal particles is further extended.

7.6. Conclusions

In this chapter, we reviewed several studies of the aggregation of Janus ellipsoids, focusing specifically on the role of two anisotropy dimensions, aspect ratio, and interaction patchiness, on particle phase behavior. We noted that the self-assembly of patchy ellipsoidal particles results in a wide variety of equilibrium structures in the gas phase regime. In addition to the micelle and vesicle phases observed in Janus sphere systems, multilayer structures are found for Janus oblate spheroids. Tubular micelles and ordered fibers are found for Janus prolate spheroids depending upon the patch arrangement on the particle surface. Moreover, additional cluster structures are obtained by varying the patch size of the spheroidal particles. By elucidating the impact of anisotropy on self-assembly, these studies have provided some new insights into programmable self-assembly [70].

Along with the aforementioned investigations of the morphology of self-assembled Janus ellipsoids, we discussed other characteristics of clustered phases. For example, a new encapsulation strategy facilitated by aggregating Janus oblate spheroids was examined. In particular, the encapsulation system, comprising Janus oblate spheroids (as encapsulating agents) and guest particles, exhibits some interesting features, such as selective encapsulations, that are dependent on guest size, interaction strength, etc. In addition, the ordered fiber formed by Janus prolate spheroids shows some intriguing mechanical properties that make the fiber a 1D shape-memory material. These characteristics may lead to specific applications in various fields, including targeted drug delivery and materials engineering. Finally, we note that the studies discussed here are mostly limited to the low-density regime. The phase behavior and self-assembly morphology of Janus ellipsoids in the high-density regime has not been studied, and other potential applications of Janus ellipsoids are yet to be explored. We hope that this chapter will provide an impetus for future work in this exciting and evolving field.

Acknowledgment

James D. Gunton would like to acknowledge support from the G. Harold and Leila Y. Mathers Foundation.

References

[1] Whitesides, G. M.; Grzybowski, B. *Science*, **295**, 2418 (**2002**).
[2] Whitesides, G. M.; Boncheva, M. *PNAS*, **99**, 4769 (**2002**).
[3] Glotzer, S. C.; Solomon, M. J. *Nat. Mater.*, **6**, 557 (**2007**).

[4] Yang, S.-M.; Kim, S.-H.; Lim, J.-M.; Yi, G.-R. *J. Mater. Chem.*, **18**, 2177 (**2008**).
[5] Pawar, A. B.; Kretzschmar, I. *Macromol. Rapid Commun.*, **31**, 150 (**2010**).
[6] Du, J.; O'Reilly, R. K. *Chem. Soc. Rev.* **40**, 2402 (**2011**).
[7] Saeanna, S.; Pine, D. J. *Curr. Opin. Colloid Interface Sci.*, **16**, 96 *(***2011***)*.
[8] Kern, N.; Frenkel, D. *J. Chem. Phys.*, **118**, 9882 (**2003**).
[9] Bianchi, E.; Largo, J.; Tartaglia, P.; Zaccarelli, E.; Sciortino, F. *Phys. Rev. Lett.*, **97**, 168301 (**2006**).
[10] Hong, L.; Cacciuto, A.; Luijten, E.; Granick, S. *Langmuir*, **24**, 621 (**2008**).
[11] Sciortino, F.; Giacometti, A.; Pastore, G. *Phys. Rev. Lett.*, **103**, 237801 (**2009**).
[12] Giacometti, A.; Lado, F.; Largo, J.; Pastore, G.; Sciortino, F. *J. Chem. Phys.*, **132**, 174110 (**2010**).
[13] Sciortino, F.; Giacometti, A.; Pastore, G. *Phys. Chem. Chem. Phys.* **12**, 11869 (**2010**).
[14] Jiang, S.; Chen, Q.; Tripathy, M.; Luijten, E.; Schweizer, K, S.; Granick, S. *Adv. Mater.*, **22**, 1060 (**2010**).
[15] Chen, Q.; K.Whitmer, J.; Jiang, S.; Bae, S. C.; Luijten, E.; Granick, S. *Science*, **331**, 199 (**2011**).
[16] Chen, Q.; Bae, S. C.; Granick, S. *Nature*, **469**, 381 (**2011**).
[17] Romano, F.; Sciortino, F. *Soft Matter*, **7**, 5799 (**2011**).
[18] Pons-Siepermann, I. C.; Glotzer, S. C. *ACS Nano*, **6**, 3919 (**2012**).
[19] Singh, J. P.; Lele, P. P.; Nettesheim, F.; Wagner, N. J.; Furst, E. M. *Phys. Rev.*, *E* **79**, 050401 (**2009**).
[20] Odriozola, G.; de J. Guevara-Rodriguez, F. *J. Chem. Phys.*, **134**, 201103 (**2011**).
[21] Odriozola, G. *J. Chem. Phys.*, **136**, 134505 (**2012**).
[22] Pawar. A. B.; Kretzschmar, I. *Langmuir*, **25**, 9057 (**2009**).
[23] Hong, L.; Jiang, S.; Granick, S. *Langmuir*, **22**, 9495 (**2006**).
[24] Preisler, Z.; Vissers, T.; Munao, G.; Smallenburg, F.; Sciortino, F. *Soft Matter*, **10**, 5121 (**2014**).
[25] Ren, B.; Ruditskiy, A.; Song, J. H. K.; Kretzschmar, I. *Langmuir*, **28**, 1149 (**2012**).
[26] Rosenthal, G.; Gubbins, K. E.; Klapp, S. H. L. *J. Chem. Phys.*, **136**, 174901 (**2012**).
[27] Hagy, M. C.; Hernandez, R. *J. Chem. Phys.*, **137**, 044505 (**2012**).
[28] Li, Z.-W.; Lu, Z.-Y.; Sun, Z.-Y.; An, L.-J. *Soft Matter*, **8**, 6693 (**2012**).
[29] Hu, J.; Zhou, S.; Sun, Y.; Fanga, X.; Wu, L. *Chem. Soc. Rev.*, **41**, 4356 (**2012**).

[30] Walther. A.; Müller, A. H. E. *Chem. Rev.*, **113**, 5194 (**2013**).
[31] Fujii, S.; Yokoyama, Y.; Miyanari, Y.; Shiono, T.; Ito, M.; ichi Yusa, S.; Nakamura, Y. *Langmuir*, **29**, 5457 (**2013**).
[32] Gunton, J. D.; Shiryayev, A.; Pagan, D. *Protein Condensation*, Cambridge University Press: Cambridge (**2007**).
[33] Liu, B.; Zhang, C.; Liu, J.; Qu, X.; Yang, Z. *Chem. Commun.*, **26**, 3871 (**2009**).
[34] Walther, A.; Drechsler, M.; Rosenfeldt, S.; Harnau, L.; Ballauff, M.; Abetz, V.; Müller, A. H. E. *J. Am. Chem. Soc.*, **131**, 4720 (**2009**).
[35] Güell, O.; Sagués, F.; Tierno, P. *Adv. Mater.*, **23**, 3674 (**2011**).
[36] Park, B. J.; Lee, D. *ACS Nano*, **6**, 782 (**2012**).
[37] Liu, Y.; Li, W.; Perez, T.; Gunton, J. D.; Brett, G. *Langmuir*, **28**, 3 (**2012**).
[38] Li, W.; Gunton, J. D. *Langmuir* **29**, 8517 (**2013**).
[39] Shah, A. A.; Schultz, B.; Kohlstedt, K. L.; Glotzer, S. C.; Solomon, M. J. *Langmuir*, **29**, 4688 (**2013**).
[40] Ruth, D.; Gunton, J. D.; Rickman, J. M.; Li, W. *J. Chem. Phys.*, **141**, 214903 (**2014**).
[41] Shah, A. A.; Schultz, B.; Zhang, W.; Glotzer, S. C.; Solomon, M. J. *Nat. Mater.*, **14**, 117 (**2015**).
[42] Xu, J.; Wang, Y.; He, X. *Soft Matter*, **11**, 7433 (**2015**).
[43] Berne, B. J.; Pechukas, P. *J. Chem. Phys.* **56**, 4213 (**1972**).
[44] Li, W.; Ruth, D.; Gunton, J. D.; Rickman, J. M. *J. Chem. Phys.*, **142**, 244705 (**2015**).
[45] Care, C. M.; Cleaver, D. J. *Rep. Prog. Phys.*, **68**, 2665 (**2005**).
[46] Wang, W.; Wang, J.; Kim, M.-S. *Compu. Aided Geomet. Des.*, **18**, 531 (**2001**).
[47] Choi, Y.-K.; Chang, J.-W.; Wang, W.; Kim, M.-S.; Elber, G. *Visualization and Computer Graphics, IEEE Transactions on* **15**, 311 (**2009**).
[48] Frenkel, D.; Smit, B. *Understanding Molecular Simulations*; Academic Press: San Diego (**1996**).
[49] Fantoni, R.; Giacometti, A.; Sciortinoc, F.; Pastore, G. *Soft Matter*, **7**, 2419 (**2011**).
[50] Sinkovits, D. W.; Barr, S. A.; Luijten, E. *J. Chem. Phys.*, **136**, 144111 (**2012**).
[51] Peyratout, C. S.; Dähne, L. *Angew. Chem. Int. Ed.*, **43**, 3762 (**2004**).
[52] Ciriminna, R.; Sciortino, M.; Alonzo, G.; de Schrijver, A.; Pagliaro, M. *Chem. Rev.* **111**, 765 (**2011**).
[53] Gouin, S. *Trends Food Sci. Technol.*, **15**, 330 *(***2004***)*.
[54] Allen, T. M.; Cullis, P. R. *Science*, **303**, 1818 (**2004**).

[55] Caruso, M. M.; Delafuente, D. A.; Ho, V.; Sottos, N. R.; Moore, J. S.; White, S. R. *Macromolecules*, **40**, 8830 (**2007**).

[56] Cho, S. H.; White, S. R; Braun, P. V. *Adv. Mater.*, **21**, 645 (**2009**).

[57] Crooks, R. M.; Zhao, M.; Sun, L.; Chechik, V.; Yeung, L. K. *Acc. Chem. Res.*, **34**, 181 (**2001**).

[58] Anal, A. K.; Singh, H. *Trends Food Sci. Technol.*, **18**, 240 (**2007**).

[59] Augustin, M. A.; Hemar, Y. *Chem. Soc. Rev.*, **38**, 902 (**2009**).

[60] Haag, R. *Angew. Chem. Int. Ed.*, **43**, 278 (**2004**).

[61] Zhang, L.; Chan, J. M.; Gu, F. X.; Rhee, J.-W.; Wang, A. Z.; Radovic-Moreno, A. F.; Alexis, F.; Langer, R.; Farokhzad, O. C. *ACS Nano*, **2**, 1696 (**2008**).

[62] Patri, A. K.; Kukowska-Latallo, J. F.; Baker Jr., J, R, *Adv. Drug Deliv. Rev.*, **57**, 2203 (**2005**).

[63] Kim, Y.-G.; Oh, S.-K.; Crooks, R. M. *Chem. Mater.*, **16**, 167 (**2004**).

[64] Dominak, L. M.; Keating, C. D. *Langmuir*, **23**, 7148 (**2007**).

[65] Rösler, A.; Vandermeulen, G. W.; Klok, H.-A. *Adv. Drug Deliv. Rev.*, **64**, 270 (**2012**).

[66] Loscertales, I. G.; Barrero, A.; Guerrero, I.; Cortijo, R.; Marquez, M.; Gañán-Calvo, A. M. *Science*, **295**, 1695 (**2002**).

[67] Chang, M.-W.; Stride, E.; Edirisinghe, M. *Soft Matter*, **5**, 5029 (**2009**).

[68] Li, W.; Liu, Y.; Brett, G.; Gunton, J. D. *Soft Matter*, **8**, 6027 (**2012**).

[69] Vallet-Regí, M.; Balas, F.; Arcos, D. *Angew. Chem. Int. Ed.*, **46**, 7548 (**2007**).

[70] Cademartiri, L.; Bishop, K. J. M, *Nat. Mater.*, **14**, 2 (**2015**).

Chapter 8

Soft, Hard, and Hybrid Janus Structures: Synthesis, Self-Assembly, and Applications — Catalytic Janus Swimmers

*S. Ebbens**,‡, *J. Howse**, *R. Archer**, *D. Gregory**, *A. Campbell*†, *and G. Dunderdale*†

*Department of Chemical and Biological Engineering, University of Sheffield, S1 3JD, UK
†Department of Chemistry, University of Sheffield, S3 7HF, UK
‡s.ebbens@sheffield.ac.uk

Abstract

This chapter describes how small-scale devices with a "two-faced" structure, often termed catalytic Janus swimmers, can be used to generate autonomous motion within fluids at small scales. The chapter starts by giving a general introduction to the field of swimming devices, and then surveys the variety of different Janus structures that have been used to produce motion. We then describe the methods that have been employed to synthesize these asymmetrical catalytic swimmers, and review the current understanding of the mechanisms that produce motion from catalytic Janus structures. The remainder of the chapter first focuses on the challenge of accurately quantifying the motion characteristics for catalytic Janus swimmers, and then describes the considerable challenge of controlling the directionality of motion, which is subject to stochastic phenomena. Finally, we survey existing demonstrations of potential applications for Janus catalytic swimmers, and discuss future directions within this research field.

Keywords: Catalysis, self-propulsion, active colloids.

8.1. Introduction

"Swimmers" is a term used to encompass the wide range of engineered nano- and micronscale materials that are able produce enhanced motion within fluidic environments [1–4]. The often quoted goals motivating the development of these synthetic swimmers are to enable a new versatile autonomous approach to microfluidic transport and to one day realize devices that can perform drug delivery and repair tasks in the body [5]. As evidence of progress in this direction, there are now numerous proof-of-concept application examples for transport applications [6–16], and a recent report of biocompatibility [17].

When initially considering the challenge of producing motion within fluids, using deformations appears to be an obvious approach. However, at small scales, fluidic properties differ from those experienced in the macroscopic world, and in particular, moving objects possess very little momentum. The concept of the difficulty of producing deformation-based swimming at small scales was first highlighted in a talk by Purcell [18]. Using a series of diagrams, this lecture considered how a simple hinged scallop can produce motion in the macroscopic aquatic environments with which we are intuitively familiar, simply by opening and closing a hinged shell (Fig. 1(a)). However, the scallop can only use this propulsion method because momentum allows it to reclose the shell without coming to a standstill. Purcell then visualized shrinking this scallop down to micron scales and explained how in this circumstance the scallops' motion would be frustrated: without momentum, the opening and closing of the hinge would simply cause the scallop to rock backward and forward. This so-called Scallop theorem is a way of expressing the reason why small-scale deformation swimmers must execute sequences of motion that are non-time-reversible.

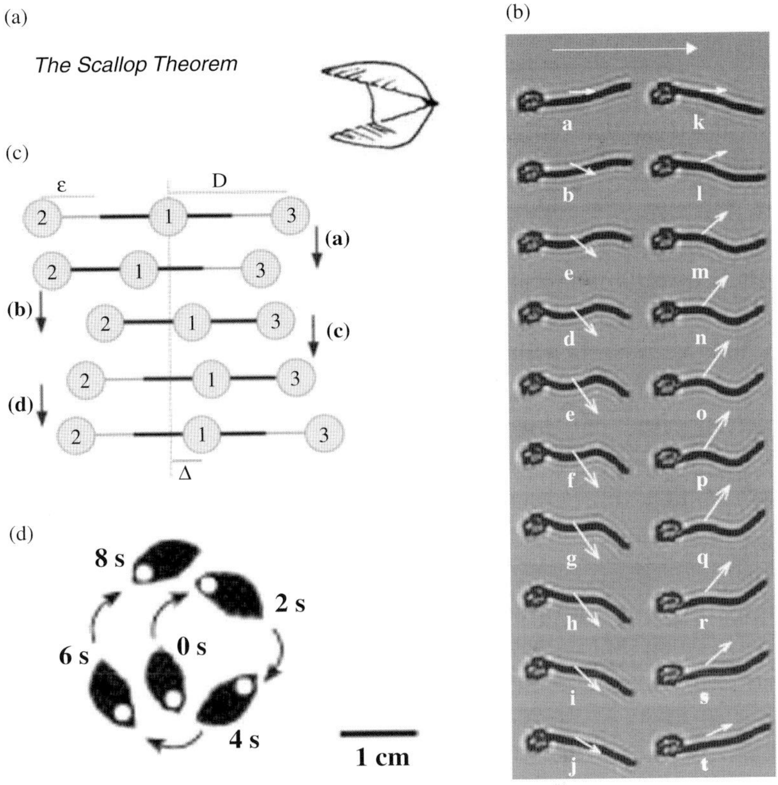

Figure 1. (a) A sketch from the Scallop theorem lecture showing the difficulties of producing deformation-driven motion at small scales. (b) Beating pattern of the motion of a magnetic flexible filament attached to a red blood cell. The time interval between each image is 5 m. The white arrows represent the externally applied magnetic field. The white arrow on the top shows the direction of motion. (c) Complete cycle of the proposed non-reciprocal motion of the simple three-body swimmer, which is composed of four consecutive time-reversal breaking stages: (a) the left arm decreases its length with the constant relative velocity, (b) the right arm decreases its length with the same velocity, (c) the left arm opens up to its original length, and finally, (d) the right arm elongates to its original size. By completing the cycle, the whole system is displaced to the right side. (d) Catalytic self-propulsion. White dots were added onto the photographs to indicate the catalyst bearing tips of the plates.

In nature, micron-scale organisms such as *Escherichia coli* rotate helical flagella to overcome this restriction and produce motion, and spermatozoa also execute non-time-reversible wave motions. However, to date, achieving the required deformations to move synthetic devices through fluids at small scales has always necessitated the application of external magnetic fields or light. As an example, an early synthetic swimmer comprised a chain of magnetic beads driven to execute spermatozoa like deformations (Fig. 1(b) [19]). Further analysis has led to theoretical postulations for the simplest swimming device that could move using a deformation-based approach. In free solution, such a device appears to be three spheres, linked by connecting material that can expand and contract. This arrangement can move through fluids if the linking material executes a prescribed series of deformations (Fig. 1(c) [20]). Since this postulation, the ability of this three-body device to produce motion has been experimentally realized, using optical manipulation to cause the three bodies to move in the required way [21]. In addition, a further scheme involving only two spheres, but requiring proximity to a planar surface has also been experimentally verified to produce motion actuated by magnetic fields [22]. While these externally driven swimmers are impressive examples, they possess somewhat limited applications due to the requirement for external actuation, and currently, it has not been possible to miniaturize, power, and synchronize materials that can autonomously actuate a time non-reversible sequence of deformations to produce motion. While hydrogel materials are able to expand and contract, and can be employed as the linking material in the three-body deformation swimmer design, a method to produce repetitive autonomous cycles of deformations does not yet exist. In this context, an alternative approach, based on the ability for catalytic reactions to produce motion, has overcome these obstacles faced by deformation swimmers and opened up the field of synthetic swimming device research.

One of the first illustrations of the key concept, which places this chapter within the remit of this book, the requirement for a Janus distribution of catalytic material in order for motion to be produced, was demonstrated by Whitesides [23]. Whitesides's paper showed that asymmetrically distributed platinum catalyst at the surface of centimeter scale thin plate-like structures produced autonomous movement along a water meniscus by generating a surface tension gradient while decomposing dissolved hydrogen peroxide fuel (Fig. 1(d)). Since this initial phenomena was reported, the general concept of Janus catalytic activity generating motion within fluidic environments containing a dissolved fuel has subsequently been elaborated upon to lead to the micron- and nano-scale catalytic Janus swimmers discussed here.

Before moving onto consider the specifics for a wide range of Janus swimmer systems, it is also important to highlight the other significant difference in fluidic environments as experienced by micron-scale devices as opposed to macroscopic objects: the dominance of Brownian phenomena. Brownian motion is most often demonstrated by observing the random motion of small-scale solid particles immersed in a fluid. These movements, resulting from the thermal energy of the surrounding fluid molecules, lead to diffusion, and so in one way provide a readily available mechanism for moving material at small scales. However, diffusive transport rate is fixed by the properties of the fluid and particle size and so can only produce motion at a limited rate, and in addition is chaotic and so unsuitable for directed transport. The small-scale swimming devices discussed throughout this chapter can produce significantly enhanced motion compared to this base level of diffusive transport; however, it should be remembered that unless otherwise confined, even the most rapidly moving small-scale object will still be undergoing Brownian translational diffusion, and that any additional propulsive transport is superimposed on this background. A further feature of Brownian motion is that all small-scale untethered objects are subject to Brownian rotation, so that their orientation is also rapidly and randomly changing. The importance of this effect for Janus catalytic swimmers will become clear as we explore the mechanisms by which these devices produce motion.

A useful way to group the majority of current synthetic catalytic swimming devices is according to their overall geometry. Three main device types have to date been investigated: nanorods, Janus spheres, and nanotubes (Fig. 2(a)). Nanorod devices comprise two contacting solid homogeneous cylindrical halves, one performing catalytic decomposition and the other inactive (Fig. 2(a) [24]). Nanorods were the first reported example of Janus catalytic swimming devices, and many studies have reported on the behavior of these devices, establishing correlations between rod composition and motion velocity [25], and investigating the effect of solution properties on speed, and also reporting collective behavior [26]. Janus sphere devices comprise one catalytically active

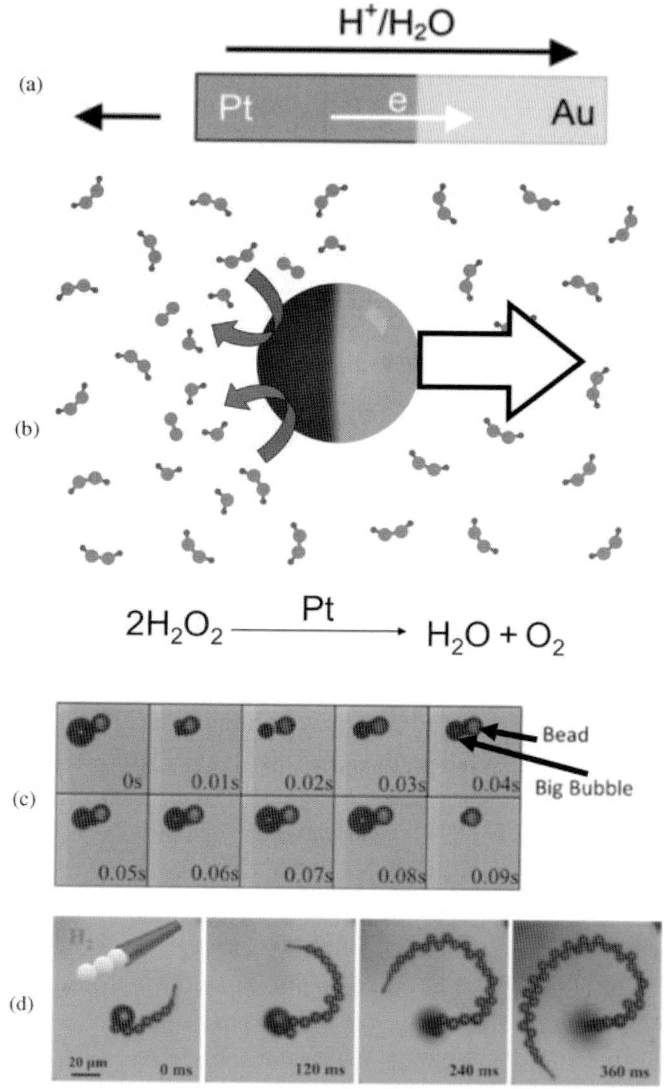

Figure 2. Major catalytic swimming device categories: (a) Schematic for a bimetallic nanorod [32] (b) Schematic of a spherical Janus active colloid. (c) Still frames from a video for a moving bubble producing spherical Janus active colloid [33]. (d) Schematic and still frames from a video for a moving bubble producing tubular active colloid [34].

hemisphere and one inactive hemisphere, and can produce motion either with or without the generation of visible bubbles depending on their size (Figs. 2(b) and 2(c) [27]). Experiments have reported motion production across a range of colloidal sizes [28,29], for different materials [30], and investigated phenomena such as self-assembly [31]. Nanotubes produce propulsion due to bubble release from one end of a tubular structure (Fig. 2(d) [6]). For nanorods and catalytic spheres, an obvious two-faced "Janus" structure is present, whereas for nanotubes, the catalyst is confined to the inside of a tube, and, for this reason, we focus our attention on the former two examples, but still draw on results for tubular swimmers where it is informative to highlight demonstrated functions that may be conferred to Janus swimmers in the future. For nanorods, a Janus structure is a critical feature, and without this feature, propulsion is not observed. As we shall discuss, this is because the mechanism driving motion for nanorods is critically reliant on two chemically distinct materials. The imperative for a Janus structure also exists for many smaller sized Janus sphere swimmers (i.e., $<10\,\mu$m), again because their motion-producing mechanism (phoresis) is reliant on asymmetry. However, larger catalytic swimmers can produce motion in fluids without the absolute requirement for a Janus structure by bubble release, but a Janus structure remains desirable in order to impart directionality to the motion.

Following this general introduction, we now review in more detail a selection of examples of reports of motion generated using a Janus catalytic structure.

As mentioned above, nanorods provided the first examples for nanoscale catalytic self-propulsion. Initial reports appeared almost simultaneously for both gold–platinum and gold–nickel rods (Fig. 3(a) [24,35]). In both cases, the catalytic reaction taking place asymmetrically across the rod was the decomposition of hydrogen peroxide into water and oxygen, by either a platinum or nickel catalyst. Since, this first observation, the hydrogen peroxide/platinum has been used to drive the vast majority of reported synthetic swimming devices, as it facilitated a rapid decomposition reaction at room temperature. These early nanorod devices were analyzed and found to produce maximum average translational velocities of up to 8 $\mu m\ s^{-1}$, and their velocity was found to increase with fuel concentration. Analysis of the energy balance for these devices showed that

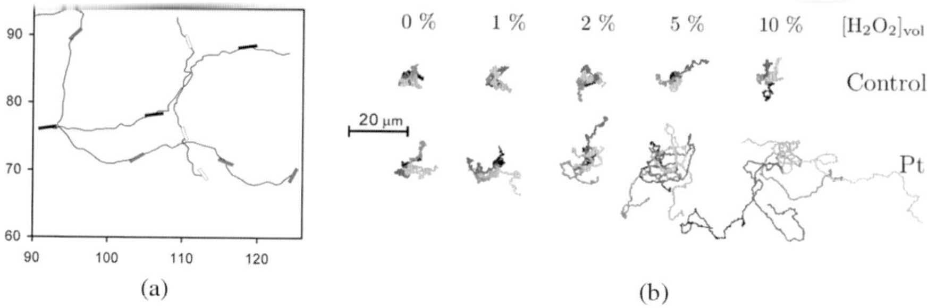

Figure 3. Trajectories for early examples of Janus catalytic swimmers (a) shows trajectories for a 2.5 μm long gold–platinum nanorods in 2.5% hydrogen peroxide fuel (b) shows trajectories for a 1.6 μm diameter platinum–polystyrene (PS) Janus sphere as a function of differing hydrogen peroxide concentrations, compared to a control pure PS sphere expected to exhibit only Brownian motion.

the proportion of chemical energy being converted into motion was low [2]. However, by changing the composition of the nanorod swimmers, it was possible to produce improvements in speed, for example, mixing carbon nanotubes into the platinum end of a nanorod increased translational speeds to 43 μm s^{-1} [36]. In addition, adding hydrazine to the hydrogen peroxide fuel also improved propulsion velocities, with a few examples observed to move at faster than 200 μm s^{-1} [36]. Another velocity improvement was found by alloying silver with gold [25]. Nanorod propulsion velocity was also shown to be dependent on temperature: at 65°C, an almost fourfold increase in nanorod propulsion speed compared to ambient conditions was reported.

The first example of a spherical Janus catalytic swimmer also exploited the platinum/hydrogen peroxide reaction, and comprised a polymeric colloid was covered on one hemisphere with the catalyst (Fig. 3(b)) [27]. This initial report demonstrated a maximum propulsion velocity of 3 μm s^{-1} and, in line with nanorod behavior, found that the propulsion velocity increased with increase in the concentration of hydrogen peroxide fuel. Since this study, Janus spheres with a range of different compositions have also been shown to produce propulsive motion. Some examples retain the platinum coating, but have varied the core colloids composition, for example using silica [37] or gold [38]. In the case of a gold core, the concentration of hydrogen peroxide required to produce an appreciable translation velocity was reduced considerably compared to the examples with an electrically insulating core. Recently an extremely small version of a solid gold/platinum Janus swimmer, with a diameter of around 30 nm, has also been shown to also produce propulsive motion [39]. Also, different catalytic layers such as iridium have been investigated and found to produce propulsion at very low levels of dissolved fuel [40].

Another area of interest that runs through the development of these devices is the quest for alternative motion-producing reactions. One area of particular focus has been to try to deploy enzymes as alternatives to metallic catalysts. An early attempt to use catalase, a very high turnover enzyme for hydrogen peroxide decomposition to power a nanorod device did not result in any detectable motion [32]. In contrast, when a larger sphere was decorated with a synthetic catalase-like complex, motion by bubble production and release was observed [41]. Similar results were also demonstrated for rod-shaped devices, using either catalase, or a combination of catalase and glucose oxidase [42,43]. More recently, a catalase-based swimmer with a Janus structure was demonstrated, also propelled by bubble generation [8]. The potential for more exotic reactions to produce motion has also been shown, for example, a polymerization-powered Janus motor was demonstrated. In this example, a Grubbs catalyst was attached to the silica side of a gold–silica Janus colloid, and the diffusion coefficient increased when the monomer norbornene was added [44]. Other Janus catalytic swimmers have not used a rigid core particle, for example, an asymmetric microcontact deposited coating of platinum nanoparticles onto a polyelectrolyte capsule resulted in a device that was propelled by bubble generation [15]. Another class of Janus swimmers relies on photocatalytic reactions occurring at the surface of a Janus colloid. One example is a gold–titanium dioxide Janus swimmer, which when activated by UV light decomposes water to result in translational propulsion at up to 25 μm s^{-1}, with a velocity that is controllable by light intensity [45]. This approach to motion production has the advantage of not requiring the addition of fuel to the fluid in which swimming is desired; however, this is at the expense of requiring actuation by external illumination.

While the details of all of these motile devices differ considerably, they share the common feature of a Janus catalyst distribution. For the remainder of this chapter, we turn our attention to considering in more detail the general experimental challenges of making and investigating these devices. We also survey the advances the field has made toward gaining control of the motion for Janus catalytic swimmers, in light of the challenges of Brownian motion outlined above, and assess how these advances align with potential future applications.

8.2. Synthesis

Adoption of the state-of-the-art methods for making catalytic Janus swimming devices surveyed elsewhere in this book has been somewhat limited, and as a result, many examples of catalytic Janus devices are currently synthesized using cumbersome approaches, which are difficult to scale. Metallic catalytic Janus nanorods are normally made using a template-based synthesis [46]. In this approach, a porous alumina template is electroplated sequentially with the two metals that comprise the different segments of the rod. The template is then dissolved to release the rods. Many spherical Janus swimmers powered by thin layers of metal such as platinum have been synthesized by exploiting the directionality of metal evaporation or sputtering from a metal target onto spherical colloids dispersed onto a planar substrate. The materials used for the starting colloids have typically been polymeric or silica spheres, either sourced commercially or synthesized, for example using the Stöber method in the case of silica to produce monodisperse devices with a readily controllable size [47]. Methods used to disperse the colloids evenly over the planar substrate include spin coating, or transfer of closely packed colloidal monolayers from a supporting liquids meniscus. Metal evaporation performed under high vacuum from either a thermal or an e-beam heated source provides conditions whereby very well-defined and controllable line of sight deposition occurs (Fig. 4). Likewise, under some conditions, sputter coating can also provide a high degree of directionality, whereas by tuning operating parameters, a more diffusive coating process where metal deposition starts to occur on surfaces that are not directly exposed to the source is possible. Even in the case of a completely diffusive coating process, the physical masking effect of the substrate on which the colloids rest has been found to prevent complete metallization and so ensure a degree of asymmetry in the coated device (Fig. 4).

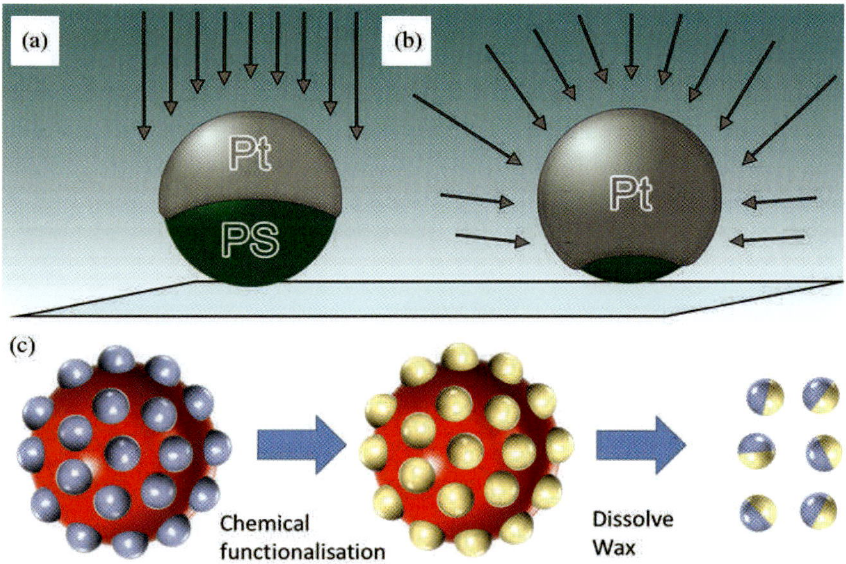

Figure 4. (a) Schematic for thermal (or e-beam) evaporation to form a Janus colloid. The coating is deposited only in regions with a line of sight to the metal source. (b) Schematic for sputter coating to produce a Janus colloid. Now the coating is less directional; however, physical masking by the substrate prevent complete coating. (c) Use of a Pickering emulsion to produce a Janus structure. The large red sphere represents the wax phase used to partially mask the colloids prior to chemical functionalization.

The evaporation method has some significant drawbacks, relying on expensive equipment, and also suffering from poor scalability due to the limited number of colloids that can be accommodated on a 2D surface, together with the requirement for these colloids to be dispersed using a non-continuous method such as spin coating. An extension to these metallization routes is to develop solution-based methods to deposit the asymmetrical catalytic coating required for motion production. One approach is to exploit Pickering emulsions, where a masking material such as wax is mixed with the colloids to be covered in catalyst (Fig. 4(c)). Under the correct conditions, this results in a suspension of larger wax particles into which the smaller colloids are partially embedded. This allows further solution chemistry to be carried out selectively on the remaining partially exposed surface of the colloids to impart asymmetrical catalytic activity. As a final stage, the colloids are released from the wax matrix. To date, this method has been employed to produce catalytic Janus bubble swimmers by selective amination of the exposed side of the embedded colloid to allow decoration with catalase [8].

8.3. Propulsion Mechanisms

The most conceptually straightforward mechanism proposed to drive catalytically powered miniaturized swimming devices is momentum transfer caused by the detachment of bubbles from the devices surface [48]. As the majority of devices are powered by decomposition reactions that generate a gaseous product, the potential for this mechanism to act is present in most reported examples. In this mechanism, the gaseous product from the catalytic reaction nucleates as bubbles on the device surface, which grow, and then eventually detach. It is during this detachment phase that the device is caused to move, and so a signature of this propulsion mechanism is "stop-start" behavior. As mentioned in the Introduction, for bubble propulsion, a Janus structure is not an essential feature for motion production, for example spheres completely covered in catalyst will also display enhanced motion [49]. While several catalytic Janus swimming devices clearly issue a trail of bubbles from their surface as they move, many other devices do not. One proposal to explain motion without obvious bubble release is that, instead, nanobubbles are being generated [48]. Nanobubbles are very hard to detect experimentally and so confirming this mechanism experimentally has proven difficult. Indeed, some devices also display examples of features inconsistent with bubble release, including a direction of travel toward the catalytically active patch. This counterintuitive direction of travel is commonly observed for bimetallic nanorods [24] and recently reported to occur under certain conditions for Janus spheres [50]. These contradictions, together with the lack of evidence for bubbles of any size being generated by many catalytic swimming devices, suggest that many devices are instead caused to move by other phenomena. Among a variety of proposals for these additional motion-generating phenomena, it has been generally accepted that phoretic mechanisms are likely to contribute to the motion for many non-bubbling devices. In general, phoresis causes an object to move through fluid due to an external gradient being present at its interface with the surrounding fluid. Phoretic phenomenon includes diffusiophoresis, where there is an interfacial gradient in solute concentration, and electrophoresis where electric field gradients are present. Conventionally, phoretic effects are demonstrated and investigated by applying an external stimulus such as an electric field to a suspension of colloids, and observing the resulting motion. However, the mechanisms proposed to move catalytic Janus swimming devices differ in that they are "self-phoretic" because the colloidal swimmer itself generates the interfacial gradient that results in motion. For example, self-electrophoresis has successfully explained many experimental observations for Janus nanorod swimmers. In this scheme, the oxidation of hydrogen peroxide fuel at the platinum end of the rod is coupled with reduction reactions at the gold end of the rod, and the resulting charge transfer due to this redox scheme induces the interfacial electric field gradient, instigating movement via electrophoresis [51]. This model has correctly predicted experimental observations including the direction of travel and magnitude of velocity (both of which can be varied according to inactive segment metal choice) [32], which are required for a conductive path between the two segments in order to produce motion [32], and reduction of swimming speed with salt concentration [32].

In the case of catalytic Janus spheres, self-diffusiophoresis due to a self-generated gradient in solute was proposed to be capable of producing enhanced motion before swimming phenomena were experimentally observed [52]. Theoretical analysis suggested that a suitable uniformly catalytically active hemisphere would generate a gradient of particles around a colloids surface, which would in turn lead to motion. The development of this model, and related simulations for dimers performed by the Kapral group [53,54], have been successful in explaining the experimental behavior for Janus devices, including the dependency of their velocity on fuel concentration and size [27,55].

However, recently it has been unexpectedly observed, both by our group and the Poon group, that catalytic Janus spheres, manufactured using the standard method of metal evaporation, show a strong velocity dependence on salt-concentration [50,56]. In both studies, it was found that adding even small amounts of salt to the fueled liquid surrounding

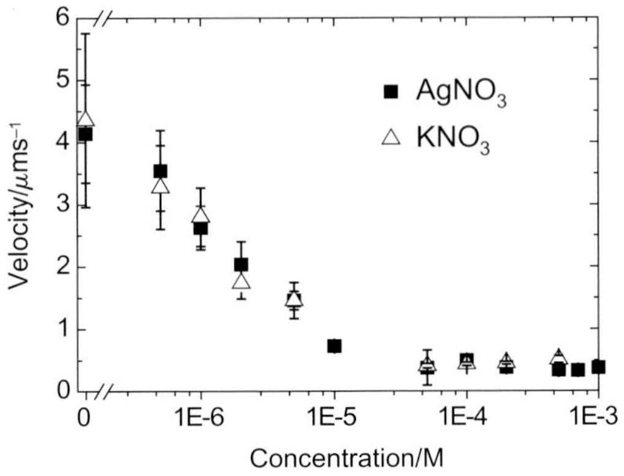

Figure 5. Effect of varying $AgNO_3$ and KNO_3 concentration on the propulsion velocity for 2 μm diameter platinum Janus sphere particles in 10% w/v H_2O_2 aqueous solutions.

the catalytically active Janus spheres significantly reduces swimming velocity, and that the magnitude of this reduction is much greater than the associated reduction in catalytic decomposition rate. For example, Fig. 5 shows the reduction in propulsion velocity observed for platinum-coated Janus catalytic swimming devices on the addition of two different salts at low concentrations. It is apparent that a rapid monotonic reduction in swimming velocity is observed, and that above 10^{-5} M the device's ability to generate propulsion has been completely quenched. As self-diffusiophoretic theory predicted that the propulsion velocity should scale with the decomposition rate of the dissolved fuel, a first glance explanation for this phenomena could be that the salts are reducing the platinum catalyst's ability to decompose peroxide fuel. To verify this proposal, Fig. 6 displays the reduction in decomposition rate relative to a salt-free solution on the addition of equivalent salt concentrations. The potassium salt shows a variable reduction in reaction rate on the addition of salt, which, however, does not correspond to the monotonic trend in velocity, or indicate complete quenching of reaction rate at 10^{-5} M concentration. Surprisingly, the silver salt showed a striking quenching of reaction rate at even low levels of addition. Taken together, it is clear that this experimental data suggests that salt is hindering the propulsion mechanism for Janus catalytic swimmers in a way that cannot be explained purely by a reduction in reaction rate.

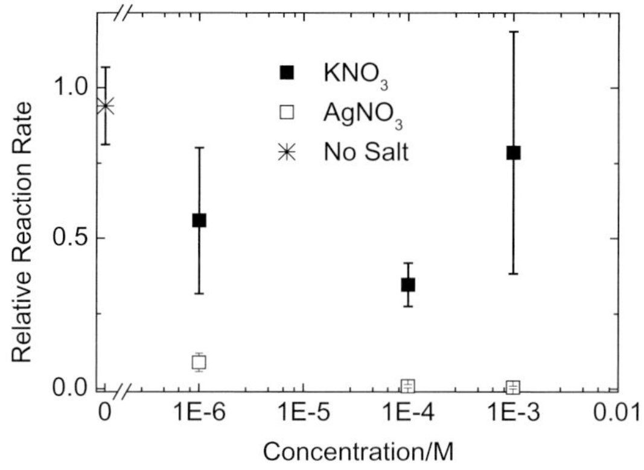

Figure 6. Relative reaction rate (k_{salt}/k_{nosalt}) as a function of salt concentration. k_{nosalt} is measured in a salt-free 10% w/v H_2O_2 solution prior to a subsequent reaction rate measurement in a salt-containing reaction mixture (k_{salt}). Hydrogen peroxide concentration was determined as a function of time using UV spectroscopy, and each experiment is summarized by a linear fit to find the initial reaction rate from the first 5 min of kinetic data. Reaction rates were measured on 1 cm^2 squares of 10 nm thick evaporated platinum. Each experiment was performed in triplicate; average ratios with the associated standard deviation are shown. The no salt data point indicates a sample that was consecutively analyzed twice in salt-free hydrogen peroxide, and so acts as a control for the remainder of the data points.

This behavior is similar to the salt-induced speed reduction for bimetallic nanorod swimming devices, which has been explained by self-electrophoresis and so suggests that the propulsion mechanism involves an electrostatic interaction. This initially surprising observation has led to two new related electrokinetic theories for Janus sphere propulsion, which both suggest that a major contribution to propulsion velocity relies on electron and proton flow induced by a variation in reaction rate across the active hemisphere [50,56]. Figure 7 shows one proposal for such a scheme, based on the existence of two separate reaction loops for the decomposition of hydrogen peroxide by platinum, one of which involves a charged species. Analysis of this kinetic scheme has shown that the electrophoretic contribution to swimming velocity, due to the charged species loop can be much larger than the diffusiophoretic contribution. A second key feature of the analysis showed that the electrophoretic contribution vanishes, if there is no variation in reaction rates across the surface of the Janus sphere.

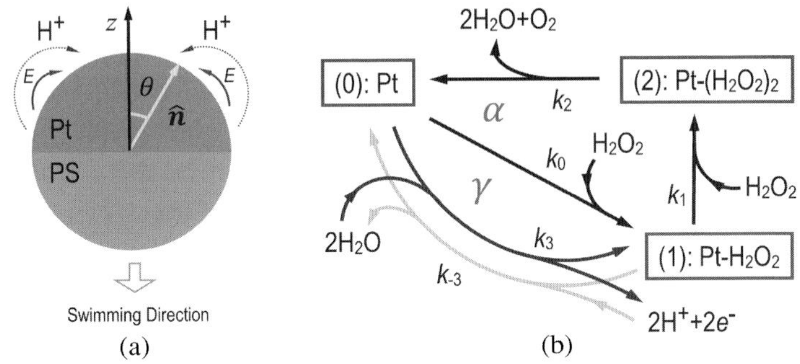

Figure 7. (a) A half Pt-coated PS Janus sphere showing the direction of flow of ions and the electric field. (b) The catalytic reaction scheme showing the competing reactions that have the topological structure of two coupled loops. Loop α is the main non-equilibrium cycle that involves only uncharged species, and loop γ is linked to the production of charged intermediates H^+, e^-.

Based on this analysis, further experiments were performed to try to ascertain any possible origin for a variation in surface reaction rate across an evaporatively produced Janus sphere. One possible source is variations in the thickness of the thin platinum layer. For example, due to the directional nature of metal evaporation from an e-beam or thermal source, a thickness gradient from pole to equator exists (Fig. 8). Separately, using experiments conducted on flat substrates, it was established that there was a strong correlation between evaporative platinum film thickness and hydrogen peroxide reaction rate over the range 0–20 nm (Fig. 8). So for a typical Janus swimmer with a nominal 10 nm thickness coating, this data suggests that a considerable gradient in reaction rate over the active hemisphere is likely, and so supports the argument that these swimmers have a large electrophoretic contribution to their swimming velocity.

We now consider the factors that may determine which mechanism dominates in a device type-bubble release or phoresis. Figure 9 displays the relationship between device size, type, propulsion velocity and the presence or absence of visible bubbles emanating from the device surface for numerous reported systems. One clear division is that rolled up catalytic nanotube swimmers all produce a prominent bubble trace, while bimetallic nanorod swimmers have not been associated with

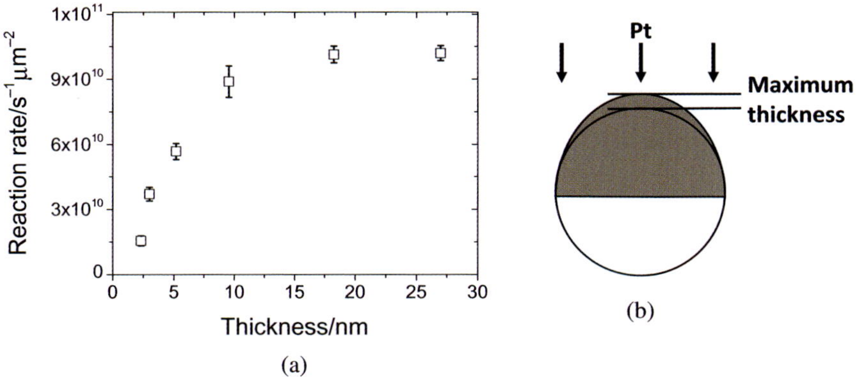

Figure 8. (a) Reaction rate for hydrogen peroxide decomposition by a series of increasingly thick evaporatively deposited platinum metal films onto flat substrates. (b) Schematic for the thickness variations across an evaporatively coated Janus catalytic swimmer.

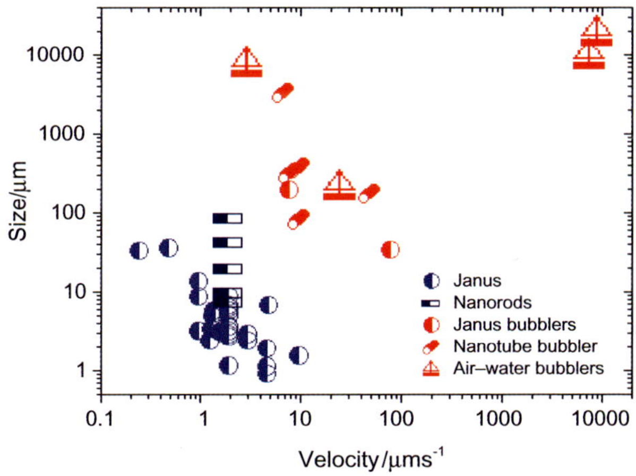

Figure 9. Summary of velocity and size for published swimming devices powered by H_2O_2 decomposition by either Pt or catalase.

Sources: Janus no bubbles [16,27,37,48,54,59,61–63]+ Unpublished Ebbens group data; Janus with bubbles [41,58]; Nanorods no bubbles [24,25,32,51,64]; nanotubes bubbling [10,65,66]; air–water devices bubbling [42,43,67–69].

visible bubble production. For the Janus sphere case, most smaller devices do not produce any visible bubbles; however, some larger examples do show bubble-driven motion. This link between size and dominant propulsion mechanism is not unexpected, as the energetic cost of nucleating bubbles is considerably increased as surfaces become more curved [57]. However, size is not the only determining factor in turning on bubble propulsion in catalytic Janus swimmers, for example, Fig. 9 shows there are two approximately 10 μm diameter Janus devices, one of which produces bubbles [58], and the other which does not [59]. These examples differ in catalytic type, with the bubble-powered swimmer being coated in platinum nanoparticles rather than a thin metallic platinum coating. This suggests that, in addition to size, higher reaction rates are potentially associated with bubble production. The observation of bubbling from the high turnover catalase/synthetic catalase covered coated Janus swimmers also supports this [8,41]. Indeed, this link has been confirmed in a study that showed how phonetically driven and bubble release drive motion can be selected by control of surface reactivity of the catalytically active region of a spherical Janus swimmer [60].

8.4. Characterizing Janus Swimmer Motion

To enable the motion of swimming devices to be analyzed and interpreted, it is necessary to deploy methods to accurately monitor the position of the small-scale devices as a function of time to obtain quantitative trajectories, and to deploy analytical methods to derive useful performance measures such as velocity and directionality of motion. This section provides a survey of commonly applied approaches.

8.4.1. Particle Tracking for Catalytic Janus Swimmers

8.4.1.1. *Microscopy*

8.4.1.1.1. 2D Particle tracking

The essence of most attempts to quantify the motion of Janus catalytic swimmers is to accurately determine the position of a fixed point on the swimming device as a function of time. The sequence of images required to achieve this goal are often obtained straightforward using video microscopy. For micron-scale devices, adequate microscopic images can be recorded using conventional transmission or reflection mode microscopy. However, when device size is reduced to less than a micron, it is often preferable to use fluorescence microscopy, which requires the colloid under investigation to be labeled with a fluorophore. This enhances signal to noise and allows smaller devices to be accurately tracked. For spherical Janus, catalytic swimming devices, the ready availability of fluorescent polymeric spheres with well-defined sizes makes it straightforward to synthesize small-scale devices suitable for fluorescent tracking.

Having recorded a series of suitable high-contrast microscopy images, it is then necessary to accurately determine the *x*,*y* coordinates (for the 2D case) for the swimming devices in each frame. This can be accomplished laboriously and with limited accuracy using manual particle tracking, where a user places a cursor on the position of the device for each video frame. However, a much better approach is to use one of the many algorithms available to automate subpixel accuracy particle tracking. One common approach uses Gaussian fitting to the reflected, transmitted, or fluorescent light intensity profile [70]. Once accurate *x*, *y* time-stamped data is obtained, the range of trajectory analysis methods discussed below can be applied.

8.4.1.1.2. 3D Particle tracking

Many eventual applications for catalytic Janus swimming devices are likely to require the ability to move in 3D throughout bulk solutions. However despite this, until recently the majority of observations made for swimming devices have been performed using conventional top-down 2D microscopy as described above. One reason for this is that some catalytic swimming devices such as metallic nanorods are very dense compared to the fluids they produce motion in, and so rapidly settle to the bottom of their containers. However, this is not the case for all devices, for example, Janus spheres with a polymeric body settle as sediments very slowly and so can be comfortably observed in bulk solutions. Consequently, it is desirable to be able to extend particle tracking to accurately obtain 3D positional data for Janus swimmers as a function of time. One method to achieve this is based on

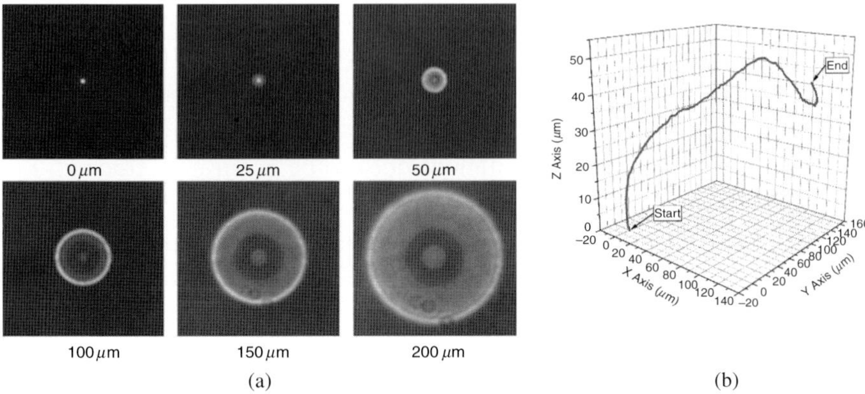

Figure 10. (a) Images of a 4.8 m diameter fluorescent Janus platinum-coated PS sphere recorded using a times 20× objective (0.4 NA). The distances below each image indicate the distance of the focal plane of the objective above the sphere. As the image is defocused from 0 µm to 200 µm the in focus image of a bright disk changes to a bright ring, the radius of which is dependent upon magnification, the sphere size and its distance from the focal plane. (b) A trajectory of a typical fluorescent Janus sphere swimming device. A sequence of images of the moving swimming device were recorded over a period of 30 s at a frame rate of 33 Hz. The (x, y, z) coordinates of the trajectory were obtained by locating the bright ring center and comparing the measured ring radius to the calibration chart for each image in the sequence.

interpreting the ring structures produced by out of focus fluorescent colloids observed with a fixed focus objective in order to estimate z-displacement relative to the focal plane [71], and has proven to be easy to apply using conventional unmodified microscopes. Figure 10 shows the essential features of this method, which relies on the change in size of the radial ring produced by a fluorescent colloid as it moves out of the focal plane. Algorithms can be employed to implement this analysis procedure rapidly and determine z-coordinates for an entire video sequence.

This approach has the significant advantage of only requiring a single frame to allow estimation of swimmer z-coordinate, which consequently allows the high frame rates, which are required to reveal the detailed trajectories. Also, this method only relies on the relative contrast of the out-of-focus colloids in single frames, rather than the absolute fluorescence intensity, and so it is resilient to quenching and blinking effects in the fluorophore. These advantages are possible at the expense of a reduced depth of field over which 3D trajectory reconstruction, and the requirement for well-separated non-overlapping colloids. Alternative approaches that could be deployed in the future for 3D trajectory acquisition include using a motorized z-stage combined with, a fast real-time image analysis feedback loop to keep a given colloid in focus.

8.4.2. Determining the Orientation for Catalytic Janus Swimmers

A further consideration for measuring the behavior of catalytic Janus swimming devices is by determining their orientation. This can be useful to provide mechanistic insights, for example to determine the direction of motion relative to the catalytically active segment. For rod-shaped devices, the orientation can simply be determined by inspection: under the correct illumination conditions, it is usually possible to distinguish the catalytic segment of the rod from the inactive segment (e.g., in a gold/platinum NR in reflection mode the different metallic hues are directly visible). For spherical devices, with no shape anisotropy, the determination of the orientation can be more difficult [72]. One approach that has proven to be effective is to utilize the optical opacity of the metallic catalytic layers that is used to power many Janus swimming devices. Before the current interest in swimming devices, it was noted that metallic layers deposited onto fluorescent colloids allowed direct observation and quantification of orientation. This is because the metal shields fluorescent emission, causing the colloid to present a "moon" like phase, which reveals its 3D orientation [73]. Serendipitously, when some Janus spherical swimming devices are observed in close proximity to a planar interface, their rotation becomes aligned and constrained about one axis so that a constant "half-moon" shape is presented (Fig. 11) [74]. This reduces the task of resolving device orientation to single axis problem, and in this case it is relatively straightforward to apply image analysis, for example via pattern matching to a model template particle with known orientation. Figure 11 shows how this approach was applied to determine a correlation between the direction of motion of a Janus catalytic swimmer and the orientation of the catalyst patch [75].

However, in other circumstances, when the particle rotation is not restricted, accurate, rapid determination of 3D orientation for these

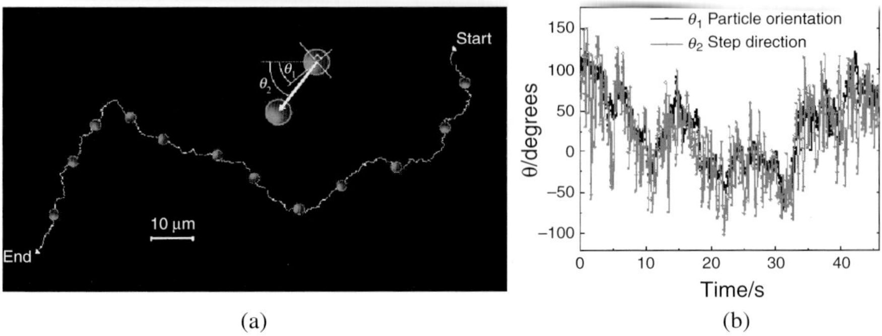

Figure 11. (a) 45 s section of trajectory for a fluorescent Janus swimmer (white) overlaid with captured images of the particle at that particular location. Inset: Definitions of particle orientation (θ_1) and step angle (θ_2) which were measured as shown relative to the horizontal dashed-line (b) Correlation between particle orientation and subsequent direction of motion (step direction).

rapidly moving devices remains a significant challenge, particularly when they are also in 3D motion within a bulk solution.

8.4.3. Trajectory Analysis

The main reason to perform analysis of catalytic swimmer trajectories is to produce quantitative measures for features of device motion that can be statistically compared to provide mechanistic insights and allow a direct comparison of performance metrics such as velocity. The simplest measures that can be extracted from x, y or x, y, z trajectories are instantaneous and average velocities, and indeed for many early reports of catalytic Janus swimmers, trajectory analysis was limited to determining these quantities. For larger devices this allows a quick comparison of performance, however, for smaller scale devices, Brownian translational diffusion becomes significant and it is necessary to attempt to decouple this motion from the velocity due to catalytic propulsion. Additionally, more advanced trajectory analysis methods also allow swimmer directionality (the degree to which a swimmer continues to move in a given direction) to be measured. In this section, a review of the principal trajectory analysis methods developed for catalytic Janus swimmers is given.

8.4.3.1. *Mean-squared displacement method*

The task of decoupling Brownian effects from motion for spherical catalytic Janus swimmers has received particular attention, and is amenable to full analysis due to the well-established correlation between swimmer orientation and the subsequent direction of motion. As shown in Fig. 11, experiments have determined that the propulsion step direction is determined by the orientation of the catalytic cap. This link has allowed equations that predict the evolution of mean-squared displacements (MSDs), ΔL^2 as a function of time interval, Δt, to be derived. MSD is a useful method to summarize and compare trajectories determined as described in Secs. 5.1.1.1 and 5.1.1.2; however, the concept of averaging the trajectory as a function of time interval Δt can cause confusion to those unaccustomed to it. To clarify this, we qualitatively describe the commonly accepted procedure: assuming a unitary time gap between the points comprising the trajectories, for $\Delta t = 1$, it is possible to calculate mean-squared displacements (i.e., $\Delta x^2 + \Delta y^2$) between frames 1 and 2, and also for frames 2 and 3, 3, and 4, 4 and 5, etc. (Fig. 12). For $\Delta t = 2$, a new set of mean-squared values are calculated, to encompass as many $\Delta t = 2$ step displacements as can be accessed from the trajectory data (i.e., frame 1–3, 2–4, 3–5, etc.). This procedure is continued for increasing Δt periods.

In this way, the number of squared displacements that can be calculated for each time interval is $n–\Delta t$, for a trajectory containing n observations of particle position. The mean of these squared displacements gives the value for ΔL^2 for a given time interval. Due to the reduction in number of experimental device position observations used in this averaging process at longer time periods, it is necessary to record significantly longer video microscopy excerpts than the time scale over which MSD will be analyzed. Bearing these considerations in mind, it is possible to attempt to fit MSD data determined from experimental trajectories to theoretical predications, which allows a decoupling of the Brownian and propulsive motion characteristics. For example, Eq. (1) shows the expected MSD evolution for a spherical Janus catalytic device with velocity v, with a Brownian translational diffusion coefficient of D, and a Brownian rotational time constant of τ_R [27].

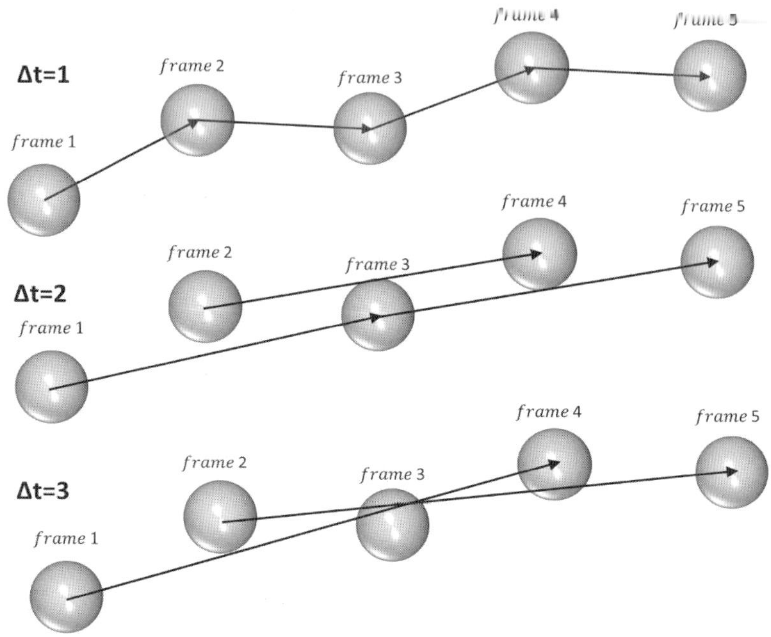

Figure 12. MSD calculation for a swimming device trajectory. The black vectors represent the displacements that are squared and then averaged as a function of increasing time step. The number of displacements that can be averaged reduces by one for each increment in time step.

$$\Delta L^2 = 4D\Delta t + \frac{V^2\tau_R^2}{2}\left[\frac{2\Delta t}{\tau_R} + e^{-\frac{2\Delta t}{\tau_R}} - 1\right]. \quad (1)$$

This equation is often simplified when fitting to experimental data for which $\Delta t \ll \tau_R$ holds true to give $\Delta L^2 = 4D\Delta t + V^2\Delta t^2$, that is, at short times relative to the Brownian rotation constant MSD evolve as a linear function of the diffusion coefficient, which is unchanged relative to that expected for pure Brownian motion, with addition of a quadratic "ballistic" contribution, which reflects MSD for the simple case of motion with constant velocity (i.e., unchanging magnitude and direction). For micron-sized particles, the rotational constant is of the order of 1–10 s ($D_r = \frac{1}{\tau_r} = kT/8\pi\eta r^3$, where k = Boltzman's' constant, T = temperature,

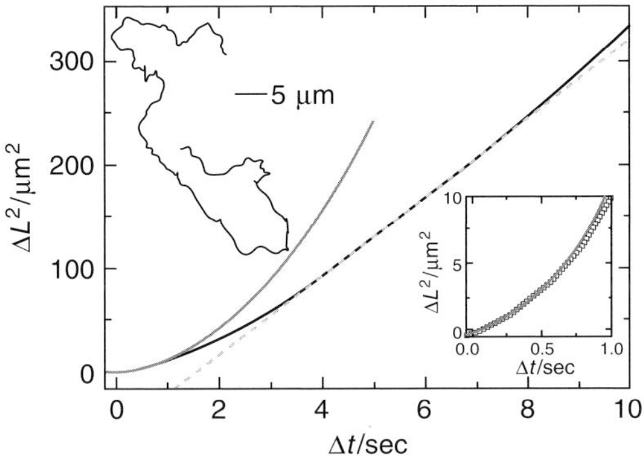

Figure 13. MSD as a function of time for the inset trajectory for a $r = 0.8$ μm Pt-coated catalytic Janus particle swimming (JPs) device in 10% H_2O. Solid line shows fit to the short-term quadratic approximation to Eq. (1), dashed line shows the fit to the long-term linear approximation. The fits yield $v = 3.1$ μm s^{-1}, $D = 0.31$ μm^2 s^{-1}, and $\tau_R = 3.9$ s.

η = viscosity, and r = radius), and so by using high frame rate video microscopy (e.g., with a frame rate between 30 and 100 Hz) it is possible to acquire sufficient trajectory data points within the period $\Delta t \ll \tau_R$ to compare experimental MSD with this prediction, and attempt to fit both the magnitude of Brownian diffusion and the propulsion velocity (Fig. 13.) Conversely, when MSD is considered over long time durations (i.e., $\Delta t \gg \tau_R$), then Eq. (1) approximates to a linear diffusive behavior, but with an enhanced diffusion coefficient $D_{\text{eff}} = D + 1/4v^2\tau_R$. Consequently, by fitting these two portions of the MSD data, or fitting Eq. (1) to the whole curve, it is possible to estimate all the relevant motion parameters for Janus catalytic swimming devices, an approach that has, for example, been used to reveal the relationship between size and propulsion velocity [29]. This analysis will also apply to Janus nanorod motion; however, in this case the rotational diffusion rate of the rod will be appear to be reduced due to asymmetrical drag forces.

However, there are many additional factors that can complicate the application of this analysis approach, as a variety of other phenomena can influence the motion of Janus devices and interfere with the isolation of catalytic propulsion magnitude. Careful consideration of these factors has not always been apparent in all published experimental reports for catalytic swimming motion. This ambiguity has, to some extent, hampered the extent to which experimental work can inform mechanistic insights and theoretical developments.

8.4.3.2. *Motion due to convection*

As well as Brownian motion, small particles may also be subjected to convection currents when observed under the optical microscope. Figure 14(a) shows the trajectory of small particles over the course of 45 s as they undergo a combination of Brownian motion and convection. Their trajectories change from globular random walks when Brownian motion dominates, to ballistic trajectories as convection becomes larger. Without due care and attention being applied to distinguish this scenario from propulsion, the potential to falsely assign this effect to catalytic propulsion is significant.

Although convective currents have to be circular over the whole of the observation chamber to maintain the same amount of liquid in all locations, at the length scales visualized by the optical microscope, convection appears to act in a single direction and at a constant velocity. Visual inspection of the trajectories of many particles undergoing convection reflects this unidirectional motion of convection, as particles migrate in straight lines at constant velocities. This contrasts with the case for small-scale catalytic swimmers, which are expected to be rotated due to Brownian rotational diffusion.

Unfortunately, distinguishing these scenarios is not possible by examining the MSD at short time intervals, as both convection and ballistic propulsion result in constant velocity motion (Fig. 14(b)). To ascertain if deviations in MSD curves are due to convection or propulsion, the MSD must be evaluated at longer time intervals. This gives time for a particle undergoing catalytic swimming to rotate and so change direction of motion, whereas convection continues to move particles in the same direction. Figure 14(c) shows the MSD of a particle propelled at 4 μm s^{-1} over time intervals up to 30 s, by convection. Only at these long time intervals does the MSD of a catalytic Janus swimmers and a particle propelled by convection differ, as rotational Brownian motion changes the direction of propulsion but not the direction of convection.

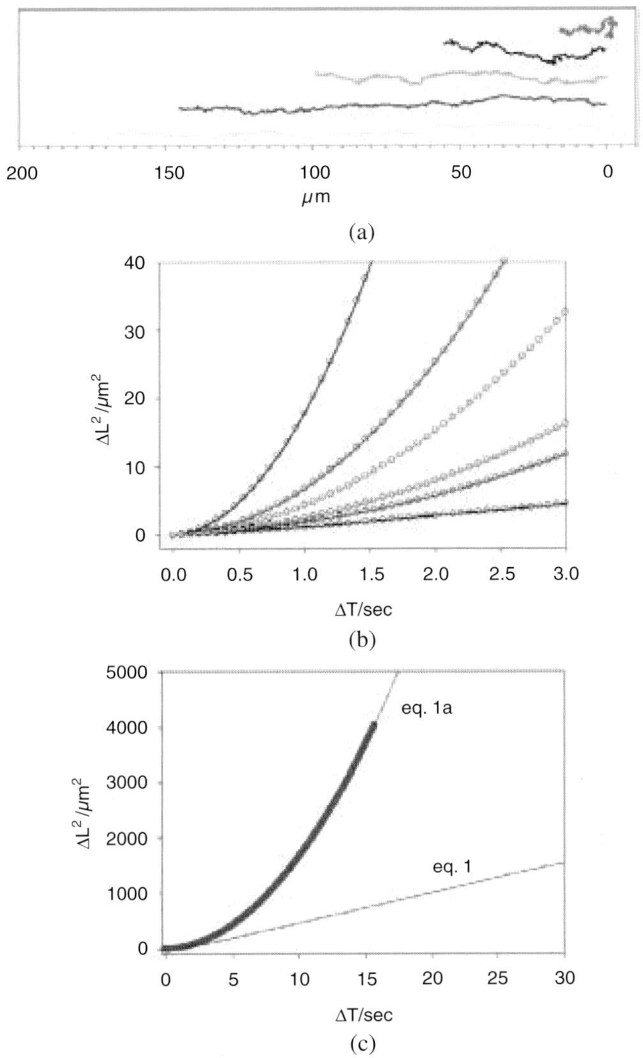

Figure 14. (a) Typical trajectories of particles undergoing a mixture of Brownian motion and convection. (b) MSD of particles at short-time periods and fit to the short-term approximation of Eq. (1). (c) MSD of particles at long time intervals, Eq. 1(a) line is a fit to the short-term approximation for Eq. (1); Eq. (1) is a fit to the full equation for a propulsive Janus swimmer.

Although this MSD approach is a valuable way to ascertain if a device is undergoing self-propulsion in a direction that corotates with its orientation, accessing this longer duration MSD data requires more experimental effort.

Sources of convection that are common in observations of Janus catalytic swimming devices include heating effects due to microscopy illumination, which can be reduced using an infrared filter and evaporation of the liquid, which can be combated by using a sealed sample cell such as a cuvette. Also, many of the chemical reactions that are responsible for the propulsion of particles cause convection by producing gas bubbles. By reducing the concentration of fuel, bubbling and convection can be reduced. Although this also reduces the velocity at which the particles are propelled. Consequently, experiments studying propulsion should seek to find an optimum fuel concentration at which high propulsion velocities and relatively low convective velocities are observed.

8.4.3.3. *Motion due to sedimentation*

Many catalytic Janus devices such as bimetallic nanorods have sufficient density that they quickly sediment through the liquid in which they are dispersed to the bottom of the observation cell. In this scenario, if the bottom surface of this cell is inclined at an angle, then particles will move sideways across the surface, in order to sediment downward and lose gravitational potential energy.

While it seems obvious that this translating motion will occur if the angle at which the surface is inclined is fairly large, in fact, even if the bottom surface of the observation chamber is inclined to an angle of only a few degrees of persistent motion can result, which could be confused with propulsion. For example, Fig. 15(a) shows the trajectory of 4 μm gold particles over 35 s on surfaces inclined at various angles. As can be seen in the figure, the incline leads to particle motion in an approximately straight line down the sloped surface. The greater the slope, the faster the particles are propelled and the greater the distance they cover. As the propulsion is due to the sloped surface, all particles in the observation cell should move in the same direction, and if they are the same mass, at the same velocity. Figure 15(b) shows the way in which this incline biases the mean-displacement steps along the direction of the incline (L). Calculating

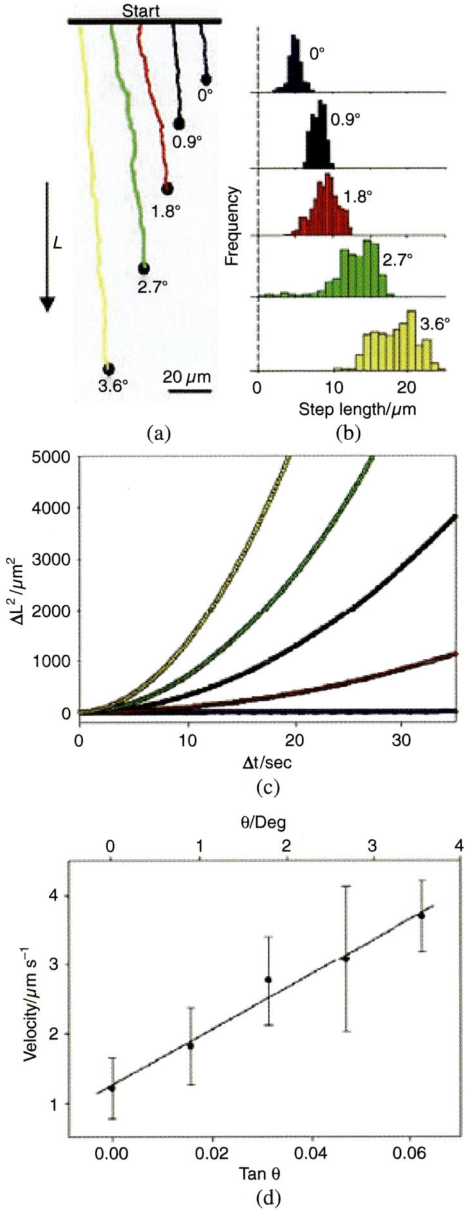

Figure 15. (a) Typical trajectories of 4 μm gold particles sedimenting across a glass surface inclined at various angles over a time period of 35 s (b) Histograms of directional displacements over time intervals of 5 s in the direction marked L (c) MSD (d) Measured velocities against tan θ, each data point is an average of at least 20 particle velocities and error bars show one standard deviation.

the MSD of these particles shows that as the angle of incline is increased, plots change from a straight line, consistent with Brownian motion, to a curved line, again consistent with the addition of a constant velocity translation. As can be seen from Fig. 15(c), even at small angles of inclination, the velocity at which particles are "propelled" is of the same order as many reports in the literature. As the intercept of the line in Fig. 15(c) is not the origin, it can be seen that the microscope stage used for these experiments was not exactly level, and indeed this can be difficult to experimentally achieve. This effect should particularly be borne in mind for experiments where directional drift in response to a chemical signal is targeted. In contrast to dense metallic devices, propulsion due to sedimentation can almost be discounted entirely for small particles, which have a low density. For example, 1 μm PS particles should sediment at less than 1 nm s^{-1} when on a slope of 1°. More care must be taken when using larger more dense particles such as 10 μm silica particles, as these will sediment at a calculated 1 μm s^{-1} on the same 1° slope.

To differentiate between catalytic propulsion and sedimentation, it is necessary to gather evidence from several sources. Sedimentation leads to all particles moving in the same direction and this direction remains constant, whereas propulsion results in particles where the orientation of the propulsion should change. These differences can be observed by plotting the trajectories of particles over relatively long times, and using MSD analysis as discussed in Sec. 5.3.2.

8.4.4. Rotational Motion

In the above discussion (Sec. 5.3.1) about MSD analysis for a propulsive spherical Janus catalytic swimmer, it was assumed that the propulsion generated by the catalytic reaction was not producing any angular velocity. However, experiments have shown that this is not necessarily the case [76], and so methods to determine the additional angular velocity introduced by catalytic propulsion are also required. With reference to the MSD method for analysis, one early paper observed that when rotational diffusion constants were fitted as described above (Sec. 5.3.1), the rotational period was shorter than that expected for Brownian motion alone. This consequently ascertained that there was a potential for enhanced angular velocity to result from catalytically driven propulsion. However, this approach does not allow a quantified value for rotational velocity to be determined [27]. A more detailed analysis of the situation of a Janus sphere moving with angular velocity, ω, in addition to translational velocity, v, and subject to Brownian motion (described by a translational diffusion coefficient D, and rotational diffusion coefficient D_r) leads to another equation for MSD Eq. (2) [31]:

$$\Delta L^2(t) = 4Dt + \frac{2\nu^2 D_r t}{D_r^2 + \omega^2} + \frac{2\nu^2(\omega^2 - D_r^2)}{(D_r^2 + \omega^2)^2} + \frac{2\nu^2 e^{-D_r^2}}{(D_r^2 + \omega^2)}[(D_r^2 - \omega^2)\cos\cdot\omega t - 2\omega D_r \sin\cdot\omega t]. \quad (2)$$

Fitting experimental MSD plots to this equation has consequently allowed full estimation of each parameter, and therefore can separate the effects of angular velocity and translational velocity for rotating catalytic devices, including the example shown in Fig. 16 [31,76].

A further approach to quantifying angular velocity for catalytic swimmers is to deploy the ability to determine particle orientation as described in Sec. 5.2. An example of this approach applied to a rapidly spinning fluorescent Janus catalytic swimmer is shown in Fig. 17. As discussed previously, when the platinum cap on a fluorescent particle is facing toward the collection objective of the microscope, it blocks the fluorescence emission. Consequently, driven catalytic rotations produce a regular modulation in fluorescence intensity. Simply by counting the period of these rotations, a direct measure of angular velocity is possible. In the depicted example, analysis using this method suggested an angular velocity of 9.7 rad s^{-1}, which was in very good agreement with the MSD trajectory fitting method applied to the corresponding trajectory.

8.4.5. Stochastic Mechanisms: Bubble Propulsion

The feature of motion for phoretic Janus catalytic swimmers that allows MSD analysis to be usefully applied is the established link between the orientation of the catalytic cap and the subsequent direction of motion. However, for other types of catalytic Janus swimming device, this link is less obvious. For example, considering a spherical catalytic Janus

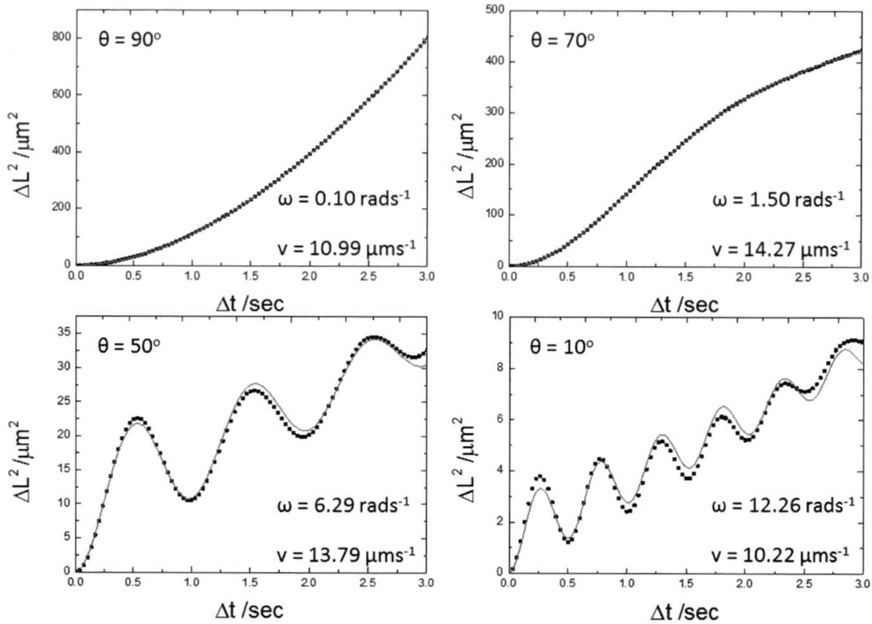

Figure 16. MSD data for Janus catalytic swimmers with varying amounts of angular velocity together with a fit to Eq. (2) (curved line).

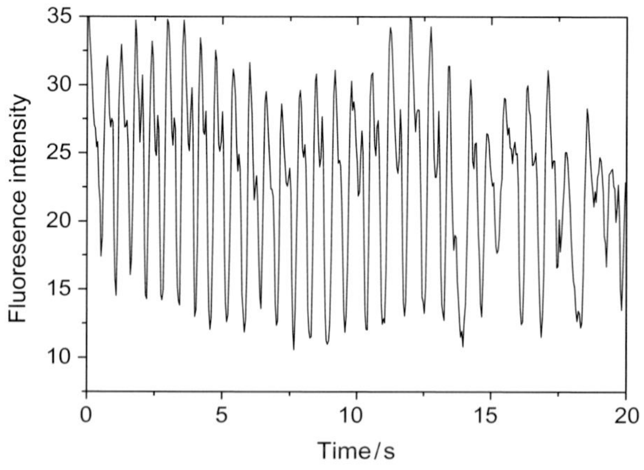

Figure 17. Fluorescence intensity as a function of time for a rapidly rotating Janus catalytic swimmer comprising a fluorescent colloid coated with a thin catalytically active layer of platinum. The period of the intensity fluctuations provides a method to estimate angular velocity.

swimmer that moves via the production of bubbles, while bubble growth and detachment may reasonably be expected to be nucleated at the surfaces covered with catalyst, rather than the inactive regions, the exact location at which bubble detachment occurs changes stochastically. This consequently makes the propulsion vector orientation and magnitude difficult to analytically relate to the devices orientation. In this situation, it is useful instead to deploy more general trajectory analysis methods such as those with demonstrated utility for related complex path problems such sperm trajectories [77] and protein motor transport [78]. Here we briefly introduce three potentially useful methods: fractal dimension, radius of gyration, and persistence length.

Fractal dimension (D_f) is a measure of the space filling properties of an object, or in this case the line representing the motion of a swimming device [79]. Unless the trajectory line completely fills the space it occupies, D_f will be smaller than the embedding dimension. Fractal dimension has been used to quantify a wide variety of systems, including aggregating diffusing particles and clusters [80], particle gels [81], bacterial colonies [82], and biopolymer gels [83]. Calculating the fractal dimension of the trajectories for catalytic Janus swimming devices can provide a number by which we can statistically compare behavior without making assumptions about the motion mechanism. For the example of a 2D trajectory, at sufficiently long time scales, a small-scale object displaying simple Brownian diffusion will have a fractal dimension of 2, while the trajectory of an object solely propelling in a straight line has a fractal dimension of 1. One promising method to obtain D_f from swimming trajectories follows the approach proposed by Katz *et al.* for the analysis of growth paths or trails for biological cells and axons moving in 2D [84]. In this treatment, the fractal dimension is given by Eq. (3)

$$D_f = \frac{\log(n)}{\log(nd/L)}, \tag{3}$$

where n is the number of trajectory segments (equivalent to the number of images in the sequence), d is the planar diameter of the trajectory, and L is the path length. The planar diameter d is found by computing the maximum distance between any two points in the trajectory. One advantage for fractal dimension analysis is that it provides a dimensionless result, and so can allow comparison for motion produced over a variety of length and time scales. For example, the fractal dimension for self-motile sperm has been reported to be around 1.3 [77] similar to that observed for an catalytic Janus bubble swimmer example where the calculation method described above was applied [49].

The mean-squared radius of gyration (R_g^2) calculates the average mean-squared distance of each point on a trajectory from a defined central fixed point, and so reflects how compact, or spread out in space a given trajectory segment is. R_g^2 can be calculated using Eq. (4), in analogy to the definition used for polymer chain analysis, where r_{mean} is chosen to be the mean position for the trajectory (i.e., the unweighted mean x, y coordinate for the trajectory segment)

$$R_{\text{g}}^2 = \frac{1}{N}\sum_{i=1}^{N}(r_i - r_{\text{mean}})^2. \tag{4}$$

Another measure for trajectory directionality is to determine the persistence length. The persistence length L_{p} denotes the distance along a set of trajectory points after which the correlation of the angle θ between consecutive segment vectors is lost; i.e., a linear set of chain segments will result in a larger L_{p} than a set of chain segments featuring curves or loops. Persistence length based analysis methods have been applied to a variety of systems including human-made polymers, biopolymers [85], and the most relevant example is the characterization of the trajectories of microtubules moving on a kinesin-coated surface [78]. L_{p} can be determined for a given trajectory by first computing the average cosine angle $\langle \cos\theta \rangle$ between vectors at all separation distances L along the chain. This can be achieved by superimposing vectors at position j on the vector at starting position i and calculating $\cos\theta$, where $j \geq i$ and i represents all starting positions along the chain (see Fig. 18). For 2D trajectories, a plot of $\langle \cos\theta \rangle$ against separation distance $L = \Delta L(j - i)$ has the form given in Eq. (5), and so fitting can be used to extract an estimate for persistence length

$$\langle \cos\theta \rangle = \mathrm{e}^{-(L/2L_P)}. \tag{5}$$

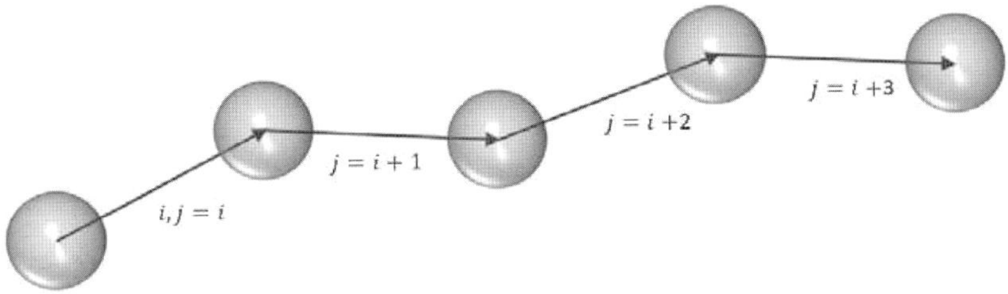

Figure 18. Swimmer trajectories are a chain of vectors. The persistence length is found by computing the average cos θ between the starting vector at position i and the superimposed vectors at position j, for all separation distances $L = \Delta L(j - i)$ along the chain and all starting positions i.

An example of the application of all three trajectory analysis methods was used to compare the trajectories for catalytic Janus bubble swimmers with a variety of coverages of the catalyst. This enabled quantification to confirm that reducing the catalytically active portion of the device produced an increase in the directionality of propulsion, as evidenced by a lower fractal dimension, greater radius of gyration, and an increased persistence length (Fig. 19).

8.4.6. Nanoscale Tracking Methods

The methods discussed above for catalytic JP tracking are applicable as long as the devices can be resolved individually; however, as the size of the device reduces, conventional microscopy becomes limited by diffraction. Two options have been applied to remove this restriction. One is to use emerging laser scattering methods that can resolve nanosized moving objects in fluids, and the second is to extract evidence for catalytic motion from signals averaged over an ensemble of particles. Examples of laser-based methods to extend resolution include the commercially available Nanosight (Malven Instruments) and ZetaView (Particle Metrix) systems. These systems use laser illumination, focused to allow the scatter from individual colloids to be visualized and recorded with a video camera. This effectively extends the ability to apply the 2D tracking methods described in Sec. 5.1.1.1 to nanometer-scale objects. Indeed, these

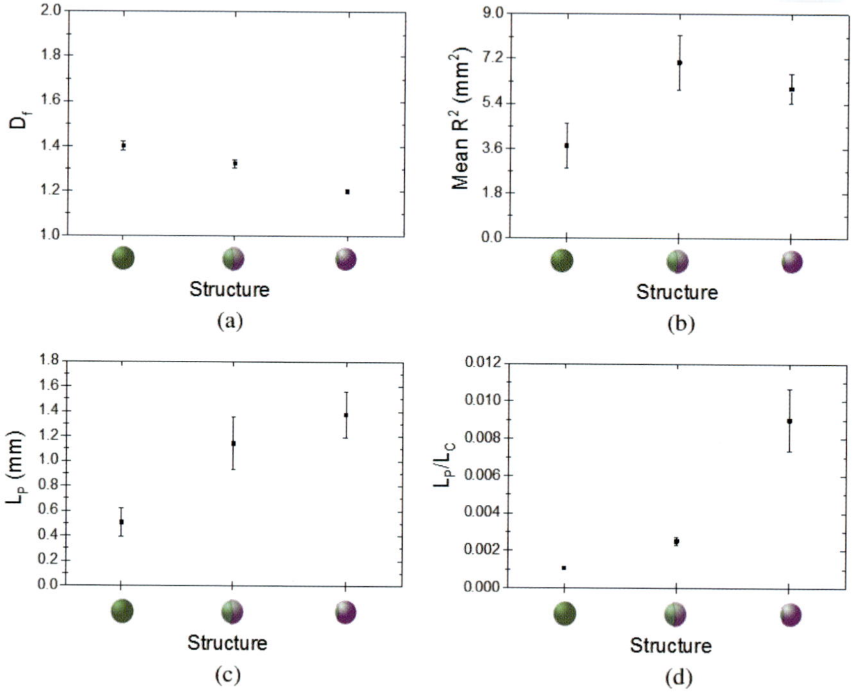

Figure 19. Results of trajectory analysis for Janus bubble swimmers with varying catalyst coverages (green depicts exposed catalyst region) (a) fractal dimension (D_f); (b) mean-squared radius of gyration (R_g^2); (c) persistence length (L_p). Panel (d) shows persistence length values normalized to total path length (LC).

instruments have been used successfully deployed to resolve trajectories for nanometer-scale platinum-loaded stomatocytes [86]. An example of the alternative ensemble approach is interpretation of data from dynamic light scattering (DLS) instrumentation, which was applied as part of the analysis procedure for nanoscale gold–platinum catalytic Janus swimming devices [39]. Also of potential future interest in this context are optical methods such as differential dynamic microscopy (DDM), which have the proven ability to quantify the motion of subresolution colloids [87]. Additionally, fluorescence-based methods such as fluorescence correlation spectroscopy (FCS) [88] and fluorescence recovery after photobleaching (FRAP) [89] can also allow the diffusion of small swimming devices labeled with fluorophores to be determined.

8.5. Motion Control

8.5.1. The Importance of Autonomy

For catalytic Janus swimming devices to be of significant utility, it is necessary to control or harness the intrinsically chaotic motion described above, which originates from Brownian phenomena and in some cases stochastic bubble release, to allow a degree of directional control. Achieving this goal will potentially lead to a new paradigm for fluidic transport so that materials can be precisely and rapidly moved and assembled at very small scales. Of the many published studies that have addressed this challenge, an apparent division in approach can be identified. One set of methods for directional control can be described as "hybrid" where an intrinsically autonomous motion producing device, such as a catalytic Janus swimmer, is steered or otherwise actuated by an external field [7]. While this approach can without doubt enable future applications, the resulting technologies are likely to be relatively complex and require additional control and monitoring infrastructure. In addition, these developments compete directly with other externally actuated methods to transport and organize material at small scales, such as those which exploit magnetic fields to move non-catalytic colloids [14]. In contrast, a second approach aims to retain full autonomy and so allow both navigation like functionality and propulsion to occur without external intervention. Here, we focus mainly on describing this later approach, motivated by the belief that pursuing this goal will lead to the most transformative deployments for catalytic Janus swimmers [5].

8.5.2. Selecting Optimum Janus Catalytic Swimmer Size

As an introduction to the theme of controlling the transport properties for catalytic Janus spheres, we initially consider the problem of selecting an optimum-sized swimming device to achieve most rapid transport over a given distance. We present analysis for the case of a phoretic Janus catalytic swimmer with a well-defined link between orientation and propulsion vector direction (see Sec. 5.2). In this scenario, swimmer size can potentially have two influences on the expected rate of diffusion away from a starting point. As described above, for a phoretic Janus catalytic swimmer, trajectory directionality is determined by the Brownian rotational diffusion coefficient, which scales as the reciprocal of device size. A consequence for this is that larger particles will show more directional trajectories as they are less susceptible to stochastic rotations changing the propulsion vectors orientation. However, changing size for catalytic Janus swimming devices has also been shown to alter the intrinsic propulsion velocity. To illustrate this, Fig. 20 shows typical trajectories for spherical catalytic swimmers with varying diameters, in each case having one entire hemisphere covered in catalyst (50% surface coverage), and all swimming in 10% w/v hydrogen peroxide aqueous solutions. These trajectories arise from the combined effects of the swimmers' intrinsic velocity imparted due to the driving catalytic reaction, together with size-dependent Brownian translational diffusion and rotation diffusion, the latter determining the propulsion direction. Qualitatively the trajectories can be seen to differ significantly. For example, the scaling of Brownian phenomena is clearly apparent, with the smaller diameter devices' rapid orientation and position randomization resulting in a torturous path, whereas at larger sizes, orientation remains almost fixed within the trajectory duration depicted. To allow further interpretation, Fig. 21 and Table 1 report the corresponding intrinsic velocities for these device sizes, and the relevant Brownian diffusion parameters. MSD analysis of many trajectories was

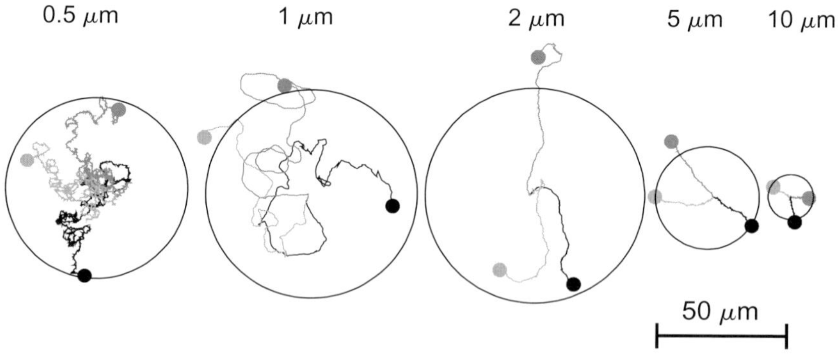

Figure 20. Typical Janus swimmer trajectories, and final positions (after 10 s) for (L-R) 0.5, 1, 2, 5, and 10 μm diameter spheres in 10% H_2O_2 fuel. Large open circles represent the corresponding predicted mean displacements after 10 s obtained from the approach outlined in the text.

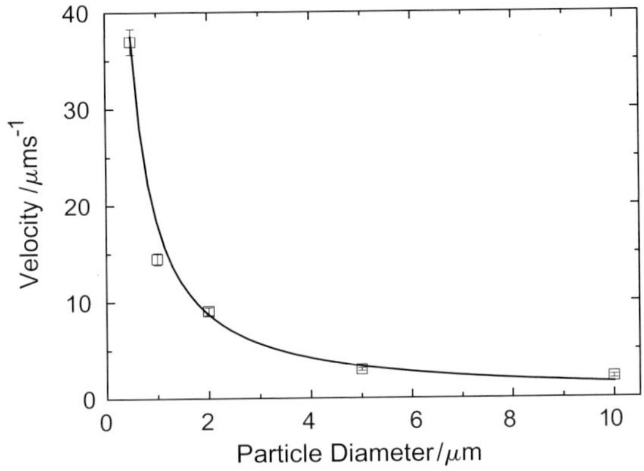

Figure 21. Average velocity vs. size data for JP swimmers.

Table 1. Parameters of motion for Janus swimmers with a hemispherical coating of platinum catalyst in 10% H_2O_2 solution.

	Calculated Brownian diffusion parameters		**Mean intrinsic velocity (μms^{-1}) (SD)**	**Expected Ballistic Run Length (mm)**
Diameter (μm)	**D (μm^{2}s^{-1})**	**τ_R (s)**		
0.5	0.86	0.097	37.0 (1.3)	0.0030
1	0.42	0.77	14.5 (0.6)	0.011
2	0.21	6.2	9.0 (0.4)	0.036
5	0.086	97	3.0 (0.1)	0.18
10	0.042	777	2.1 (0.1)	0.60

used to establish these intrinsic velocity values as described in Sec. 5.3.1 [29]. It is apparent that the intrinsic velocity of the devices increases dramatically at the smaller device size (the scaling approximates a $1/R$ dependency over this size range).

Revisiting the trajectories (Fig. 20), with knowledge of the quantified intrinsic velocities illustrates more clearly the significant role the magnitude of Brownian effects play in determining the paths and resulting transport rate for swimming devices. For example, it is clear that while the 500 nm diameter JP has the fastest intrinsic velocity (37 μms^{-1}); this does not result in the furthest unconstrained transport from the origin over the time period considered, for the trajectories depicted. This is due to the rapid Brownian rotation that this particle undergoes. In contrast, the 2 μm diameter particle, despite its significantly slower propulsion, exhibits trajectories that appear to produce the largest mean transport distance from the starting position, due to the two orders of magnitude slower Brownian rotation rate at this increased size (Table 1). Above this diameter, while the Brownian rotation rate reduces further, the drop in intrinsic velocity results in shorter transport lengths. These trajectories hence appear qualitatively to exhibit an optimum size for maximum transport rate over a given distance/duration, due to the interplay of Brownian and intrinsic velocity scaling. A derived parameter can be used to summarize the character of the trajectories shown in Fig. 20: the expected "ballistic run length," which represents the average distance a given sized swimmer, will cover before changing orientation. This can be simply estimated by multiplying the rotational time constant by the intrinsic velocity (Table 1). Expected run lengths reach millimeters for 5 μm diameter swimmers and above. It should be noted that this complex behavior is uniquely introduced in this system by the coupling of the translational propulsive velocity direction with Brownian rotation diffusion [75]. In the absence of a propulsive force, the expected distance traveled is simply determined monotonically by the Brownian translational diffusion coefficient, which scales as $1/R$, so smaller particles are expected to travel further in a given time period, with trajectories that are completely decoupled from Brownian rotation effects.

To generalize the analysis of the size dependency of transport rate, a model to predict the statistical displacements of hemi-spherical catalytic swimmers as a function of size and time is required. This is possible by making use of the previously derived full analytical expression for MSDs Eq. (1) [27]. Substituting the appropriate values from the intrinsic velocity scaling relationship (Fig. 21) into Eq. (1), together with corresponding D and τ_R values from Table 1 allows the MSDs for spherical swimmers to be modeled as a function of time and diameter.

Expected mean displacements calculated for the device sizes discussed here are shown in Fig. 22(b). These graphs effectively represent the expected behavior that would occur in a virtual race between different diameter swimmers. The points at which the displacement curves cross

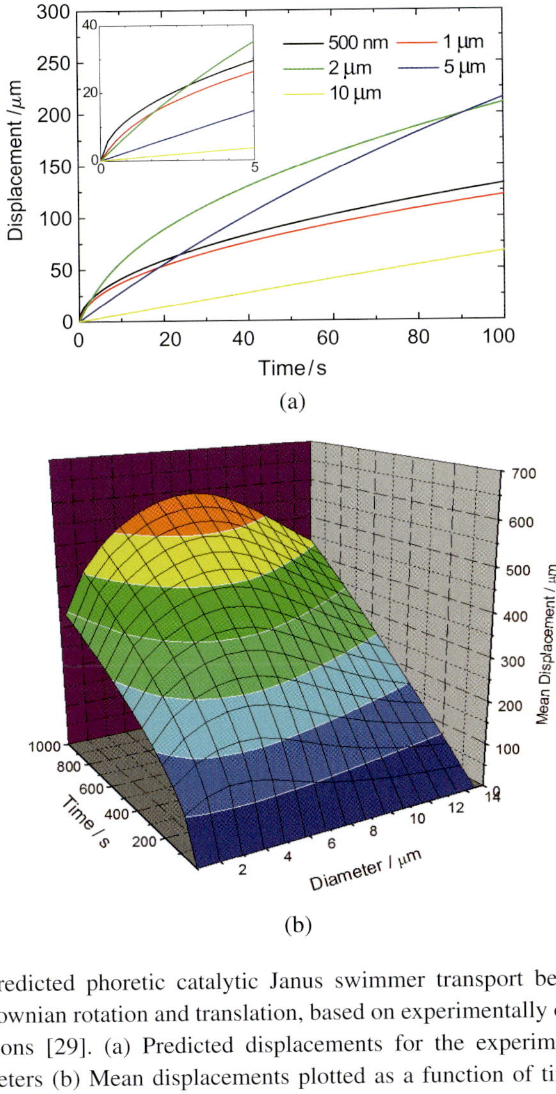

Figure 22. Predicted phoretic catalytic Janus swimmer transport behavior under the influence of Brownian rotation and translation, based on experimentally observed intrinsic velocity variations [29]. (a) Predicted displacements for the experimentally measured swimmer diameters (b) Mean displacements plotted as a function of time and swimmer diameter.

each other show how the racers' position and leader changes in a complex manner with duration. This unique behavior reflects the combination of the swimmers intrinsic velocity differences, combined with the degree of ballistic and diffusive behavior shown by their trajectories at a particular point in the time period. A snapshot from these predicted displacements is overlaid against the experimental trajectories shown in Fig. 20, and are in good agreement. This numerical approach can also be used to extrapolate the transport rate of symmetrical Janus swimmers in general as a function of size and duration (Fig. 22(b)).

The prominent ridge feature seen in these predicted mean displacements shows that there is a unique optimum device size for fastest autonomous transport over a given distance from the device starting position. It is also apparent that for longer distance transport tasks, the optimum swimming device size increases. This implies that the high intrinsic velocities possessed by the smaller devices (e.g., 37 μm s^{-1} for 500 nm diameter Janus swimmers) cannot be harnessed effectively for long-distance autonomous transport tasks. However, while intrinsic velocities are slower at larger sizes, they remain sufficient to dominate Brownian translational diffusion and so produce directed motion at least up to 10 μm diameters, while the correspondingly slow rotational rates allow autonomous motion without deviation of course (Fig. 20). This analysis suggests that the most rapid autonomous transport over the useful distance of 0.5 mm will be achieved using a 6.6 μm diameter device, and also that a slightly slower 9 μm diameter device can cover this distance within its expected ballistic path length, with very little loss of orientation. In summary, it is apparent that for transport using Janus catalytic swimmers that when moving in the absence of any guidance mechanism, size can be used to dramatically change the trajectory character.

8.5.3. Control of Rotational Motion

It has become clear that there are application areas and fundamental investigations that are enabled by the ability to produce swimming devices with well-defined spiraling and spinning trajectories, rather than intrinsically purely translational motion. This has motivated the development of methods to produce rotational propulsion. As an example of the demonstrated potential applications for rotating devices, micron-sized spinning colloids have been shown to influence the directional growth of neurons *in vitro* [90], induce cell death by bursting lysosomes through the generation of shear forces [91], and enhance surface protein binding rates [92]. However, these demonstrations lacked the feature of autonomy, requiring external actuation of the rotational behavior, achieved via applying circularly polarized light with angular momentum to an optically trapped bead, using dynamic magnetic fields to rotate superparamagnetic iron oxide nanoparticles, and the rotation of magnetic colloids using a quadrupole magnet, respectively. Rapid rotation has also been generated by the assembly of arrays of nanowire rotors onto patterned nanomagnetic bearings, actuated by quadrupole microelectrodes, giving controlled rotations at over 18,000 rpm [93]. In addition, ultrasonically actuated gold nanorods can rotate at kilohertz speeds around their long axis [94]. However, the lack of autonomy limits potential applications, due to the complexity of the field-generating apparatus. This motivates the quest to produce autonomous rotations based on catalytic fuel. Realizing these rotating devices would also allow interesting phenomena proposed for self-motile rotor/translators to be experimentally accessed. These include chiral diffusion effects [95], and predications for reversal of intrinsic rotation direction if placed in a confined geometry [96], together with proposals for novel chiral sorting methods (Fig. 23) [97]. An extra motivation for investigating autonomous rotation is provided by the potential for devices that can transition from linear running to rapid rotation to emulate the chemotaxis behavior displayed by bacteria in order to respond to local stimuli gradients [98].

Two approaches to achieve this goal of generating well-defined autonomous rotational velocity using catalytic Janus swimmers are described: one utilizing the ability to control the geometry of the catalytic patch, and the second the self-assembly of individually non-rotating translationally propulsive swimmers.

8.5.3.1. *Glancing angle metal deposition for control of Janus particle angular velocity*

Breaking the rotational symmetry for the hemispherical active layer in Janus catalytic spheres has been investigated as a potential route to impart well defined spin, or rotational velocity. The electrokinetic mechanism for Janus swimmers introduced in Sec. 4 proposes that catalyst thickness variations from the hemispheres' pole to the equator cause reactivity variations, which drive fluid pumping around the swimmer body [56]. Consequently, modifying this flow by removing sections of the reactive cap may provide a route to induce rotations. One method to break the cap

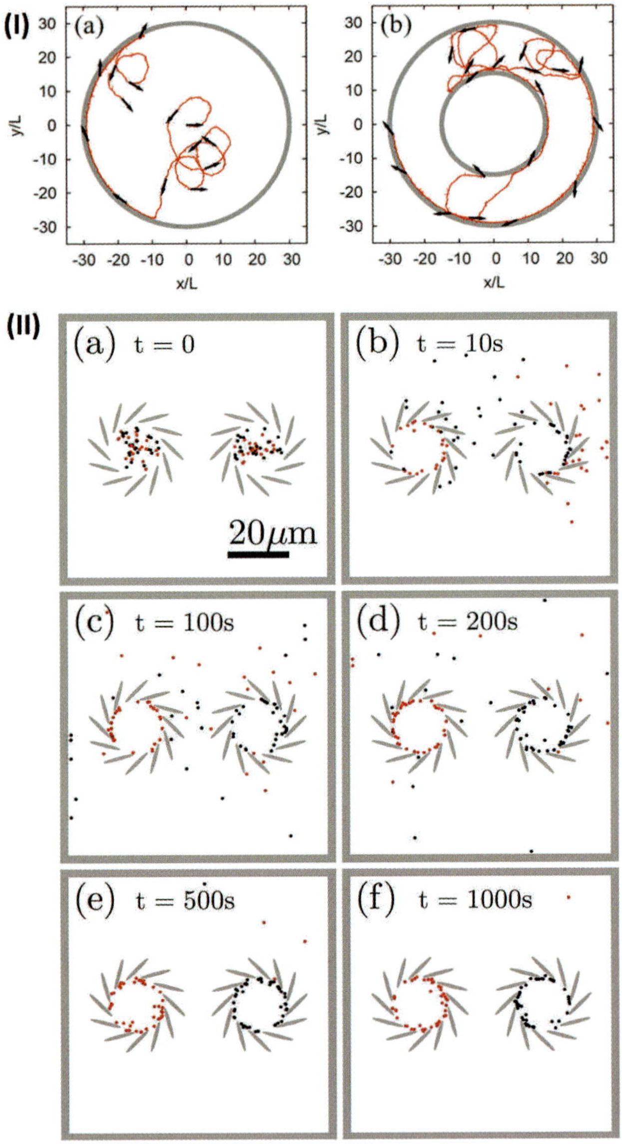

Figure 23. (I) Simulated trajectories for rod-shaped swimmers with angular velocity within (a) a petri dish like confinement and (b) in a ring-like confinement. (II) Simulation for the separation of levogyre and dextrogyre microswimmers. (a) At t = 0 s, a racemic mixture of microswimmers (R = 1,000 nm) is released inside two chiral flowers with opposite chirality. (b)–(f) As time progresses, the levogyre (red symbols) microswimmers are trapped in the left chiral flower and the dextrogyre (black symbols) ones in the right chiral flower.

symmetry is to use a glancing angle metal evaporation technique developed by Pawar and Kretzschmar to deposit the active platinum catalyst onto colloids dispersed onto a planar substrate [99]. In this scheme, colloids are assembled into close packed colloidal crystals and exposed to a directional metal vapor from a well-defined glancing angle. The oblique deposition angle allows colloids to exert shadowing effects on each other, breaking the symmetry through the deposited hemisphere to a variable extent controlled by the glancing angle, and thereby allowing a high level of control over cap shape and coverage.

Figures 24(a) and 24(b) illustrates why for any shadowing to occur, colloids must be close together. For a sparse coverage, regardless of the glancing angle a rotationally symmetrical hemisphere of catalyst is expected to be deposited. Figure 24(c) depicts the possible shadowing effects and principal parameters for glancing angle metal evaporation, which are the glancing angle, θ and also the in plane orientation of the colloidal crystal, α. The effect of reducing θ is clear, and the rotational symmetry of the active hemisphere is broken. Decreasing θ also reduces the total coverage of the deposited metal due to increased shadowing. Varying the crystal orientation angle α for a fixed θ value has a more subtle effect: changing the symmetry and position of the cusp-like features that define the cap perimeter according to the alignment of neighboring colloids in the crystal arrangement. Figures 25(a)–25(e) show top down back-scattered scanning electron microscope (SEM) images for small regions of colloidal crystal after metal deposition at a series of decreasing glancing angles to allow comparison with the geometrically predicted behavior. In these image regions covered with platinum metal generate bright contrast. As glancing angle decreases, the distribution of platinum metal across each colloid becomes visibly asymmetric, indicating shadowing has indeed occurred. This asymmetry is aligned across the entire colloidal crystal, and is orientated with respect to the direction of the metal source relative to the colloidal crystal orientation (marked by gray arrows). At lower glancing angles, the asymmetry is increasingly pronounced as more of the directional metal vapor is shadowed by neighboring colloids. While Figs. 25(a)–25(e) depict single crystal regions of colloidal crystal, in fact, the colloidal crystal used in this experiment was polycrystalline over larger regions. Consequently, when the catalytic

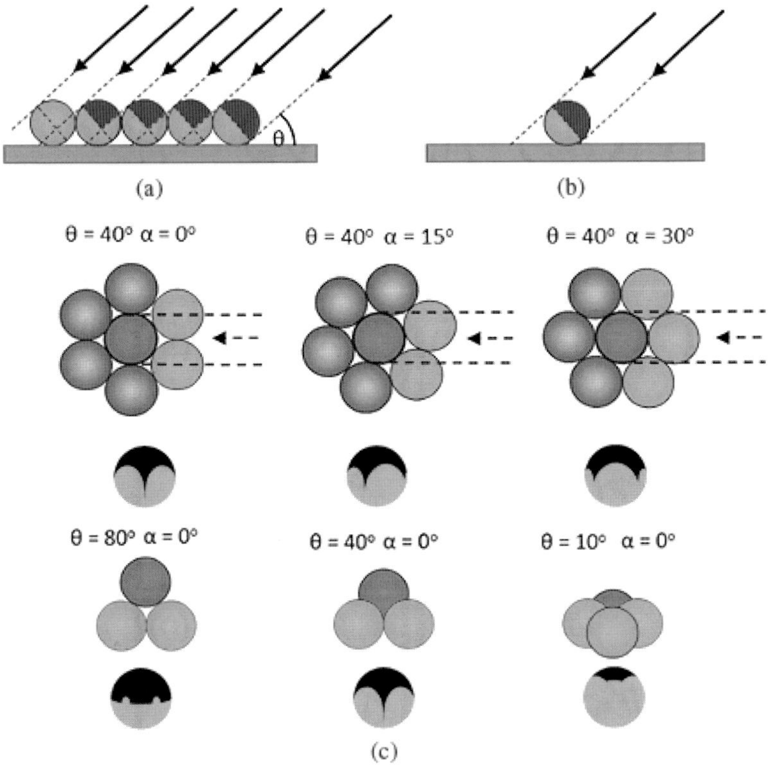

Figure 24. (a) Glancing angle evaporation onto a colloidal crystal modifies the cap shape as the glancing angle (θ) is varied due to the potential for neighboring colloids to shadow the line of sight path of the metal (black shading, arrows indicate direction of metal source). (b) If colloidal coverage is sufficiently sparse, changing the glancing angle, θ, will not affect the deposited cap shape, which will remain hemispherical. (c) Top section: At a fixed glancing angle, the orientation of the colloidal crystal (α) relative to the metal source will produce variations in the deposited cap shape. The resulting deposited metal pattern on the dark gray sphere is shown below, light gray spheres are responsible for the shadowing effects. Bottom section: The effect of varying glancing angle for a fixed crystal orientation is depicted. As the glancing angle is reduced, the patch coverage is reduced, and the perimeter shape is also altered.

swimmers were resuspended from the substrate into solution for further SEM observation or to allow observations of catalytic propulsion, this resulted in many different crystalline domains being sampled, leading to the crystal orientation angle α being uncontrolled and randomized.

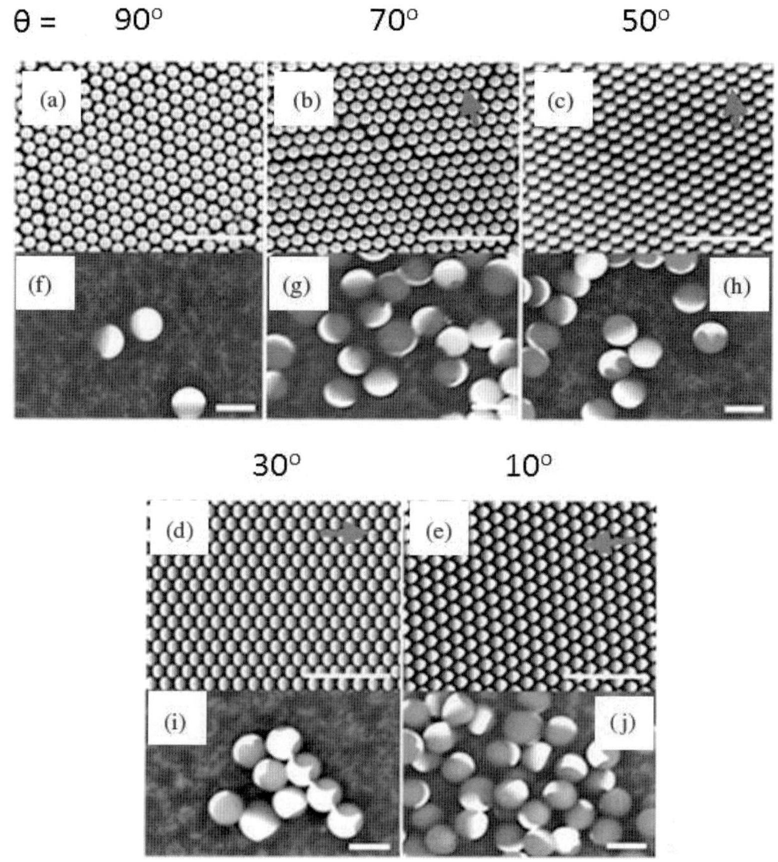

Figure 25. Backscattered SEM images following glancing angle platinum deposition onto a colloidal crystal of 1.9 μm diameter carboxylated PS particles. (a)–(e), Pt coated colloidal crystals after evaporation at angles of 90°, 70°, 50°, 30°, and 10°, (a)–(e) respectively, where the scale bar represents 10 μm. Gray arrows are used to mark the direction of the incoming platinum vapor where it was apparent. (f)–(j) colloids prepared at glancing angles 90°, 70°, 50°, 30°, and 10° after dispersal and redeposition of the colloidal crystals shown in (a)–(e), the scale bars represent 2 μm.

To further verify the perimeter shape of the cap, colloidal crystals prepared at different glancing angles were dispersed into water and then redeposited onto a planar substrate and examined again by SEM. This procedure served to randomize the viewing angle for each colloid and so allowed inspection of the entire shape of the catalytically coated region. Figure 25(f) shows these randomly orientated colloids from the normal glancing angle ($\theta = 90°$) sample. A smooth unbroken equatorial cap boundary is observed for this case, consistent with the expectation that normal evaporation will not produce any shadowing effects. Figures 25(g)–25(j) allow the perimeter of the catalytic cap to be assessed for samples prepared at decreasing glancing angles. At a glancing angle of 70°, small deviations in the equatorial cap perimeter line are seen for some particles, indicating that near neighbors have shadowed the metal deposition (Fig. 25(g)). Other particles lying at different randomized orientations display smooth equatorial perimeters. This is as expected, because, each individual JP will have one side facing toward the metal source which is shadowed by neighboring colloids, whereas the other side of the caps shape is masked by the particle itself. As the glancing angle is decreased, the deviations in perimeter on the side facing the evaporation source become more pronounced, however again some of the smoother, trailing edge perimeter can be seen at other orientations. These perimeter features show good agreement with the expected variations shown in Fig. 24(c).

Having verified that the glancing angle procedure is breaking the rotational symmetry for the platinum cap as expected, we now consider the resulting swimming behavior for a series of colloids prepared at different glancing angles. Figure 26 shows *x*–*y* trajectories for catalytic Janus colloids as a function of glancing angle, determined using 2D microscopy as described in Sec. 5.1.1.1. For normal ($\theta = 90°$) deposition, the trajectories resemble those shown above for similar sized Janus swimmer colloids (Fig. 20), and clearly exhibit enhanced motion significantly exceeding the displacements observed for unfueled colloids in water. The trajectories show linear runs, followed by direction changes due to Brownian phenomena, which we are now familiar with (Sec. 6.2). However, as the glancing angle decreases, a dramatic change in trajectory character is observed: increasingly tight spiraling is induced. This suggests that the asymmetry introduced by neighboring colloid shadowing effects is resulting in the introduction of increasing amounts of angular propulsive velocity. Using the MSD fitting approach described in Sec. 5.4 (Eq. (2)), many trajectories for each batch of colloids were analyzed to determine the mean translational and rotational angular velocity for each glancing angle (Fig. 27). Considering average rotational velocity (Fig. 27(a)), an increase

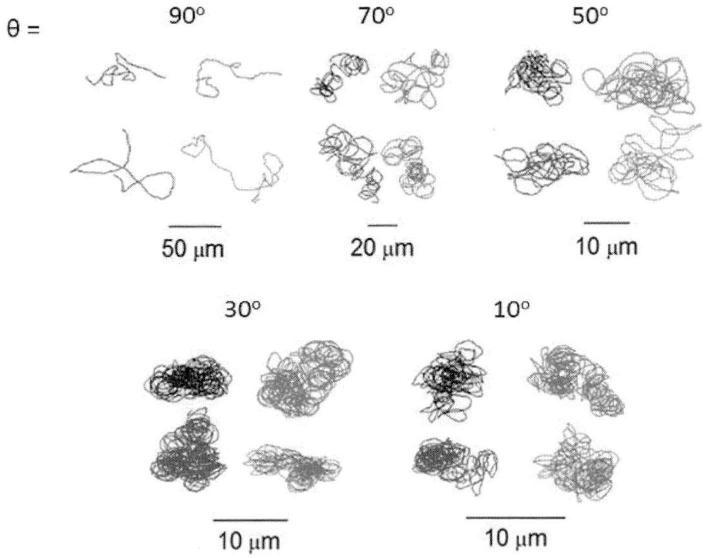

Figure 26. Representative x–y plot trajectories (30 s) for 1.9 μm diameter Janus colloids redispersed into 10% H_2O_2 solutions after platinum evaporation onto a colloidal crystal at a range of glancing angles.

in ω with decreasing glancing angle from 1.16 to 16.46 rad s^{-1} (corresponding to 0.25 to 2.62 Hz rotational frequency) peaking at $\theta = 20°$ is observed. In comparison, samples with a sparse colloidal distribution over the planar substrate (obtained by spin-coating a low volume fraction colloidal suspension) where no shadowing is expected, give a mean ω of 1.32 (±0.28) rads^{-1} (0.21 Hz) and the glancing angle makes no significant change to ω.

The average translational velocity for JPs resuspended from a colloidal crystal is 16.48 (±4.40) μms^{-1}, while sparse-coverage spin-coated samples give a mean value of 12.28 μms^{-1} (±0.69). For the spin-coated samples, translational velocity appears to be unaffected by the glancing angle, as expected due to their symmetrical hemispherical cap. However, significantly higher translational velocities are observed for some of the colloidal crystal samples, particularly for glancing angles in the range 20°–40° (Fig. 27(b)). Due to the variations in translational velocity as a function of glancing angle for colloidal crystal samples, it is instructive to

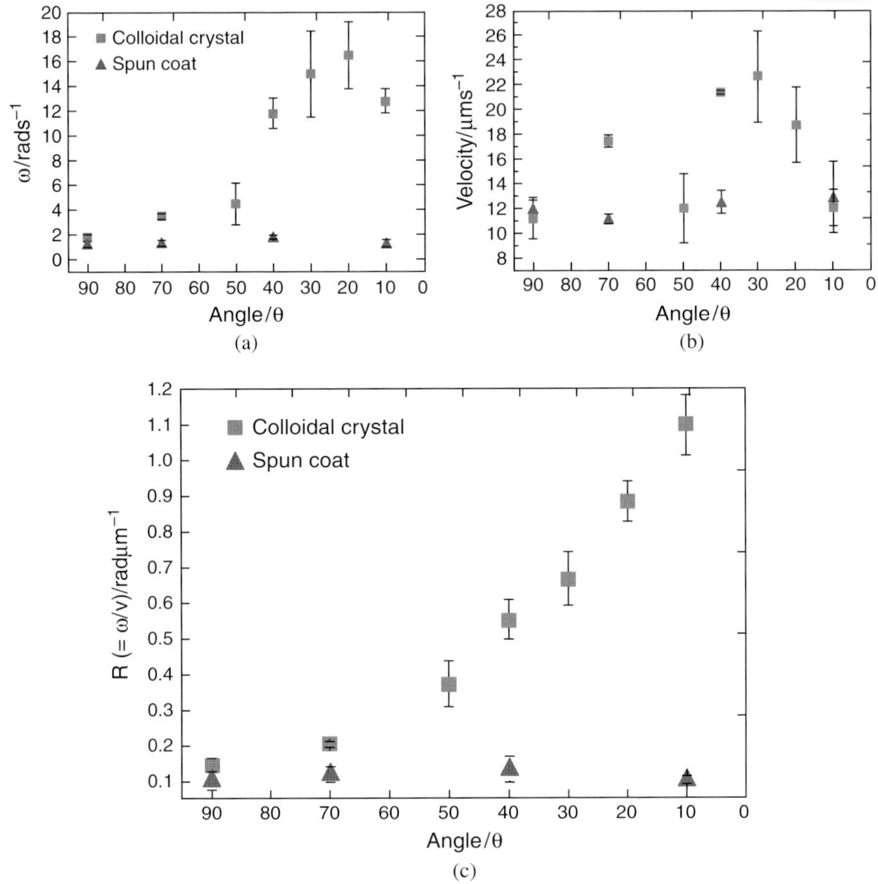

Figure 27. (a) Angular velocity (ω) profile of microswimmers prepared from colloidal crystals (square) and sparsely spun coat colloids (triangle) for a range of angles from 0 to 80° θ. (b) Velocity (v) profile of microswimmers prepared from colloidal crystals (square) and sparsely spun-coat colloids (triangle) for a range of angles from 90° to 10° θ. (c) Ratio (R) between ω and v for microswimmers prepared from colloidal crystals (square) and sparsely spun-coat colloids (triangle) for a range of angles (θ) from 90° to 10°.

also consider the derived parameter, $R = \frac{\omega}{v}$. R represents the number of radians turned by a given Janus colloid per unit translational distance traveled, and so gives a measure of how tightly spinning the trajectory of an average colloid prepared under a given condition is, irrespective of the absolute magnitudes of ω and v. This is useful to ascertain the relative production of translational and rotational velocity for a given batch of colloids.

Figure 27(c) shows that R increases monotonically with a reduction in the glancing angle θ, whereas for spun-coat samples there is no link between R and θ. This demonstrates the potential to use the glancing angle deposition method to reproducibly prepare batches of Janus colloids with a well-defined angular trajectory character.

Considering in more detail the potential origin for the modifications to the platinum cap shape caused by changing the glancing angle, and the degree of observed spin, we first recall the two proposed motions producing phoretic mechanisms for spherical catalytic Janus swimmers (Sec. 4). The original mechanism, self-diffusiophoresis, was developed from a theoretical analysis of the propulsion produced from rotationally symmetrical distributions of catalyst with uniform activity [100]. As such, there is no existing framework to predict the self-diffusiophoretic propulsion character for the rotationally asymmetrically catalyst shapes produced by glancing angle deposition. Further to this, the experiments reporting the salt dependency for Janus catalytic swimmers suggest that, in fact, an electrokinetic mechanism may dominate the observed propulsion [56]. This mechanism suggests that a flow of protons may be generated across the catalytically active cap driven by variations in catalytic activity. As shown in Fig. 8, the reactivity at the surface of platinum patches is proportional to deposited thickness (over the range 0–15 nm). For the case of a rotationally symmetrical platinum hemisphere, the thickness varies symmetrically from the pole (thickest) to the equator (thinnest), and so we can expect a higher reaction rate at the pole, and a low rate at the equator. Based on the electrokinetic mechanism, this will serve to produce a correspondingly symmetrical propulsion thrust, resulting in pure translational motion without any angular velocity for a perfect hemisphere. Figures 28(a) and 28(b) displays the thickness variations for this case calculated using a line-of-sight geometric model assuming parallel incident "rays" of platinum are produced by the metal evaporation source. Using this basis, it is now possible to consider the effect of the variations in cap shape induced by changing glancing angle. Figure 28(e) shows the geometrically determined thickness variations as a function of glancing angle for a fixed colloidal orientation ($\alpha = 0°$). It is clear that shadowing from neighboring colloids within the crystal increasingly breaks the rotational symmetry of the deposited cap. Consequently, based on the argument above, the predicted electrokinetic propulsion force, indicated by

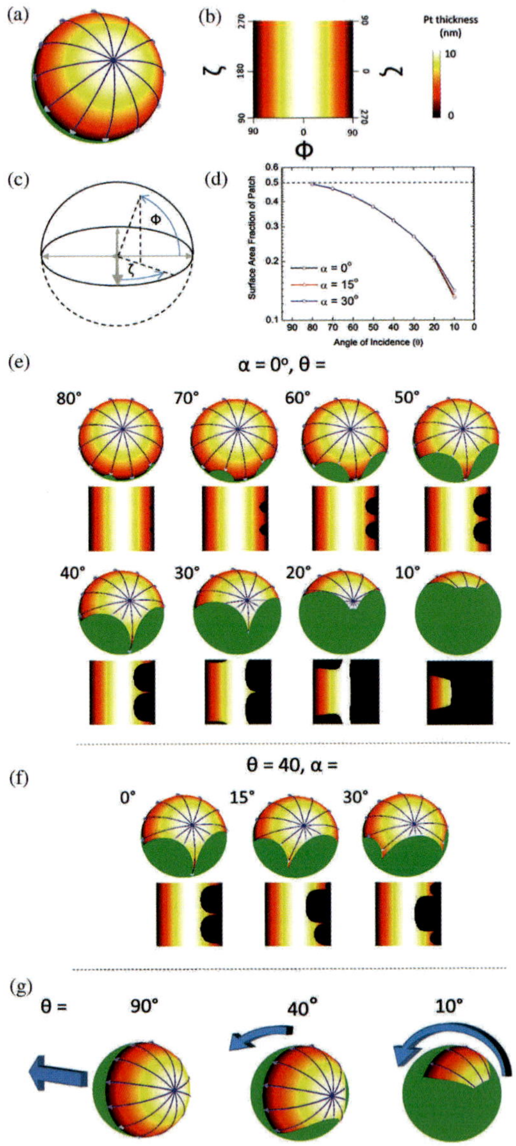

Figure 28. (a) 3D model of a Janus swimmer without shadowing. (b) Cap shape and thickness diagram represented in unwrapped 2D format using altitude and azimuthal coordinates and color representation plot of platinum cap thickness from 0 to 10 nm. (c) Schematic representation of the altitude ϕ and azimuthal ξ angles. (d) Graphical representation of overall patch area coverage with θ for alpha (0°, 15°, and 30°).

blue arrows, is increasingly unbalanced across the cap, as the symmetry of the thickness gradient around $\phi = 0°$ is broken. This explanation is consistent with the observations of increasing amounts of angular velocity being produced in the $\theta = 90°$ to $10°$ range. Also of note is the predicted effect of glancing angle on total coverage of platinum (Fig. 28(d)). It can be seen that the surface area fraction of catalyst reduces considerably over the glancing angle range considered; however, the experiments above did not report a significant systematic variation in translational velocity. The remaining parameter that affects the shape of the Janus cap and variations in thickness is the orientation of colloidal crystal. While this was not controlled in the experiments discussed here, it is experimentally feasible to produce colloidal crystals with well-defined orientations, for example by using electrostatic methods. Figure 28(f) illustrates how variations in the crystal orientation would subtly alter the symmetry of the cap shape for a given glancing angle. It appears that this could result in a modification to the axis of rotation, by breaking the symmetry about $\xi = 0°$. The crystal orientation can be seen to have very little effect on the total catalyst coverage. Finally, Fig. 28(g) pictorially summarizes the way in which breaking the rotational symmetry for the Janus cap may be producing the observed varying amounts of angular velocity, based on the mechanistic and geometric arguments outlined above.

8.5.3.2. *Self-assembly*

A second approach to generate well-defined trajectories with both translational and rotational velocity was discovered serendipitously during observations of agglomerate formation from individual catalytic Janus swimming devices. If solutions containing separate swimming devices are stored under the correct conditions, collisions between the isolated units will occur, and, in some cases, result in the production of stable agglomerates containing two or more devices. Each agglomerate shows a fixed

Figure 28 (*Continued*) (e) 3D and 2D representation of the cap (same format as a and (b) for $\theta = 80°$ to $10°$ at $\alpha = 0°$. (f) 3D and 2D representation of the cap (same format as a and b) for $\alpha = 0°$, $15°$ and $30°$ at $\theta = 40°$. (g) Schematic representation of changing cap shape.

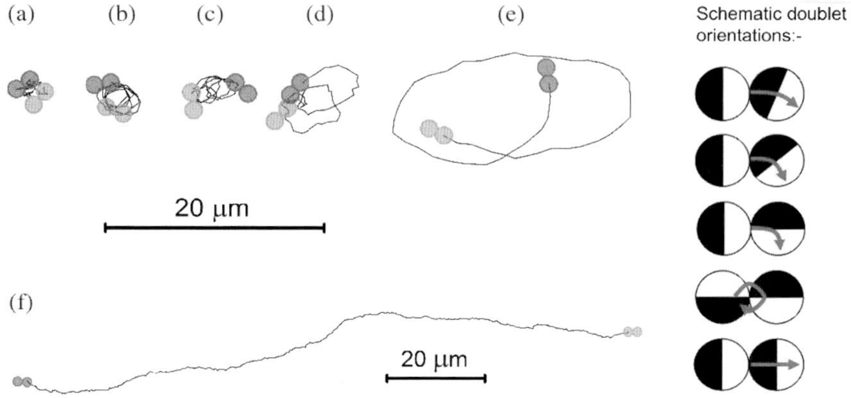

Figure 29. Trajectories of 25 s for a range of Janus doublets in solution: (a) Janus doublet in pure water (b)–(f) Janus doublets in 10% H_2O_2.

trajectory type; however, examination of a large number of agglomerates reveals a rich variety of behaviors. Figure 29 compares trajectories for a range of doublet swimmers in the presence of hydrogen peroxide fuel, and a control doublet in water. The addition of fuel to the doublets causes them to cover longer paths in a given time period, indicating propulsion. The trajectories observed vary from a linear translator, to a range of spirals and tighter rotations. Again, MSD analysis can be performed to determine the range of translational and angular velocities (v = 2–6 μms^{-1}; ω = 0–1.4 rad s^{-1}), showing that the glancing angle method discussed above is able to access considerably higher maximum angular velocities (16 rad s^{-1}).

The explanation for the behavior of each Janus doublet lies in considering the various orientations in which the individually propulsive units could couple, RHS (Fig. 29). It becomes clear that perfect alignment of the catalytic cap would produce a non-rotating swimmer of the type shown in Fig. 29(f); however, it is much more likely that caps are not perfectly aligned, which will produce a combination of both translational and rotational thrust, as shown in Figs. 29(b) and 29(c). Indeed, the frequency distribution for the different trajectory types qualitatively supports this analysis, whereby most doublets were observed to rotate, and linear translators were rare (out of all doublets observed (>100) only one linear translator was seen) corresponding to low probability of this alignment occurring. In addition to doublets, higher order aggregates of up to six JPs, in a variety of morphologies also formed, and were found to show a rich variety of rotational behavior.

Other reports have used alternative schemes to introduce angular velocity into catalytic swimming devices. One approach is to utilize shape asymmetry, for example it has been shown that L-shaped swimming devices can generate an asymmetric velocity dependent torque, resulting in circular trajectories [101]. This study required light actuation to induce device motion via localized heating at a metallic strip coated on the polymeric device body, but the principle could also be applied to autonomous devices. Also, in some early reports, self-propulsive nanorods were observed to undergo tight rotations; however, this was ultimately found to be due to them becoming randomly stuck to substrate defects and was not well controlled [35]. However, nanorods exhibiting fast autonomous rotation have been subsequently manufactured using a complex trimetal arrangement that allowed hydrodynamic interactions between pairs of rotors to be observed [102].

8.5.4. Magnetic Steering

As we have seen above, Brownian rotational diffusion is the origin of the randomization of trajectories for catalytic Janus swimmers moving in free solutions. Therefore, a way to impose directionality on catalytic Janus colloids with a well-defined link between their orientation and propulsion direction (such as phoretic catalytic JPs and nanorods) is to generate a constraining torque with sufficient energy to prevent Brownian rotation. In this scenario, the propulsion vector will be fixed and aligned to the direction of the constraint and allow the swimmer to be steered. A method to achieve this involves incorporating a magnetic component into the device, and then using a magnetic field, to align the device and impose an energetic penalty for rotation away from the aligned configuration, which is significantly larger than the available thermal energy [16,103]. This approach has been observed in nature, where directional migration determined by the earths local geomagnetic field is exhibited by some bacteria, and is made possible due to them containing iron-rich structures. An early example of the application of this method to control catalytic devices was shown for nanorods, with additional magnetized sections inserted [64]. An

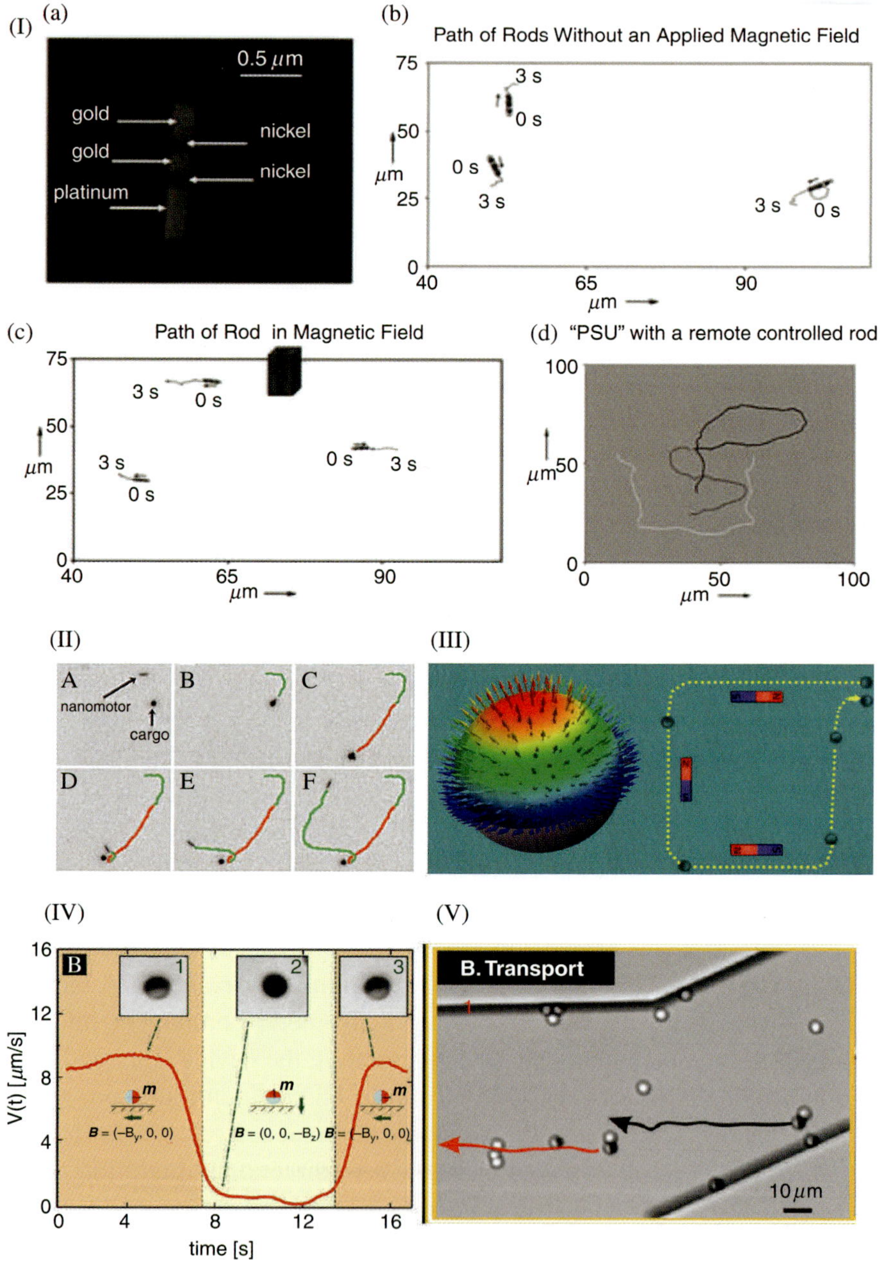

Figure 30. (I) (a) SEM image of a magnetically steerable rod. (b) and (c) The effect of applying an external magnetic field to the trajectory of the rods. (d) Demonstration of the manipulation of magnetic nanorods to follow desired trajectories by changing the

effective approach was to use a "stripped" pattern of single domain nickel segments, to allow magnetization perpendicular to the long axis of the rod (Fig. 30(I)). By applying a sufficiently strong magnetic field to a suspension of these devices moving in hydrogen peroxide fuel, it was shown that their orientation, and trajectories could be arbitrarily aligned, and that by dynamically changing the orientation of the field, the device could be made to travel along well defined path (Fig. 30(I)). This ability to accurately steer nanorods was subsequently used to allow the capture of a colloidal cargo, move it along a well-defined path, and finally release the cargo by causing the transporter rod to execute a rapid direction change (Fig. 30(II)) [103]. Spherical Janus catalytic swimmers have also been directed using a similar strategy to produce directionality [16]. In this case, a multilayer stack of cobalt/platinum was used to achieve the required magnetic properties. This structure allowed a coercive magnetic field to impart a well-defined magnetization vector to each Janus colloid localized on a planar support (Fig. 30(III)). Janus colloids prepared in this way then demonstrated both the ability to be guided along 2D paths by external magnetic fields, as for nanorods, but additionally, when moving close to a substrate could be stopped and started by temporarily orientating the platinum cap so that it was facing the substrate (Fig. 30(IV)). Also, demonstrations for cargo transport were shown using these magnetically controlled spherical swimmers (Fig. 30(V)).

8.5.5. Gravitaxis

While the use of magnetism provides an extremely useful way to manipulate devices, this is not an autonomous approach, as it relies on external

Figure 30. (*Continued*) orientation of the external magnetic field. (II) Sequential optical microscopy images of an en-route cargo pickup, transport, and release. Green and red traces show the path traveled by the rod and cargo loaded rod respectively (A, B), manipulation and pickup (C), transport (D), release (E + F) continued rod motion after release (III) (LHS). Schematic for the magnetization vectors for the spherical Janus swimmer discussed in the text (RHS) demonstration of control of orientation and trajectory control for this spherical Janus swimmer (IV). Controlled motion of the JPs can be achieved by orienting external magnetic field either in-plane (deterministic propulsion with a certain velocity, regions 1 and 3) or out-of-plane (particle stops, region 2) (V) catalytic motors transport colloidal cargo through the microfluidic channel.

field manipulation. However, it has recently been shown that interactions of Janus catalytic swimming devices with the Earth's gravitational field can also provide an aligning torque. The idea that mass asymmetry due to the asymmetrical metallic coating on Janus swimming devices can orientate "swimming" behavior stems from consideration of biological entities such as algae. Simple swimming cells have been shown to navigate upward by virtue of an asymmetrical mass distribution alone, a phenomena termed "gravitaxis" [104]. The mechanism for this effect was subject to debate, but it is now established that these cells are simply bottom heavy, a feature which allows them to orientate their flagella-driven motion to consistently move upward. A key feature of this behavior is that it demonstrates that it is possible for such cells to show gravity-guided behavior without having a sophisticated gravity-sensing mechanism to control their flagella. As autonomous sensing and response mechanisms are very challenging to implement in synthetic systems, this suggests the potential for a biomimetic alternative approach.

Figure 31 shows representative sideways on 2D and 3D trajectories for different sizes of Janus catalytic spheres. The 3D trajectories were obtained as described in Sec. 5.1.1.2. Some of the differences between the depicted trajectories as a function of size, such as the reduction in the rate of Brownian rotation, rate leading to straighter trajectories are consistent with the behavior discussed in Sec. 6.2. However, considering the distribution of the trajectories with respect to the direction of gravity, qualitatively, it is apparent that the isotropy of motion in the z-axis, in which gravity acts, is lost at larger particle sizes and replaced by a pronounced bias toward trajectories where the swimmers move upward. Examination of the larger-sized sphere trajectories in 3D allows the isotropy along all axes to be assessed and makes it clear that isotropic behavior remains in the xy plane but is lost in the xz plane. This data provides strong evidence that a synthetic gravitaxis mechanism is being turned on as particle size increases. A qualitative explanation for this size dependency is that for larger particles, the total platinum caps' mass is significantly larger due to the larger hemispherical surface area, and this mass is positioned further from the center of the Janus swimmer, resulting in a larger torque to constrain the Brownian rotations of the particles, which are limited by the available thermal energy.

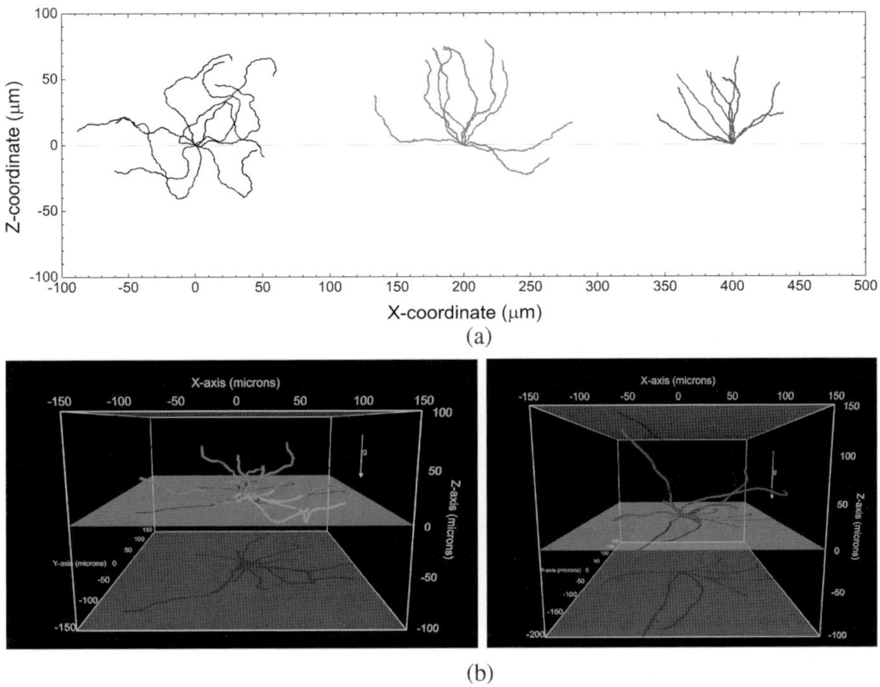

Figure 31. (a) 2D Swimming trajectories for representative Janus spheres observed in the (x,z) plane for $r = 0.95\ \mu m$ (black), $r = 1.55\ \mu m$ (light gray), and $r = 2.4\ \mu m$ (dark gray) spheres. The swimmers were suspended in a 10 wt.% solution of H_2O_2. Predicted mean path lengths are 19, 59, and 151 μm, respectively. Gravity is acting downward along the z-axis. (b) 3D swimming trajectories for representative Janus spheres; $r = 1.55\ \mu m$ (left) and $r = 2.4\ \mu m$ (right).

Starting from this qualitative interpretation, it is possible to use Boltzmann statistics to create a simple numerical model for the expected azimuthal angle distribution, $P(\theta_A)$, for an asymmetrically platinum-coated catalytic Janus sphere. In order to do this, an expression for the energetic cost of rotating a given Janus sphere from its lowest energy position ($\theta_A = 0°$, i.e., the cap is pointing downwards) is required. Equation (6) gives such an expression, for the additional gravitational energy required to rotate a Janus sphere to a given angle

$$E_{\text{grav}}(\theta_A) = mgl[1 - \cos(\theta_A)], \tag{6}$$

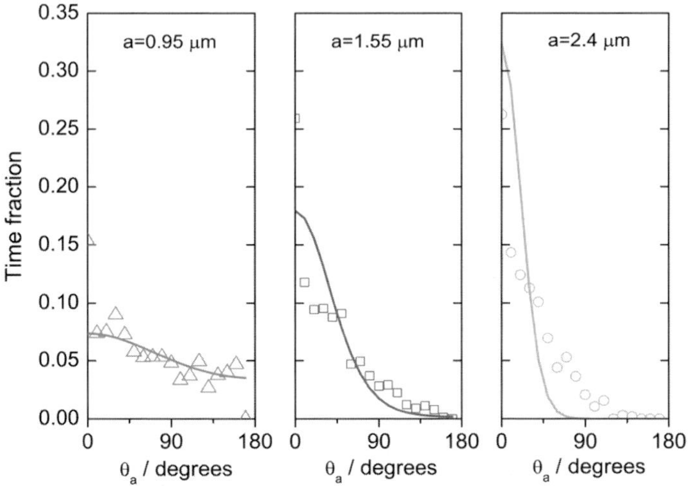

Figure 32. Scatter graphs showing the experimental azimuthal angle, $\theta\nu$ distribution. Solid line is $P(\theta\nu)$ estimated from a Boltzmann distribution for the gravitational energy cost for rotating a Janus cap of platinum.

where m is the experimentally determined mass of the cap, and l is the distance of the center of mass of the ellipsoidal shell from the center of rotation of the Janus spheres. Applying Boltzmann statistics, the expected residency time at a given angle is then given by Eq. (7),

$$P(\theta_A) = mg[1 - \cos\theta_A]. \tag{7}$$

Figure 32 shows a comparison of these quantitative predictions for azimuthal angle to distributions of the catalytic JPs azimuthal angle determined from trajectory analysis [61]. It is clear that a good agreement is obtained, demonstrating that the mass anisotropy for catalytic Janus swimmers is the origin of the observed gravitational alignment.

This model indicates that an increased degree of autonomous gravitationally aligned directional control can be achieved by increasing the mass of the platinum catalyst, which would be most easily accomplished via an increase in coating thickness. However, there will be a limits in this scenario as increasing the coating mass will also increase the sedimentation rate, and the swimming particles must be capable of exceeding this in order to move upward. Despite this, as an indication of the scope for gravitactic control, at $r = 1.55$ μm Janus swimmer with a 50 nm thick Pt coating is predicted to spend 85% of its time moving upward with a gravitational alignment that remains within 20° of vertical and be capable of producing a 1 μm s^{-1} translational rate.

Some new applications may be enabled by optimizing and exploiting this easily experimentally accessible gravitaxis behavior. For example, small-scale "lifts" transporting cargo could be constructed and made to switch between lifting and sedimenting by a changing a stimuli that determines propulsion speed, for example, dissolved fuel concentration. A similar scenario could be used to effect controlled release of active colloids from gravitational confinement within a flow, where a solution stimulus causes colloid's to "activate" and escape a gravitational well. Additionally, separation methods may be envisaged, for example catalytic and non-catalytic colloids will separate vertically due to gravitactic propulsion, while differently sized swimmers displaying differing amounts of gravitaxis may also be separated. As a further possibility, it is known that algae produce flow patterns due to their gravitactic accumulation at the top surface of a fluid, which causes sinking and rising columns due to local densification, and a similar strategy may allow synthetic devices to self-organize at sufficiently high volume fractions. Guidance by gravitaxis could also be exploited to rapidly concentrate colloids from a 3D volume to a 2D interface, and open the potential to execute sequential processes in different regions of an autonomous microfluidics device comprising suitable vertically arranged chambers, or inclined surfaces. Furthermore, the degree of gravitaxis can be potentially be altered by cargo attachment, due to size, density, and shape changes, allowing dynamic responses in the navigation behavior.

As an alternative approach to gravitaxis, in a second study, shape anisotropy was also found to result in upward-biased motion [105]. In this example, the swimming mechanism relied on laser illumination and so was not autonomous, however, deploying this phenomena to direct chemically powered devices should be straightforward. In summary, the overall utility of gravitaxis is apparent, as it provides a mechanism to break the otherwise isotropic motion for catalytic Janus colloids and impart a well-defined and controllable propulsion direction without the need to apply controlling fields.

8.5.6. Boundary Steering

An alternative emerging route to overcome the randomization of catalytic Janus swimmer trajectories is to exploit interactions with physical structures. Again, drawing an analogy with nature, self-motile cells have been observed to display a range of interactions with surfaces, resulting in them accumulating at planar interfaces [106], and when exposed to more sophisticated patterns, showing rectification and sorting behaviors according to motion characteristics [107]. In a similar fashion, the interactions between active colloids and patterned surfaces have recently been considered. For example, in two independent studies both bimetallic Janus nanorods [108] and Janus spheres [109] have displayed circling behavior around the base of the colloidal obstacles. Following these studies, the phenomeology that leads to boundary steering has been studied in more detail for the specific case of a catalytic Janus swimmer. The starting point for this investigation was considering in more detail the observation briefly alluded to in Sec. 5.2, that Brownian rotation for catalytic Janus swimmers are quenched by proximity to planar interfaces about one specific axis perpendicular to the plane.

Using the method described in Sec. 5.2, observing the fluorescence emission from a fluorophore-labeled polymer core for Janus catalytic swimmers across a range of sizes allows their asymmetry to be directly visualized as various "Moon phases", and in turn reveals the orientation. Figure 33(c) displays still frames from fluorescence videos recorded for Janus catalytic swimmers both in water, and while undergoing enhanced translational motion in hydrogen peroxide. For the smaller particles ($a = 1\ \mu$m, $a = 1.55\ \mu$m) in water, Brownian rotations about both rotational axis (azimuthal and in-plane) are occurring. However, for the largest particle size observed in water, an unchanging "full" phase of fluorescence emission is observed. As the microscopic observation and illumination are occurring from above, bright phases correspond to a "cap down" orientation where the fluorescence emission is not impeded at all by the platinum cap (azimuthal angle, θ close to 0°). This can be understood following the discussion in Sec. 6.5 as evidence that the gravitational resistance to rotation from the lowest energy $\theta = 0°$ orientation due to the asymmetric mass distribution is preventing thermal reorientation, and as expected, this effect becomes more noticeable for the largest sized sphere. A further key feature for the data recorded in the absence of fuel is that no correlation between particle orientation in 3D and subsequent Brownian displacement direction could be found, reflecting physical expectation of the independence of Brownian translational and rotational diffusion for a non-propulsive

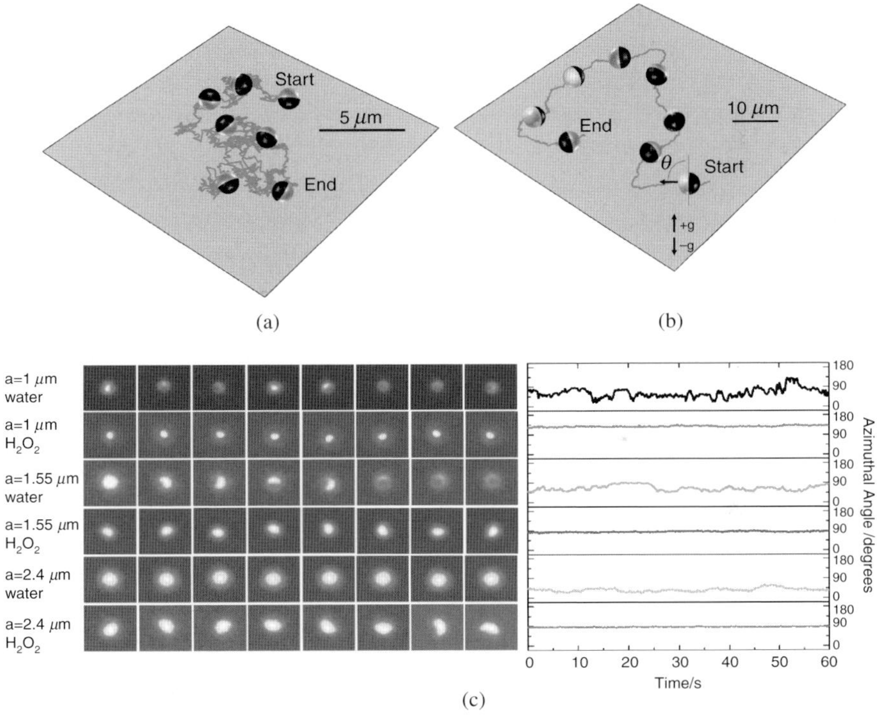

Figure 33. (a) and (b) Schematic 3D orientation and experimental trajectories (45 s duration) for an a = 1 μm platinum-polymer JP in (a) water settled under gravity against a planar glass substrate; colloid and (b) 10% hydrogen peroxide solution at the planar surface of a rectangular glass cuvette. The azimuthal angle, θ, is defined. (c) LHS: Selection of frames from fluorescent microscopy videos (15 μm × 15 μm field of view) for fluorescent platinum-PS Janus spheres of varying radii, near a planar interface in water, and in 10% aqueous hydrogen peroxide solutions. RHS: Azimuthal angle, θ, as a function of time for typical JPs determined from fluorescent microscopy videos.

particle, (orientations shown schematically in Fig. 33(a)). However, when still frames in water are compared to those recorded in the presence of 10% hydrogen peroxide fuel, it is apparent that at all particle sizes, each catalytic Janus swimmer presents a constant hemi-spherical "half moon" shape during propulsion. This suggests that instead of undergoing Brownian rotation, or gravitational confinement, the azimuthal angle, θ, now remains close to 90°, i.e., the particles' propulsion vector is orientated so that it is parallel to the substrate. It is, however, apparent from the selected frames that Brownian rotation about the out-of-plane axis does still occur (orientation shown schematically in Fig. 33(b), and this phenomena is also visible from the data shown in Fig. 11).

This observation motivated an examination of catalytic Janus swimmers moving in geometries where the propelling particles can interact with more than one solid interface at the same time to see if an additional axis of rotation can be quenched, thus providing a method to control both the orientation and resulting propulsive trajectory.

Figures 34(a) and 34(b) depict two geometries where multiplanar interactions are possible: a planar surface with a vertical side wall and a rectangular groove. Figure 34(c) shows a $r = 1.55$ μm catalytic Janus swimmer moving across a planar interface at the bottom of a glass cuvette. While only in proximity to one planar surface, the particles trajectory is subject to unrestricted Brownian rotation about the axis perpendicular to the planar surface. Figure 34(d) (inset) shows the same Janus swimmer at a later time, immediately before it encounters the side-wall of the cuvette. At the point at which the JP reaches the curved perimeter of the cuvette, it is clear that rotational diffusion about the axis perpendicular to the bottom of the cuvette also ceases, and that the orientation of the JP is now confined to be aligned to the second vertical interface, resulting in a trajectory that is steered by the vertical cuvette side wall. Figure 34(d) main image shows how this phenomena can result in Janus swimmers following curved boundaries for considerable distances (centimeter-scale direction motion was observed). Figure 34(e) illustrates this behavior for a linear section of square wall boundary, and a detail where the fixed alignment relative to the steering boundary for the Janus swimmer is clearly visible. In this inset, the fluorescence images have the resolution to reveal that the Janus swimmer is slightly canted away from the side of the cuvette as it moves.

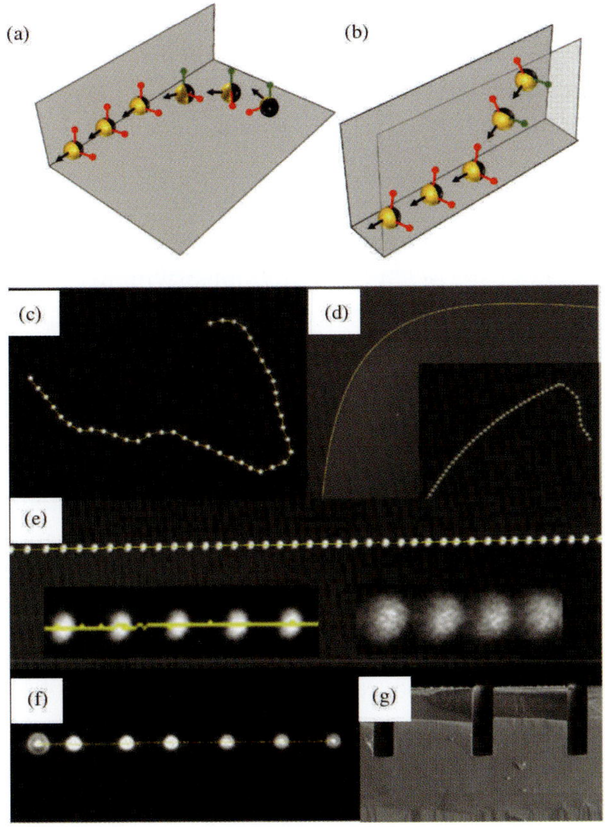

Figure 34. (a) and (b) Schematics of JPs encountering multiplanar geometries. Red axis indicates forbidden rotations due to proximity to a plane; green axis indicates unquenched axis of rotation: (a) JP encountering a planar edge while moving along a 2D surface, expected to result in Brownian rotational quenching about two orthogonal axes. (b) JP confined within a square groove; parallel vertical walls confine the rotational diffusion about one axis; however, if the particle descends to the base of the groove, it is confined about two orthogonal axes. (c)–(f) Overlaid still frames from fluorescence microscopy videos with equal time gaps: yellow line shows complete trajectory, green line shows location of vertical cuvette walls, and red arrows indication direction of motion: (c) $r = 1.55$ μm JP (10% H_2O_2) moving at the bottom of a rectangular glass cuvette a long way away from the edges. (d) $r = 1.55$ μm Janus colloid (10% H_2O_2) moving along the curved edge of a glass cuvette — inset shows a colloid reaching the cuvette boundary. (e) $r = 2.4$ μm Janus colloid (10% H_2O_2) moving along the straight edge of a glass cuvette — left-hand inset shows a magnified region, right-hand inset shows a "stuck" aligned agglomerate formed at the cuvette boundary. (f) $r = 2.4$ μm Janus colloid (10% H_2O_2) moving within a rectangular section groove (width = 8.75 μm). (g) SEM image of a section of the rectangular grooves (widths 7–9 μm) used in f.

Figure 34(d) (right-hand side inset) shows what happens when a Janus swimmer is occasionally stopped by small blemishes at the edge of the cuvette, and then additional colloids encounter the "stuck" object, resulting in "pile-ups" of aligned colloids. This provides a potential route to produce orientated multicomponent swimming devices of the type described in Sec. 6.3.2.

The second geometry investigated was a deep rectangular channel. Figure 34(f) shows $r = 2.4$ μm catalytic Janus swimmer moving confined within a 9 μm channel. In this case, achieving complete rotational diffusion control requires the Janus swimmer to interact with the bottom surface of the well, in addition to the two parallel walls (Fig. 34(b)). In this experiment, grooves with a variety of widths were investigated, and $r = 2.4$ μm colloids were observed to be rotationally quenched within channels with widths ranging from 7 to 9 μm. The rectangular groove structure was made using conventional silicon soft lithography (see cross-section in Fig. 34(g)), and so there is clear potential to direct Janus swimmers in this way along increasingly complex channel designs, realizing a new form of substrate-based autonomous microfluidics. A similar range of phenomena to those discussed here have also been observed by the Sanchez group, however, in this case the topographical features used for guidance were steps and edges with heights smaller than the radius of the Janus colloid [110].

8.5.7. Chemotaxis

An additional strategy for autonomous directional control of catalytic Janus colloids is inspired by the ability of swimming organisms such as *E. coli* to display chemotaxis i.e., show spatially biased motion in response to differences in the local solution composition, such as a gradient in concentration of a dissolved component [111]. For *E. coli*, chemotaxis is used to find nutrients or avoid damaging chemicals. To achieve this function, *E. coli* are able to temporally sense the levels of the stimulus chemical and then adjust their "flagella's" motion to execute running or tumbling trajectories as required. While this complex sensing and response mechanism is not currently viable for synthetic devices, there are several examples of Janus catalytic swimming devices displaying broadly analogous behavior, resulting in accumulations in response to a gradient in the fuel molecules that drive their motion. The benefit for applications afforded by this chemical sensing approach is the potential for Janus catalyst colloids to autonomously follow signals to, for example, allow delivery of cargo in a particular region or accumulate to perform a specific task. The first example of chemotaxis for active colloids was reported for Janus bimetallic nanorods by the Sen group in 2007 (Fig. 35(I)) [26]. However, more recently chemotaxis has also been investigated in more detail in well-defined fuel gradients established using a microfluidic channel. This approach was used to show that both Janus catalytic colloids and tubular swimmers moved toward higher fuel concentration regions (Fig. 35(II)) [112]. In addition, the so-called match-stick shaped swimmers with a spherical head made from manganese oxide attached to a silica rod, providing the Janus structure, were also observed to move up fuel concentration gradients (Fig. 35(III)) [113]. Surprisingly, all these experiments, displaying up fuel gradient motion are actually in disagreement with predictions for the direction of chemotaxis scenario made by simulations which suggest migration toward low-fuel regions [98], and the exact mechanism for these phenomena is unclear. A recent review article by Golestanian explores the different scenarios for chemotaxis like phenomena in more detail, and interested readers are directed toward this [114].

To expand on this fuel concentration gradient driven motion control, it will be desirable to enable chemotaxis stimulated by other signals. Solution properties including conductivity [50] and certain small molecules [65] are known to modulate propulsion velocity, and so these are also candidates for a chemotaxis response. Indeed, one very recent report has demonstrated that metal organic framework swimmers powered by peptide assembly can be autonomously directed toward a target based on the interaction of the propulsion mechanisms with local pH gradients [115]. An alternative proposed approach to enable a potentially more rapid response to gradients in a constant fuel environment is to use size changes in a responsive material incorporated within the active colloid [98]. As we have seen above, in Sec. 6.2, size changes can alter both the velocity and trajectory for Janus catalytic swimmers. Using this as a starting point, a simulation for the behavior of a catalytic Janus swimmer made from a responsive hydrogel material in a gradient that would cause it to change size was constructed. To make this simulation realistic, the size changes for a real pH-responsive hydrogel particle (polymethacrylic acid, PMAA)

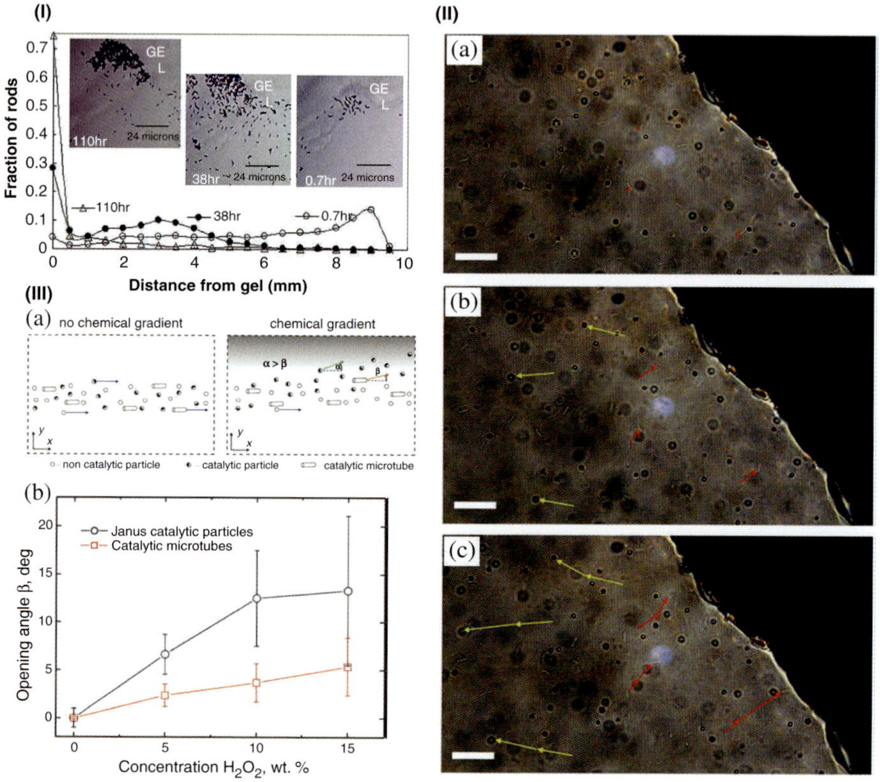

Figure 35. (I) Chemotaxis of bimetallic platinum–gold nanorods reported in Ref. [50]. The gel is at the top of the micrograph inserts, and is soaked in hydrogen peroxide. The main figure shows the distribution of nanorods with respect to distance from the gel at various times after the start of the experiment. (II) Optical micrographs showing a mixture of chemically "inert" PS microspheres and self-propelled "matchstick" catalytic silica–manganese oxide rods in the presence of a hydrogen peroxide gradient (H_2O_2 well is on the right of the images) at (a) $t = 0$ s, (b) $t = 35$ s, (c) $t = 70$ s. Red vectors show rod displacement between each image, whereas yellow vectors indicate motion of microspheres between each image. Scale bar = 10 μm. (III) Comparison of spherical Janus catalytic swimmers and microtubes in a chemical gradient generated using a microfluidic device. (a) Chemotactic behavior of Janus sphere and tubular catalytic motors is quantified by determining the opening angles (α, β). (b) Opening angle of catalytic micromotors in different concentrations of hydrogen peroxide solution injected by one of the microfluidic channels. Black plot JPs and red plot tubular microjets.

were determined experimentally. A simulation containing a periodic variation in pH that would cause a Janus catalytic swimmer made from this material to almost double in size as it moved between a low pH and high pH zone was constructed. The velocity change due to size modulation for each device in the ensemble simulation was determined by a reciprocal relationship (i.e., $\nu \propto \frac{1}{R}$) based on the previously discussed data (Fig. 21), and the rotational constant was calculated from the swimmer radius (time constant $\propto \frac{1}{R^3}$). The qualitative experience for each device within the simulation was consequently that high pH zones would result in a size increase leading to slower propulsive velocities and a reduced rate of trajectory randomization. Simulations started with an initially uniform concentration of swimmers, and then the distribution was allowed to evolve in response to the pH gradients. Figure 36(a) shows the time evolution of the probability density function (PDF) for device positions in the periodic pH function. It is apparent that the catalytic swimming devices show a statistical accumulation at the high pH (size expanded) zone, and that equilibrium is reached after around 60 min. A control simulation in which the same size changing catalytic devices were moved only by Brownian motion, with no propulsion, showed no response to the same pH pattern over an equivalent time scale, proving that this statistical accumulation behavior is reliant on propulsion. This simulation suggests that catalytic Janus swimmers made from materials that can produce a size change in response to a pH gradient will exhibit a significant and rapid stimulus directed accumulation, and so this could provide a route for deployment as an autonomous navigation strategy. To further understand the roles of the modulation of intrinsic propulsion velocity due to the size, and the modulation of rotational diffusion, it was possible to run additional simulations, where each factor was separately studied. Figure 36(b) shows the PDF evolution in the case where rotational rate remains constant and only the device velocity responds to pH-induced size changes. In this case, accumulation within the high pH region also evolves during the simulation, with equilibrium being reached more rapidly compared to the full-response simulation. However, the extent of variation induced in the PDF due to the pH gradient appears rather less. Figure 36(c) shows the evolution of PDF in the converse situation where the rotational rate changes but propulsion velocity is held constant. It is clear that this

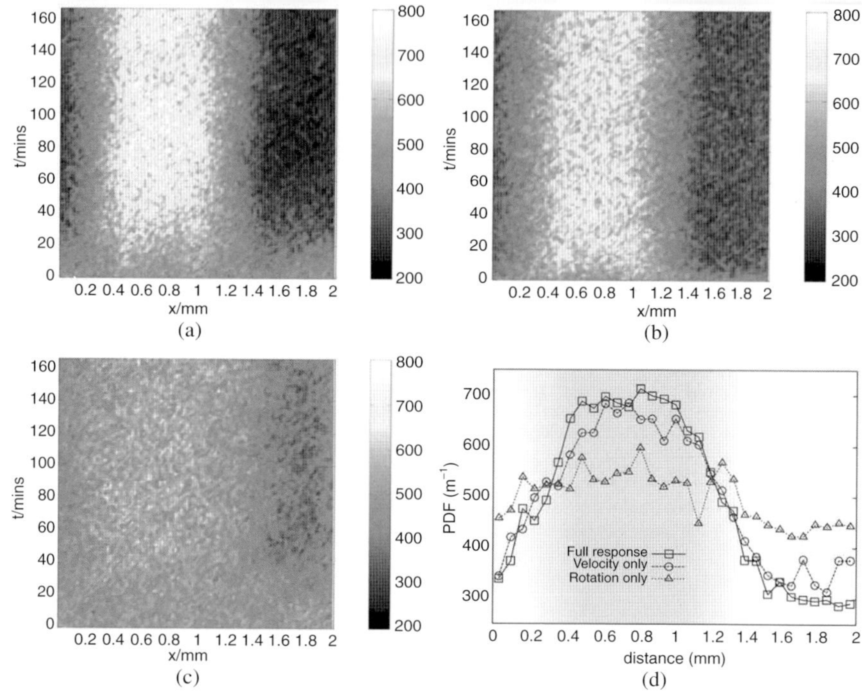

Figure 36. Temporal evolution of swimming device densities in response to a 1D periodic pH. (a) Full response. (b) Velocity only response (rotation rate fixed at the value for a 1.1 mm radius swimmer). (c) Rotation rate only response (velocity fixed at 5.68 mm s^{-1}, the value for a 1.1 mm radius swimmer). (d) Comparison of probability distribution function at equilibrium for the three response scenarios. The shaded region indicates pH concentration (darker shading, higher pH).

scenario also produces a PDF enhancement in the high pH regions, reaching equilibrium after around 60 min; however, the extent of enrichment is significantly less than for the other two cases. Figure 36(d) compares the equilibrium PDF profiles in each of these cases (full response, velocity only, rotation only) and confirms the full response simulation produces the most significant particle accumulation within the high pH stripes (pH concentration is depicted with shading in Fig. 36(d)). With this information, it is apparent that the overall statistical response (Fig. 36(a)) appears to be due to the combined effect of both the size change-induced velocity modulation, and rotational rate variation, which function together to concentrate swimmers in the expanded "large size" high pH region. The significant and rapid velocity response contribution to this accumulation is not unexpected: the swimmers move more slowly when they are in the expanded high pH regions, and so spend more time in this zone. However, the extrarotational modulation contribution to the PDF enhancement, also causing accumulation in high pH regions, is less easily visualized. Additional simulations suggested that size-changing swimmers trying to escape a high pH region are "turned around", as they move toward lower pH regions due to the increase in rotational rate.

It is also interesting to consider which of the three scenarios simulated would match the behavior for realistic size-changing swimming devices. Due to adhesion issues, it appears that successful size changing catalytic devices may require the driving metallic catalyst to be localized on a rigid, non-responsive part of the bead, or a separate rigid propulsion generating module, which would break the symmetry of the devices, and so alter the velocity dependence on expansion. In this case, the degree of velocity modulation may be less than that used here. However, the potential to use more responsive hydrogel materials, for example at four times the swelling ratio of the hydrogel, has recently been reported and could lead to more dramatic responses than those used in the simulation. The simulations also show that even if the velocity modulation effect is removed entirely, size changing allows a degree of statistical control over swimmer distribution.

8.5.8. Clustering

Another potential mechanism to control Janus catalytic swimming devices is to harness the interactions between individual devices, for example to produce collective behavior such as the formation of clusters. This differs from the scenario discussed above, where rigid assemblies of Janus catalytic swimmers were investigated (Sec. 6.3.2), in that the clusters discussed here are often dynamic and the individual colloids do not become permanently joined. One example of this behavior has been reported for spherical gold JPs half-coated in platinum that can propel themselves at moderate velocities (3 μm s^{-1}) in low concentrations of hydrogen peroxide (0.1%) [38]. This low fuel concentration allows high number densities of particles to be investigated without the oxygen bubbles evolved during the catalytic decomposition of peroxide causing significant convection. Clustering of these particles was observed, at volume fractions of particles between 3% and 50%. For example, at a volume fraction of 5% clusters formed within a few seconds, which themselves are able to also collectively translate at low speeds. These clusters were reported to be dynamic, with JPs joining and leaving the clusters.

Another system capable of propulsion using low concentrations of hydrazine fuel has been reported. These spherical iridium-based Janus microswimmers [40] are capable of propelling themselves in concentrations of hydrazine as low as 10^{-7}%. The iridium particles were also found to aggregate into clusters, especially in the larger of the two sizes synthesized (4.74 μm diameter). A further chemical propulsion system displaying clustering uses particles powered by the phase separation of water/lutidine mixtures triggered by an external light source that induces localized heating [116]. This again overcomes the problem of significant unwanted oxygen bubble formation often found when using the platinum/hydrogen peroxide propulsion system. Without illumination, the particles undergo only Brownian motion and were found to have a homogeneous distribution for all the volume fractions investigated (0.1–0.4). However, when illuminated and undergoing propulsion, clusters formed (Fig. 37(a)). The cluster size increased with increasing velocity, and as before, the particles dynamically joined and left the clusters. The JPs clusters were observed to evolve, merging together or splitting back up into smaller clusters. Once the particles were clustered, if the illumination was turned off, the particles redispersed to a homogeneous distribution.

The mechanisms for these cluster formation phenomena is the subject of debate, but proposals include a generic particle density dependent instability, a chemotaxis like effect due to self-generated concentration gradients, and a "self-trapping" effect where the cluster longevity is due to the time taken for the propulsion vectors of opposingly orientated colloids to rotate to allow them to resume motion (Fig. 37(b)). In the cases above, clustering of particles was turned on by features of the Janus swimmer propulsion. However, propulsion can also be used to break up clusters of particles. As an example, bimetallic nanorods were induced to form clusters by the application of ultrasound, but when the ultrasound was turned off, catalytic propulsion caused the rapid dissipation of the particle swarms [117].

The issue of collection motion phenomena in catalytic Janus swimming devices has also been examined from a theoretical perspective. Saha

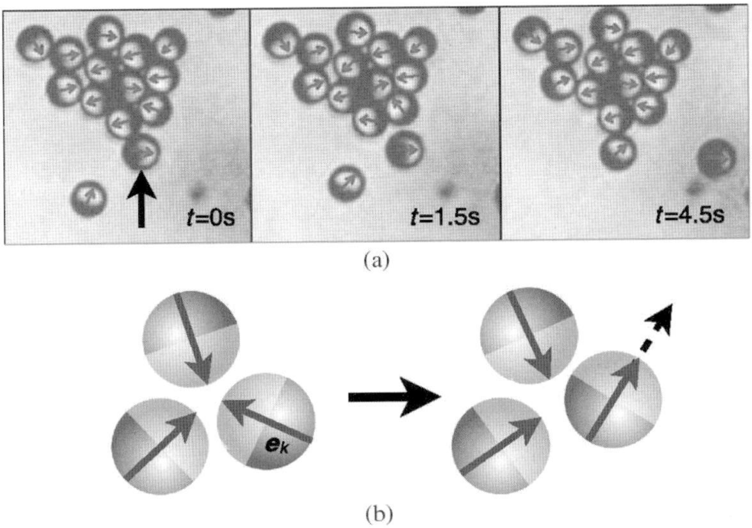

Figure 37. Cluster formation in propelled JPs. (a) Consecutive micrographs of a cluster, showing the direction of propulsion (gray arrows). The forward path of particles in the cluster is blocked until they rotate to point outward. (b) Schematic of the self-trapping mechanism showing how particles can become free by waiting for their orientations to change due to rotational diffusion. From Ref. [116].

et al. presented a comprehensive analytical approach describing possible ensemble structure formation for catalytic swimmers due to interactions with fuel and product molecule gradients [114]. This work describes the variety of methods by which catalytic Janus swimmers can respond to concentration gradients, and then applies this understanding to the case of many interacting particles. Two phenomena are proposed: at low reactant concentration clumping and patterning is possible, while at high reactant concentration, analogs of electrostatic and gravitational effects are possible such as Debye screening, microphase separation, plasma oscillations, and gravitational collapse (Fig. 38).

It is clear from this analysis that there is future potential for harnessing the phenomena discussed in this section to achieve a rich variety of autonomous control over Janus catalytic swimmer motions, and that this will be a key enabling method to allow the proposed applications for these devices to be realized, as discussed in the following section.

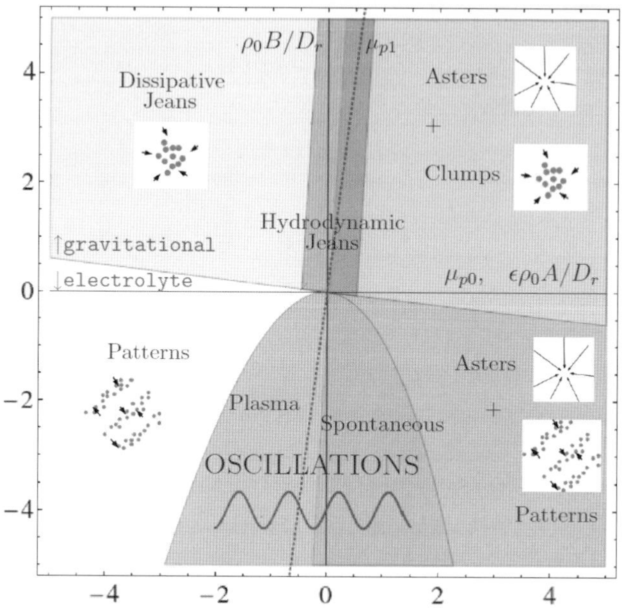

Figure 38. The phase diagram in the *reaction-limited* regime (with abundant fuel) shows a variety of possible experimental paths that can be explored in sequences of experiments on particles designed with suitable mobility coatings.

8.6. Applications

8.6.1. Catalytic Swimmers with Functionality

In order for catalytic swimmers to be of real-world utility, it will often be necessary for the devices to be able to perform a function in combination with the rapid autonomous transport phenomena discussed above. Many proposed future applications for catalytic Janus swimmers center on transport, and consequently, the ability to attach cargo, ranging from small molecules to micron-sized components, at the surface of catalytic Janus swimmers has received much attention. In this context, a key feature for the Janus structure in many swimming devices is that the inactive, non-catalytic surface of the device remains available for further chemical functionalization. As examples of progress, both bimetallic nanorod swimmers and Janus catalytic swimmers have been shown to be able to transport colloidal cargo (Fig. 30) [16,103]. However, these transport demonstrations have not been autonomous, as cargo release and pickup required external guidance by magnetic fields. A recent further example for cargo transport function from the Sanchez group shows the potential to exploit the porosity of catalytic silica Janus swimmers to transport small molecules. In a two-stage process, it was demonstrated that small molecules could be diffusively loaded into exposed pores, and that propulsive transport could be used in conjunction with physical guidance to concentrate Janus colloids at an enhanced rate in a well-defined location. Following this transport demonstration, it was also shown that sustained release of the small molecule cargo from the pores could be achieved [28]. While clearly showing future potential, accomplishing the cargo capture and release processes autonomously in synergy with propulsive transport was not possible. It was suggested that using responsive materials to control the release behavior from the pores could address this in the future. To illustrate future potential for Janus catalyst functionality, related tubular devices have shown the ability to dynamically attach specific biomolecules to their surfaces [118]. In addition to transport, more easily obtainable functions that rely solely on the propulsion generation capacity are emerging, for example, bubble propulsive nanotubes have been shown to enhance mixing rates considerably due to a combination of bubble generation and enhanced motion [119].

8.6.2. Current Application Areas

While applications for catalytic Janus swimmers are still in the developmental stage, here we review the general application areas in which early demonstrations of device utility is being reported. We do not limit this discussion to Janus swimming devices, but instead broaden the remit to consider reports for the whole family of swimming devices, as there is a high degree of potential to apply the strategies demonstrated for tubular device, to Janus swimmers in the future.

8.6.2.1. *Environmental applications*

One application area that has received attention as a potential arena within which Janus catalytic swimmers can serve a useful function is environmental remediation. The idea for this deployment is that the swimming devices can perform tasks such as neutralization of a harmful chemical, utilizing the benefit of their enhanced motion to increase the frequency of contact with the contaminants and so speed up the remediation process. This application does not require deployment of advanced motion control methods, as it is likely to be feasible to rely on stochastic motion and achieve targeting simply by manually releasing the devices at the site of contamination. However, the ability to produce motion without the addition of external fuel molecules is likely to be important. One approach to pursue this goal has been to move away from catalytic propulsion, and instead utilize a conventional chemical reaction between part of the swimming device and a component of the medium in which propulsion is sought. As an example, a JP made from a manganese core, masked with an overlayer of gold was able to produce motion via hydrogen bubble expulsion [120]. This reaction was persistent when chloride ions were present, which made motion within sea-water viable. In a demonstration of remediation potential, an additional photocatalytic overlayer of titanium dioxide was added to the Janus structure. This layer was shown to be able to decompose an anthrax-like contaminant, and nerve agent simulants. When compared to functional colloids operating without propulsion, it was established that enhanced motion resulted in improved solution mixing and also promoted self-cleaning of the functional surface, leading to a much increased rate of decontamination (Fig. 39(a)) [121]. A second example using a similar manganese-based JP employed a hydrophobic coating functionalized onto a gold outer layer to collect oil droplets, as a potential route to clean up for oil spillages at sea (Fig. 39(b)) [120]. This study used magnets to steer the Janus swimmer toward the oil droplet. In both of these examples, while the use of a chemical, rather than catalytic reaction produced Janus colloid motion without the addition of fuel molecules, a surfactant additive was required to instigate motion, in order to facilitate bubble nucleation and release. The requirement for surfactants is not, however, a general limitation for all bubble propulsive Janus swimming devices, as in contrast, platinum-based bubble propelled Janus swimmers decomposing hydrogen peroxide only require the addition of fuel in order to produce motion in aqueous environments, so there appears to be the potential to combine these features in the future [60].

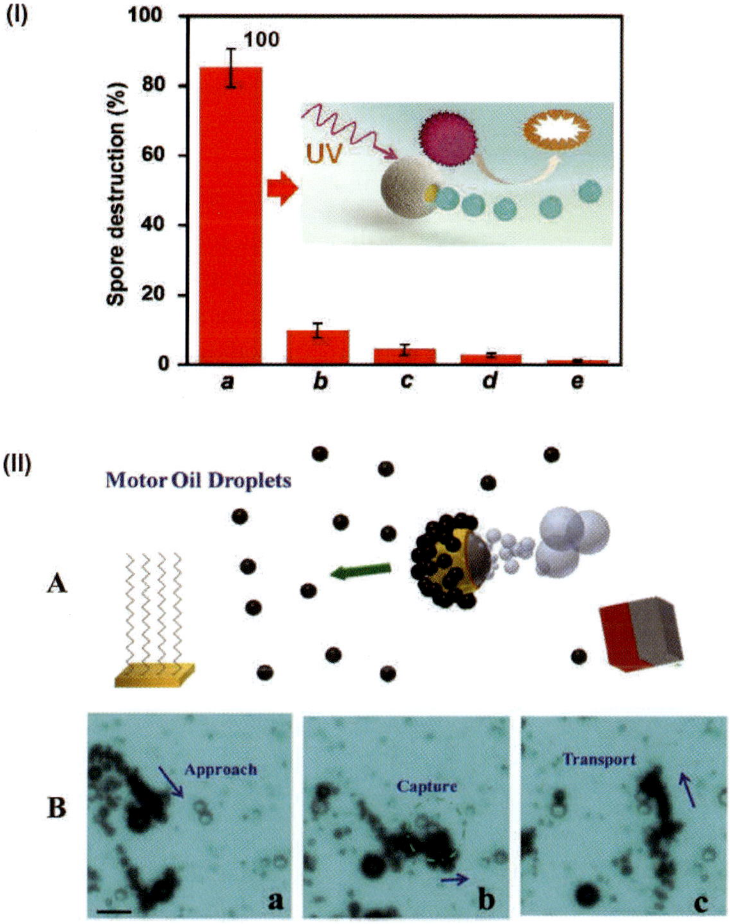

Figure 39. Remediation by active colloids. (I) Schematic: a titanium dioxide (gray) coated manganese core (yellow) particle produces hydrogen bubbles at an uncoated pore, resulting in fuel free bubble propulsion. The functional titanium layer decomposes anthrax spores when exposed to UV light. Graph shows the autonomous remediation function efficiency for (a) the situation where colloids are illuminated and producing propulsion and (b) no light (c) moving motors without a functional layer (d) UV light alone (e) static motors with UV light (II) A: Schematic of oil droplet collection by an active colloid, again a manganese core (gray) produces fuel free motion while a gold surface functionalized with an alkane thiol gathers oil droplets. B: Still frames from video microscopy showing that magnetic guidance is used to direct the colloid toward oil droplets, which are then captured and transported.

Environmentally relevant functions have also been demonstrated for related catalytic nanotube swimmers, in one case using Fenton oxidation to decompose organic contaminants [122].

8.6.2.2. *Lab-on-a-chip applications*

Lab-on-a-chip devices comprise small-scale interconnected chambers and pipes, together with other miniaturized instrumentation, allowing low volumes of solution to be used for complex analytical processes. The goals for this technology include enabling rapid medical diagnosis and the ability to detect environmental contamination events at the "point-of-care". In order to execute multiple step analysis methods within current devices, the motion of the fluid sample through miniature tubing must be controlled. This is only possible by deploying additional components that are external to the fluidic path, resulting in devices that are complex and require external power. Instead, it is possible that catalytic JPs may be deployable in this situation, to enable the transport of analytes within miniaturized fluid networks without requiring additional external apparatus. This approach is often termed static microfluidics, as in contrast to other methods the fluid will not be required to flow, but rather the swimming devices will carry attached cargo through the static fluidic environment. By no longer needing to generate and create small-scale flow, lab-on-a-chip devices can be significantly simplified, which in turn will increase their utility. It is apparent that to make use of Janus catalytic swimmers beneficial in this context, the developing capabilities for autonomous guidance and motion control described in Sec. 6 will be important. To illustrate the state of play for this application area, a recent study has deployed non-Janus, nanotube swimming devices for a lab-on-a-chip proof-of-concept demonstration [123]. In this example, the swimming devices were transferred between different connected chambers using magnetic guidance to allow a sequence of chemical operations used in common immunoassay protocols to be performed. To enable this, antibodies were attached to the surface of the propulsive devices to selectively identify specific proteins or cells, and then the captured analyte was carried by the devices to an additional chamber where a labeling operating using a second recognition component completed the assay. Encouraging for the development of similar applications for Janus catalytic swimmers, these biological binding events took place within a

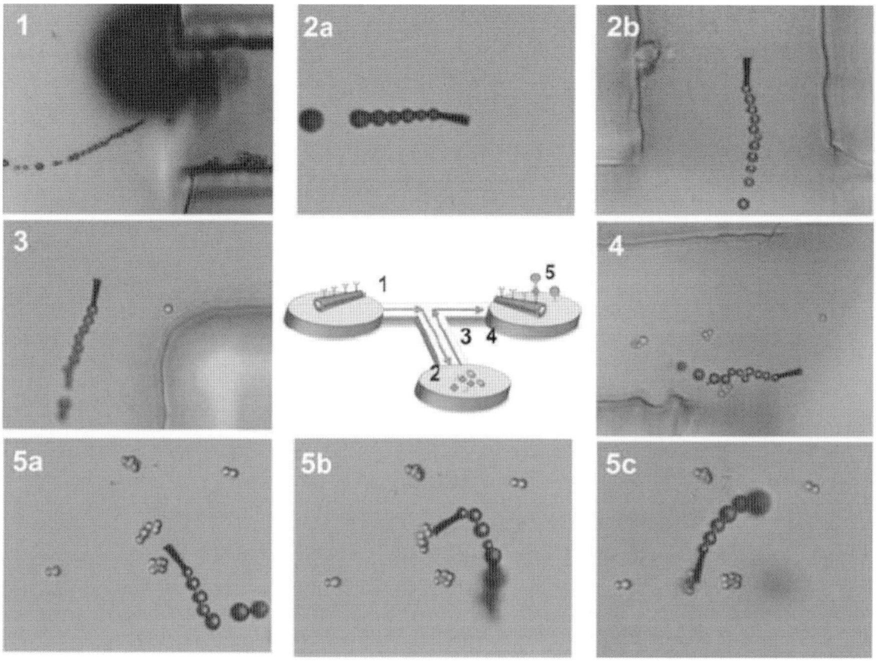

Figure 40. Transfer of an antibody-functionalized nanotube through a microfluidic device (shown centrally). (1) nanotube leaves initial reservoir, (2a) nanotube moves within a reservoir containing both target and other protein molecules, and then (2b) leaves to enter a third chamber (3 + 4). (5) Shows beads tagged with a second antibody complementary to the target protein attaching to the nanotube, confirming selective pickup and labeling of the target. All chemistry and propulsion were autonomous in this example, but magnetic fields were used to guide the device though the interconnecting channels.

fluid containing the hydrogen peroxide fuel necessary to support propulsion motion, and the rate of protein binding was enhanced by the motion of the "roving" analyte collector (Fig. 40).

8.6.2.3. *In vivo applications*

One of the original motivations for the catalytic Janus swimmer field was to enable futuristic visions of small-scale devices moving within the human body to deliver drugs to specific locations or effect repairs. As a measure of progress toward this end goal, recently, early examples for the potential to use swimming devices within living animals have been published. In one example, hydrogen bubble driven tubular swimmers made from zinc and a polymeric matrix were administered orally to mice [124]. It was found that some propulsion was generated within the mouse stomach, and that this was sufficient to embed the swimmers into the stomach lining, which improved their residence time compared to an inactive control experiment. The study used gold nanoparticles as a proxy for a possible therapeutic agent incorporated into the body of the swimming devices, and showed that there was an enhanced retention of this delivery vector in the stomach lining compared to that observed for conventional oral delivery.

Other potential future-envisaged *in vivo* applications for Janus catalytic swimmers could involve them functioning within the bloodstream, for example to repair arteries. To assess if such applications will be viable, and also to move closer to the ability to implement lab-on-a-chip medical analysis, catalytic propulsion within fluids containing blood components has been investigated. In one report, it was found that blood proteins considerably reduced the mobility for Cu/Pt microtubular motors, due to fouling of the catalytic surface and solution viscosity [125]. However, more promisingly, Fe/Pt microtubular motors were found to be able to produce motion at physiological temperatures in diluted serum and red-blood-cell-containing solutions [126].

Despite the range of navigation capabilities surveyed in Sec. 6, these strategies have not yet been applied for an *in vivo* application. However, in the future phenomena such as chemotaxis may be able to be used to enable accumulation of devices in response to the many solution gradients found within the body, and flows present in the body may also provide an additional method to orientate asymmetrical JPs.

8.7. Future Directions

One current feature for Janus catalytic devices that requires more attention is the fact that at least two mechanisms producing motion — bubble release and phoretic gradients — exist, and that each mechanism has distinct features both in terms of tolerance for different fluid environments, and the trajectories that are produced. When comparing trajectories, it is apparent that while spherical Janus swimmers that move by bubble release show some relationship between catalyst distribution and directionality, their

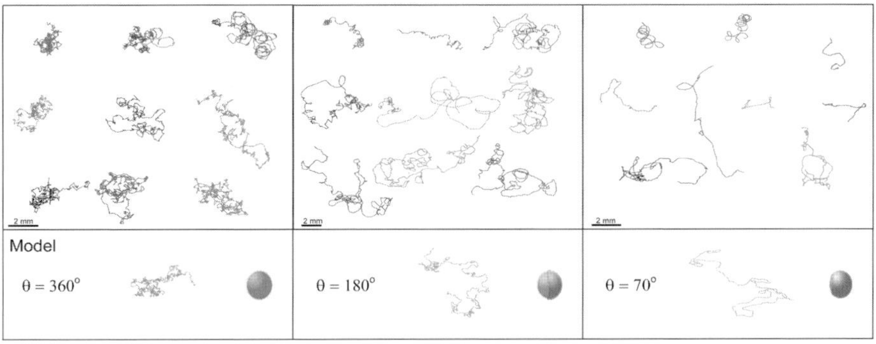

Figure 41. Trajectories of 40 s duration for typical bubble propulsive colloids recorded in 10% w/v hydrogen peroxide (left to right: symmetrical activity, Janus activity, and pore activity).

motion is overall highly chaotic, and does not have a strong link to the orientation of the catalytic patch (Fig. 41) [49]. The lack of directionality will make spherical Janus bubble swimmers harder to control, when compared to phoretic Janus swimmers, where a large variety of routes for motion control have been demonstrated. Against this, bubble-propulsive swimmers produce significantly higher velocities, of the order of hundreds of microns per second compared to tens of microns per second. Considering differences in tolerance for solutions, as we have discussed, bubble swimmers can need surfactant additives in order to produce motion, but are not sensitive to high salt environments, which signposts an improved future potential for operating within biofluids, whereas phoretic Janus swimmers, which do not produce visible bubbles, show propulsion velocities that are rapidly quenched by high ionic strength solutions. To take the development of Janus swimming devices forward, it is clear that ideally combining the features for both of these mechanisms would be desirable to result in salt tolerant, extremely rapid, easy to autonomously direct devices.

Another interesting possibility is the potential to use Janus swimming devices that can switch between bubble propulsion and phoretic propulsion. The trajectory and velocity of such devices would be expected to change considerably at such a transition point, which could be useful for some transport applications. For a given device size, changing fuel concentration provides a route to vary the rate of catalytic reaction, so for example in the case of the hydrogen peroxide/platinum system the rate of oxygen production per device can be increased by adding more hydrogen peroxide. Bubble nucleation is encouraged by higher local oxygen production rates [60], and so it is possible that for a suitably sized device, transition between these two mechanisms could be mediated by fuel concentration. Given that some motion control phenomena such as boundary steering appear to be unique features for the phoretic mechanism, it may be possible in this way to turn on and off particular features of swimming devices motion guidance, for example allowing for guidance by boundaries between interconnected chambers in a fluidic device transitioning to bubble-driven rapid motion within a higher fuel concentration region of a microfluidic device.

A further general challenge for the field of catalytic swimmers is to overcome the current reliance on hydrogen peroxide fuel for many of the existing experimental demonstrations. Ideally, motion could be produced without the need for any additional fuel. As discussed above, there has been some progress in this respect, for example, short-lived bubble propulsion can be generated by chemical reactions, within suitable solutions without adding fuel. An intermediate goal would be to generate motion from fuel molecules that are less likely to interact with components of the solution in which motion is desired. Enzymes offer an attractive route to exploit in this context, due to the wide range of decomposition reactions that could potentially generate motion. Indeed, recently colloids symmetrically functionalized with urease and catalase were found to undergo enhanced diffusion, without the generation of visible bubbles [127]. The phenomenology of this effect is different from any of the mechanisms discussed above, and was thought to be potentially due to local temperature variations induced by the exothermic enzymatic reactions. Previously it has also been shown to be possible to generate bubble propulsion from a glucose fuel using a combination of glucose oxidase and catalase [43].

This chapter has described the numerous methods that have been demonstrated to control the motion for catalytic swimming devices being deployed separately, rather than being used in tandem. This reflects the compartmentalization of these developments at this stage of the fields' maturity. A further feature is that many of the application examples, such as performing an immunoassay with a tubular swimmer, revert to using magnetic guidance, rather than incorporating autonomous guidance procedures. It is clear that there is much potential in the near future to start combining the various guidance methods detailed in Sec. 6, to enable autonomous applications, fully benefiting from the essential feature of catalytic swimming: motion production at small scales without external actuation. The toolkit of motion control methods suggest that complex 3D manipulations for swimming catalytic colloids are within the realms of possibility. Gravitaxis provides a mechanism for vertical motion control, chemotaxis-like chemical gradient steering can provide additional bias, while boundary steering provides another new route to achieving directionality. A future challenge for the field is to deploy these mechanisms together to, for example, allow autonomous transfer of a mobile sensor between the different solutions required to execute an immunoassay. Some specific scenarios in which these guidance mechanisms could be used together can be imagined. For example, the path of a swimmer at a junction in a physically guided microfluidic devices could be determined by the bias imparted by a chemical stimulus, providing a "switch" like functionality. Similarly, inclined physically patterned substrates could be used to harness the combined effects of boundary steering and gravitaxis bias. Gravitaxis also provides a method to control the vertical transit rate of a swimming colloid within free solution and could be used, for example, to allow collection of an analyte cargo from solution, and concentration to the top surface of a container, where physical guidance could provide a method for further manipulation, for example localization at a specific sensor region.

While this chapter has focused mainly on experimental results for Janus swimming devices, this field has also inspired a considerable amount of theoretical and modeling research. An example of the interplay between these two approaches, the initial concept for spherical Janus swimmer devices, in fact, originated from a theoretical proposal [52]. Continuing developments in the Janus swimmer field will also clearly benefit from theory and modeling. In particular, theory and modeling allow investigations for scenarios that are currently experimentally hard to realize, such as high volume fractions of catalytic Janus swimmers, where gas evolution and fuel depletion can limit observations [128]. These theoretical insights will inform experimentalists and provide direction to ensure the continued development of this exciting field toward enabling future applications.

References

[1] Wang, W.; Duan, W.; Ahmed, S.; Mallouk, T. E.; Sen, A. *Nano Today*, **8**(5), 531 (**2013**).

[2] Wang, W.; Chiang, T.-Y.; Velegol, D.; Mallouk, T. E. *J. Am. Chem. Soc.*, **135**(28), 10557 (**2013**).

[3] Kapral, R. *J. Chem. Phys.*, **138**(2), 020901 (**2013**).

[4] Ebbens, S. J.; Howse, J. R. *Soft Matter*, **6**(4), 726 (**2010**).

[5] Ebbens, S. *Mater. Today*, **15**(7–8), 294 (**2012**).

[6] Solovev, A. A.; Sanchez, S.; Pumera, M.; Mei, Y. F.; Schmidt, O. G. *Adv. Funct. Mater.*, **20**(15), 2430 (**2010**).

[7] Baraban, L.; Tasinkevych, M.; Popescu, M. N.; Sanchez, S.; Dietrich, S.; Schmidt, O. G. *Soft Matter*, (**2012**).

[8] Simmchen, J.; Baeza, A.; Ruiz, D.; Esplandiu, M. J.; Vallet-Regí, M. *Small*, **8**(13), 2053 (**2012**).

[9] Palacci, J.; Sacanna, S.; Vatchinsky, A.; Chaikin, P. M.; Pine, D. J. *J. Am. Chem. Soc.*, 8 (**2013**).

[10] Solovev, A. A.; Xi, W.; Gracias, D. H.; Harazim, S. M.; Deneke, C.; Sanchez, S.; Schmidt, O. G. *ACS Nano*, **6**(2), 1751 (**2012**).

[11] Patra, D.; Sengupta, S.; Duan, W.; Zhang, H.; Pavlick, R. A.; Sen, A. *Nanoscale*, **5**(4), 1273 (**2013**).

[12] Peyer, K. E.; Tottori, S.; Qiu, F.; Zhang, L.; Nelson, B. J. *Chemistry*, **19**(1), 28 (**2013**).

[13] Fischer, P.; Ghosh, A. *Nanoscale*, **3**(2), 557 (**2011**).

[14] Mair, L. O.; Evans, B.; Hall, A. R.; Carpenter, J.; Shields, A.; Ford, K.; Millard, M.; Superfine, R. *J. Phys. D Appl. Phys.*, **44**(12), 125001 (**2011**).

[15] Wu, Y.; Wu, Z.; Lin, X.; He, Q.; Li, J. *ACS Nano*, **6**(12), 10910 (**2012**).

[16] Baraban, L.; Makarov, D.; Streubel, R.; Mönch, I.; Grimm, D.; Sanchez, S.; Schmidt, O. G. *ACS Nano*, **6**(4), 3383 (**2012**).

[17] Khim Chng, E. L.; Zhao, G.; Pumera, M. *Nanoscale*, **6**(4), 2119 (**2014**).

[18] Purcell, E. M. *Am. J. Phys.*, **45**(1), 3 (**1977**).

[19] Dreyfus, R.; Baudry, J.; Roper, M. L.; Fermigier, M.; Stone, H. A.; Bibette, J. *Nature*, **437**(7060), 862 (**2005**).

[20] Najafi, A.; Golestanian, R. *Phys. Rev. E.*, **69**(6), 062901 (**2004**).

[21] Leoni, M.; Kotar, J.; Bassetti, B.; Cicuta, P.; Lagomarsino, M. C. *Soft Matter*, **5**, 472 (**2009**).

[22] Tierno, P.; Golestanian, R.; Pagonabarraga, I.; Sagués, F. *Phys. Rev. Lett.*, **101** (November), 218304 (**2008**).

[23] Ismagilov, R. F.; Schwartz, A.; Bowden, N.; Whitesides, G. M. *Angew. Chemie*, **114**(4), 652 (**2002**).
[24] Paxton, W. F.; Kistler, K. C.; Olmeda, C. C.; Sen, A.; St Angelo, S. K.; Cao, Y.; Mallouk, T. E.; Lammert, P. E.; Crespi, V. H. *J. Am. Chem. Soc.*, **126**(41), 13424 (**2004**).
[25] Demirok, U. K.; Laocharoensuk, R.; Manesh, K. M.; Wang, J. *Angew. Chem., Int. Ed.*, **47**(48), 9349 (**2008**).
[26] Hong, Y.; Blackman, N.; Kopp, N.; Sen, A.; Velegol, D. *Phys. Rev. Letts.*, **99**(17), 1 (**2007**).
[27] Howse, J.; Jones, R.; Ryan, A.; Gough, T.; Vafabakhsh, R.; Golestanian, R. *Phys. Rev. Letts.*, **99**(4), 8 (**2007**).
[28] Ma, X.; Hahn, K.; Sanchez, S. *J. Am. Chem. Soc.*, 150410110204000 (**2015**).
[29] Ebbens, S.; Tu, M.-H. M.; Howse, J. R. J. J. R.; Golestanian, R.; Mei-Hsien Tu. *Phys. Rev. E.*, **85** (2 Pt, 1), 020401 (**2012**).
[30] Ke, H.; Ye, S.; Carroll, R. L.; Showalter, K. *J. Phys. Chem. A*, **114**, 5462 (**2010**).
[31] Ebbens, S.; Jones, R. A. L.; Ryan, A. J.; Golestanian, R.; Howse, J. R. *Phys. Rev. E.*, **82**(1), 6 (**2010**).
[32] Wang, Y.; Hernandez, R. M.; Bartlett, D. J.; Bingham, J. M.; Kline, T. R.; Sen, A.; Mallouk, T. E. *Langmuir*, **22**(25), 10451 (**2006**).
[33] Manjare, M.; Yang, B.; Zhao, Y. P. *Phys. Rev. Lett.*, **109** (September), 1 (**2012**).
[34] Gao, W.; Uygun, A.; Wang, J. *J. Am. Chem. Soc.*, **134**, 897 (**2012**).
[35] Fournier-Bidoz, S.; Arsenault, A. C.; Manners, I.; Ozin, G. a. *Chem. Commun.* (*Camb*). No. 4, 441 (**2005**).
[36] Laocharoensuk, R.; Burdick, J.; Wang, J. *ACS Nano*, **2**(5), 1069 (**2008**).
[37] Ke, H.; Ye, S.; Carroll, R. L.; Showalter, K. *J. Phys. Chem. A*, **114**(17), 5462 (**2010**).
[38] Theurkauff, I.; Cottin-Bizonne, C.; Palacci, J.; Ybert, C.; Bocquet, L. *Phys. Rev. Lett.*, **108**(26), 268303 (**2012**).
[39] Lee, T. C.; Alarcón-Correa, M.; Miksch, C.; Hahn, K.; Gibbs, J. G.; Fischer, P. *Nano Lett.*, **14**(5), 2407 (**2014**).
[40] Gao, W.; Pei, A.; Dong, R.; Wang, J. *J. Am. Chem. Soc.*, **136**(6), 2276 (**2014**).
[41] Vicario, J.; Eelkema, R.; Browne, W. R.; Meetsma, A.; La Crois, R. M.; Feringa, B. L. *Chem. Comms.*, **No. 31**, 3936 (**2005**).
[42] Pantarotto, D.; Browne, W. R.; Feringa, B. L. *Chem. Comms.*, **No. 13**, 1533 (**2008**).
[43] Mano, N.; Heller, A. *J. Am. Chem. Soc.*, **127**(33), 11574 (**2005**).

[44] Pavlick, R. A.; Sengupta, S.; McFadden, T.; Zhang, H.; Sen, A. *Angew. Chemie Int. Ed.* **50**(40), 9374 (**2011**).
[45] Dong, R.; Zhang, Q.; Gao, W.; Pei, A.; Ren, B. *ACS Nano*, acsnano., 5b05940 (**2015**).
[46] Martin, B. R.; Dermody, D. J.; Reiss, B. D.; Fang, M.; Lyon, L. A.; Natan, M. J.; Mallouk, T. E. *Adv. Mater.*, **11**(12), 1021 (**1999**).
[47] Stöber, W.; Fink, A.; Bohn, E. *J. Colloid Interface Sci.*, **26**(1), 62 (**1968**).
[48] Gibbs, J. G.; Zhao, Y.-P. *App. Phys. Letts.*, **94**(16), 163104 (**2009**).
[49] Gregory, D. A.; Campbell, A. I.; Ebbens, S. J. *J. Phys. Chem. C.*, 150610185912009 (**2015**).
[50] Brown, A.; Poon, W. *Soft Matter*, 4016 (**2014**).
[51] Paxton, W. F.; Baker, P. T.; Kline, T. R.; Wang, Y.; Mallouk, T. E.; Sen, A. *J. Am. Chem. Soc.*, **128**(46), 14881 (**2006**).
[52] Golestanian, R.; Liverpool, T.; Ajdari, A. *Phys. Rev. Letts.*, **94**(22), 1 (**2005**).
[53] de Buyl, P.; Kapral, R. *Nanoscale*, **5**(4), 1337 (**2013**).
[54] Valadares, L. F.; Tao, Y.-G.; Zacharia, N. S.; Kitaev, V.; Galembeck, F.; Kapral, R.; Ozin, G. A. *Small*, **6**(4), 565 (**2010**).
[55] Ebbens, S.; Tu, M.; Howse, J.; Golestanian, R. *Phys. Rev. E, 020401*, 1 (**2012**).
[56] Ebbens, S.; Gregory, D. A.; Dunderdale, G.; Howse, J. R.; Ibrahim, Y.; Liverpool, T. B.; Golestanian, R. *EPL*, **106**(5), 58003 (**2014**).
[57] Fletcher, N. H. *J. Chem. Phys.*, **29**(1958), 572 (**1958**).
[58] Wilson, D. A; de Nijs, B.; van Blaaderen, A.; Nolte, R. J. M.; van Hest, J. C. M. *Nanoscale*, **5**(4), 1315 (**2013**).
[59] Ebbens, S.; Tu, M.-H.; Howse, J. R.; Golestanian, R. *Phys. Rev. E. Stat. Nonlin. Soft Matter Phys.*, **85**(2 Pt, 1), 020401 (**2012**).
[60] Wang, S.; Wu, N. *Langmuir* (**2014**).
[61] Campbell, A. I.; Ebbens, S. J. *Langmuir*, **29**(46), 14066 (**2013**).
[62] Palacci, J.; Cottin-Bizonne, C.; Ybert, C.; Bocquet, L. *Phys. Rev. Lett.*, **105**(8), 1 (**2010**).
[63] ITheurkauff. *PRL*, **108**(26), 1 (**2012**).
[64] Kline, T. R.; Paxton, W. F.; Mallouk, T. E.; Sen, A. *Angew. Chem., Int. Ed.*, **44**(5), 744 (**2005**).
[65] Zhao, G.; Sanchez, S.; Schmidt, O. G.; Pumera, M. *Nanoscale*, **5**(7), 2909 (**2013**).
[66] Gao, W.; Sattayasamitsathit, S.; Uygun, A.; Pei, A.; Ponedal, A.; Wang, J. *Nanoscale*, **4**(7), 2447 (**2012**).

[67] Sanchez, S.; Solovev, A. A.; Mei, Y.; Schmidt, O. G. *J. Am. Chem. Soc.*, **132**(38), 13144 (**2010**).

[68] Ismagilov, R. F.; Schwartz, A.; Bowden, N.; Whitesides, G. M. *Angew. Chem., Int. Ed.*, **41**(4), 652 (**2002**).

[69] Gu, Y.; Sattayasamitsathit, S.; Kaufmann, K.; Vazquez-Duhalt, R.; Gao, W.; Wang, C.; Wang, J. *Chem. Commun.* (*Camb*). **49**(66), 7307 (**2013**).

[70] Yildiz, A.; Selvin, P. R. *Acc. Chem. Res.*, **38**(7), 574 (**2005**).

[71] Peterson, S. D.; Chuang, H.-S.; Wereley, S. T. *Meas. Sci. Technol.*, **19**(11), 115406 (**2008**).

[72] Anthony, S. M.; Yu, Y. *Anal. Methods* **No. 2004**, 7020 (**2015**).

[73] Behrend, C. J.; Anker, J. N.; McNaughton, B. H.; Brasuel, M.; Philbert, M. A.; Kopelman, R. *J. Phys. Chem. B*, **108**(29), 10408 (**2004**).

[74] Das, S.; Garg, A.; Campbell, A. I.; Howse, J. R.; Sen, A.; Velegol, D.; Golestanian, R.; Ebbens, S. J. *Nat. Commun.*, **6** (8999) (**2015**).

[75] Ebbens, S. J.; Howse, J. R. *Langmuir*, **27**(20), 12293 (**2011**).

[76] Archer, R. J.; Campbell, A. I.; Ebbens, S. *Soft Matter*, **11**, 6872 (**2015**).

[77] Mortimer, S. T.; Swan, M. A.; Mortimer, D. *Hum. Reprod.*, **11**(5), 1049 (**1996**).

[78] Van Den Heuvel, M. G. L.; Bolhuis, S.; Dekker, C. *Nano Lett.*, **7**, 3138 (**2007**).

[79] Mandelbrot, B. B. *American Journal of Physics.*, 1983, p, 286.

[80] Witten, T. A.; Sander, L. M. *Phys. Rev. Lett.*, **47**(19), 1400 (**1981**).

[81] Poon, W. C. K.; Pirie, A. D.; Pusey, P. N. *Faraday Discuss.*, **101**, 65 (**1995**).

[82] Vicsek, T.; Cserző, M.; Horváth, V. K. *Physica A: Statistical Mechanics and its Applications*. August, 1990, pp. 315–321.

[83] Takahashi, A.; Kita, R.; Shinozaki, T.; Kubota, K.; Kaibara, M. *Colloid Polym. Sci.*, **281**(9), 832 (**2003**).

[84] Katz, M. J.; George, E. B. *Bull. Math. Biol.*, **47**(l), 273 (**1985**).

[85] Li, D. C.; Banon, S.; Biswal, S. L. *Soft Matter*, **6**(17), 4197 (**2010**).

[86] Wilson, D. A.; Nolte, R. J. M.; van Hest, J. C. M. *Nat. Chem.*, **4**(4), 268 (**2012**).

[87] Reufer, M.; Martinez, V. a; Schurtenberger, P.; Poon, W. C. K. *Langmuir* (**2012**).

[88] Banachowicz, E.; Patkowski, A.; Meier, G.; Klamecka, K.; Gapiński, J. *Langmuir* (**2014**).

[89] Lopez, a; Dupou, L.; Altibelli, A; Trotard, J.; Tocanne, J. F. *Biophys. J.*, **53**(6), 963 (**1988**).

[90] Wu, T.; Nieminen, T. A.; Mohanty, S. K.; Miotke, J.; Meyer, R. L.; Rubinsztein-Dunlop, H.; Berns, M. W. *Nat. Photonics*, **6**(1), 62 (**2011**).

[91] Apoptosis, C.; Zhang, E.; Kircher, M. F.; Koch, X. M.; Eliasson, L.; Goldberg, S. N., **No. 4**, 3192 (**2014**).
[92] Van Reenen, A.; De Jong, A. M.; Prins, M. W. J. *J. Phys. Chem. B*, **117**(5), 1210 (**2013**).
[93] Kim, K.; Xu, X.; Guo, J.; Fan, D. L. *Nat. Commun.*, *5*, 3632 (**2014**).
[94] Balk, A. L.; Mair, L. O.; Mathai, P. P.; Patrone, P. N.; Wang, W.; Ahmed, S.; Mallouk, T. E.; Liddle, J. A.; Stavis, S. M. *ACS Nano*, **8**(8), 8300 (**2014**).
[95] Nourhani, A.; Lammert, P. E.; Borhan, A.; Crespi, V. H. *Phys. Rev. E*, **87**(5), 050301 (**2013**).
[96] van Teeffelen, S.; Zimmermann, U.; Löwen, H. *Soft Matter*, **5**(22), 4510 (**2009**).
[97] Mijalkov, M.; Volpe, G. *Soft Matter*, **9**(28), 6376 (**2013**).
[98] Ebbens, S. J.; Buxton, G. A.; Alexeev, A.; Sadeghi, A.; Howse, J. R. *Soft Matter*, **8**(11), 3077 (**2012**).
[99] Pawar, A. B.; Kretzschmar, I. *Langmuir*, **24**(22), 355 (**2008**).
[100] Golestanian, R. *Phys. Rev. Lett.*, **102**(18), 188305 (**2009**).
[101] Kümmel, F.; Ten Hagen, B.; Wittkowski, R.; Buttinoni, I.; Eichhorn, R.; Volpe, G.; Löwen, H.; Bechinger, C. *Phys. Rev. Lett.*, **110**(19), 198302 (**2013**).
[102] Wang, Y.; Fei, S. T.; Byun, Y. M.; Lammert, P. E.; Crespi, V. H.; Sen, A.; Mallouk, T. E. *J. Am. Chem. Soc.*, **131**(29), 9926 (**2009**).
[103] Burdick, J.; Laocharoensuk, R.; Wheat, P. M.; Posner, J. D.; Wang, J. *J. Am. Chem. Soc.*, **130**(26), 8164 (**2008**).
[104] Kam, V.; Moseyko, N.; Nemson, J.; Feldman, L. J. *Int. J. Plant Sci.*, **160**(6), 1093 (**1999**).
[105] ten Hagen, B.; Kümmel, F.; Wittkowski, R.; Takagi, D.; Löwen, H.; Bechinger, C. *Nat. Commun.*, **5**, 4829 (**2014**).
[106] Elgeti, J.; Kaupp, U. B.; Gompper, G. *Biophysj*, **99**(4), 1018 (**2010**).
[107] Galajda, P.; Keymer, J.; Chaikin, P.; Austin, R. *J. Bacteriol.*, **189**(23), 8704 (**2007**).
[108] Takagi, D.; Palacci, J.; Braunschweig, A. B.; Shelley, M. J.; Zhang, J. *Soft Matter I*(1), 3 (**2014**).
[109] Brown, A. T.; Vladescu, I. D.; Dawson, A.; Vissers, T.; Schwarz-Linek, J.; Lintuvuori, J. S.; Poon, W. C. K. *arXiv:1411.6847*, 7 (**2015**).
[110] Simmchen, J.; Katuri, J.; Uspal, W. E.; Popescu, M. N.; Tasinkevych, M.; Sánchez, S. *Nat. Commun.*, **7** (May, 2015), 10598 (**2016**).
[111] Berg, H. C. *Annu. Rev. Biophys. Bioeng.*, **4**(00), 119 (**1975**).

[112] Baraban, L.; Harazim, S. M.; Sanchez, S.; Schmidt, O. G. *Angew. Chemie — Int. Ed.*, **52**(21), 5552 (**2013**).
[113] Morgan, A. R.; Dawson, A. B.; Mckenzie, H. S.; Skelhon, T. S.; Beanland, R.; Franks, H. P. W.; Bon, S. A. F. *Mater. Horizons*, **1**, 65 (**2014**).
[114] Saha, S.; Golestanian, R.; Ramaswamy, S. *Phys. Rev. E — Stat. Nonlinear, Soft Matter Phys.*, **89**, 1 (**2014**).
[115] Ikezoe, Y.; Fang, J.; Wasik, T. L.; Shi, M.; Uemura, T.; Kitagawa, S.; Matsui, H. *Nano Lett.*, 150526144524005 (**2015**).
[116] Buttinoni, I.; Bialké, J.; Kümmel, F.; Löwen, H.; Bechinger, C.; Speck, T. *Phys. Rev. Lett.*, **110**(23), 238301 (**2013**).
[117] Wang, W.; Duan, W.; Zhang, Z.; Sun, M.; Sen, A.; Mallouk, T. E. *Chem. Commun.*, **51**(6), 1020 (**2015**).
[118] Wang, J. *Lab Chip*, **12**(11), 1944 (**2012**).
[119] Orozco, J.; Jurado-Sánchez, B.; Wagner, G.; Gao, W.; Vazquez-Duhalt, R.; Sattayasamitsathit, S.; Galarnyk, M.; Cortés, A.; Saintillan, D.; Wang, J. *Langmuir*, **30**(18), 5082 (**2014**).
[120] Gao, W.; Feng, X.; Pei, A.; Gu, Y.; Li, J.; Wang, J. *Nanoscale*, **5**, 4696 (**2013**).
[121] Li, J.; Singh, V. V; Sattayasamitsathit, S.; Orozco, J.; Kaufmann, K.; Dong, R.; Gao, W.; Jurado-Sanchez, B.; Fedorak, Y.; Wang, J.; Diego, S.; Fellow, M. C. *ACS Nano*, **8**(11), 11118 (**2014**).
[122] Soler, L.; Magdanz, V.; Fomin, V. M.; Sanchez, S.; Schmidt, O. G. *ACS Nano*, **7**(11), 9611 (**2013**).
[123] García, M.; Orozco, J.; Guix, M.; Gao, W.; Sattayasamitsathit, S.; Escarpa, A.; Merkoçi, A.; Wang, J. *Nanoscale*, **5**(4), 1325 (**2013**).
[124] Gao, W.; Dong, R.; Thamphiwatana, S.; Li, J.; Gao, W.; Zhang, L.; Wang, J. *ACS Nano.*, **No. 1**, 117 (**2015**).
[125] Wang, H.; Zhao, G.; Pumera, M. *Chem. — A Eur. J.*, **19**, 16756 (**2013**).
[126] Soler, L.; Martínez-cisneros, C.; Swiersy, A.; Sánchez, S.; Schmidt, O. G. *Lab Chip*, **13**(22), 4299 (**2013**).
[127] Dey, K. K.; Zhao, X.; Tansi, B. M.; Méndez-Ortiz, W. J.; Córdova-Figueroa, U. M.; Golestanian, R.; Sen, A. *Nano Lett.*, **15**(12), 8311 (**2015**).
[128] Elgeti, J.; Winkler, R. G.; Gompper, G. *Reports Prog. Phys.*, **78**(5), 056601 (**2015**).

Chapter 9

Janus Particles for Biomedical Applications

*Yi Yi, Kwahun Lee, Lucero Sanchez, and Yan Yu**,†

Department of Chemistry, Indiana University,
Bloomington, IN 47405, USA
*yy33@indiana.edu

Abstract

Recent years have witnessed a rapid development of novel Janus structures and exploration of their biomedical applications. Unlike conventional homogeneous particles, the surface or structural anisotropy of Janus particles (JPs) allows different or even incompatible functionalities to be combined within a single particle unit. As a result of such anisotropy, novel biomedical applications were made possible. Recent advances include synergistic drug delivery, cell targeting, multimodal bioimaging, biosensing, self-propelled therapeutic motors, and various combinations of these capacities. These examples, which are highlighted in this chapter, are likely to be just the beginning of a new era of JP-based therapeutics. This chapter also discusses the challenges that must be overcome to realize the many unique opportunities presented by JPs.

Keywords: Janus particles, dual-imaging, multifunctional, bioanalytical, therapeutic delivery.

† Corresponding author

9.1. Introduction

"The whole is greater than the sum of its parts."

Aristotle

Synthetic nanoparticles are an innovation that has transformed numerous biomedical fields. The emergence of multifunctional nanomaterials is of particular importance, since when many parts work together to perform multiple functions, their coordinated action can produce entirely new synergistic effects. The potential therapeutic use of multifunctional nanoparticles appears extremely promising for applications combining functions such as drug delivery, diagnostic imaging, cell therapy, and molecular sensing [1–5]. A single nanoparticle can be designed to use chemical cues to deliver drugs to a specific target region in the body, while also monitoring their release and efficacy. The inherent magnetic, electric, and optical properties of nanomaterials can also facilitate additional applications [4–7]. However, creating multifunctional nanomaterials is not a simple matter of assembling different parts into one. The base structure of a nanoparticle must be capable of carrying the multifunctional units. More importantly, the assembly process for the different building blocks must not compromise the function of the others. Some functions, such as cell targeting, molecular sensing, and *in vivo* imaging, interfere with one another when they are combined on the same surface. It is thus desirable to keep these functions spatially separated while combining them in the same structural unit. JPs are a possible solution to this design challenge. The surface of JPs is anisotropic and divided into multiple compartments. Sensing and targeting functions can thus be performed by spatially separated domains. Drugs of different water solubilities can coexist in different compartments of a single JP. Drug carriers can also contain a compartment for delivering chemotherapy drugs and another for delivering reagents to reduce drug resistance. This provides more than just an efficient way to combine multiple applications; it allows entirely new functions to be performed due to synergetic effects between the individual units. The aim of this chapter is to highlight recent advances in the biomedical uses of JPs. The new applications include synergistic drug delivery, cell targeting, multimodal bioimaging, biosensing, remote control of biological functions, self-propelled therapeutic motors, and combinations of these capacities.

Some unique opportunities presented by JPs, and some challenges that must be overcome to realize their broad application will be discussed.

9.2. Biomedical Applications of Janus Particles

9.2.1. Drug Delivery and Release

9.2.1.1. *Synergistic codelivery of drugs*

When therapeutic agents are delivered in a spatiotemporally coordinated fashion, new synergistic effects emerge. In combination therapy, drugs, RNA, and DNA may be codelivered. This emerging strategy has been shown to enhance the therapeutic efficacy of medications [8,9]. For instance, the efficacy of chemotherapeutic drugs in cancer treatment is often limited by the drug resistance of tumor cells. A promising solution to this problem is to codeliver small interfering RNAs (siRNAs) that knock out the genes that facilitate drug resistance. However, siRNAs are polymers with many charged regions, which make them hydrophilic, whereas chemotherapeutic drugs are usually small hydrophobic molecules. Effectively combining substances with different water solubilities in a single carrier has posed a major challenge. Compartmentalized JPs offer a potential solution to this obstacle [10–19]. For example, if the polarity of each compartment of a JPs can be independently adjusted to match the polarity of the drugs, therapeutic reagents that differ in water solubility can be loaded, transported, and delivered with a single carrier.

The creation of polymeric Janus beads that contained two compartments was first reported in 2005 [10]. The particles were fabricated using an electrohydrodynamic cojetting process. In this method, parallel streams of two polymer solutions become electrically charged by exposure to a strong electrical field. The streams deform into strings of micrometer-sized droplets due to interfacial tension. The resulting droplets contain both polymer solutions, one in each compartment. This fabrication approach allows the synthesis of JPs that are difficult to make in other ways (Fig. 1(a)). By way of the cojetting fabrication process, it is possible to control the composition of each compartment of the particle. For example, particles have been made with portions composed of biocompatible and biodegradable polymers [20]. Biologically active

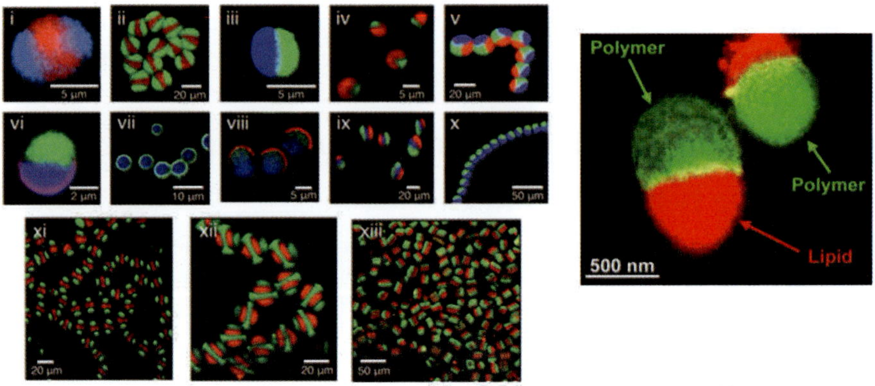

Figure 1. (a) Fluorescence images of representative multicompartmental polymer particles made from the electrohydrodynamic cojetting. Reprinted with permission from Ref. [20]. Copyright (2011) Wiley-VCH Verlag GmbH & Co. KGaA. (b) Fluorescence image of bicompartmental Janus nanoparticles (JNPs) containing a PLGA phase labeled with FITC (shown in green) and a lipid phase labeled with DiR (shown in red). Reprinted with permission from Ref. [24]. Copyright (2014) American Chemical Society.

reagents, such as dyes or drugs, can be loaded specifically into each structural compartment. Unfortunately, particles produced via this technique are typically larger than 200 nm and polydisperse in size, which may limit their usefulness for drug delivery. Compartmentalized JPs have also been made using other methods including microfluidics,[7] fluidic nanoprecipitation [21], and seeded-emulsion polymerization.[8] Particles containing up to seven compartments have been made by a method combining electrified cospinning and microsectioning [22,23].

The promise of compartmentalized JPs for the codelivery of drugs or gene vectors has been demonstrated in several recent studies. In a study by Garbuzenko *et al.*, bicompartmental JPs were synthesized to codeliver the hydrophobic drug curcumin and the hydrophilic drug doxorubicin for treating lung cancer [24]. The JPs consisted of a relatively hydrophilic poly (lactic-*co*-glycolic acid) (PLGA) compartment and a more hydrophobic lipid compartment to encapsulate the two drugs separately (Fig. 1(b)). The JPs loaded with both drugs were found to be more effective in inhibiting lung tumor progression than nanoparticles loaded with either curcumin or doxorubicin alone. This codelivery strategy could be beneficial for treating cancer cells that develop resistance to a single chemotherapeutic agent over time and also diseases that require synergistic functions of two or more drugs.

Independent control over the release of drugs from the compartments of single JPs can be achieved with external stimuli such as pH [14,15], light [25], and oxidative stress [25]. Lahann *et al.* used the different degradation rates of polymers to construct JPs that release drugs at differential rates (Fig. 2(a)). The base material for the particles was poly(acrylamide-*co*-acrylic acid) (p(AAm-*co*-AA)), but poly(ethylene oxide) (PEO) was mixed into the base material for one hemisphere and dextran into that for the other. Both compartments were stable at pH 3.0, but only the PEO-containing compartment degraded at pH 7.4. This differential degradation is useful for releasing orally administrated drugs at physiological pH after passing through the low pH environment of the stomach [14]. The same design concept is also applicable to other polymeric materials [19,26]. With properly designed chemistry, differential drug release has also been achieved using heat or light. A combination of heat and light was used by Zhao *et al.* for a two-step controlled release system based on JPs (300 nm) containing two mesoporous silica compartments (Fig. 2(b)) [26]. One side was a core/shell/shell structured silica (SiO_2) nanosphere. The other compartment was a periodic mesoporous organosilica (PMO) single-crystal nanocube. The two hemispheres differed in their mesoporous structure, pore size (2.1 nm for nanosphere, 3.5–5.5 nm for nanocube), and surface chemistry (hydrophobic for the nanosphere and hydrophilic for the nanocube). Those differences rendered such a Janus nanocomposite ideal for loading drugs of different molecular weights and water solubilities. As a proof-of-concept demonstration, the hydrophobic drug molecule paclitaxel and the hydrophilic drug doxorubicin were loaded into separate compartments and then sealed with azobenzene (light sensitive) and 1-tetradecanol (heat sensitive), respectively. The drugs were selectively released using heat or light.

In addition for codelivery and differential release from compartments, the anisotropic surfaces of JPs can be further functionalized for specific cell targeting. Coating ligands onto only one side of JPs leads to unidirectional binding of the particles to human endothelial cells.[7] The synergistic

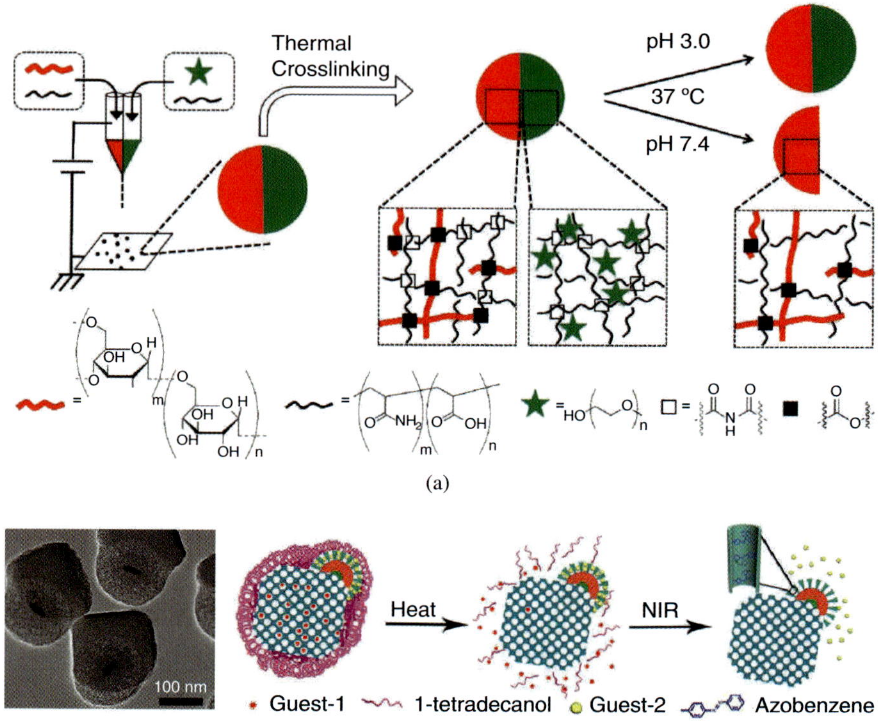

(a)

(b)

Figure 2. Independent control of drug release from bicompartmental JPs. (a) Schematic illustration of JPs containing two compartments with different degradation rates. Reprinted with permission from Ref. [14]. Copyright (2012) Wiley-VCH Verlag GmbH & Co. KGaA. (b) (Left) TEM image showing JPs that consist of an upconversion nanoparticle (UCNP)/SiO_2/$mSiO_2$ nanosphere and POM mesoporous JPs. (Right) Schematic illustration of the heat and NIR-triggered drug release from the JPs. Reprinted with permission from Ref. [26]. Copyright (2014) American Chemical Society.

yet spatially separated presentation of multiple ligands was also made possible using Janus microdisks consisting of three compartments with different surface chemistries [18]. In a recent study, Janus nanocylinders (50–100 nm) were developed to achieve targeted codelivery of drugs to breast cancer cells [17]. The nanocylinders consisted of a polyethyleneimine (PEI) compartment for endosome targeting and a poly(D,L-lactide-*co*-glycolic acid) (PLGA) compartment for encapsulating drugs that inhibit chemokine signaling of triple negative breast cancer cells. The optical and magnetic properties of the building blocks in compartmentalized particles have also been used to incorporate more capabilities into the drug-delivery JPs. For example, bicompartmental nanoparticles were synthesized by Lahann and coworkers for the delivery of siRNA and the imaging of cancer cells [15]. The PLGA compartment of the nanoparticles contained dye molecules for fluorescence imaging, whereas the PEI hydrogel compartment released encapsulated siRNA at low pH.

9.2.1.2. *Combining drug delivery with other biomedical functions*

The versatility of JPs allows them to be used to perform other functions in coordination with the delivery of drugs to a targeted site in the body. Since the ability to perform these additional functions is a consequence of the geometry of the particle, and its method of production, these additional functions are discussed here in relation to the shape of the particle.

Janus spheres. Spherical particles are the most common, because droplets minimize surface tension compared with other shapes of the same volume. Gold (Au), in the form of nanoparticles or thin films, is a commonly used material in the production of JPs capable of performing multiple functions, because gold surfaces have unique chemical, optical, and photothermal properties [27,28]. Au-thiol chemistry is versatile and enables easy surface conjugation of biomolecules [29]. The plasmonic properties of gold surfaces significantly enhance signals from Raman spectroscopy, which can therefore sensitively detect molecules adhering to them. This approach is known as surface-enhanced Raman spectroscopy (SERS) [30,31]. Gold nanoparticles also produce heat when excited by near infrared (NIR) light, making them useful for photothermal therapy [32]. Rayleigh scattering from gold nanoparticles also renders them useful as non-bleaching labels for optical imaging of cells [33,34]. Hsieh *et al.* reported a study on using JPs with gold caps for drug delivery, SERS/fluorescence imaging, and protein targeting [35]. To create the JPs, fluorescent polystyrene nanoparticles (250–500 nm) were coated on one hemisphere with a thin layer of gold using metal evaporation. The fluorescent hemisphere of the particles was conjugated with anti-CD44 for specific targeting of cancer cells that overexpress CD44 proteins. The gold surface was modified with cleavable disulfide linkers (biotin-SS-NHS-sulfo) for further conjugation of drug molecules. The disulfide linkers were cleaved in the cell cytoplasm to release the drug. Meanwhile, these JPs also enabled simultaneous SERS and fluorescent imaging in cells. In another notable example, polymeric JPs half-coated with gold showed promise for combined chemo- and photothermal treatment of rheumatoid arthritis [36]. The JPs were made by depositing a layer of Au on one side of PEGylated PLGA particles that were pre-loaded with the drug molecule methotrexate. Arginine-glycine-aspartic acid (RGD) tripeptides for targeting specific cell types were then conjugated onto the Au surfaces. *In vivo* NIR absorbance imaging, enabled by the gold cap on JPs, indicated that the JPs were selectively accumulated in the inflamed joints in arthritic mouse models. NIR radiation on the gold caps of the JPs also generated local heating and degraded PLGA, which led to the release of the drug. The synergistic execution of chemo- and photothermal therapies was effective in damaging inflamed tissue, which cannot be done with using either therapy alone.

Sliver, like gold, is a plasmonic material that can enable SERS detection of JPs [37]. For example, real-time SERS monitoring of drug release was achieved by using JPs that consisted of a mesoporous silica rod on one side and a silver (Ag) nanoparticle on the other side [38]. The drug release was triggered by lowering pH. By using magnetically responsive Fe_3O_4 nanoparticles instead of gold, Wang *et al.* prepared polymeric JPs (~300 nm) that consisted of a silica cap with embedded Fe_3O_4 nanoparticles for tumor cell targeting, drug release, and magnetic hyperthermia therapy (Fig. 3) [39]. The pH-sensitive hydrazone linker was used to conjugate the drug doxorubicin on the silica hemisphere for pH-triggered release.

Dumbbell-like Janus particles. Dumbbell-shaped JPs are typically formed by merging two separate particles that have different material properties into one. This strategy is not specific to a given type of particle materials, so it theoretically allows JPs to be tailor-made with the desired multifunctionality. Particles made via this method can be as small as 50 nm and monodisperse in size distribution because the size of the

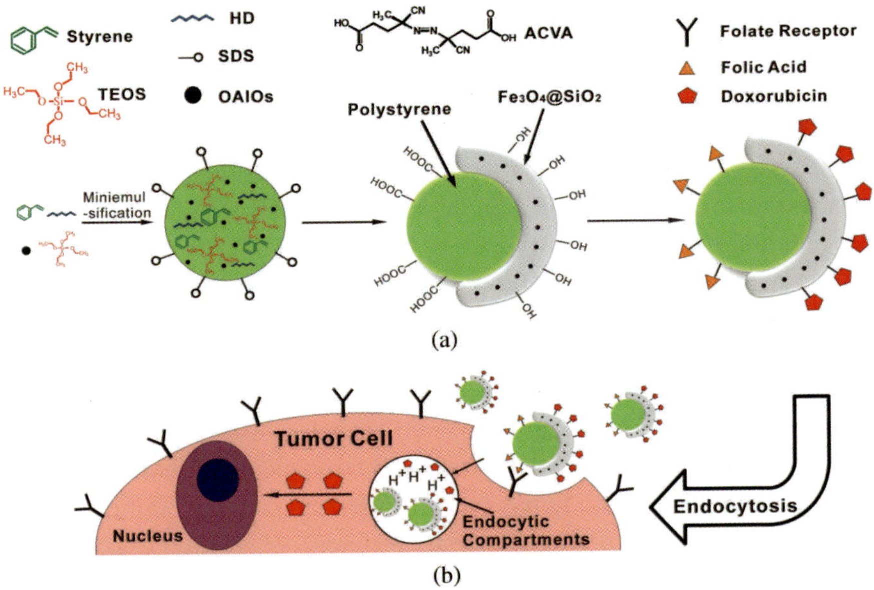

Figure 3. Schematic illustration of the synthesis (a) and targeted drug release (b) of JPs. The polystyrene-based nanoparticles are partially coated with a SiO_2 shell doped with magnetically responsive Fe_3O_4 nanoparticles. Surface-attached drug molecules are released by the pH-triggered dissociation of hydrazone linkers. The magnetic response of Fe_3O_4 nanoparticles enables magnetic hyperthermia therapy. Reprinted with permission from Ref. [39]. Copyright (2013) American Chemical Society.

constituent particles is independently controlled. However, the major technical challenge of preparing dumbbell-like JPs is attaining precise control over the stoichiometry of the constituent particles. Sun *et al.* synthesized monodisperse dumbbell-like Au/Fe_3O_4 JPs (2–8 nm for Au and 4–20 nm for Fe_3O_4) for targeted drug delivery to breast cancer cells [40,41]. The anticancer therapeutic drug, cisplatin, was selectively coupled onto the Au surface. Ligands specific for HER2 receptors, which are overexpressed on cancer cells, were conjugated onto the Fe_3O_4 particle surface to achieve cell-specific targeting. *In vitro* cell study showed that the cisplatin-loaded nanoparticles specifically targeted HER2 positive Sk-Br3 cancer cells and were internalized by the cells through receptor-mediated endocytosis. Release of the drug was triggered by the low pH inside endosomes. Because of its high specificity in targeting and drug release, this drug delivery system required only half of the usual concentration of chemotherapeutic drug to induce cancer cell death. The photothermal properties of the gold nanoparticles in dumbbell-like Au/mesoporous silica JPs has also been exploited to achieve simultaneous chemotherapy and photothermal therapy [42]. Gold nanoparticles generate intense local heat upon exposure to NIR radiation. In addition to encapsulating drugs, the mesoporous silica side of the JPs was also functionalized with tumor-targeting lactobionic acid (LA). *In vivo* studies showed that the synergistic chemo- and photothermal therapies were more effective in suppressing tumor growth in mice models than therapies using free drug molecules or JPs without NIR radiation.

In addition to drug release triggered by external stimuli, such as pH, light, heat, or ultrasound, enzyme-triggered drug release is another interesting strategy. A recent study has reported the application of dumbbell-like JPs for the triggering of drug release by the presence of two enzymes [43]. The JP contained a mesoporous silica nanoparticle (97 ± 15 nm) linked to a gold nanoparticle (20 nm) (Fig. 4). The pores of the silica served to encapsulate the drug, and the gold facilitated surface attachment of enzymes. The enzymes glucose oxidase and esterase were used in that study. The encapsulating pores of the mesoporous silica nanoparticles were capped by β-cyclodextrin (β-CD). β-CD is pH-responsive. It opens the pores at low pH to release the encapsulated drugs. When proper enzyme substrates were encountered, glucose oxidase and esterase functioned to produce gluconic acid and butyric acid, respectively. The generation of the two acidic products lowered pH locally and triggered the opening of the β-CD caps to release the encapsulated drug molecules. The JP device was also made to self-reset by incorporating another enzyme, urease, onto the gold nanoparticle. Urease catalyzes the hydrolysis of urea to CO_2 and NH_3, leading to an increase in pH, which shuts off the release of the drug.

Multisegmented nanorods. Rod-shaped particles, compared to spherical ones, are more effective at entering cells, which makes them potentially advantageous for some applications that take place inside cells [44]. Nevertheless, the majority of the studies on multisegmented nanorods have been focused on the biosensing and imaging applications, which will be discussed in the next section. Salem *et al.* reported a few studies that

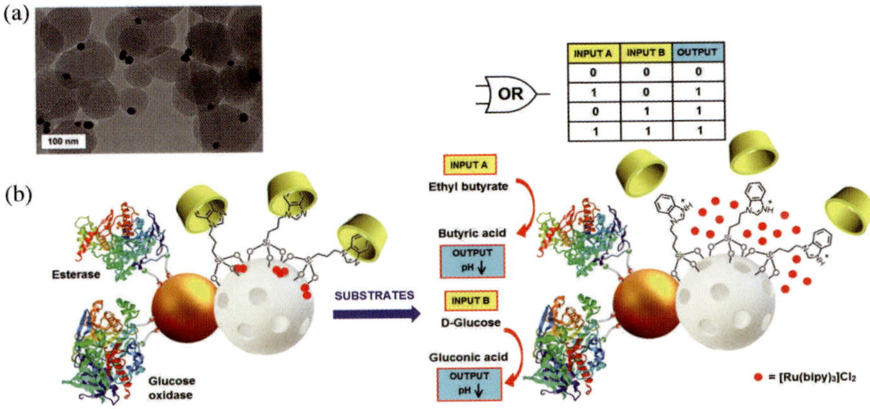

Figure 4. Enzyme-triggered drug releases from dumbbell-like JPs. (a) TEM image showing the JPs that consist of a mesoporous silica nanoparticle on one side and gold nanoparticle on the other. (b) Schematic illustration of drug release triggered by the enzymatic reactions of esterase and glucose oxidase with their respective substrates. The two reactions generate two acidic products: gluconic acid and butyric acid. The reduced pH triggers the detachment of the β-CD and thus opens up the pores for drug release from the mesoporous silica nanoparticles. Reprinted with permission from Ref. [43]. Copyright (2014) American Chemical Society.

demonstrate the potential of Janus nanorods for gene delivery. The Janus nanorods were 100 nm in diameter and contained segments of Au and nickel (Ni), each 100 nm in length [45,46]. DNA plasmids were conjugated to the surface of the Ni segment, and the cell-targeting ligand transferrin was tethered to the surface of the Au segment (Fig. 5). The transfection efficiency of these dual-functional Au/Ni Janus rods was enhanced many-fold by comparison with nanorods functionalized with DNA plasmids alone. This was likely due to the high targeting specificity of transferrin.

9.2.1.3. *Micro/nanomotors as self-propelled drug delivery vehicles*

Self-mobility has been investigated as a possible means to facilitate a drug delivery nanoparticle's ability to reach its target within the body.

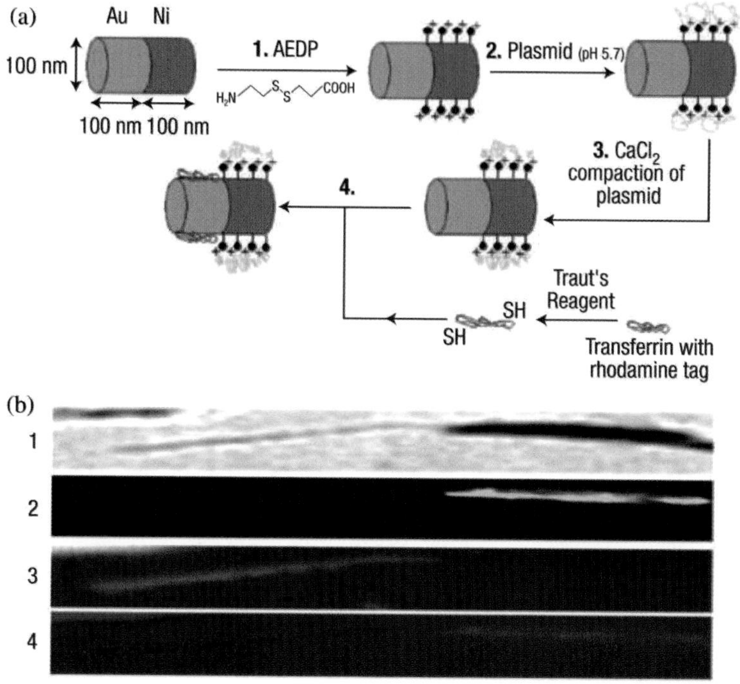

Figure 5. Janus nanorods for gene delivery. (a) Schematic illustration showing the composition and selective surface functionalization of the Janus Au/Ni nanorods. DNA plasmids are attached on the surface of the Ni block and the cell-targeting ligand transferrin is tethered on the Au block. (b) Wide-field and fluorescence images showing the Au/Ni nanorod 20 μm in length (Image 1), dye-labeled transferrin on the nanorod (Image 2), dye-labeled DNA plasmids on the nanorod (Image 3), and an overlay image of Images 2 and 3. Reprinted with permission from Ref. [45]. Copyright (2003) Nature Publishing Group.

Micro/nanomotors are synthetic particles that move autonomously. They are inspired by bacteria, which locomote by means of flagella [47,48]. Two sorts of self-propulsion have so far been demonstrated for nanoparticles. The first is chemical propulsion by catalytic reactions that generate gases [49]. The second is propulsion by an applied external force, such as a ultrasound [50–55], magnetic fields [56–60]. or a combination of both [61]. Micro/nanomotors are promising candidates for a variety of important medical and non-medical applications, including intelligent biomedicine [49, 62–64], efficient energy storage [65], water remediation, and environmental monitoring [66,67]. The first reported nanomotors were designed based on Janus Au/Pt nanorods. The nanomotors were propelled by oxygen gas generated by the catalytic reaction between the platinum (Pt) segment and an aqueous hydrogen peroxide (H_2O_2) solution [68]. Following this groundbreaking study, significant progress has been made in the development of Janus and heterostructured nanomotors as platforms for novel biomedical applications [52–54,69–71]. Most studies reported so far have been focused on the fabrication of micro/nanomotors and demonstration of techniques. The range of possible propulsion mechanisms and the tremendous potential of self-propelled particles in biomedical applications so far remain largely unexplored. A few possibilities that have been investigated illustrate the feasibility of self-propelled nanodevices for drug delivery. They are discussed below.

Chemically powered micro/nanomotors. Among the many catalytic reactions used for providing a power source for micro/nanomotors, the decomposition of aqueous H_2O_2 is the mostly commonly used. In the presence of a suitable catalyst, such as Pt or silver (Ag), H_2O_2 rapidly decomposes into water and oxygen, with oxygen propelling the locomotion of the nanoparticle. This design strategy has been utilized for developing nanomotors for drug delivery [69,72]. For example, Xuan *et al.* presented a self-powered drug delivery system based on JPs that consisted of mesoporous silicate nanoparticles (75 nm) with chromium (Cr)/Pt caps [71]. The anti-cancer drug, doxorubicin hydrochloride, was encapsulated inside the mesoporous silica. To generate gases to propel the nanomotors, 0.2% H_2O_2 was added to the cell culture medium to react with the Pt coating on the particles. The self-propulsion of the JPs was shown to enhance the efficiency of drug delivery into HeLa cancer cells. However, this work also illustrates a major roadblock to the biomedical use of micro/nanomotors. The external power source used, H_2O_2, is toxic to living cells. This issue of cytotoxicity limits the choice of fuels and reactions for nanomotors that must function in a biological environment.

To avoid the use of toxic chemicals as fuels for nanomotor propulsion, Wang *et al.* developed micromotors powered by water. The micromotors were composed of gallium (Ga)–aluminum (Al) alloy microspheres (20 μm) partially coated with titanium (Ti) [73].

Aluminum reduces water to produce hydrogen, which acts as a propellant gas. This reaction is typically hindered by the rapid formation of an oxide passivation layer on the Al surface. To prevent this, Ga was added to form an alloy with Al. The ejection of hydrogen bubbles propelled the directional locomotion of the micromotors. This micromotor system was shown to function in a variety of biological media, including cell culture media, phosphate-buffered saline, and human serum. Mou *et al.* later demonstrated that the metal–water reaction can also be used to propel magnesium (Mg)/Pt Janus micromotors (20 μm) [74]. In this study, the H_2 gas that supplied motive force was generated from the reaction between Mg and water. The reaction also produces $Mg(OH)_2$, which has the effect of passivating the surface and preventing further reaction between Mg and water. The investigators used a strategy similar to that used by Wang *et al.* to prevent this passivating reaction. They added aqueous $NaHCO_3$ to the medium. Follow-up work from the same group demonstrated the potential of using Mg/Pt micromotors to actively transport and deliver drugs in biological media [75]. The Pt surface of the Mg/Pt micromotor was coated with a 50 μm layer of poly(*N*-isopropylacrylamide) (PNIPAM) in order to encapsulate and release anti-cancer drugs (Fig. 6). The ability to use micromotors for *in vivo* drug delivery was recently demonstrated for the first time with the successful delivery of a drug to the stomach of mice [76]. The micromotors were tubular and were made up of an exterior layer

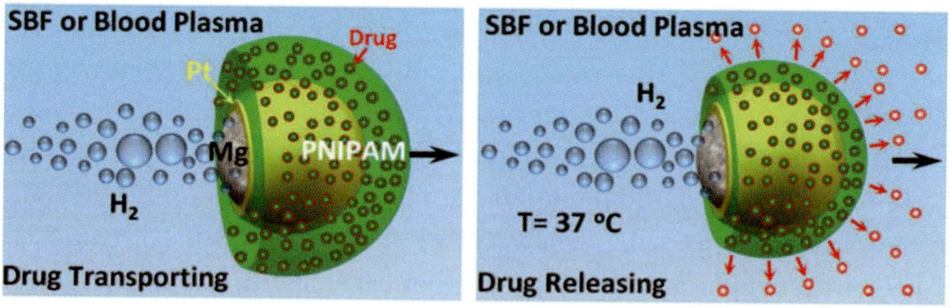

Figure 6. Schematic illustration of self-propelled Janus micromotors for drug transport and release in biological media. The reaction between Mg and water generates H_2 gas that propels the micromotors. Drug molecules were encapsulated in the PNIPAM layer (shown in green) and released upon the change of temperature. Reprinted with permission from Ref. [75]. Copyright 2014 American Chemical Society.

of poly(3,4-ethylenedioxythiophene) (PEDOT) and an interior layer of zinc. The zinc layer converts protons in the acidic stomach fluid to hydrogen gas. A PEDOT polymer is a non-cytotoxic material used to prolong the hydrogen producing lifetime the zinc layer. This study demonstrated that acid-driven propulsion of micromotors enhances the binding and retention of the particles as drug carriers.

Another strategy to avoid using toxic fuels, such as H_2O_2, in micro/nanomotor systems is to combine chains of catalytic reactions in which a toxic fuel chemical is produced as an intermediate from harmless substances via a reaction step, but is immediately eliminated by subsequent reaction steps. This strategy has been demonstrated using enzymatic reactions. As noted above, the degradation of H_2O_2 by catalase into water and oxygen can be used to power micro/nanomotors [77], but a relatively high concentration of H_2O_2 fuel is required to generate sufficient propulsive force for locomotion. This problem can be eliminated by producing H_2O_2 at the point of use by an enzymatic reaction using glucose oxidase. Glucose oxidase catalyzes the oxidation of glucose found in living tissue into H_2O_2 and D-glucono-δ-lactone. By combing catalase and glucose oxidase on one hemisphere of JPs, only the non-toxic glucose instead of H_2O_2 was required as external fuel [63]. The Janus micromotors were even shown to exhibit enhanced propulsion, due to the reaction gradient from the asymmetric coating of enzymes on one hemisphere of the JPs. A similar idea was demonstrated in another study, in which JP nanomotors were coated with urease, glucose oxidase and catalase on one hemisphere [62]. Urease catalyzes the hydrolysis of urea into CO_2 and NH_3, providing propulsive forces in addition to those from the reactions of glucose oxidase and catalase. Besides enzymes, the motility of particle-adhered bacteria can also be harnessed as a form of power source for micromotors [78].

Fuel-free micro/nanomotors. Another approach to avoiding the cytotoxicity issue of many chemical fuel-powered micro/nanomotor systems is to design fuel-free motors that are able to move under the power of external influences, such as ultrasound and magnetic fields. Micromotors (5.8 μm in length, 200 nm in diameter) containing Ni and Ag segments were guided by a magnetic field to load and transport drug-encapsulating PLGA microparticles to *in vitro* HeLa cancer cells (Fig. 7) [58]. Multiple

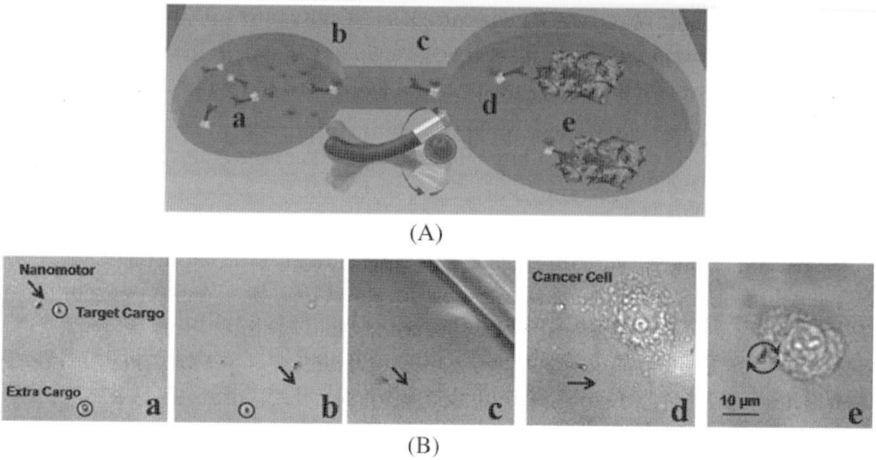

Figure 7. Fuel-free micromotors for drug delivery. (A) Schematic illustration of the magnetic control of Ni/Ag micromotors for drug delivery to HeLa cells. (B) A micromotor is controlled to capture, transport, and deliver drug-loaded PLGA microparticles. Arrows indicate the moving direction of the micromotor. Reprinted with permission from Ref. [58]. Copyright 2011 Wiley-VCH Verlag GmbH & Co. KGaA.

external forces have been combined to achieve precise control of the movements of micro/nanomotors. Wang *et al.* reported a study in which Au/Ni/Au triblock nanorods (0.8/0.2/0.8 μm in length, 250 nm in diameter) were propelled by ultrasound and guided by a magnetic field to capture and transport bacteria [52]. The two Au segments were selectively functionalized with lectin and anti-protein A antibody to make them capable of binding to *Escherichia coli* and *Staphylococcus aureus* bacteria, respectively. The Ni segment rendered the nanorod motors magnetically responsive. Polymeric or mesoporous compartments can also be linked to one end of Au/Ni/Au nanorods to achieve pH- or light-triggered drug release after the nanomotors successfully reach their cell targets [52,54]. The Mallouk group also combined ultrasound and magnetic fields to control the locomotion of Au/Ni/Ru motors for purpose of targeting HeLa cancer cells *in vitro* [53]. They demonstrated for the first time that nanorods motors propelled by low-power ultrasound could operate inside living cells [55]. Most studies so far have focused on the fabrication of micro/nanomotors, with only a relative few dealing with biological applications. Nevertheless, these pioneering studies have blazed a path toward realizing many novel biomedical applications of synthetic motors.

9.2.1.4. *Janus protein constructs for drug delivery*

Non-traditional JPs have been made by the assembly of proteins. As naturally occurring build blocks, proteins offer advantages including uniformity of size and composition, high biocompatibility, and the capacity for site-specific structural modification. Janus protein nanocages (9 nm) have been engineered via topographically selective modifications of their surfaces [79]. This approach is similar to the methods used to generate more conventional JPs by protecting portions of particle surfaces from modification. Protein cages were immobilized on solid substrates and their top exposed surface was modified. To demonstrate the microbial targeting capacity of Janus protein cages, ligands specific for the microbial pathogen, *S. aureus*, were attached to one side of the Janus protein cages via a streptavidin–biotin linkage (Fig. 8(a)) [80]. The protein cages are useful for targeted drug delivery and bioimaging because they can encapsulate cargo and undergo site-specific modifications of their surfaces. A separate study has demonstrated that Janus protein cages can be created via the surface functionalization of single sites on virus-based nanoparticles (Fig. 8(b)) [81]. Weil *et al.* fused a transporter protein with a cargo protein using a non-covalent iminobiotin–avidin linker [82]. These were, in effect, Janus protein complexes (Fig. 8(c)). Studies indicated that these Janus protein complexes were able to enter into cells easily while preserving the enzymatic activity of the proteins, and could release cargo proteins in the acidic tumor environment due to the selective dissociation of the iminobiotin linkers under those conditions.

9.2.1.5. *Janus particle-enabled cellular backpacks*

Certain cells, such as immune cells, naturally migrate to specific tissues. This homing property of cells can be harnessed to carry and deliver immunoregulatory drugs for cancer treatment [83,84]. In 2008, Rubner *et al.* first reported the creation of a disk-like polymeric Janus microparticles that were able to attach to the surface of living immune cells without affecting their normal functioning [85]. These "cellular backpacks" were composed of multiple layers of polyelectrolytes that were held together by hydrogen bonding and electrostatic attraction [85,86]. The "cellular backpacks" consisted of three polymer layers with distinctive functions: a cell-adhesive layer to anchor the backpack on the cell surface, a payload layer to

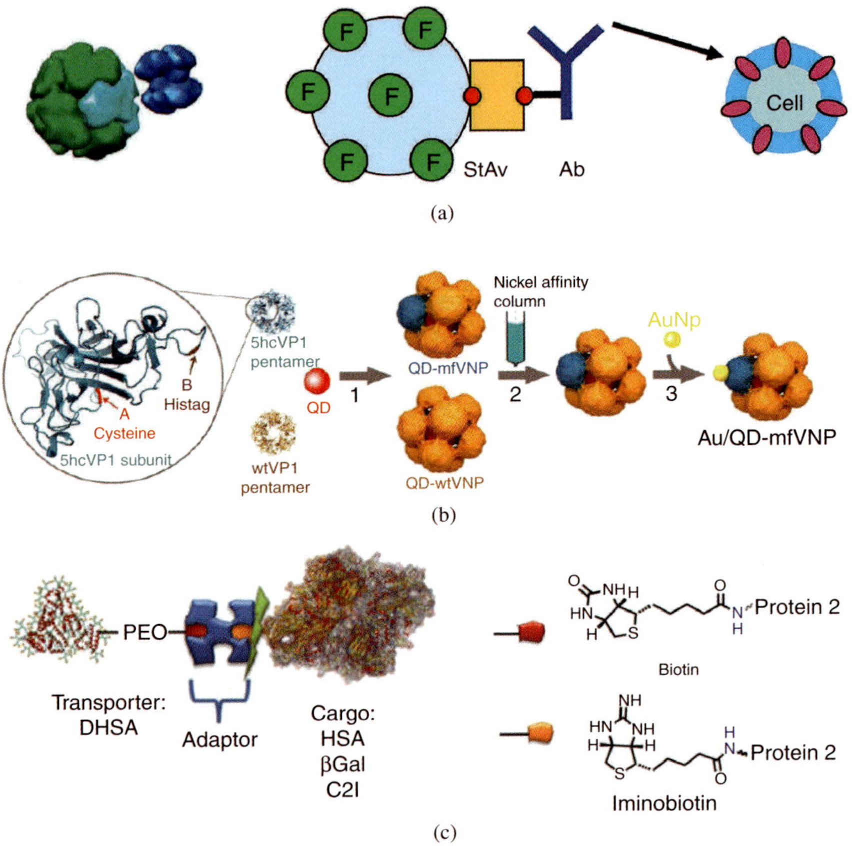

Figure 8. Schematic illustration of representative Janus protein constructs. (a) A Janus protein cage based on streptavidin (StAv)-functionalized LiDps (the DNA-binding protein from Listeria innocua), which is conjugated with a biotinylated mAb for targeting *S. aureus*; (b) A Janus protein cage based on monofunctionalized cysteine-containing virus-based nanoparticle of simian virus 40. The Janus protein cages were prepared in three steps. Step 1: Coassembly of 5hcVP1, wtVP1 and quantum dots (QDs). Step 2: Isolation and purification of Janus protein nanocages by nickel affinity chromatography and sucrose density gradient centrifugation. Step 3: Au nanoparticle binding assay to confirm the Janus structure of the protein cages; (c) pH-responsive Janus supramolecular fusion proteins from the assembly of an avidin-modified transporter protein, dendronized human serum albumin (DHSA), with different cargo proteins. The dimer of DHSA–avidin was made by toposelective phase immobilization using biotin–PEO–maleimide as the linker. Cargo proteins were coupled to DHSA-avidin through pH-sensitive avidin–iminobiotin interaction. At low pH, cargo proteins were released from the supramolecular protein complexes due to the protonation of iminobiotin. Reprinted with permission from Ref. [80]. Copyright (2009) (a), Ref. [81]. Copyright (2013) (b), and Ref. [82]. Copyright (2011) (c) American Chemical Society.

encapsulate therapeutic cargos, and a magnetic region doped with iron oxide (Fe_3O_4) nanoparticles to allow remote manipulation, *in vivo* tracking, and easy separation of cells (Fig. 9(a)). Because of their unique flattened anisotropic shape, these "cellular backpacks" were also resistant to phagocytosis, a clearance mechanism used by immune cells to remove foreign particles. Following this pioneering work, additional studies have been carried out to explore the application of this strategy in cell-mediated

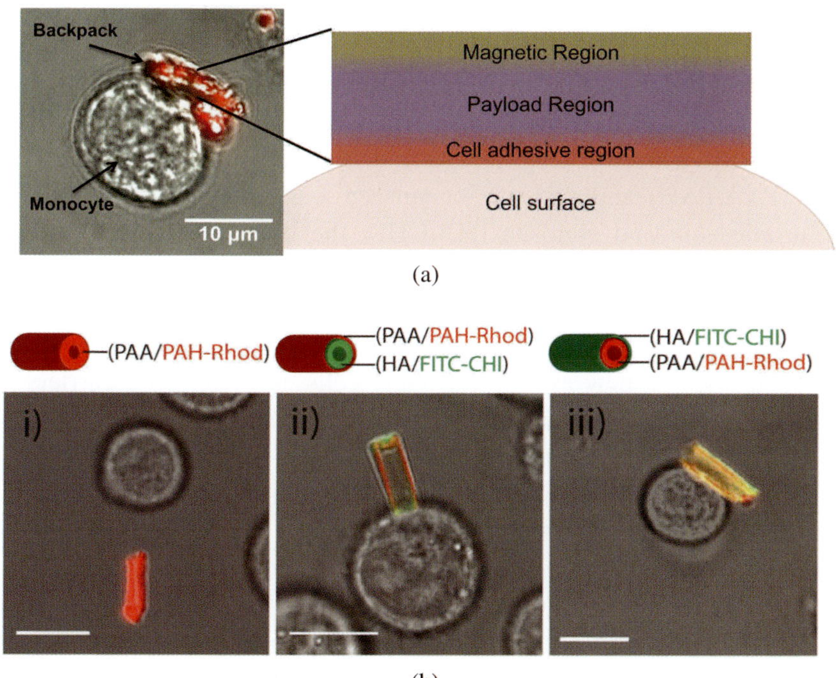

Figure 9. (a) An overlaid bright-field and fluorescence image showing a "cellular backpack" (shown in red) adhered to a mouse monocyte cell. The "cellular backpack" contained three layers for cell adhesion, drug loading, and magnetic response, respectively. (b) Confocal images showing the controlled orientation of three different microtubes on the surfaces of B-cells: (i) microtubes consisted of the cell-repulsive polymer (PAA/PAH-Rhod) 40; (ii) microtubules with cell repulsive (PAA/PAH-Rhod) 40 as the exterior layer and cell-adhesive (HA/FITC-CHI) 40 as interior; (iii) microtubules with cell-adhesive (HA/FITC-CHI) 4 as exterior and cell-repulsive (PAA/PAH-Rhod) 40 as interior. Scale bar 10 μm. Reprinted with permission from Ref. [89]. Copyright 2015 (a) and Ref. [90]. Copyright 2013 (b) Wiley-VCH Verlag GmbH & Co. KGaA.

therapy [87–89]. For example, monocytes with "cellular backpacks" were employed to deliver drugs to inflamed tissues [87]. The monocytes retained their natural ability to migrate to inflammation sites and delivered the drug-loaded backpacks with high targeting specificity. Cargos other than drug molecules, such as liposomes, have also been included in the layer-by-layer structure of "cellular backpacks". This further broadens the potential applications of this approach [89]. In order to control the orientation in which "cellular backpacks" bind to cells, the Rubner group developed hollow tubular particles whose orientation, upon binding to cells, was controlled by the surface chemistry of the particles. The microtubes were composed of layers of polymers that were either cell-adhesive or cell-repulsive [90]. The sequence of these polymer layers determined the orientation of the microtubes on the cell surface. When the exterior surface was a cell-repulsive layer of poly(acrylic acid) and poly(allylamine hydrochloride) (PAA/PAH) and the interior surface was a cell-adhesive layer of hyaluronic acid and chitosan (HA/CHI), the microtubes predominantly attached to B-cells from the ends (Fig. 9(b) (ii)). In contrast, the inverse arrangement of PAA/PAH and HA/CHI caused the microtubes to adhere to cells from the side (Fig. 9(b), (iii)). Overall, the "cellular backpack" approach opens a new avenue for using cells as living drug carriers for therapies that target cancer cells with high specificity [91,92].

9.2.2. Bioimaging

Since the invention of X-ray imaging for visualizing the internal anatomy of a human being in 1895, the development and application of non-invasive imaging techniques has been an important focus of biomedical research. Imaging modalities have been utilized as diagnostic tools for studying the progression of a disease, as monitoring tools for evaluating the efficacy of a therapeutic treatment, or as visualization guides for surgery. On the cellular and molecular scale, imaging techniques are also essential for applications such as understanding the biodistribution and working mechanisms of drug molecules. Being able to directly visualize proteins and subcellular structures is critical for the fundamental understanding of cellular functions. The commonly used bioimaging techniques, from magnetic resonance imaging (MRI) to optical and ultrasound imaging, differ significantly in their resolution, sensitivity, and their applicable environment. Multiple imaging modalities are often combined to offset the limitations of each technology. This is known as multimodality imaging. Nanoparticles have been widely used as multimodality imaging agents because multiple imaging probes can be combined into a single particle structure [93,94]. The large surface area of nanoparticles is available for conjugating ligands to ensure that imaging nanoprobes accumulate selectively at the target site. Imaging nanoprobes can also incorporate other biomedical functions, such as hyperthermia therapy. Whenever several functions are incorporated into single nanoparticles, there is a potential for interference between functionalities. Materials suitable for hyperthermia therapy may not provide the surface chemistry suitable for the conjugation of targeting ligands. As for other nanoparticle applications, this problem can be addressed by spatially separating functions on the particle surface by using JPs.

Fluorescent, magnetic, and plasmonic nanoparticles are those most commonly used for bioimaging. Therefore, the recent development of JPs as multimodality imaging nanoprobes has been mostly focused on dye-doped inorganic nanoparticles [95], quantum dots [96–102]. Au nanoparticles [28,103,104], and superparamagnetic iron oxide nanoparticles (SPION) [105–107] as the basic building blocks. Here, we summarize these studies based on the different types of imaging.

9.2.2.1. *Multimodality imaging*

Fluorescence bioimaging provides unique advantages over other techniques. It is highly sensitive and non-invasive. More importantly, it provides high spatial and temporal imaging resolutions, which are crucial for applications that require real-time monitoring, such as evaluating cell responses to therapeutic treatments. Optically anisotropic Janus nanospheres that emit different colors of fluorescence from their two hemispheres were developed by Lahann *et al.* as imaging probes [11]. The JPs, made via electrified cojetting of two aqueous solutions, contained two compartments labeled fluorescently with either a green dye fluorescein isothiocyanate (FITC) or a red dye tetramethylrhodamine isothiocyanate (TRITC), respectively (Fig. 10). Only the TRITC-labeled hemisphere was coated with biotin for further

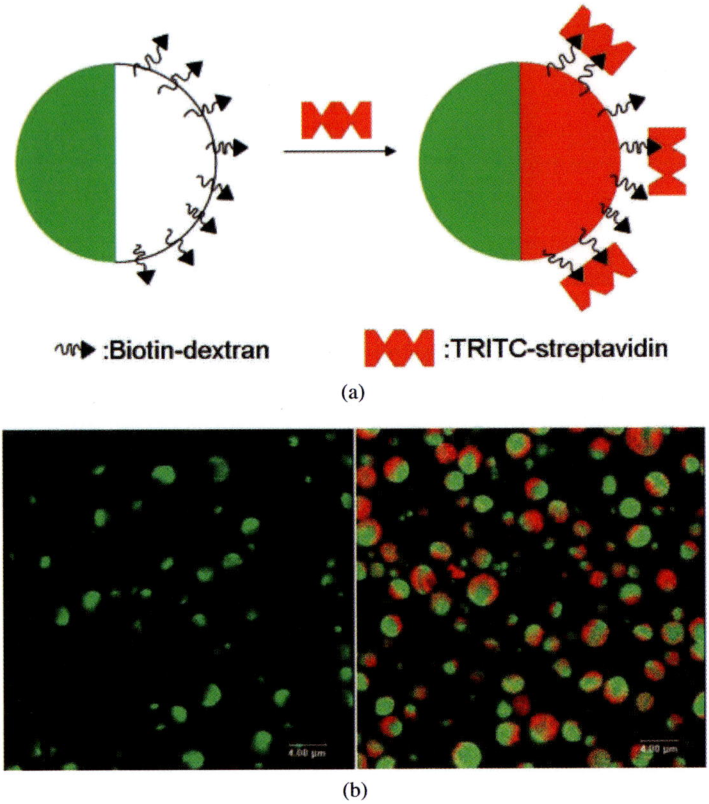

Figure 10. (a) Schematic illustration showing dextran-based JPs as optically anisotropic imaging probes. One hemisphere of the JPs contained FITC (shown in green) and the other hemisphere was coated with TRITC-conjugated streptavidin (shown in red). (b) Confocal fluorescence images showing the dual-labeled JPs. Reprinted with permission from Ref. [11]. Copyright 2007 American Chemical Society.

conjugation of cell-targeting ligands. The distinctive colors of the two hemispheres of the JPs allow the orientation of the particles upon binding to cells to be directly visualized [12].

QDs are widely used imaging nanoprobes owing to their excellent photostability, high quantum yield, and the large selection of colors available for imaging [108–110]. QDs are often combined with magnetic nanoparticles to enable simultaneous fluorescence and magnetic imaging. As shown by Ying *et al.*, QDs and magnetic nanoparticles can be combined with well-defined stoichiometry into dumbbell-like JPs [111]. To make the particles, cadmium selenide (CdSe) nanospheres (4–5 nm) were grown on one side of Fe_2O_3 nanoparticles (8–10 nm). The entire composite particle was subsequently coated with a layer of SiO_2 and then conjugated with poly(ethylene glycol) (PEG). The PEG layer was used to render the nanoparticles water soluble and to reduce non-specific adsorption of proteins. This study was mostly focused on the synthesis of magnetic-QD JPs. Using the same basic idea, Pt-conjugated CdSe QDs were linked to guanine-conjugated iron oxide magnetic nanoparticles [112]. The optical properties of QDs enabled real-time cell imaging, whereas the magnetic properties of the iron oxide nanoparticles enabled MRI imaging. The specificity of cell- and tissue-imaging can be improved by coating the surfaces of imaging probes with targeting ligands. As shown by Sun *et al.,* cell-specific MRI and optical reflection imaging were achieved by using a single type of JPs [113]. The JPs were dumbbell-shaped Au/Fe_3O_4 nanoparticles (3–8 nm for Au and 18–25 nm for Fe_3O_4) (Fig. 11). The Fe_3O_4 side not only enabled MRI, but was also coated with epidermal growth factor receptor (EGFR) antibody, which enabled it to target human epithelial carcinoma cells. Meanwhile, the high reflectance of the gold compartment enabled optical imaging of cancer cells without fluorescence labels. Fluorescence imaging can also be enabled by materials other than QDs. For example, Tremel *et al.* developed Au/ZnO/SiO_2 Janus nanorods for enabling simultaneous dark-field, two-photon and fluorescence imaging [114]. The nanorods were ≈60 nm in length, ≈10 nm in diameter, and had a matchstick-like shape. The "head" of each nanorod was a gold nanosphere and the "tail" was a ZnO rod covered with a silica shell. The gold nanoparticle enables dark-field light scattering imaging, the photoluminescence of ZnO enhances two-photon imaging at 832 nm, and the silica shell was doped with dye molecules for fluorescence imaging. Although the Au head was left unmodified in the study, it can be further functionalized with cell-targeting ligands for highly selective bioimaging of a particular type of cell or tissue.

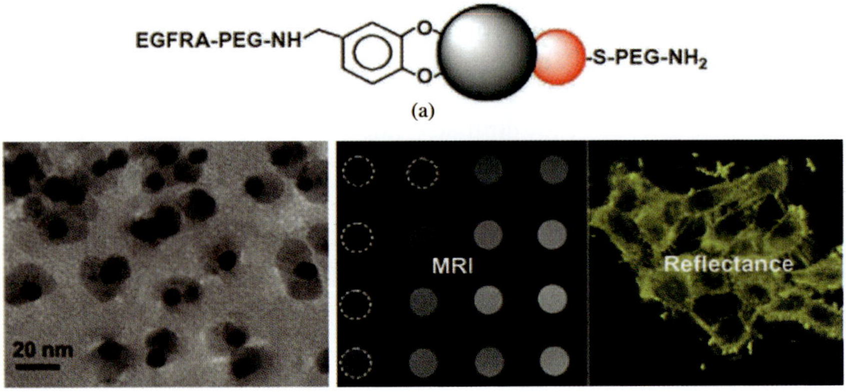

Figure 11. (a) Schematic illustration of a dumbbell-shaped Janus Au/Fe_3O_4 nanoparticle. The gold nanoparticle (shown in red) was functionalized with HS-PEG-NH2 via thiol-Au chemistry. The Fe_3O_4 nanoparticle (shown in gray) was coated with the EGFR antibody to target human epithelial carcinoma cells. (b) TEM image (left) and MRI image (middle) of surface-functionalized Janus Au/Fe_3O_4 nanoparticles; reflectance image (right) of human epithelial carcinoma cells (A431) labeled with Janus Au/Fe_3O_4 nanoparticles. Reprinted with permission from Ref. [113]. Copyright (2008) Wiley-VCH Verlag GmbH & Co. KGaA.

9.2.2.2. *Combining multimodal imaging and other biomedical applications*

Optical imaging combined with magnetic control of living cells. JPs containing magnetically responsive compartments allow both magnetic imaging and remote control of biological systems [115]. In a study by Gao *et al.*, CdSe/Fe_3O_4 JPs that were internalized by cells into endosomes were pulled to aggregate on one side of the cell under a weak magnetic field [116]. Such spatial manipulation is useful for precisely controlling protein gradients inside living cells and manipulating cell polarity. Polymeric Janus nanospheres (100–300 nm) for fluorescent imaging and magnetolytic therapy were developed using a similar concept [117]. One hemisphere of a JP was doped with SPION and the other hemisphere consisted of fluorescently labeled polymers. In addition to enabling fluorescence imaging, these JPs were also twisted under a spinning magnetic field to generate mechanical forces that were sufficient to damage the cell membrane and induce cell death. Magnetic manipulation of cells has also been realized with JPs of other shapes or materials. In the study by Kilinc *et al.* Fe/Au nanorods were used for combined cell targeting, magnetic hyperthermia, and magnetic imaging (Fig. 12) [118]. The Au segments were coated with ligands to target cancer cells and the Fe block enables MRI. Meanwhile, the magnetic response of the Fe segment under a magnetic field induces cell death by mechanically perturbing normal cell signaling. Other magnetically responsive materials, such as MnO, have also been used in place of iron oxides to develop magnetic JPs. For instance, dumbbell-like Au/MnO/SiO_2 nanoparticles have been synthesized [119]. The nanoparticles were composed of an Au nanoparticle

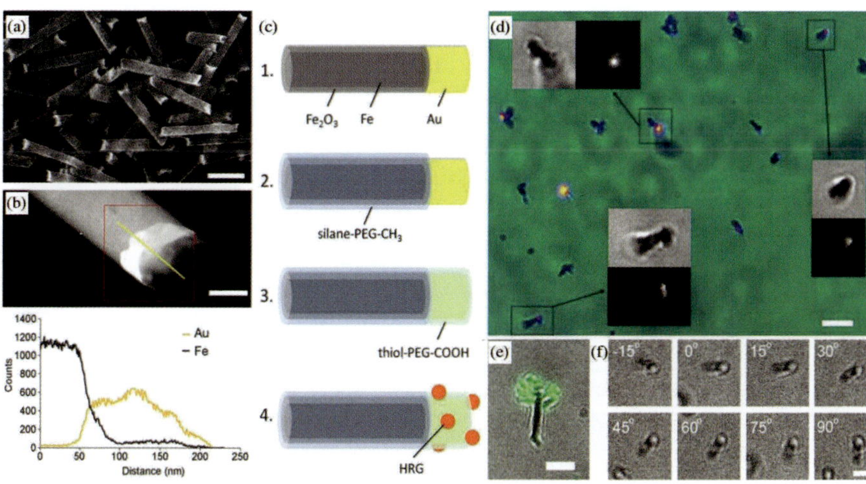

Figure 12. (a) and (b) From top to bottom: Scanning electron image, TEM image, and energy-dispersive X-ray spectroscopy analysis of Janus Fe/Au nanorods; (c) Schematic illustration showing the composition and surface functionalization of Janus Fe/Au nanorods; (d) The pseudo-colored fluorescence image showing streptavidin-conjugated Janus Fe/Au nanorods after incubation with fluorescein–biotin. It was superposed with phase contrast transmission optical image; (e) The superimposed fluorescence and optical image showing Janus Fe/Au nanorods coated with fluorescein-DNA; (f) Phase contrast images showing rotation of a streptavidin-conjugated nanorod on a PEG–biotin surface in an external magnetic field. Scale bars = 1 μm (a), 100 nm (b), 5 μm (d), and 2 μm (e) and (f). Reprinted with permission from Ref. [118]. Copyright (2015) Wiley-VCH Verlag GmbH & Co. KGaA.

(2–8 nm) on one side and a MnO nanoparticle (10–30 nm) coated with a fluorescent silica shell on the other. The nanoparticles are superparamagnetic, fluorescent, and two-photon active at 970 nm, which makes them useful for simultaneous multimodality imaging and magnetic control of cells. In addition, both the gold nanoparticles and the SiO_2 coating are available for use by conjugating biomolecules to them. This endows them with additional functions such as the ability to target specific cell types.

Plasmonic imaging and photothermal therapy. The surface plasmon resonance of noble metal nanoparticles can be harnessed for special plasmonic imaging methods. Unlike fluorescent emitters, metal nanoparticles are extremely photostable, which is important for long-time observations. At the same time, the absorption enhancement at the plasmon resonance frequency also enhances the photothermal response of metal nanoparticles [120]. The ability of metal nanoparticles to generate intense localized heat under optical illumination is important for thermally related applications such as photothermal therapy. JPs that contain noble metal nanoparticles as their building blocks therefore are useful for plasmonic imaging and photothermal therapy, in addition to other functions [121,122]. Tadpole-shaped JPs that contain gold nanoparticles were developed for near-infrared (NIR) phototherapy and NIR imaging in cells (Fig. 13) [123]. The head of each "tadpole" nanoparticle was a silica nanoparticle (≈165 nm)

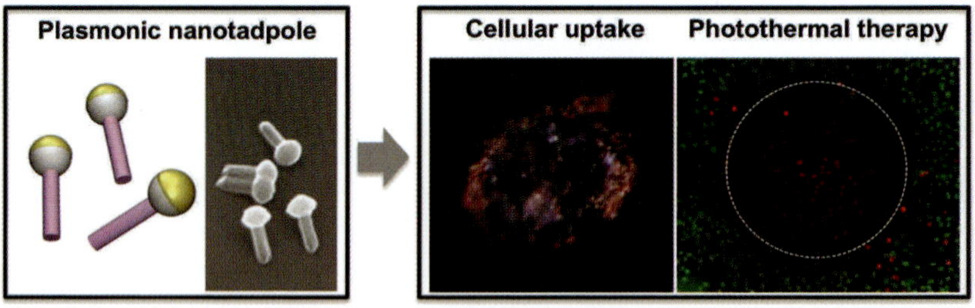

Figure 13. (Left) Schematic illustration and TEM image showing tadpole-like JPs consisting of a head of a silica nanoparticle half-coated with Au and a tail of plasma-treated polystyrene. (Middle) Dark field microscopy image showing HeLa cells after internalizing tadpole-like JPs. (Right) Fluorescence image showing the dead HeLa cells after photothermal treatment. Reprinted with permission from Ref. [123]. Copyright 2014 Royal Society of Chemistry.

that was half-coated with Au to enable NIR imaging and photothermal therapy. The other half of the silica nanoparticle was rendered hydrophobic, while the "tail" of each "tadpole" nanoparticle was composed of hydrophilic polystyrene. This amphiphilicity was designed to enhance the cell entry of the particles. The Janus "tadpoles" were shown to require lower radiation intensities for NIR imaging and phototherapy when compared to Au-capped SiO_2 nanospheres.

Quantitative fluorescence imaging for rotational dynamics measurements. The majority of studies employing JPs for bioimaging, as summarized above, are concerned with qualitative imaging of cells and intracellular organelles. In contrast, only a few groups have employed JPs as tools for making quantitative measurements, such as measurements of rotational dynamics in living systems. The Kopelman group was the first to apply optically asymmetric JPs that were half-fluorescent and half-dark to probe rotation inside living cells [124]. These JPs were made by coating thin films of metal onto one hemisphere of fluorescent beads via the metal vapor deposition method. The particles were named modulated optical nanoprobes (MOONs). Because only one hemisphere of these particles is fluorescent, the shape and intensity of the fluorescent profile changes as the particle rotates, just as the Earth's moon changes its phase from "new" to "full" as the spatial relationship between the hemisphere illuminated by the sun and the Earth changes over the course of a month. Even without resolving the shape of the fluorescent hemisphere, its intensity fluctuations alone provide a means of reliably measuring its out-of-plane rotation. Based on the known relationship between particle rotation and local viscosity, Kopelman and coworkers were able to demonstrate the use of MOONs as probes for measuring local viscosity inside living cells. The Granick group later reported an image processing algorithm for measuring rotation around two of the three possible rotational axes of JPs by resolving the shape and intensity of the fluorescent hemisphere of single JPs [125,126]. The tracking precision of this technique decreases for smaller JPs and may not be applicable for particles smaller than a few hundred nanometers due to the technical challenge of resolving the shape of the fluorescent hemisphere optically. This single-particle rotational tracking technique has also not been applied in biological systems. Yu *et al.* reported a method that takes advantage of the optical asymmetry of "triblock" JPs to precisely measure their in-plane rotation (Fig. 14) [127]. The JPs, prepared through a "sandwiched" microcontact printing procedure, displayed two patches of molecules of different fluorescence colors on the two poles. Once the centroid of each fluorescent patch is identified using a single-particle tracking algorithm, a vector connecting centers of the two patches on single JPs reveals the in-plane orientation of the particle, and the center of this vector indicates the center of the particle.

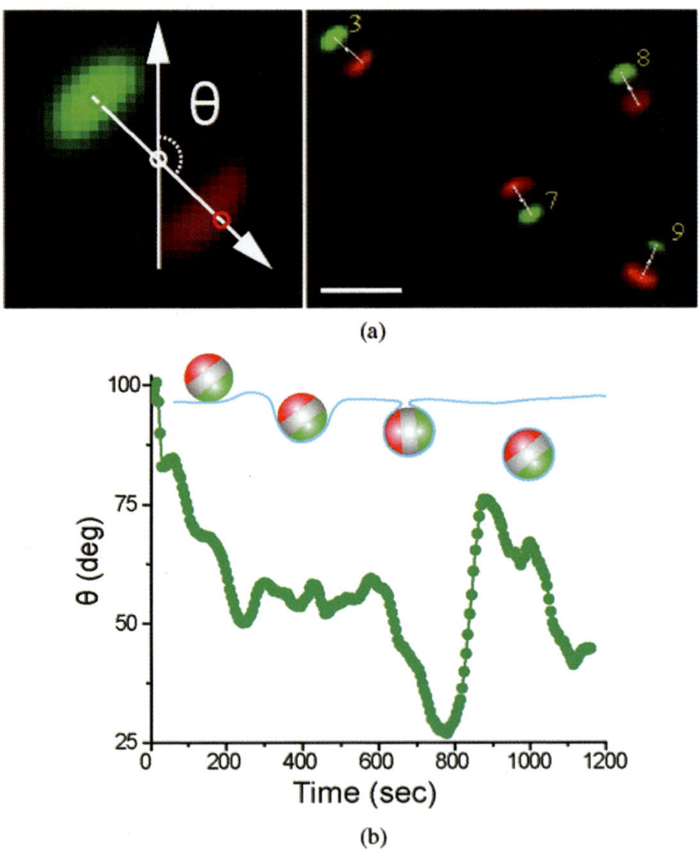

Figure 14. (a) Single-particle tracking of the in-plane rotational angle (θ) of JPs displaying two patches of proteins of different fluorescence. (b) A representative plot showing θ as a function of time of a particle as it entered a macrophage cell. Reprinted with permission from Ref. [127]. Copyright 2015 Royal Society of Chemistry.

With this method, it was shown that during cell internalization, particles exhibit heterogeneous rotational dynamics.

9.2.3. Janus Particles for Biosensing

9.2.3.1. *Biosensing enabled by the surface anisotropy of Janus particles*

Nanoparticles have been extensively studied as sensors for detecting biomolecules. Plasmonic nanoparticles, in particular, are widely used for label-free detection [103,104,128,129]. Molecules or particles that come into close proximity to nanoparticle sensors cause changes in their plasmons, including a peak shift in the surface plasmon resonance spectrum, effects on SERS, and quenching or enhancement of fluorescence emission of dyes. It is also desirable to use nanoparticle biosensors in a cell- or tissue-specific manner. That usually requires that specific ligands be attached to their surfaces. However, attaching molecules to the surfaces of nanoparticles can interfere with their sensing functions. The surface anisotropy of JPs solves this problem by providing a surface that spatially separates the targeting and detection functions. This advantage of JPs was demonstrated by Wu *et al.* who employed JPs for simultaneous targeting and SERS sensing inside tumor cells (Fig. 15) [130]. The JP sensors were made by coating a thin gold film on one hemisphere of polystyrene nanoparticles. The gold cap was roughened by oxygen plasma etching for two reasons. First, the roughened surface provides many hotspots for SERS. Second, the increased surface area due to surface roughness allowed for high-capacity collection of analytes. Anti-HER-2 antibodies were functionalized only on the exposed polystyrene surface, to enable selective targeting of breast cancer cells without affecting the SERS detection localized on the gold surface. The capacity of fluorescence imaging can be further incorporated into such JPs by rendering the polymeric hemisphere fluorescent [35]. The surface anisotropy of JPs has also been exploited to couple enzymatic reactions into a single biosensor unit. A representative application is the colorimetric detection of glucose [131]. Conventionally, this involves two steps. The first is the conversion of glucose by the enzyme glucose oxidase into

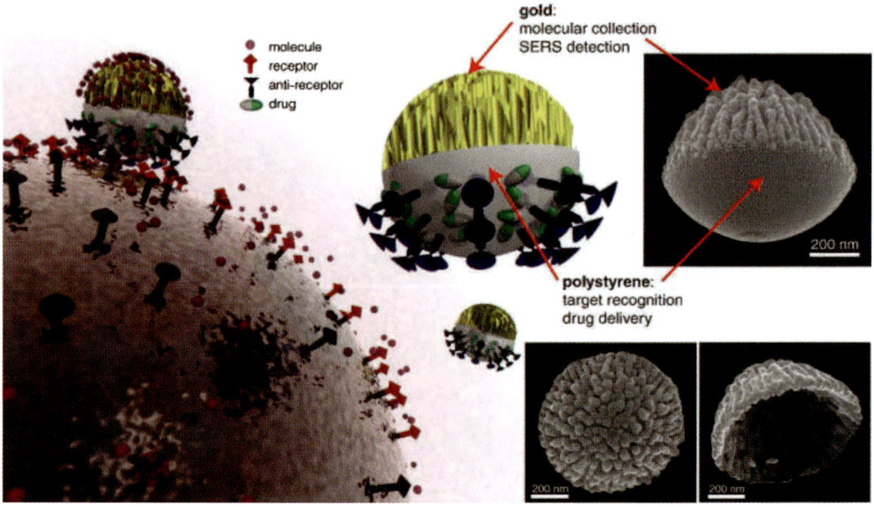

Figure 15. Schematic illustration showing Janus Au/polystyrene nanoparticles for cancer cell targeting, SERS sensing, and drug delivery. Insets are SEM images showing the JPs before and after the polystyrene hemisphere was etched away. Reprinted with permission from Ref. [130]. Copyright 2010 Wiley-VCH Verlag GmbH & Co. KGaA.

gluconic acid and H_2O_2. The second is the reaction of H_2O_2 with probe molecules to generate products of distinct colors. The two reaction steps were coupled on the surface of single γ-Fe_2O_3/SiO_2 JPs, with one reaction on each hemisphere. Specifically, glucose oxidase was conjugated onto the silica side of the particles for the glucose conversion step, whereas the Fe_2O_3 side catalyzes the oxidation of tetramethylbenzidine (TMB) by H_2O_2 into a blue product. The nanoparticles were easily separated from the bulk solution by applying a magnetic field due to the magnetic response of the Fe_2O_3 hemisphere.

The non-detecting region of two-faced JP sensors has also been employed as a kind of barcode for multiplex bioanalysis. The first reported example used striped multiblock microrods for multiplexed immunoassays [132,133]. The microrods consisted of segments of metals (Pt, Pd, Ni, Co, Ag, Cu, and Au). The sequence of metal segments and their individual lengths can be precisely controlled through the sequential electrochemical deposition process that was used to fabricate the microrods. Because different metals emit different colors in reflectance optical microscopy, microrods with unique combination of metal segments were easily distinguishable when mixed, so that different analytes can be simultaneously detected in one solution mix. The Doyle group has done extensive studies aimed at developing polymeric JPs for simultaneous biosensing and barcoding applications [134–138]. They made compartmentalized JPs using continuous-flow photolithography. The particles displayed arrays of dots in one compartment. The combination of dot arrays served as barcodes. The other compartment was conjugated with probe molecules for detecting analytes. After the capture of analytes on biosensors, the next step is to efficiently separate the sensors from the bulk solution. This has been achieved by incorporating magnetically responsive materials into the design of the Janus biosensors. The most commonly used magnetic materials are nickel or iron oxide nanoparticles. Triblock Au/Ni/Au nanorods have been fabricated for the detection and separation of protein mixtures (Fig. 16) [139,140]. For selective detection of protein mixtures, the Au segments were functionalized with nitrostreptavidin for capturing biotinylated proteins,

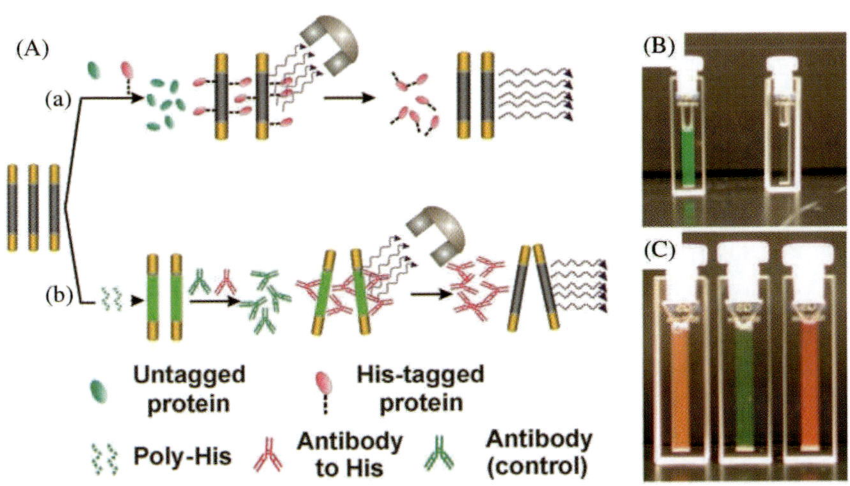

Figure 16. (A) Schematic illustration showing the separation of His-tagged proteins from untagged proteins (route a) and the separation of anti-histidine antibodies from other antibodies (route b). (B) Solutions of fluorescein-tagged poly-His before (green) and after (clear) exposure to triblock Au/Ni/Au nanorods; (C) Color of solutions changes from orange to green, indicating removal the red-dye-labeled His-tagged proteins upon exposure to the nanorods. Color change from green to red indicates the release of the red-dye labeled proteins. Reprinted with permission from Ref. [139]. Copyright (2004) Wiley-VCH Verlag GmbH & Co. KGaA.

and the Ni block specifically bound polyhistidine-tagged proteins. The magnetic response of the nickel block also allowed the nanorods to be separated using a magnet.

9.2.3.2. *Janus micro- and nanomotors as active sensors*

In addition to their usefulness for drug delivery, self-propelled motor particles also hold great promise as active sensors. With conventional biosensors, detection relies on the passive diffusion of analytes. In contrast, biosensors with micro- and nanomotors, like bacteria, can potentially actively search for analytes. The Wang group developed a series of biosensing techniques based on self-propelled Janus motors [141–143]. For instance, a nucleic acid detection assay was developed based on the Ag^+-catalyzed propulsion of nanomotors in the H_2O_2 fuel solution (Fig. 17). In this assay, Au/Pt nanomotors were coated with a capture sequence that hybridizes with a particular part of the target DNA. A detection sequence that partially hybridizes with the remaining part of the target DNA was attached to silver nanoparticles [144]. A successful capture of the target DNA by both particles thus leads to the conjugation of silver nanoparticles to the Au/Pt nanomotors. This conjugation was found to speed up the mobility of the Au/Pt nanomotors, due to the additional propulsive force from the silver ions generated when silver nanoparticles were exposed to the H_2O_2 produced by the motors. The acceleration of the nanomotors increased with the concentration of DNA, so that the distance traveled provided a means to quantitatively measure the nucleic acid concentration. Micromotors have also been shown to be capable of capturing chemical wastes that would otherwise be challenging or dangerous to remove [145]. Self-propelled Janus micromotors also function like miniature stir bars to enhance fluid mixing in immunoassays. By this means, they can improve the overall detection efficiency of conventional microarray-based immunoassays [146].

9.2.4. Non-traditional Biomedical Applications of Janus Particles

Research on the biomedical applications of JPs has been heavily focused on drug delivery, bioimaging, and biosensing, as discussed above.

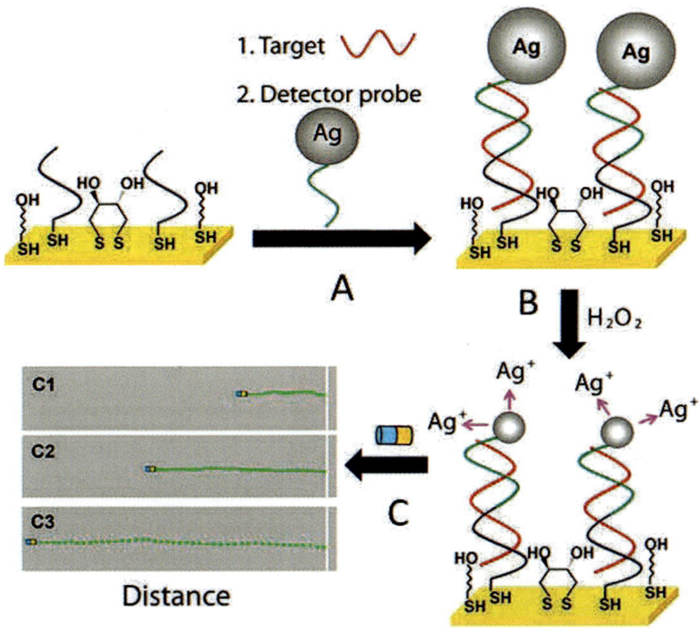

Figure 17. Schematic illustration showing mechanisms of using Au/Pt nanomotors for nucleic acid detection. Step A: Hybridization of the target DNA with the surface-tethered capture DNA and the detection DNA bound on Ag nanoparticles; Step B: Release of Ag^+ from dissolution of silver nanoparticle in the peroxide fuel; Step C: Locomotion of the nanomotors propelled by Ag^+ ions. The distance traveled by the Au–Pt nanomotor (C1, C2, and C3) within a given time increases with the increase of the target DNA concentration. Reprinted with permission from Ref. [144]. Copyright (2010) Nature Publishing Group.

However, a handful of studies have also demonstrated other novel applications of JPs. One interesting idea involves the use of their surface anisotropy to manipulate cell signaling. When a cell communicates with its environment and other cells, its membrane receptors first must bind to their specific target ligands and then trigger the specified cascades of enzymatic reactions inside the cell, which exercise control of specific cell functions, such as cell migration. It is known that membrane receptors must cluster together to initiate this cell signaling process. Yu and coworkers have demonstrated that the spatial segregation of ligands on JPs can be used as an effective way to control the clustering of cell receptors and to subsequently control the immune response of T cells (Fig. 18) [147]. JPs

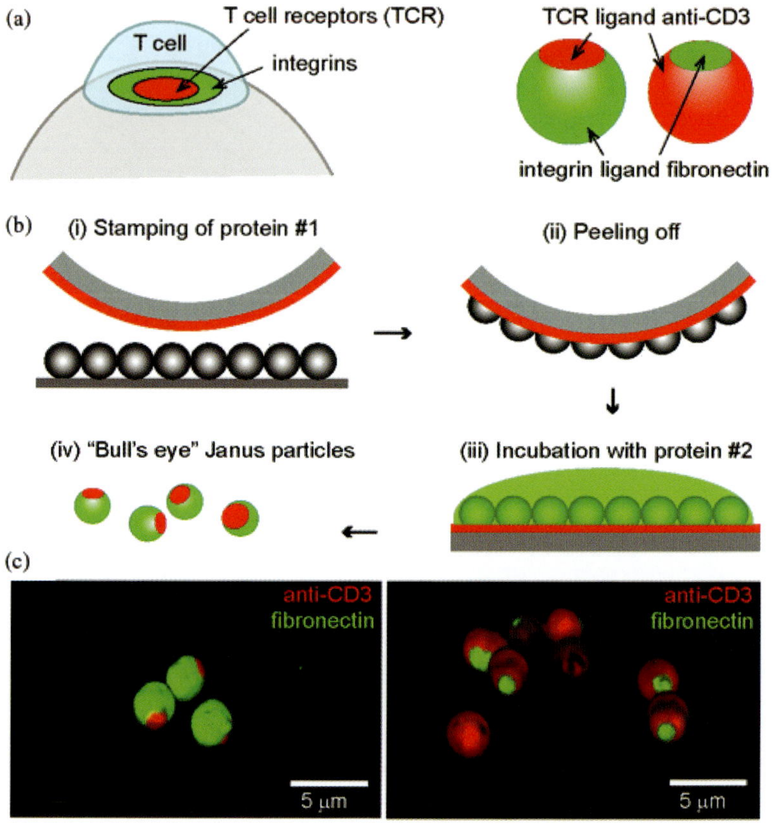

Figure 18. JPs for immune modulation. (a) Patches of proteins on JPs resemble the "bull's eye" pattern of proteins at the interface between a T cell and an antigen-presenting cell. (b) Schematic illustration of the microcontact printing method used to create the JPs displaying protein patterns. (c) Fluorescence images showing the JPs with opposite patterns of proteins. Scale bars: 5 μm. Reprinted with permission from Ref. [147]. Copyright 2014 American Chemical Society.

were created to display separate patches of two ligands, one for binding T cell receptors and one for cell adhesion. This was intended to mimic the protein pattern found at the cell junction between a T cell and an antigen-presenting cell. It was shown that T cells were stimulated by the JPs to different levels depending on the ligand patterns they presented. The stimulation of T cells is a critical step in cancer immunotherapy, which has proven difficult to achieve through other means. This study demonstrated the ability to use the surface anisotropy of JPs to modulate T cell responses. The surface anisotropy of JPs has also been employed to induce selective adhesion of cells for creating bioactuators [148]. In this design, cardiac muscle cells were attached onto only one side of polymeric Janus rods. The elastic rods bend in one direction as the cells contract spontaneously. The achievement of controlled contraction would allow the flexible rods to work as actuators for micromachines.

JPs have also been used for antimicrobial purposes. They can, for example, act as support for antimicrobial reagents. Matchstick-shaped Fe_3O_4/SiO_2 Janus rods coated with Ag nanospheres have been shown to bind to bacteria and inhibit their growth [149]. The magnetic response of the Fe_3O_4 compartment enabled efficient removal of the captured bacteria. In a different strategy, Janus nanomotors were employed to carry enzymes that damage the bacterial cell wall. The nanomotors were powered by an acoustic field and contained a porous gold compartment for encapsulating lysozymes. It was shown that these nanomotors were >30-fold more effective in killing *Micrococcus lysodeikticus* and *E. coli* bacteria than was using lysozyme alone.

Some other interesting applications have been demonstrated with non-traditional JPs. One example is the use of large Janus hydrogel particles as carriers of mammalian cells for potential applications, including constructing artificial tissues [150]. One hemisphere of the particles was used to encapsulate cells, and the other hemisphere contained magnetic nanoparticles to enable remote manipulation of the cell–particle complexes. In another study, Janus cellular spheroids, which are cell aggregates, have been created as building blocks to fabricate scaffold-free tissues [151]. These cell aggregates are similar in morphology to synthetic JPs, but much larger in size. They contain a patch of aggregated Fe_3O_4 nanoparticles on one side that enabled the magnetic control over the morphology and fusion and the spheroids.

9.3. Concluding Remarks

JPs are unique in that they consist of multiple building blocks that differ in chemical, physical, and material properties. This is in stark contrast with conventional micro- and nanoparticles that have a uniform structure, or at least, a uniform surface coating. In the decade since their first synthesis, the surface and structural anisotropy of JPs has been shown to be of tremendous functional usefulness. It allows combining multiple functions into a single particle and thus enables entirely novel biomedical applications that would be impossible with uniform particles. The applications demonstrated so far have included drug delivery, bioimaging, biosensing, cell and tissue targeting, and miniature active motors. These applications are biomedically relevant to both therapy and diagnostics. They are likely to be just the beginning of a new era of JP-based therapeutics.

Many obstacles still stand in the way of realizing the vast potential of this unique particulate system. One major challenge is the lack of any basic understanding of the interactions between JPs and biological systems. So far, studies have been heavily focused on the basics of how to fabricate biologically active JPs and proof-of-concept demonstrations of the relevant techniques. It has been observed in many such studies that JPs exhibit superior properties compared to conventional uniform particles, but little is known about the working mechanisms underlying those observations. Such mechanistic understanding is necessary to optimize the design of JPs. With the rapidly increasing interest in these particles, it is also essential to properly assess their potential impacts on living systems, negative as well as positive. Such an assessment likewise requires a fundamental understanding of how the surface and structural anisotropy of JPs affects biological functions. Even though extensive empirical studies have been done to elucidate the potential health impact of uniform particles [152–163], findings from those studies have not yielded a theory that would allow one to predict the much more complex behavior of anisotropic particles over the broad range of circumstances that might potentially be encountered. Both experimental and some simulation studies have already indicated that the surface anisotropy of JPs gives rise to a new and complex spectrum of phenomena, from disruption of cell membranes to altered cell entry behavior [164–168]. Through investigation is needed to elucidate the general principles and mechanisms governing the interactions between JPs and biological systems and thereby understand how and why this complex spectrum of phenomena arises.

Besides this central challenge, another problem is the lack of simple strategies for the synthesis of JPs that can be readily adopted by researchers across the range of disciplines that will potentially be needed to understand and use them. The few easy-to-follow methods are only suitable for synthesizing particles larger than a few hundred nanometers. Preparation of smaller Janus nanoparticles, which are potentially more useful as therapeutics, often involves much more complex synthetic procedures. Many of these methods provide low yields of particles. Currently, such factors greatly limit the broad application of JPs. Given the rapid advancement in JP synthesis, this challenge is likely to be resolved in the near future.

References

[1] Sanvicens, N.; Marco, M. P. *Trends Biotechnol.*, **26**, 425 (**2008**).
[2] Bao, G.; Mitragotri, S.; Tong, S. *Annu. Rev. Biomed. Eng.*, **15**, 253 (**2013**).
[3] Jia, F.; Liu, X. P.; Li, L. H.; Mallapragada, S.; Narasimhan, B.; Wang, Q. *J. Control. Release*, **172**, 1020 (**2013**).
[4] Lee, D. E.; Koo, H.; Sun, I. C.; Ryu, J. H.; Kim, K.; Kwon, I. C. *Chem. Soc. Rev.*, **41**, 2656 (**2012**).
[5] Kim, J.; Piao, Y.; Hyeon, T. *Chem. Soc. Rev.*, **38**, 372 (**2009**).
[6] Gao, J. H.; Gu, H. W.; Xu, B. *Acc. Chem. Res.*, **42**, 1097 (**2009**).
[7] Hao, R.; Xing, R. J.; Xu, Z. C.; Hou, Y. L.; Gao, S.; Sun, S. H. *Adv. Mater.*, **22**, 2729 (**2010**).
[8] Gadde, S. *MedChemComm.*, **6**, 1916 (**2015**).
[9] Lee, J. H.; Nan, A. *J. Drug Deliv.*, **2012**, 17 (**2012**).
[10] Roh, K.-H.; Martin, D. C.; Lahann, J. *Nat. Mater.*, **4**, 759 (**2005**).
[11] Roh, K.-H.; Yoshida, M.; Lahann, J. *Langmuir*, **23**, 5683 (**2007**).
[12] Yoshida, M.; Roh, K.-H.; Lahann, J. *Biomaterials*, **28**, 2446 (**2007**).
[13] Yoshida, M.; Roh, K.-H.; Mandal, S.; Bhaskar, S.; Lim, D.; Nandivada, H.; Deng, X.; Lahann, J. *Adv. Mater.*, **21**, 4920 (**2009**).
[14] Hwang, S.; Lahann, J. *Macromol. Rapid Comm.*, **33**, 1178 (**2012**).
[15] Misra, A. C.; Bhaskar, S.; Clay, N.; Lahann, J. *Adv. Mater.*, **24**, 3850 (**2012**).
[16] Sokolovskaya, E.; Yoon, J.; Misra, A. C.; Bräse, S.; Lahann, J. *Macromol. Rapid Comm.*, **34**, 1554 (**2013**).
[17] Misra, A. C.; Luker, K. E.; Durmaz, H.; Luker, G. D.; Lahann, J. *Biomacromolecules*, **16**, 2412 (**2015**).
[18] Rahmani, S.; Saha, S.; Durmaz, H.; Donini, A.; Misra, A. C.; Yoon, J.; Lahann, J. *Angew. Chem. Int. Ed.*, **53**, 2332 (**2014**).

[19] Rahmani, S.; Ross, A. M.; Park, T.-H.; Durmaz, H.; Dishman, A. F.; Prieskorn, D. M.; Jones, N.; Altschuler, R. A.; Lahann, J. *Adv. Healthc. Mater.*, **5**, 94 (**2016**).
[20] Lahann, J. *Small*, **7**, 1149 (**2011**).
[21] Xie, H.; She, Z.-G.; Wang, S.; Sharma, G.; Smith, J. W. *Langmuir*, **28**, 4459 (**2012**).
[22] Bhaskar, S.; Lahann, J. *J. Am. Chem. Soc.*, **131**, 6650 (**2009**).
[23] Bhaskar, S.; Hitt, J.; Chang, S.-W. L.; Lahann, J. *Angew. Chem. Int. Ed.*, **48**, 4589 (**2009**).
[24] Garbuzenko, O. B.; Winkler, J.; Tomassone, M. S.; Minko, T. *Langmuir*, **30**, 12941 (**2014**).
[25] Sokolovskaya, E.; Rahmani, S.; Misra, A. C.; Bräse, S.; Lahann, J. *ACS Appl. Mater. Interfaces*, **7**, 9744 (**2015**).
[26] Li, X.; Zhou, L.; Wei, Y.; El-Toni, A. M.; Zhang, F.; Zhao, D. *J. Am. Chem. Soc.*, **136**, 15086 (**2014**).
[27] Saha, K.; Agasti, S. S.; Kim, C.; Li, X.; Rotello, V. M. *Chem. Rev.*, **112**, 2739 (**2012**).
[28] Zhou, W.; Gao, X.; Liu, D.; Chen, X. *Chem. Rev.*, **115**, 10575 (**2015**).
[29] Hakkinen, H. *Nat. Chem.*, **4**, 443 (**2012**).
[30] Willets, K. A.; Van Duyne, R. P. *Annu. Rev. Phys. Chem.*, **58**, 267 (**2007**).
[31] Anker, J. N.; Hall, W. P.; Lyandres, O.; Shah, N. C.; Zhao, J.; Van Duyne, R. P. *Nat. Mater.*, **7**, 442 (**2008**).
[32] Jain, P. K.; Huang, X. H.; El-Sayed, I. H.; El-Sayed, M. A. *Acc. Chem. Res.*, **41**, 1578 (**2008**).
[33] Murphy, C. J.; Gole, A. M.; Stone, J. W.; Sisco, P. N.; Alkilany, A. M.; Goldsmith, E. C.; Baxter, S. C. *Acc. Chem. Res.*, **41**, 1721 (**2008**).
[34] Schultz, D. A. *Curr. Opin. Biotechnol.*, **14**, 13 (**2003**).
[35] Hsieh, H.-Y.; Huang, T.-W.; Xiao, J.-L.; Yang, C.-S.; Chang, C.-C.; Chu, C.-C.; Lo, L.-W.; Wang, S.-H.; Wang, P.-C.; Chieng, C.-C.; Lee, C.-H.; Tseng, F.-G. *J. Mater. Chem.*, **22**, 20918 (**2012**).
[36] Lee, S. M.; Kim, H. J.; Ha, Y. J.; Park, Y. N.; Lee, S. K.; Park, Y. B.; Yoo, K. H. *ACS Nano*, **7**, 50 (**2013**).
[37] Yang, Y.; Matsubara, S.; Xiong, L.; Hayakawa, T.; Nogami, M. *J. Phys. Chem. C*, **111**, 9095 (**2007**).
[38] Shao, D.; Zhang, X.; Liu, W.; Zhang, F.; Zheng, X.; Qiao, P.; Li, J.; Dong, W.-f.; Chen, L. *ACS Appl. Mater. Interfaces*, **8**, 4303 (**2016**).
[39] Wang, F.; Pauletti, G. M.; Wang, J.; Zhang, J.; Ewing, R. C.; Wang, Y.; Shi, D. *Adv. Mater.*, **25**, 3485 (**2013**).
[40] Xu, C.; Wang, B.; Sun, S. *J. Am. Chem. Soc.*, **131**, 4216 (**2009**).

[41] Yu, H.; Chen, M.; Rice, P. M.; Wang, S. X.; White, R. L.; Sun, S. *Nano Lett.*, **5**, 379 (**2005**).
[42] Zhang, L.; Chen, Y.; Li, Z.; Li, L.; Saint-Cricq, P.; Li, C.; Lin, J.; Wang, C.; Su, Z.; Zink, J. I. *Angew. Chem. Int. Ed.*, **55**, 2118 (**2016**).
[43] Díez, P.; Sánchez, A.; Gamella, M.; Martínez-Ruíz, P.; Aznar, E.; de la Torre, C.; Murguía, J. R.; Martínez-Máñez, R.; Villalonga, R.; Pingarrón, J. M. *J. Am. Chem. Soc.*, **136**, 9116 (**2014**).
[44] Champion, J. A.; Mitragotri, S. *Proc. Natl. Acad. Sci. U.S.A.*, **103**, 4930 (**2006**).
[45] Salem, A. K.; Searson, P. C.; Leong, K. W. *Nat. Mater.*, **2**, 668 (**2003**).
[46] Salem, A. K.; Hung, C. F.; Kim, T. W.; Wu, T. C.; Searson, P. C.; Leong, K. W. *Nanotechnology*, **16**, 484 (**2005**).
[47] Wang, H.; Pumera, M. *Chem. Rev.*, **115**, 8704 (**2015**).
[48] Zhang, L.; Abbott, J. J.; Dong, L.; Kratochvil, B. E.; Bell, D.; Nelson, B. J. *Appl. Phys. Lett.*, **94**, 064107 (**2009**).
[49] Sánchez, S.; Soler, L.; Katuri, J. *Angew. Chem. Int. Ed.*, **54**, 1414 (**2015**).
[50] Kagan, D.; Benchimol, M. J.; Claussen, J. C.; Chuluun-Erdene, E.; Esener, S.; Wang, J. *Angew. Chem. Int. Ed.*, **124**, 7637 (**2012**).
[51] Wang, W.; Castro, L. A.; Hoyos, M.; Mallouk, T. E. *ACS Nano*, **6**, 6122 (**2012**).
[52] Garcia-Gradilla, V.; Orozco, J.; Sattayasamitsathit, S.; Soto, F.; Kuralay, F.; Pourazary, A.; Katzenberg, A.; Gao, W.; Shen, Y.; Wang, J. *ACS Nano*, **7**, 9232 (**2013**).
[53] Ahmed, S.; Wang, W.; Mair, L. O.; Fraleigh, R. D.; Li, S.; Castro, L. A.; Hoyos, M.; Huang, T. J.; Mallouk, T. E. *Langmuir*, **29**, 16113 (**2013**).
[54] Garcia-Gradilla, V.; Sattayasamitsathit, S.; Soto, F.; Kuralay, F.; Yardımcı, C.; Wiitala, D.; Galarnyk, M.; Wang, J. *Small*, **10**, 4154 (**2014**).
[55] Wang, W.; Li, S.; Mair, L.; Ahmed, S.; Huang, T. J.; Mallouk, T. E. *Angew. Chem. Int. Ed.*, **53**, 3201 (**2014**).
[56] Ghosh, A.; Fischer, P. *Nano Lett.*, **9**, 2243 (**2009**).
[57] Gao, W.; Sattayasamitsathit, S.; Manesh, K. M.; Weihs, D.; Wang, J. *J. Am. Chem. Soc.*, **132**, 14403 (**2010**).
[58] Gao, W.; Kagan, D.; Pak, O. S.; Clawson, C.; Campuzano, S.; Chuluun-Erdene, E.; Shipton, E.; Fullerton, E. E.; Zhang, L.; Lauga, E.; Wang, J. *Small*, **8**, 460 (**2012**).
[59] Tottori, S.; Zhang, L.; Qiu, F.; Krawczyk, K. K.; Franco-Obregón, A.; Nelson, B. J. *Adv. Mater.*, **24**, 811 (**2012**).
[60] Baraban, L.; Streubel, R.; Makarov, D.; Han, L.; Karnaushenko, D.; Schmidt, O. G.; Cuniberti, G. *ACS Nano*, **7**, 1360 (**2013**).

[61] Li, J.; Li, T.; Xu, T.; Kiristi, M.; Liu, W.; Wu, Z.; Wang, J. *Nano Lett.*, **15**, 4814 (**2015**).

[62] Gao, W.; Wang, J. *Nanoscale*, **6**, 10486 (**2014**).

[63] Chałupniak, A.; Morales-Narváez, E.; Merkoçi, A. *Adv. Drug Deliv. Rev.*, **95**, 104 (**2015**).

[64] Guix, M.; Mayorga-Martinez, C. C.; Merkoçi, A. *Chem. Rev.*, **114**, 6285 (**2014**).

[65] Singh, V. V.; Soto, F.; Kaufmann, K.; Wang, J. *Angew. Chem. Int. Ed.*, **54**, 6896 (**2015**).

[66] Srivastava, S. K.; Guix, M.; Schmidt, O. G. *Nano Lett.*, **16**, 817 (**2016**).

[67] Soler, L.; Sanchez, S. *Nanoscale*, **6**, 7175 (**2014**).

[68] Paxton, W. F.; Kistler, K. C.; Olmeda, C. C.; Sen, A.; St. Angelo, S. K.; Cao, Y.; Mallouk, T. E.; Lammert, P. E.; Crespi, V. H. *J. Am. Chem. Soc.*, **126**, 13424 (**2004**).

[69] Kagan, D.; Laocharoensuk, R.; Zimmerman, M.; Clawson, C.; Balasubramanian, S.; Kang, D.; Bishop, D.; Sattayasamitsathit, S.; Zhang, L.; Wang, J. *Small*, **6**, 2741 (**2010**).

[70] Ma, X.; Hahn, K.; Sanchez, S. *J. Am. Chem. Soc.*, **137**, 4976 (**2015**).

[71] Xuan, M.; Shao, J.; Lin, X.; Dai, L.; He, Q. *ChemPhysChem.*, **15**, 2255 (**2014**).

[72] Sundararajan, S.; Sengupta, S.; Ibele, M. E.; Sen, A. *Small*, **6**, 1479 (**2010**).

[73] Gao, W.; Pei, A.; Wang, J. *ACS Nano*, **6**, 8432 (**2012**).

[74] Mou, F.; Chen, C.; Ma, H.; Yin, Y.; Wu, Q.; Guan, J. *Angew. Chem. Int. Ed.*, **52**, 7208 (**2013**).

[75] Mou, F.; Chen, C.; Zhong, Q.; Yin, Y.; Ma, H.; Guan, J. *ACS Appl. Mater. Interfaces*, **6**, 9897 (**2014**).

[76] Gao, W.; Dong, R.; Thamphiwatana, S.; Li, J.; Gao, W.; Zhang, L.; Wang, J. *ACS Nano*, **9**, 117 (**2015**).

[77] Wu, Y.; Lin, X.; Wu, Z.; Möhwald, H.; He, Q. *ACS Appl. Mater. Interfaces*, **6**, 10476 (**2014**).

[78] Stanton, M. M.; Simmchen, J.; Ma, X.; Miguel-López, A.; Sánchez, S. *Adv. Mater. Interfaces*, **3**, 1500505 (**2015**).

[79] Kang, S.; Suci, P. A.; Broomell, C. C.; Iwahori, K.; Kobayashi, M.; Yamashita, I.; Young, M.; Douglas, T. *Nano Lett.*, **9**, 2360 (**2009**).

[80] Suci, P. A.; Kang, S.; Young, M.; Douglas, T. *J. Am. Chem. Soc.*, **131**, 9164 (**2009**).

[81] Li, F.; Chen, Y.; Chen, H.; He, W.; Zhang, Z.-P.; Zhang, X.-E.; Wang, Q. *J. Am. Chem. Soc.*, **133**, 20040 (**2011**).

[82] Kuan, S. L.; Ng, D. Y. W.; Wu, Y.; Förtsch, C.; Barth, H.; Doroshenko, M.; Koynov, K.; Meier, C.; Weil, T. *J. Am. Chem. Soc.*, **135**, 17254 (**2013**).

[83] Zachos, T.; Diggs, A.; Weisbrode, S.; Bartlett, J.; Bertone, A. *Mol. Ther.*, **15**, 1543 (**2007**).

[84] Pelinkovic, D.; Lee, J. Y.; Engelhardt, M.; Rodosky, M.; Cummins, J.; Fu, F. H.; Huard, J. *Tissue Eng.*, **9**, 143 (**2003**).

[85] Swiston, A. J.; Cheng, C.; Um, S. H.; Irvine, D. J.; Cohen, R. E.; Rubner, M. F. *Nano Lett.*, **8**, 4446 (**2008**).

[86] Zhang, P.; Liu, Y.; Xia, J.; Wang, Z.; Kirkland, B.; Guan, J. *Adv. Healthc. Mater.*, **2**, 540 (**2013**).

[87] Anselmo, A. C.; Gilbert, J. B.; Kumar, S.; Gupta, V.; Cohen, R. E.; Rubner, M. F.; Mitragotri, S. *J. Control. Release.*, **199**, 29 (**2015**).

[88] Xia, J.; Wang, Z.; Huang, D.; Yan, Y.; Li, Y.; Guan, J. *ACS Appl. Mater. Interfaces*, **7**, 6293 (**2015**).

[89] Polak, R.; Lim, R. M.; Beppu, M. M.; Pitombo, R. N. M.; Cohen, R. E.; Rubner, M. F. *Adv. Healthc. Mater.*, **4**, 2832 (**2015**).

[90] Gilbert, J. B.; O'Brien, J. S.; Suresh, H. S.; Cohen, R. E.; Rubner, M. F. *Adv. Mater.*, **25**, 5948 (**2013**).

[91] Doshi, N.; Swiston, A. J.; Gilbert, J. B.; Alcaraz, M. L.; Cohen, R. E.; Rubner, M. F.; Mitragotri, S. *Adv. Mater.* **23**, H105 (**2011**).

[92] Swiston, A. J.; Gilbert, J. B.; Irvine, D. J.; Cohen, R. E.; Rubner, M. F. *Biomacromolecules*, **11**, 1826 (**2010**).

[93] Sharma, P.; Brown, S.; Walter, G.; Santra, S.; Moudgil, B. *Adv. Colloid Interface Sci.* **123**, 471 (**2006**).

[94] Bigall, N. C.; Parak, W. J.; Dorfs, D. *Nano Today*, **7**, 282 (**2012**).

[95] Peng, F.; Su, Y.; Zhong, Y.; Fan, C.; Lee, S.-T.; He, Y. *Acc. Chem. Res.*, **47**, 612 (**2014**).

[96] Chan, W. C. W.; Nie, S. *Science*, **281**, 2016 (**1998**).

[97] Bruchez, M.; Moronne, M.; Gin, P.; Weiss, S.; Alivisatos, A. P. *Science*, **281**, 2013 (**1998**).

[98] Chan, W. C. W.; Maxwell, D. J.; Gao, X.; Bailey, R. E.; Han, M.; Nie, S. *Curr. Opin. Biotechnol.*, **13**, 40 (**2002**).

[99] Walling, M.; Novak, J.; Shepard, J. R. E. *Int. J. Mol. Sci.*, **10**, 441 (**2009**).

[100] Wegner, K. D.; Hildebrandt, N. *Chem. Soc. Rev.*, **44**, 4792 (**2015**).

[101] Mukherjee, A.; Shim, Y.; Myong Song, J. *Biotechnol. J.*, **11**, 31 (**2016**).

[102] Zhou, J.; Yang, Y.; Zhang, C.-y. *Chem. Rev.*, **115**, 11669 (**2015**).

[103] Talley, C. E.; Jackson, J. B.; Oubre, C.; Grady, N. K.; Hollars, C. W.; Lane, S. M.; Huser, T. R.; Nordlander, P.; Halas, N. J. *Nano Lett.*, **5**, 1569 (**2005**).

[104] Dasary, S. S. R.; Singh, A. K.; Senapati, D.; Yu, H.; Ray, P. C. *J. Am. Chem. Soc.*, **131**, 13806 (**2009**).
[105] Lee, N.; Yoo, D.; Ling, D.; Cho, M. H.; Hyeon, T.; Cheon, J. *Chem. Rev.*, **115**, 10637 (**2015**).
[106] Veiseh, O.; Gunn, J. W.; Zhang, M. *Adv. Drug Delivery Rev.*, **62**, 284 (**2010**).
[107] Hsing, I. M.; Xu, Y.; Zhao, W. *Electroanalysis*, **19**, 755 (**2007**).
[108] Gao, X.; Cui, Y.; Levenson, R. M.; Chung, L. W. K.; Nie, S. *Nat. Biotechnol.*, **22**, 969 (**2004**).
[109] Larson, D. R.; Zipfel, W. R.; Williams, R. M.; Clark, S. W.; Bruchez, M. P.; Wise, F. W.; Webb, W. W. *Science*, **300**, 1434 (**2003**).
[110] Michalet, X.; Pinaud, F. F.; Bentolila, L. A.; Tsay, J. M.; Doose, S.; Li, J. J.; Sundaresan, G.; Wu, A. M.; Gambhir, S. S.; Weiss, S. *Science*, **307**, 538 (**2005**).
[111] Selvan, S. T.; Patra, P. K.; Ang, C. Y.; Ying, J. Y. *Angew. Chem. Int. Ed.*, **46**, 2448 (**2007**).
[112] Lee, J.; Hwang, G.; Hong, Y. S.; Sim, T. *Analyst*, **140**, 2864 (**2015**).
[113] Xu, C.; Xie, J.; Ho, D.; Wang, C.; Kohler, N.; Walsh, E. G.; Morgan, J. R.; Chin, Y. E.; Sun, S. *Angew. Chem. Int. Ed.*, **47**, 173 (**2008**).
[114] Susewind, M.; Schilmann, A.-M.; Heim, J.; Henkel, A.; Link, T.; Fischer, K.; Strand, D.; Kolb, U.; Tahir, M. N.; Brieger, J.; Tremel, W. *J. Mat. Chem. B*, **3**, 1813 (**2015**).
[115] Dobson, J. *Nat. Nanotechnol.*, **3**, 139 (**2008**).
[116] Gao, J. H.; Zhang, W.; Huang, P. B.; Zhang, B.; Zhang, X. X.; Xu, B. *J. Am. Chem. Soc.*, **130**, 3710 (**2008**).
[117] Hu, S.-H.; Gao, X. *J. Am. Chem. Soc.*, **132**, 7234 (**2010**).
[118] Kilinc, D.; Lesniak, A.; Rashdan, S. A.; Gandhi, D.; Blasiak, A.; Fannin, P. C.; von Kriegsheim, A.; Kolch, W.; Lee, G. U. *Adv. Healthc. Mater.*, **4**, 395 (**2015**).
[119] Schick, I.; Lorenz, S.; Gehrig, D.; Schilmann, A.-M.; Bauer, H.; Panthöfer, M.; Fischer, K.; Strand, D.; Laquai, F.; Tremel, W. *J. Am. Chem. Soc.*, **136**, 2473 (**2014**).
[120] Lee, K.-S.; El-Sayed, M. A. *J. Phys. Chem. B*, **110**, 19220 (**2006**).
[121] Liu, M.; Chao, J.; Deng, S.; Wang, K.; Li, K.; Fan, C. *Colloids Surf., B*, **124**, 111 (**2014**).
[122] Sotiriou, G. A.; Hirt, A. M.; Lozach, P.-Y.; Teleki, A.; Krumeich, F.; Pratsinis, S. E. *Chem. Mater.*, **23**, 1985 (**2011**).
[123] Kang, H.; Kim, S.-H.; Yang, S.-M.; Park, J.-H. *J. Mater. Chem. B*, **2**, 6462 (**2014**).

[124] Behrend, C. J.; Anker, J. N.; McNaughton, B. H.; Brasuel, M.; Philbert, M. A.; Kopelman, R. *J. Phys. Chem. B*, **108**, 10408 (**2004**).

[125] Anthony, S. M.; Hong, L.; Kim, M.; Granick, S. *Langmuir*, **22**, 9812 (**2006**).

[126] Anthony, S. M.; Kim, M.; Granick, S. *Langmuir*, **24**, 6557 (**2008**).

[127] Sanchez, L.; Patton, P.; Anthony, S. M.; Yi, Y.; Yu, Y. *Soft Matter*, **11**, 5346 (**2015**).

[128] Qian, X. M.; Nie, S. M. *Chem. Soc. Rev.*, **37**, 912 (**2008**).

[129] Cao, Y. C.; Jin, R.; Mirkin, C. A. *Science*, **297**, 1536 (**2002**).

[130] Wu, L. Y.; Ross, B. M.; Hong, S.; Lee, L. P. *Small*, **6**, 503 (**2010**).

[131] Lu, C.; Liu, X.; Li, Y.; Yu, F.; Tang, L.; Hu, Y.; Ying, Y. *ACS Appl. Mater. Interfaces*, **7**, 15395 (**2015**).

[132] Nicewarner-Peña, S. R.; Freeman, R. G.; Reiss, B. D.; He, L.; Peña, D. J.; Walton, I. D.; Cromer, R.; Keating, C. D.; Natan, M. J. *Science*, **294**, 137 (**2001**).

[133] Nicewarner-Peña, S. R.; Carado, A. J.; Shale, K. E.; Keating, C. D. *J. Phys. Chem. B*, **107**, 7360 (**2003**).

[134] Pregibon, D. C.; Toner, M.; Doyle, P. S. *Science*, **315**, 1393 (**2007**).

[135] Bong, K. W.; Chapin, S. C.; Doyle, P. S. *Langmuir*, **26**, 8008 (**2010**).

[136] Yuet, K. P.; Hwang, D. K.; Haghgooie, R.; Doyle, P. S. *Langmuir*, **26**, 4281 (**2010**).

[137] Appleyard, D. C.; Chapin, S. C.; Srinivas, R. L.; Doyle, P. S. *Nat. Protoc.*, **6**, 1761 (**2011**).

[138] Lee, J.; Bisso, P. W.; Srinivas, R. L.; Kim, J. J.; Swiston, A. J.; Doyle, P. S. *Nat. Mater.*, **13**, 524 (**2014**).

[139] Lee, K.-B.; Park, S.; Mirkin, C. A. *Angew. Chem. Int. Ed.*, **43**, 3048 (**2004**).

[140] Oh, B.-K.; Park, S.; Millstone, J. E.; Lee, S. W.; Lee, K.-B.; Mirkin, C. A. *J. Am. Chem. Soc.*, **128**, 11825 (**2006**).

[141] Esteban-Fernández de Ávila, B.; Martín, A.; Soto, F.; Lopez-Ramirez, M. A.; Campuzano, S.; Vásquez-Machado, G. M.; Gao, W.; Zhang, L.; Wang, J. *ACS Nano*, **9**, 6756 (**2015**).

[142] Wang, J.; Gao, W. *ACS Nano*, **6**, 5745 (**2012**).

[143] Singh, V. V.; Wang, J. *Nanoscale*, **7**, 19377 (**2015**).

[144] Wu, J.; Balasubramanian, S.; Kagan, D.; Manesh, K. M.; Campuzano, S.; Wang, J. *Nat. Commun.*, **1**, 36 (**2010**).

[145] Singh, V. V.; Martin, A.; Kaufmann, K.; D. S. de Oliveira, S.; Wang, J. *Chem. Mater.*, **27**, 8162 (**2015**).

[146] Morales-Narváez, E.; Guix, M.; Medina-Sánchez, M.; Mayorga-Martinez, C. C.; Merkoçi, A. *Small*, **10**, 2542 (**2014**).

[147] Chen, B.; Jia, Y.; Gao, Y.; Sanchez, L.; Anthony, S. M.; Yu, Y. *ACS Appl. Mater. Interfaces*, **6**, 18435 (**2014**).

[148] Yoon, J.; Eyster, T. W.; Misra, A. C.; Lahann, J. *Adv. Mater.*, **27**, 4509 (**2015**).

[149] Zhang, L.; Luo, Q.; Zhang, F.; Zhang, D. M.; Wang, Y. S.; Sun, Y. L.; Dong, W. F.; Liu, J. Q.; Huo, Q. S.; Sun, H. B. *J. Mater. Chem.*, **22**, 23741 (**2012**).

[150] Zhao, L. B.; Pan, L.; Zhang, K.; Guo, S. S.; Liu, W.; Wang, Y.; Chen, Y.; Zhao, X. Z.; Chan, H. L. W. *Lab Chip*, **9**, 2981 (**2009**).

[151] Mattix, B. M.; Olsen, T. R.; Casco, M.; Reese, L.; Poole, J. T.; Zhang, J.; Visconti, R. P.; Simionescu, A.; Simionescu, D. T.; Alexis, F. *Biomaterials*, **35**, 949 (**2014**).

[152] Hutter, E.; Boridy, S.; Labrecque, S.; Lalancette-Hébert, M.; Kriz, J.; Winnik, F. M.; Maysinger, D. *ACS Nano*, **4**, 2595 (**2010**).

[153] Rabolli, V.; Thomassen, L. C. J.; Princen, C.; Napierska, D.; Gonzalez, L.; Kirsch-Volders, M.; Hoet, P. H.; Huaux, F.; Kirschhock, C. E. A.; Martens, J. A.; Lison, D. *Nanotoxicology*, **4**, 307 (**2010**).

[154] Hulander, M.; Lundgren, A.; Berglin, M.; Ohrlander, M.; Lausmaa, J.; Elwing, H. *Int. J. Nanomedicine*, **6**, 2653 (**2011**).

[155] Greish, K.; Thiagarajan, G.; Herd, H.; Price, R.; Bauer, H.; Hubbard, D.; Burckle, A.; Sadekar, S.; Yu, T.; Anwar, A.; Ray, A.; Ghandehari, H. *Nanotoxicology*, **6**, 713 (**2012**).

[156] Arvizo, R. R.; Miranda, O. R.; Thompson, M. A.; Pabelick, C. M.; Bhattacharya, R.; Robertson, J. D.; Rotello, V. M.; Prakash, Y. S.; Mukherjee, P. *Nano Lett.*, **10**, 2543 (**2010**).

[157] El Badawy, A. M.; Silva, R. G.; Morris, B.; Scheckel, K. G.; Suidan, M. T.; Tolaymat, T. M. *Environ. Sci. Technol.*, **45**, 283 (**2011**).

[158] Schaeublin, N. M.; Braydich-Stolle, L. K.; Schrand, A. M.; Miller, J. M.; Hutchison, J.; Schlager, J. J.; Hussain, S. M. *Nanoscale*, **3**, 410 (**2011**).

[159] Park, Y.-H.; Bae, H.; Jang, Y.; Jeong, S.; Lee, H.; Ryu, W.-I.; Yoo, M.; Kim, Y.-R.; Kim, M.-K.; Lee, J.; Jeong, J.; Son, S. *Mol. Cell. Toxicol.*, **9**, 67 (**2013**).

[160] Moyano, D. F.; Goldsmith, M.; Solfiell, D. J.; Landesman-Milo, D.; Miranda, O. R.; Peer, D.; Rotello, V. M. *J. Am. Chem. Soc.*, **134**, 3965 (**2012**).

[161] Saha, K.; Moyano, D. F.; Rotello, V. M. *Mate. Horiz.*, **1**, 102 (**2014**).

[162] Zuo, G.; Huang, Q.; Wei, G.; Zhou, R.; Fang, H. *ACS Nano*, **4**, 7508 (**2010**).

[163] Kennedy, A. J.; Hull, M. S.; Steevens, J. A.; Dontsova, K. M.; Chappell, M. A.; Gunter, J. C.; Weiss, C. A. *Environ. Toxicol. Chem.*, **27**, 1932 (**2008**).

[164] Gao, Y.; Yu, Y. *Langmuir*, **31**, 2833 (**2015**).

[165] Gao, Y.; Yu, Y. *J. Am. Chem. Soc.*, **135**, 19091 (**2013**).

[166] Alexeev, A.; Uspal, W. E.; Balazs, A. C. *ACS Nano*, **2**, 1117 (**2008**).

[167] Anna C. Balazs, O. K. A. A. *Macromol. Theory Simul.*, **18**, 11 (**2009**).

[168] Ding, H.-m.; Ma, Y.-q. *Nanoscale*, **4**, 1116 (**2012**).

Chapter 10

Janus Particles as Novel Building Blocks for Active Functional Surfaces and Interfaces

Alla Synytska[*] *and Alina Kirillova*

Leibniz-Institut für Polymerforschung Dresden e.V., Hohe Str. 6, 01069 Dresden, Germany
Technische Universität Dresden, Fakultät Mathematik und Naturwissenschaften, 01062 Dresden, Germany
*synytska@ipfdd.de

Abstract

Janus particles (JPs), named after the two-faced Roman god, are a unique class among the multifunctional patchy particles, combining two distinct chemical/physical functionalities at their opposite sides. The exclusive chemical and geometrical asymmetry of the JPs provides the possibility to address complex self-assembled architectures and materials inaccessible for homogeneous building blocks. In this chapter, we will highlight the recent developments in the employment of JPs as novel building blocks for active multifunctional materials. First, a short introduction into the synthesis, assembly, and applications of the JPs will be presented, covering a broad variety of the state-of-the-art strategies and techniques. The challenges faced in the topic of JPs and the subsequent opportunities will be discussed. The main focus, however, will be placed on the application of JPs for active surfaces and interfaces, providing a review on the diversity of up-to-date studies aimed at different applications. In terms of active interfaces, the stabilization efficiency of the JPs at interfaces can be combined with the ability of interface structuring/functionalizing. Due to the high interfacial activity of the JPs and their favorable orientation, the inner and outer sides of the stabilized interfaces (e.g., emulsion droplets) can be furnished with active compounds and catalysts, or feature different optical appearances and other properties, yielding active functional interfaces. In terms of active surfaces, JPs can be utilized as advanced building blocks to construct materials with controlled chemical and topographical heterogeneity as well as wettability and adhesion. Especially, combining both chemical and topographical heterogeneity, such materials may produce synergistic effects favorable for specific applications. Finally, our critical view on the applicability and rational design of multifunctional materials based on the JPs will be suggested [153], closing with the future perspectives and visions.

Keywords: Janus particles, hybrid Janus particles, active surfaces, multi-functional coatings, active interfaces, materials based on Janus particles.

10.1. Janus Particles: Novel Building Blocks with Two Faces

10.1.1. Introduction and Historical Perspective

Asymmetry, the absence or violation of symmetry, is very widely seen in nature, having evolved at many levels of organization from individual cells to entire organs and body shapes (Fig. 1) [153]. Going deeper onto the molecular level, the asymmetrical structure of amino acids determines the asymmetrical architectures of proteins, whereas the amphiphilic nature of asymmetric lipids results in their self-organization into bilayers, yielding vesicles, liposomes, or membranes for cells and bacteria. Furthermore, molecular asymmetry is crucial for the design of synthetic molecules, surfactants, liquid crystals, or block copolymers, which further self-assemble into micelles, vesicles, polymersomes, as well as sophisticated liquid crystalline, and block copolymer bulk phases [1]. Therefore, asymmetry at the building block level can lead to complex assembled constructs with hierarchical architectures favorable for various fields of application.

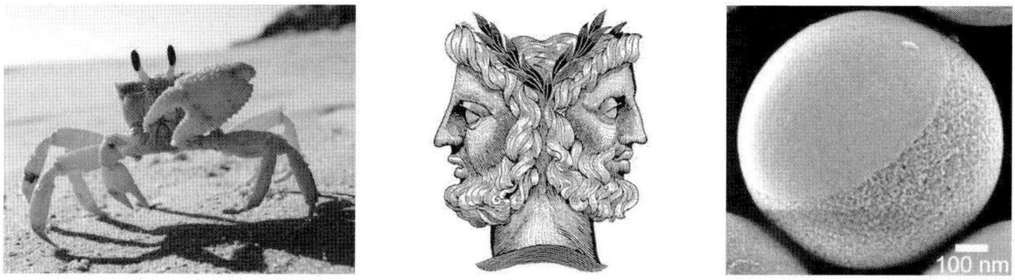

Figure 1. The concept of JPs: (from left to right) asymmetry in nature on an example of a crab with differently sized claws (ask.com), the Roman god Janus with two distinct faces (creationoutreach.com), and an scanning electron microscope (SEM) image of a synthesized JP (own work).

On the other hand, colloidal particles, which are small objects suspended in continuous medium in particulate form with sizes at least in one dimension being in the range from several nanometers to a few micrometers, may be considered as building blocks for advanced materials [2]. Such particles can provide the engineered materials with rich functionality, such as optical activity, mechanical strength, conductivity, etc. For instance, periodic arrays of homogeneous colloids (colloidal crystals) exhibit unique photonic properties [3]. The recent breakthroughs in the particle synthesis have led to a remarkable variety of building blocks with different shapes, compositions, patterns, and functionalities [4]. Among these building blocks, asymmetric patchy particles have gained significant attention in the recent years due to their advanced anisotropy, which significantly contrasts with the isotropic spherical colloids, providing the materials assembled from patchy particles with different macroscopic properties [5].

Janus particles (JPs) are a unique class among the multifunctional patchy particles, combining two distinct chemical/physical functionalities at their opposite sides (Fig. 1). These particles were named after the two-faced Roman god Janus, the god of beginnings and endings, gates, transitions, time, and passages (Fig. 1). The term "Janus beads" was first coined by Casagrande *et al.*, describing glass spherical particles with one hydrophilic and one hydrophobic hemisphere, and the behavior of these beads at oil/water interfaces [6]. The amphiphilic solids were prepared by protecting one hemisphere with a cellulose varnish and chemically treating the other one with octadecyltrichlorosilane. In 1991, Pierre-Gilles de Gennes addressed the JPs in his Nobel lecture, pointing out their potential and significance [7]. He mentioned the unique property of the amphiphilic Janus grains to form dense films at interfaces, while allowing chemical exchange to occur through the gaps between the particles, identifying it by terming that "the skin can breathe". Furthermore, the exclusive chemical and geometrical asymmetry of the JPs provides the possibility to address complex self-assembled architectures and materials inaccessible for homogeneous building blocks.

Significant progress has been made in the past two decades concerning the fabrication of JPs through the development of numerous synthetic strategies, the understanding of their advanced physical properties, and their potential fields of applications [1]. Originally, a considerable challenge was the advancement of synthetic procedures to break the symmetry of existing particles, or to directly synthesize the JPs with a high yield and monodispersity. However, extensive advances have been made to attain various types of JPs, ranging from spherical, disk-shaped, or cylindrical to dumbbell-like JPs, and Janus capsules/vesicles [1]. Along with improved pathways to synthesize the JPs, increasing interest has been focused on their unusual self-assembly behavior, as well as on their possible fields of applications, such as solid surfactants, sensors, elements for electronic paper, or biomedical applications.

10.1.2. Design, Assembly, and Applications of Janus Particles

Using the newly developed synthetic strategies, JPs with diverse functionalities, well-defined shapes, and various architectures were fabricated with sizes throughout the nano- to microscopic length scale [1]. However, despite the numerous routes to construct different kinds of JPs, the successful preparation of well-defined multifunctional JPs with advanced physical properties on a large scale is still an important issue in the field. According to their physical properties, the JPs can be classified into soft (organic, polymer-based), hard (inorganic, metallic), and hybrid organic/inorganic JPs.

Soft JPs: Soft, organic JPs represent the largest group of JPs, including unimolecular JPs such as Janus dendrimers [8–12], or Janus-type heteroarm star copolymers [13] and cylindrical polymer brushes [14], biological and biomimetic JPs [15–19], self-assembled polymeric Janus micelles

[20–29], as well as larger microscale polymeric JPs [30–40]. The smallest unimolecular JPs are typically prepared using classical organic and polymer synthesis [8–14], while biomimetic JPs can be accessed through protein engineering, and controlled masking and modification [15–19]. Very effective tools for the generation of polymeric Janus micelles are block copolymer self-assembly in solution (Fig. 2(a)) [20–23], as well as self-assembled block terpolymer bulk structures [24–29], (Sub-) micrometer-sized polymeric JPs can be fabricated using a diversity of well-defined approaches, including phase separation in confined volumes [30–32,40], seeded emulsion polymerization [33,34], electrohydrodynamic cojetting [35,36,41], and microfluidics [37–39,42].

Hard JPs: Hard, inorganic JPs typically comprise either (sub-) micrometer-sized metal oxide particles (silica, titanium dioxide) partly coated with metals or inorganic nanoparticles (NPs), or functionalized (bi-) metallic Janus NPs. The most widespread strategy for the preparation of hard (sub-)micrometer-sized JPs is the partial temporary masking of one side of the particles using a temporary immobilization template. This strategy includes the modification of the exposed surfaces, typically via metal or NP deposition, of particle monolayers immobilized on planar substrates (Fig. 2(b)) [43–51], as well as the toposelective modification of particles in Pickering emulsions [52–55]. Inorganic Janus NPs, usually comprising Au or Fe_3O_4 NPs, can be fabricated through controlled immobilization and subsequent selective modification of metal NPs [56,57], or through the generation of bimetallic dumbbell-shaped NPs [58,59].

Hybrid JPs: Hybrid, organic/inorganic JPs are a particularly interesting class of JPs, which possess advanced multifunctionality due to their hybrid structure, combining the responsiveness of polymers with mechanical stability of inorganic particles. The most common design in the case of hybrid JPs is an inorganic core particle toposelectively covered with either one or two distinct polymer shells. The same techniques as in the case of soft and hard JPs can be applied with necessary modifications for the synthesis of hybrid JPs, such as phase separation in miniemulsions [61] seeded emulsion polymerization (Fig. 2(c)) [60,62,63], microfluidics [64], as well as toposelective modification of particle monolayers [65], or particles in Pickering emulsions [66–69]. Additionally, controlled/living radical polymerization techniques can be used in the fabrication process

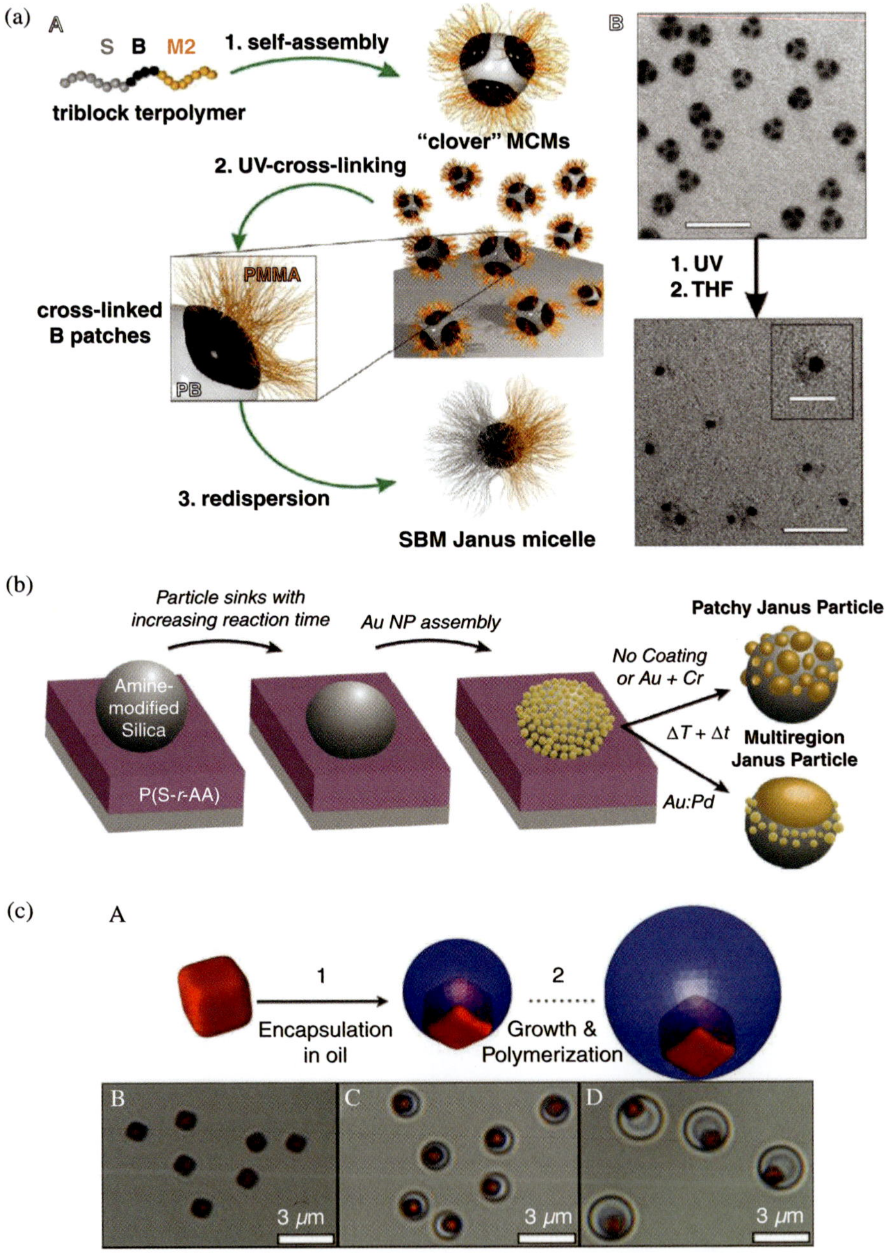

Figure 2. Synthetic strategies for the fabrication of (a) soft, (b) hard, and (c) hybrid JPs. (a): Soft JP synthesis via self-assembly of triblock terpolymers into multicompartment micelles (MCMs), subsequent cross-linking of the compartments and redispersion in THF

of hybrid JPs via simultaneous biphasic grafting of different polymer brushes onto the opposite sides of the core particles [67–71]. For instance, hybrid hairy JPs of different (sub-)micrometer sizes can be prepared via a combination of "grafting from" and "grafting to" approaches [68,71]. First, a wax-water pickering emulsion is formed, stabilized by amino-modified silica particles, where one side of the particles is masked, and the exposed one is modified with an initiator for atomic transfer radical polymerization (ATRP, "grafting from" approach), which is performed after the subsequent dissolution of the wax (Fig. 3). The "grafting to" approach is then used to couple carboxy-terminated polymer chains with the amino groups on the opposite particle side in the melt, yielding responsive hybrid hairy JPs.

Self-assembly of JPs: The understanding and programming of the colloidal particle self-assembly process is of great fundamental interest for the creation of novel materials, which may possess advanced properties that are not present in the individual building blocks. As JPs are a unique class of building blocks, combining chemical/physical anisotropy, their assembly can lead to a variety of sophisticated structures inaccessible to their homogeneous counterparts [1,72]. Therefore, diverse experimental and theoretical studies were dedicated to the self-assembly of various JPs, including the self-assembly of soft nanoscale JPs (Fig. 4(a)) [11,24,25,73], amphiphilic (Fig. 4(b)) [74–76], or dipolar [54,68,77–80]. Janus colloids, and the assembly of JPs in external electric (Fig. 4(c)) [81–84] or magnetic (Fig. 4(d)) [60,85–89] fields. Nevertheless, a deeper insight into the assembly mechanisms of the JPs is needed in order to translate these mechanisms toward the production of real materials.

Figure 2. (*Continued*) (a); corresponding transmission electron microscope (TEM) images of the "clover" MCMs and the derived JPs with OsO_4 staining: polystyrene (PS) gray, polybutadiene (PB) black, poly(methyl methacrylate) (PMMA) is not visible due to electron beam degradation, scale bars are 200 and 50 nm in the inset. (b): Schematic representation of the self-assembled formation of patchy and multiregion hard JPs using a temporary immobilization template. (c): Schematic diagram showing the synthetic steps for the preparation of hybrid JPs with a single magnetic patch via seeded emulsion polymerization (a); encapsulation and growth followed by optical microscopy (B–D). Reproduced with permission from Refs. [23, 47, 60]. Copyright (2010), (2012) American Chemical Society.

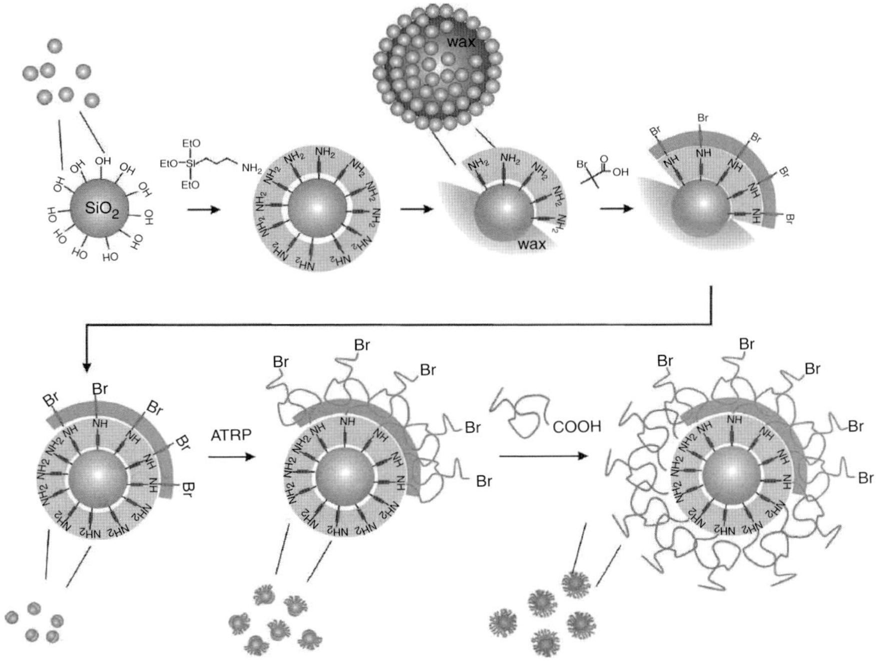

Figure 3. Schematic illustration of the hybrid hairy JP synthesis via "grafting from"/"grafting to" approaches. Reproduced with permission from Ref. [68]. Copyright (2008) American Chemical Society.

Applications of JPs: One of the most fascinating applications of JPs arises from their enhanced surface activity compared to homogeneous particles. Therefore, JPs are perfect candidates for the stabilization of multiphasic fluid mixtures, such as emulsions (Fig. 5(a)) [31,90], or polymer blends (Fig. 5(b)) [91,92]. Besides the stabilization effects of JPs, the possibility to use them for structuring and functionalizing of interfaces is very promising for special applications. The use of JPs for active interfaces as well as surfaces will be discussed in detail in the forthcoming chapters. Briefly, in terms of interfaces, the JPs can be applied to prepare functional emulsion droplets [93], self-assembled liquid mirrors [94], liquid marbles [50], and for interfacial catalysis [95], while in terms of surfaces JPs can be used for water-repellent textiles [96], ultra-hydrophobic coatings [97], electronic paper displays [98–100], and biological applications [101–103]. Taking advantage of their asymmetric character,

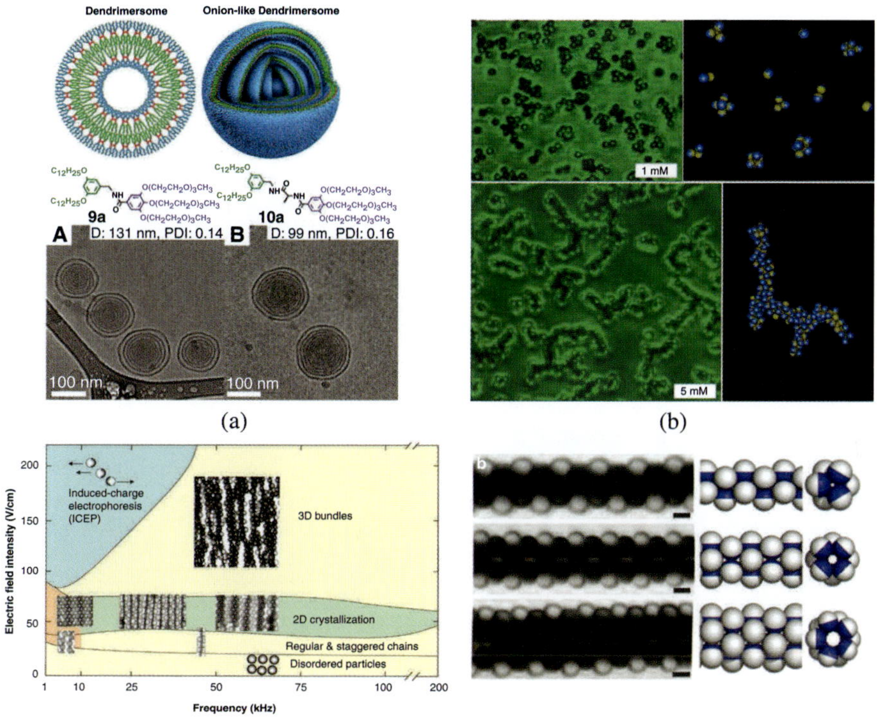

Figure 4. Self-assembly of JPs. (a): Schematic representation of a single-bilayer dendrimersome self-assembled from Janus dendrimers and a multibilayer onion-like dendrimersome (upper part); representative cryo-TEM images (A, B) of onion-like dendrimersomes self-assembled by the injection of a THF solution of Janus dendrimers (9a or 10a) in water. (b): Clusters formed from amphiphilic Janus colloids (1 μm diameter) at different salt concentrations — epifluorescence images (left) and Monte Carlo computer simulations (right; blue and yellow colors represent charged and hydrophobic hemispheres, respectively). Small clusters in equilibrium with one another are formed at lower salt concentrations (1 mM KNO_3, upper part), and long, branched wormlike strings — at higher salt concentrations (5 mM KNO_3, lower part). (c): JP assembly in electric fields — phase diagram as a function of the AC electric field intensity versus field frequency revealing five different regions. (d): JP assembly in magnetic fields — observed images (left) and corresponding models (right; side and end views) of microtubes parallel to the precession axis formed via synchronized self-assembly using magnetic Janus colloids. Reproduced with permission from Refs. [11,75,81,87]. Copyright (2008) American Chemical Society, (2012) Nature Publishing Group, (2014) National Academy of Sciences. (Ref. [11] (PNAS) to be requested).

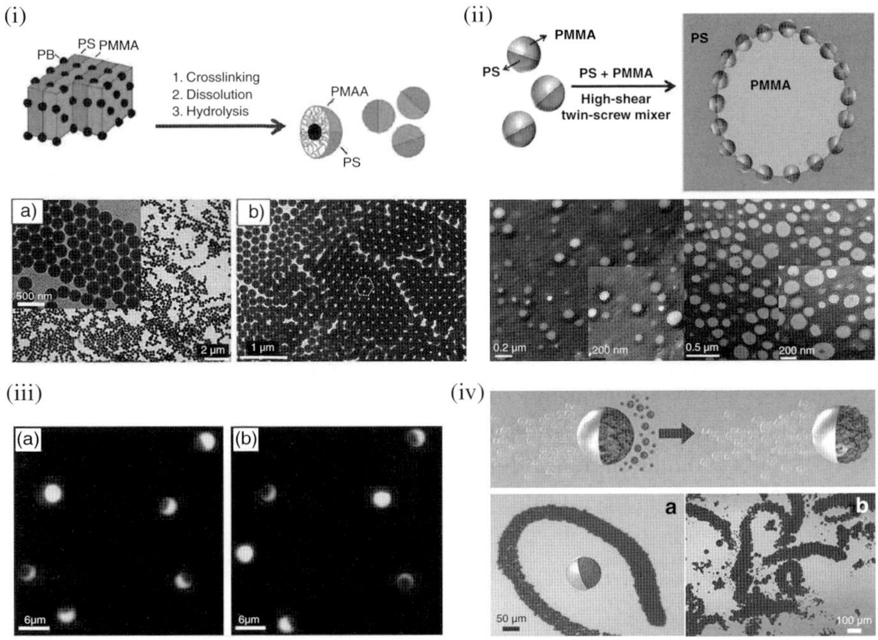

Figure 5. Applications of JPs. (i) Amphiphilic PS/poly(methacrylic acid) (PS/PMAA) JPs applied as emulsifiers for an emulsion polymerization. Template-assisted synthesis of the JPs (upper part); TEM images (lower part) of the monodisperse PS latexes formed with a JP stabilizer content of (a) 0.5 and (b) 4 wt.%. (ii) PS/PMMA JPs applied as compatibilizers for polymer blends. Schematic representation of JPs and their adsorption at the blend interface of a PS/PMMA blend (upper part); TEM images (lower part) obtained for 10 wt.% JPs in an 8/2 PS/PMMA blend (left) and for 20 wt.% JPs in a 6/4 PS/PMMA blend (right). (iii) JPs applied as modulated optical nanoprobes (MOONs). Fluorescence from six different 3.4 μm aluminum-capped Brownian MOONs individually rotating through the phases of the moon at two different times. (iv) Self-propelled Janus micromotors for water purification. Schematic of the activated carbon/Pt Janus micromotor and its "on-the-fly" decontamination use (upper part; dark hemisphere — spherical activated carbon, light gray area — catalytic Pt hemisphere patch); time-lapse images (lower part) illustrating the motor propulsion over a 4 s period in (a) ultrapure water, and (b) 70% seawater. Reproduced with permission from Refs. [90,91,106,110]. Copyright (2005) Elsevier, (2008), (2015) Wiley-VCH Verlag GmbH & Co. KGaA, (2008) American Chemical Society.

individual JPs can be applied as modulated optical nanoprobes (Fig. 5(c)) [104–106], or self-propelled micro/nanomotors (Fig. 5(d)) [107–110]. Clearly, the true potential of JPs lies in their unique asymmetry and multifunctionality, distinguishing them from other structures, and allowing to access new advanced materials.

10.1.3. Challenges and Opportunities

As already mentioned in the previous sections, important breakthroughs have been made concerning the synthesis, assembly, and applications of the JPs. As a result, the number of contributions on the topic of "Janus particles" is rapidly growing (Fig. 6(a)), indicating the importance of this field of research. Nevertheless, considering the number of contributions regarding the topic of JPs in general (Fig. 6(a)), by far less contributions are being published regarding the applications of the synthesized JPs toward rational material design (Fig. 6(b)). This undoubtedly indicates on the need of a more extensive research in the field of possible applications of the JPs based on their unique asymmetric features.

One of the most appealing and beneficial properties of the JPs is their high interfacial activity, which results in the ability of JPs to adsorb on the interfaces more efficiently than their homogeneous analogs. Therefore, one of the main application directions for the JPs is the stabilization of different interfaces, for example, in emulsions, foams, or polymer blends. Statistically, almost half of the contributions regarding the applications of the JPs are dedicated to interfaces (Fig. 7(a)). Most of these studies apply the JPs as stabilizing agents at different fluid–fluid interfaces. However,

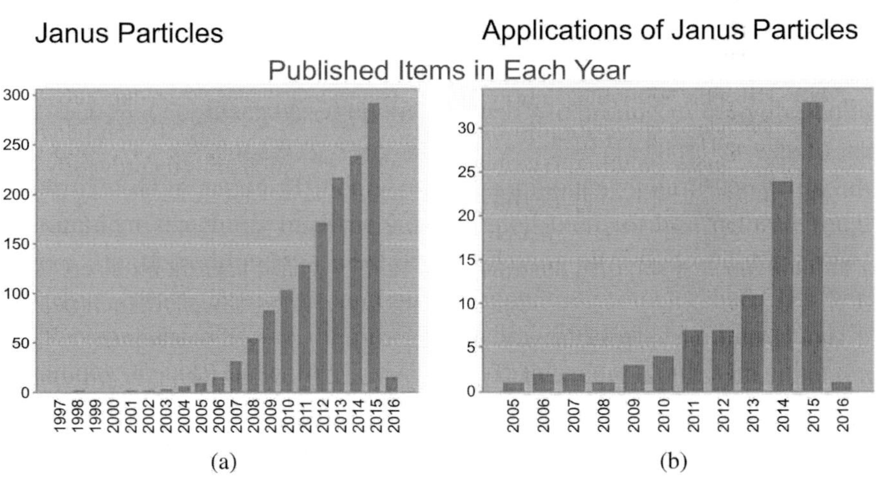

Figure 6. Number of contributions concerning the topic JPs in general (a) as well as the applications of JPs (b) published each year over the last 20 years (according to ISI web of science, 22.02.2016) [111].

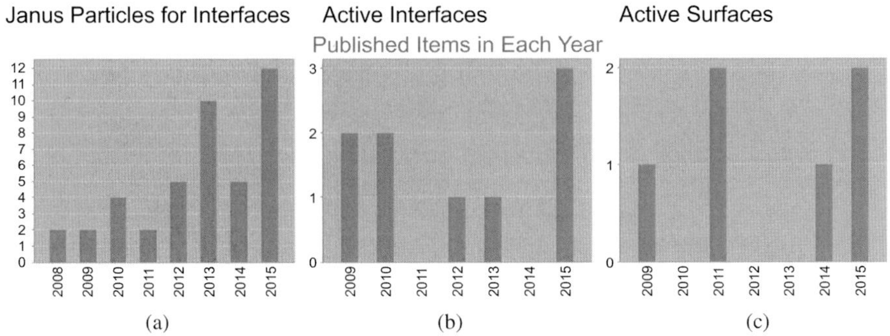

Figure 7. Number of contributions concerning the applications of JPs for interfaces in general (a), active interfaces (b), and active surfaces (c), published each year over the last 20 years (according to ISI web of science, 22.02.2016) [111].

aside from the consideration of simply the stabilization efficiency of JPs at interfaces, a very promising property is also the ability of interface structuring/functionalizing with JPs toward *active functional interfaces*. Due to the high interfacial activity of JPs and their favorable orientation, the inner and outer sides of the stabilized interfaces (e.g., emulsion droplets) can be furnished with active compounds and catalysts, or feature different optical appearances and other properties. Some work has already been done in this direction, and contributions regarding the JPs for active interfaces are starting to appear in the recent years (Fig. 7, middle). Nevertheless, the number of such contributions is very small, especially considering the number of contributions on the topic of JPs in general as well as their applications (Fig. 6). Hence, this pinpoints a very promising and beneficial area for the application of JPs, where their interfacial activity can be combined with their multifunctionality.

Apart from active functional interfaces, surfaces equipped with JPs are another advantageous field of research, where the multifunctionality of JPs may lead to a synergistic effect and thus enhanced desired features of the resulting material. *Active functional surfaces* are a very important and promising topic that can be extended to the production of real materials in, for instance, the coating technology. Typically, mono-functional surfaces (e.g., hydrophilic, hydrophobic) are being studied in this regard, but more and more attention is being drawn to the heterogeneous surfaces, which combine the favorable properties inherent to the corresponding monofunctional ones. Consequently, structuring of surfaces with JPs can provide them with rich functionality and heterogeneity on different scales required to address specific applications. In the recent years, several contributions on the topic of JPs for active surfaces emerged (Fig. 7, right), though their number is almost negligible compared to the plentitude of contributions on the topic of JPs in general, and even to the much smaller number of contributions dedicated to their applications (Fig. 6). Correspondingly, this points out another appealing area of the JPS applications.

In this chapter, the application of JPs for active interfaces and active surfaces will be discussed in detail, showing the state-of-the-art and the benefits of the JPs for the corresponding topics.

10.2. Active Functional Interfaces

10.2.1. Interfacial Activity of JPs

The ability of solid particles to adsorb at interfaces such as gas–liquid or liquid–liquid is of great interest due to its diverse potential applications in the stabilization of foams and emulsions, respectively. Particularly, a Pickering emulsion, first described by S. U. Pickering in 1907, is a two-phase system consisting of oil and water, where the oil/water interface is stabilized by solid particles (for example, colloidal silica) binding to the interface via purely physical interactions [112]. The type of the emulsion (oil-in-water or water-in-oil) and its coalescence stability is dictated by the hydrophobicity of the particles, which can be varied by coating them with different materials or to different extents. As a result, particles of intermediate hydrophobicity are the most effective for stabilizing submicrometer-sized oil or water droplets. Such particles typically have different ratios of hydrophobic and hydrophilic groups on the surface and exhibit different wettabilities, but in each case the overall wettability is uniform over the particle surface [113].

JPs possess wettability anisotropy, that is, amphiphilicity inherent to low-molecular-weight surfactants or block copolymers, and can be used to stabilize Pickering emulsions during longer periods of time and under experimentally more stressful conditions (temperature changes or shear effects) than the homogeneous particles (HPs) [113]. Contrary to HPs, amphiphilic JPs can exhibit high interfacial activity regardless of their

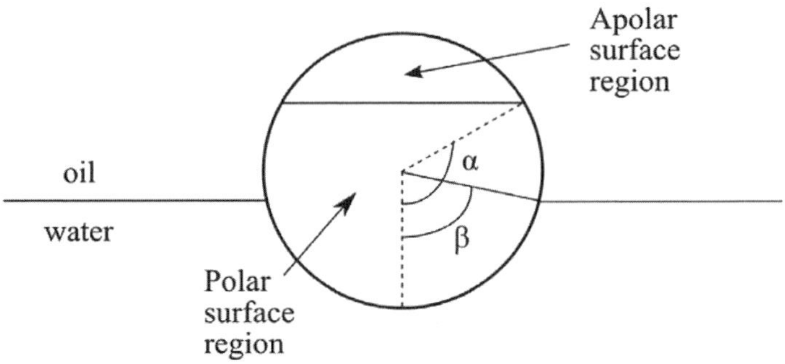

Figure 8. Geometry of a JP within an oil–water interface. The relative areas of the polar and apolar particle surface regions are parameterized by the angle α. The immersion depth of the particle in the oil/water interface is parameterized by the angle β. Reproduced with permission from Ref. [113]. Copyright (2001) American Chemical Society.

amphiphilicity degree, simply due to the spatial separation of the different wettability regions [114]. Accordingly, it is crucial to properly evaluate the colloidal stability in bulk and the interfacial activity of JPs to develop their favorable applications.

A theoretical comparison between the desorption energies of particles with uniform wettability (HPs) and JPs adsorbed at an oil–water interface was accomplished by Binks and Fletcher [113]. The amphiphilicity degree was varied from zero to the maximum value. The amphiphilicity of JPs can be tuned by the variation of both the Janus balance (Fig. 8, relative areas of the polar and apolar domains, α) and the magnitude of the difference between the θ_A and θ_P, which correspond to the equilibrium contact angles of the two distinct domains (A — apolar, P — polar) of the particle adsorbed at the oil–water interface. Zero amphiphilicity correlated to HPs corresponds to either $\alpha = 0°/180°$, or $(\theta_A - \theta_P) = 0°$. On the other hand, the highest degree of amphiphilicity is expected when $\alpha = 90°$ and $|\theta_A - \theta_P| = 180°$. The total surface free energy E of a JP at an oil–water interface as a function of the angle β, characterizing the immersion depth of the particles (Fig. 8), can be expressed as

For $\beta \leq \alpha$

$$E(\beta) = 2\pi R^2[\gamma_{AO}(1+\cos\alpha) + \gamma_{PO}(\cos\beta - \cos\alpha) + \gamma_{PW}(1-\cos\beta) - \tfrac{1}{2}\gamma_{OW}(\sin^2\beta)], \tag{1}$$

For $\beta \geq \alpha$

$$E(\beta) = 2\pi R^2[\gamma_{AO}(1+\cos\beta) + \gamma_{AW}(\cos\alpha - \cos\beta) + \gamma_{PW}(1-\cos\alpha) - \tfrac{1}{2}\gamma_{OW}(\sin^2\beta)]. \quad (2)$$

Here, R is the particle radius, and γ_{AO}, γ_{PO}, γ_{AW}, γ_{PW}, and γ_{OW} are the interfacial tensions of the apolar domain–oil, polar–oil, apolar–water, polar–water, and oil–water interfaces, respectively.

The contact angles θ_A and θ_P correspond to the equilibrium angles given by the Young's equation as follows:

$$\cos\theta_A = \frac{\gamma_{AW} - \gamma_{AO}}{\gamma_{OW}}, \quad (3)$$

$$\cos\theta_P = \frac{\gamma_{PW} - \gamma_{PO}}{\gamma_{OW}}. \quad (4)$$

The average contact angle of the JP is weighted by the relative areas of the polar and apolar domains as follows:

$$\theta_{\text{average}} = \frac{\theta_A(1+\cos\alpha) + \theta_P(1-\cos\alpha)}{2}. \quad (5)$$

The immersion angle β corresponding to the minimum surface energy configuration of the JPs depends on the relative magnitudes of α, θ_A, and θ_P. There are three possibilities in this case:

$$\begin{aligned} &\text{For } \alpha < \theta_A < \theta_P; && \text{then } \beta = \theta_A, \\ &\text{For } \theta_A < \alpha < \theta_P; && \text{then } \beta = \alpha, \\ &\text{For } \theta_A < \theta_P < \alpha; && \text{then } \beta = \theta_P. \end{aligned} \quad (6)$$

Equations (1) and (2) together with the inequalities in Eq. (6) allow the calculation of the minimum surface energy of the adsorbed particle, E_{ads}. The surface energy of the particle located entirely in either the oil (E_{oil}) or the water (E_{water}) phase is given by Eqs. (1) and (2) with β set to either 0° (for the particle in oil) or 180° (for the particle in water). The surface activity of the (Janus) particle was calculated by Binks and Fletcher [113] as the desorption energy, which is defined as the free energy required to desorb the particle from the interface into either the oil or the water phase, $(E_{oil}-E_{ads})$ or $(E_{water}-E_{ads})$, whichever is the lowest. The two most important conclusions drawn from this theoretical study are: (1) the desorption energies of HPs may be increased threefold by maximizing the amphiphilicity of JPs; (2) contrary to HPs, JPs retain their strong adsorption even for average contact angles of 0° and 180°.

Following this study, further contributions to theory and simulations of JPs at mostly fluid–fluid interfaces appeared, highlighting their unique interfacial properties and behaviors that are not observed in HPs [114–117], and investigating, for instance, the adsorption of JPs to curved interfaces [118], the influence of the JP shape on their orientation and surface activity at interfaces [119], the influence of the Janus balance [120], or the behavior of magnetic JPs at interfaces interacting with an external magnetic field [121]. Experimentally, the surface activity of JPs at fluid–fluid interfaces can be determined using two different methods: the classical Langmuir balance, or the pendant drop tensiometry [114]. Generally, the important parameters established in all theoretical and experimental studies that impact the interfacial activity of JPs are shape, morphology, charge, composition, spatial distribution of the distinct domains in a JP, as well as spreading solvent, and colloidal stability [114].

Clearly, the theoretical predictions and experimental model works emphasizing the enhanced interfacial activity of the JPs also led to the attempts to apply the JPs as stabilizers in emulsions, foams, or polymer blend systems [1]. In terms of liquid–liquid emulsion interfaces, the first real application study was based on successfully employing polymeric Janus micelles as Pickering emulsifiers for the conventional emulsion polymerization of styrene and *n*-butyl acrylate (Fig. 5(a)) [90]. Following this study, various contributions appeared on the efficient use of JPs for the stabilization of mostly oil–water interfaces (Figs. 9(a) and 9(b)) [31,61,69,122–127]. Apart from liquid–liquid interfaces, the behavior of JPs in homopolymer mixtures or within block copolymers and the subsequent polymer blend compatibilization is another promising area for the application of JPs. In the first experimental contribution, the efficient compatibilization of PS/PMMA

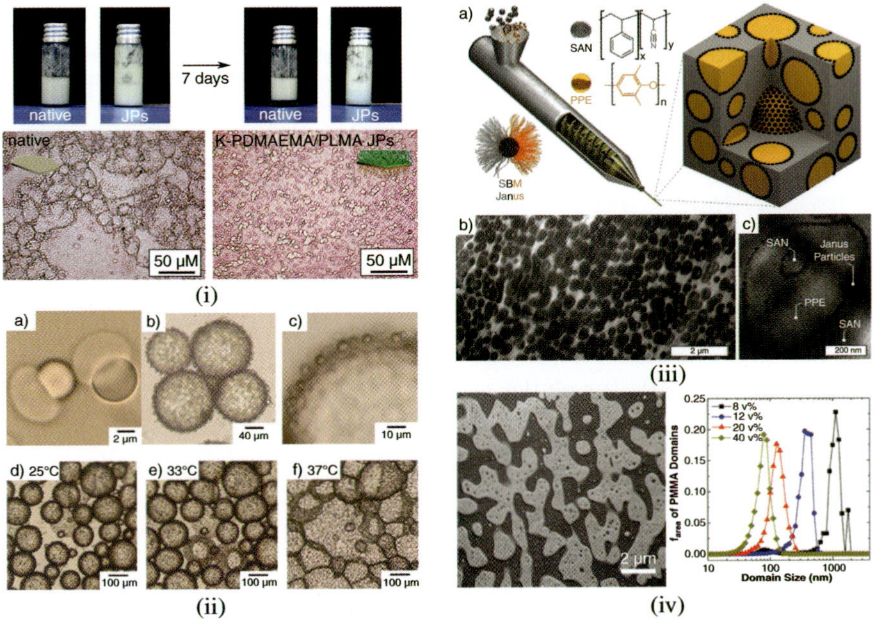

Figure 9. Applications of JPs for the stabilization of emulsions (a) and (b) and polymer blends (c) and (d). (i) Stabilization of a water–oil emulsions with native kaolinite particles and poly(2-dimethylaminoethyl methacrylate)/poly(lauryl methacrylate) (PDMAEMA/PLMA) core–shell JPs. Upper part: optical images of the emulsions directly after preparation and after seven days, showing the destabilization of the native kaolinite-stabilized emulsion, whereas the JP-stabilized emulsion remains stable. Lower row: optical micrographs revealing the difference in the droplet size in the native kaolinite-stabilized emulsion (left) and the JP-stabilized emulsion (right). (ii) Optical micrographs of (a) mushroom-like PMMA/P(S-BIEM)-*g*-PDMAEMA (S-BIEM = styrene-2-(2-bromoisobutyryloxy)ethyl methacrylate) JPs, (b) and (c) 1-Octanol-in-water emulsion droplets stabilized by the JPs, (d)–(f) the emulsion stabilized by the JPs taken at the same visual field and at a rising rate of 5°C/min showing temperature control over the emulsion stability. (iii) Processing of polymer blends using JPs as compatibilizers, (a) during extrusion of the polymer melt, PS/PMMA JPs (black dots) compatibilize and stabilize poly(2,6-dimethyl-1,4-phenylene ether) (PPE, yellow) droplets within the poly(styrene-*co*-acrylonitrile) (SAN, gray) matrix, (b) and (c) TEM images of the compatibilized blend at different magnifications showing a small drop morphology, whereas the JPs are exclusively located at the interface. (iv) TEM micrograph (left) of the PS/PMMA blend (solvent-cast film) with an 8% PS/PMMA JP loading and a histogram plot (right) of the area-weighted fraction of PMMA domains as a function of domain size (chord length), as determined by image analysis, showing the change in domain size with varying JP loadings. Reproduced with permission from Refs. [69,127–129]. Copyright (2014), (2015) American Chemical Society.

polymer blend systems was demonstrated using PS/PMMA JPs under high shear conditions in a twin-screw minimixer, which is rather similar to the real conditions in polymer processing (Fig. 5(b)) [91]. The following experimental studies and simulations also discussed the beneficial effects of JPs for the compatibilization of polymer blend systems (Figs. 9(c) and 9(d)) [92,128–131].

On the other hand, the enhanced interfacial activity of the JPs can be combined with their inherent multifunctionality and used for the structuring and functionalizing of the stabilized interfaces. As already mentioned in the previous section, some studies have emerged by this time concerning the topic of JPs for active interfaces (Fig. 7(b)), which will be discussed in more detail in the following section.

10.2.2. Interface Design

Due to the favorable orientation of JPs at interfaces and their strongly decreased rotational diffusion, the inner and outer sides of the emulsion droplets can be selectively modified to present optically different properties, or equipped with catalysts and other compounds. Another promising example of an active interface, where JPs can be applied with great potential, is the electronic paper technology.

10.2.2.1. *Optically active emulsion droplets*

In terms of JP-loaded emulsion droplet interfaces featuring optically different appearances at their inner and outer sides, Sailor and Link demonstrated a selective orientation of bifunctional porous Si smart-dust particles containing a hydrophobic green mirror and a hydrophilic red mirror at the interface between an organic liquid and water according to their polarity resulting in a homogeneously colored emulsion droplet (Fig. 10(a)) [93]. The amphiphilic micrometer-sized Janus plates were prepared by electrochemically etching two discrete porous dielectric mirrors into Si, one on top of the other, chemically modifying the first mirror by hydrosilylation and the second one by mild thermal oxidation, and subsequent fracturing of the films into small particles, each particle containing the necessary components to allow self-assembly, spectroscopic identification, and chemical sensing (Fig. 10(a)). These

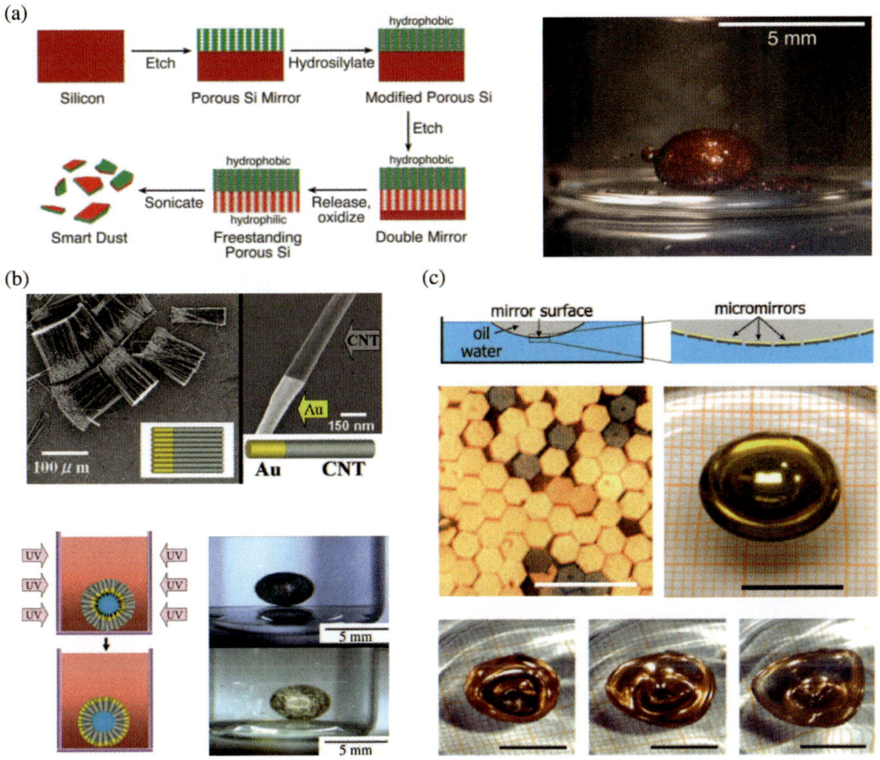

Figure 10. (a) Synthesis of bi-functional smart-dust particles (left) and an optical image of their assembly at the interface of a dichloromethane droplet in water (right). (b) SEM images and schematics (insets) of the CNT–Au hybrid nanowires (upper part) and schematics with optical images showing the photoinduced surface modification of the hybrid nanowire assembly using UV irradiation (lower part). (c) Self-assembled liquid mirror and micromirror organization — schematics (upper part), optical microscopy (middle left) and optical image (middle right) of the micromirror assembly at the oil–water interface with the gold face oriented toward the oil phase; frame captures (450 ms between frames) of an acoustically excited liquid mirror (lower part, scale bars: 1 cm). Reproduced with permission from Refs. [93,94,132]. Copyright 2003 National Academy of Sciences, (2008), (2009) American Chemical Society (Ref. [93] (PNAS) to be requested).

particles spontaneously align at an organic liquid–water interface, with the hydrophobic green side oriented toward the organic phase and the hydrophilic red side toward the water phase (Fig. 10(a)). Sensing is accomplished when liquid at the interface infuses into the porous mirrors, inducing predictable shifts in the optical spectra of both mirrors.

In a similar context, Ajayan *et al.* produced hybrid inorganic multi-segmented nanowires with hydrophobic CNT tails and hydrophilic metal nanowire heads, which allowed the assembly and manipulation of massive ordered structures such as oil–water emulsion droplets in solution (Fig. 10(b)) [132]. The hybrid Au–CNT and Au–Ni–CNT structures were grown using AAO templates by a combination of electrodeposition and chemical vapor deposition methods. Owing to their amphiphilic nature, the assembly of the hybrid nanowires is sensitive to the environment and they can convey the environment change through the color change induced by the alternation in their assembly. For instance, the functionalization of the Au surface with a perfluorinated thiol rendered it more hydrophobic, resulting in its facing the interior of an emulsified dichloromethane (DCM) droplet, yielding a black droplet due to the CNTs pointing to the surrounding water (Fig. 10(b)). However, upon photoinduced cleavage of the thiol, the hydrophobic–hydrophilic balance between the Au and CNT segments is changed, and the black sphere flips back to the one with the golden surface due to the CNTs now pointing to the DCM phase (Fig. 10(b)). Such reconstruction of the assembled structure via photoinduced surface modification may be effectively used in applications such as smart delivery systems. Moreover, the incorporation of a Ni segment in the hybrid nanowires allows them to be manipulated using a magnetic field.

In another approach, Krupenkin *et al.* employed the self-assembly of selectively gold-functionalized hexagonal Si/SiO_2 micromirrors (Janus tiles), produced using photolithographic techniques, on the surface of an oil droplet to create a concave liquid mirror (Fig. 10(c)) [94]. The hydrophobic gold sides of the Janus tiles were facing predominantly toward the oil phase because of the different wetting properties of the two sides. The best results with regard to reflectivity and alignment of the micromirrors were observed using pentyl benzoate as the oil phase: more than 80% of the micromirrors demonstrated proper orientation. The focal length and axial position of the liquid mirror could be controlled using electrowetting. Moreover, the mirror was mechanically robust and retained its integrity even at high levels of vibrational excitation of the interface (Fig. 10(c)). Ultimately, the presented approach may enable a wide range of new optofluidic devices such as a 3D projector, i.e., a light source, which can scan an image onto a moving, non-planar focal surface.

10.2.2.2. *Flexible barriers and liquid marbles*

A highly flexible and impregnable water-repelling interface was created by Kim *et al.* using the self-assembly of porous and rough Janus microspheres with superhydrophobic and hydrophilic faces [50]. The Janus microspheres were prepared utilizing photopolymerized emulsion droplets of an ethoxylated trimethylolpropane triacrylate (ETPTA) resin decorated with arrays of silica particles as a template. The silica particles were then removed by a wet-etching process to create cavity arrays, and superhydrophobicity as well as additional porosity and the subsequent asymmetric character were achieved by applying reactive ion etching with sulfur hexafluoride. The fabricated monodisperse Janus microspheres formed a hexagonal array at the air–water interface, providing a flexible barrier against coalescence, as water droplets freely floated on top of them, while the hydrophilic surfaces of the JPs were strongly held and aligned along the interface. Moreover, a membrane of Janus microspheres maintained its integrity under a dynamic disturbance induced by a hydrophilic glass stick. Additionally, the JPs were employed to make liquid marbles, which consisted of a water droplet completely covered with Janus microspheres with their hydrophilic surfaces directed toward the center of the droplet. The resulting liquid marbles even sustained handling with tweezers. Eventually, such JP-stabilized monolayers may have great potential in size-dependent semipermeable membranes along interfaces between immiscible fluids, buoys for water floating micromachines, and various superhydrophobic coatings.

In a similar context, Zhao *et al.* created JPs with multiple features, such as different wettability, tailored arrangement of structural colors, and magnetism in distinct compartments of the JPs (Fig. 11) [133]. The JPs were generated using microfluidic emulsification followed by emulsion polymerization and surface treatment (Fig. 11(a)). Colloidal crystal solutions of monodisperse silica or α-Fe_2O_3 NPs in an ETPTA resin were used as the dispersed oil phases for the emulsions. Because of the periodic arrangement of these NPs, the prepared ETPTA resins displayed iridescent colors due to the diffraction of light, whereas encapsulating magnetic NPs into one hemisphere of the JPs made them magnetically responsive. Amphiphilicity was achieved by surface-treating the JPs, resulting in a

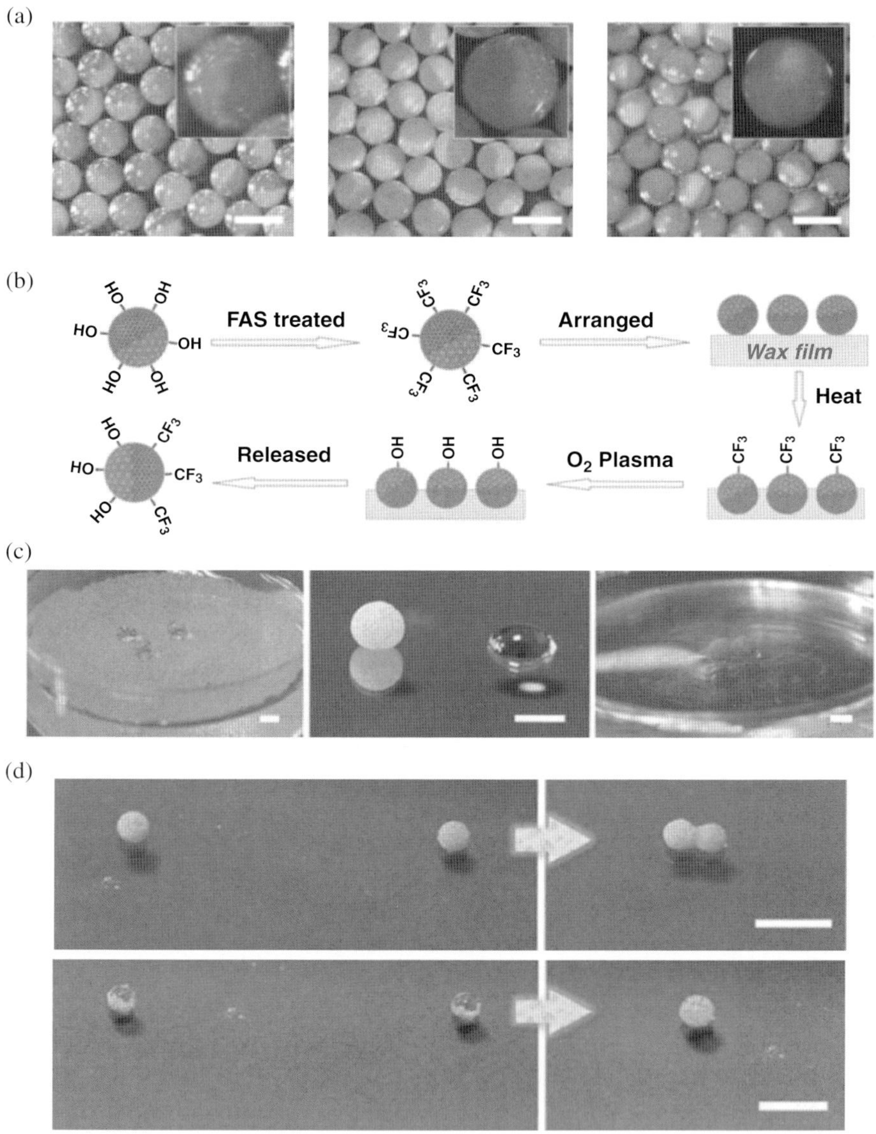

Figure 11. (a) Photographs of the fabricated JPs (scale bars: 100 μm); (b) schematic illustration of the procedure for the spatial variation of the JP surface wettability; (c) photographs of water droplets sitting on a monolayer of JPs (left), and JP-coated liquid marbles placed on a glass slide (middle) or on a water surface (right), scale bars are 2 mm; (d) forced coalescence of liquid marbles under magnetic force, resulting in a dumbbell-shaped marble (upper part), or a single spherical liquid marble (lower part). Reproduced with permission from Ref. [133]. Copyright (2013) American Chemical Society.

hydrophilic plasma-treated colored surface and a hydrophobic fluoroalkyl-silane-modified magnetic surface (Fig. 11(b)). The fabricated JPs formed a monolayer at the air–water interface and oriented themselves according to the wettability of their sides, providing a flexible barrier for preventing the coalescence of water droplets (Fig. 11(c)). Furthermore, liquid marbles were created by coating 5 μL water droplets with JPs, which were stable enough to be handled by tweezers and remained intact even after they were transferred onto a hydrophilic glass substrate or a water surface (Fig. 11(c)). Apart from their stability, the liquid marbles could be manipulated using a magnetic field on both flat and curved surfaces, thus triggering the coalescence of the droplets (Fig. 11(d)). Moreover, the encapsulation of colloidal crystal arrays in the JPs resulted in different structural colors that could be used for barcoding the particle-decorated liquid marbles.

10.2.2.3. *Electronic paper display technology*

Electronic paper (e-paper) displays are a rapidly expanding technology having unique features, including low energy consumption, high visibility, flexibility, thinness, and light weight. Different types of displays are already known as e-paper, such as electrophoretic, electrochromic, electrowetting, cholesteric liquid crystal displays, etc. [134]. Among these, a twisting ball display is based on bichromal (black/white) JPs having electrically anisotropic properties dispersed in a transparent silicone elastomer sheet, which is sandwiched between two parallel electrodes. A liquid-filled cavity exists around each JPs, allowing it to rotate in response to an electrical field. Due to the different charge properties at the opposite hemispheres of the JPs, one polarity of field will cause the white hemispheres to face the viewing window, while the other polarity will allow the black hemispheres to be seen. Thus, such Janus particle-based displays may be considered as active functional interfaces.

The first twisting ball display, called Gyricon, was developed in the 1970s by Sheridon [135]. Since that time, different improvements have been made by Sheridon *et al.* in the JP quality, production technique, and performance [136,137]. The spinning disk method was developed to produce the JPs by introducing molten white and black wax-like plastics at its opposite sides. They flowed to the edge of the disk, where they formed ligaments or small jets that subsequently broke up into bichromal Janus balls and solidified. The reflective light e-paper display based on these black/white Gyricon beads exhibited diffuse reflectance of more than 18% and a contrast ratio of more than 6:1, while the viewing angle approached that of paper [99].

Later, Nisisako *et al.* developed a microfluidic coflow system to produce monodisperse polymeric bicolored JPs having asymmetric charge distribution and pigments of carbon black and titanium oxide at their opposite sides, which were synthesized outside the fluidic module by thermal polymerization of the generated Janus droplets (Fig. 12(a)) [98]. Electrical actuation of the fabricated JPs was demonstrated by preparing a display panel and placing electrodes on the top and bottom of a monolayer of particles. The particles turned to orient their black hemispheres to the negatively charged panel and vice versa, whereas by reversing the electric field gradient, the particles flipped (Fig. 12(a)).

In a further contribution, Yang *et al.* synthesized electro-responsive photonic Janus balls with black and structural color regions based on an optofluidic approach (Fig. 12(b)) [100]. The JPs were fabricated via emulsification of two suspensions, one of them containing pure silica particles in an ETPTA resin and the other one containing silica particles in ETPTA along with carbon black particles. As a result, the silica particles self-organized in ETPTA into an fcc lattice on one hemisphere producing structural colors, whereas the other hemisphere was black due to the carbon black pigments. To demonstrate the electrical actuation of the photonic Janus balls, a twisting ball display scheme was used, showing the flipping of JPs in response to an AC electric field (Fig. 12(b)). The structural colors (in this case red and green) arising from the non-planar fcc lattice showed favorable optical properties such as a viewing-angle-independent reflection color. Such photonic Janus balls can be used as color pigments in reflection-mode displays (including e-paper), by creating the pixels with RGB-colored Janus balls.

Apart from electrical actuation, a magnetically switchable display based on magnetic-fluorescent Janus supraballs was fabricated by Chen *et al.* [138]. Ionomer-based monodisperse Janus supraballs with superparamagnetic NPs confined in one hemisphere and quantum dots (QDs) localized in the other hemisphere were synthesized via the microfluidic approach, realizing fluorescent and magnetic responsive performances

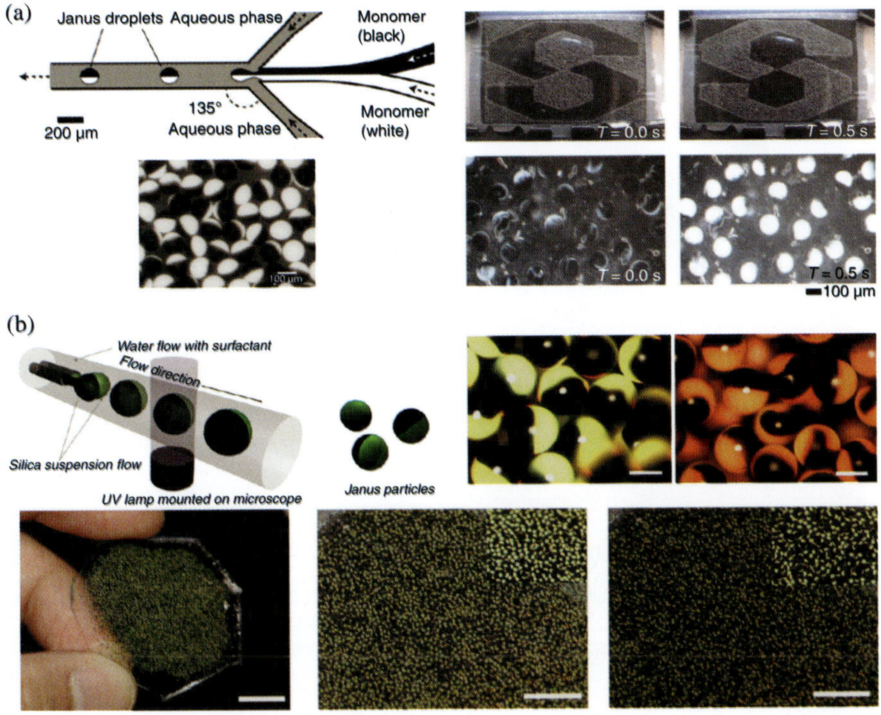

Figure 12. (a) Schematic of the Janus droplet generation using microfluidics and an optical microscopy image of the bicolored JPs (left); electrical actuation of the synthesized spheres (right; lower row — magnified top view of the display panel). (b) Schematic of the photonic Janus ball synthesis using optofluidics and optical microscopy images of the JPs (upper row); images of the JPs embedded in a flexible PDMS matrix and of the aligned JPs with their green or black side up (lower row; scale bars are 5 mm). Reproduced with permission from Refs. [98, 100]. Copyright (2006), (2008) Wiley-VCH Verlag GmbH & Co. KGaA.

within one JP (Fig. 13(a)). A magnetoresponsive fluorescent switch was developed to achieve free-writing on a rotating Janus bead panel under an external magnetic field. By varying the direction of the field, the alternation of the upward orientation of Fe_3O_4 NP-loaded hemispheres (OFF state) and QD-loaded hemispheres (ON state) could be enabled (Fig. 13(a)). When the Fe_3O_4 NP-dropped hemispheres face upward, the panel shows reddish-brown color under daylight and black color under UV light, whereas when the QD-polymer hemispheres orient upward, the

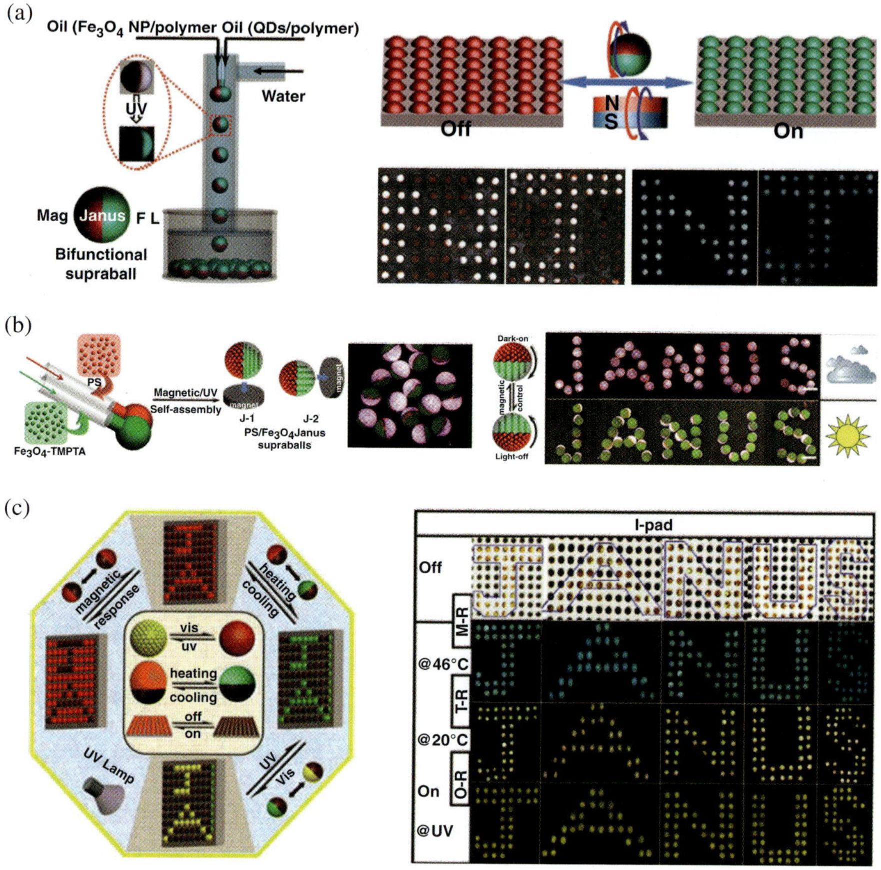

Figure 13. (a) Schematic representation of a microfluidic system for the synthesis of Janus supraballs (left) and of a fluorescent switch of the JPs controlled by varying the direction of the magnetic field (right) with optical images of the bead display under daylight (left) and under UV irradiation (right). (b) Schematic illustration of the PS/Fe_3O_4 JP fabrication and their optical microscopy image (left); photographs (right) of the PS hemispheres upside under weak light intensity (red) and Fe_3O_4 hemispheres upside under strong light intensity (green), scale bars are 500 μm. (c) The applications of JPs as switches and sensor displays (left); I-pad prepared from multifunctional JPs showing a temperature response and a reaction-induced response (right). Reproduced with permission from Refs. [138,140,141]. Copyright (2011) Wiley-VCH Verlag GmbH & Co. KGaA, 2014 Royal Society of Chemistry, 2015 American Chemical Society.

panel displays white color under daylight and bright blue emission under UV light (Fig. 13 (a)). The same group has further extended their approach and developed a triphase microfluidic-directed self-assembly to construct Janus colloidal photonic crystal (CPC) supraparticles with controllable and predictable shape, which exhibited superparamagnetism due to the encapsulated magnetic NPs and a photonic bandgap in two distinct hemispheres [139]. Such Janus CPC supraparticles demonstrate "dark" and "light" switchable behaviors under an external magnetic field and, more importantly, each particle can act as an independent pixel unit in the multipixel array, which can be micromanipulated by a magnetic needle, enabling rewritable and color-tunable photonic patterns. Furthermore, the synthetic procedure for the Janus CPC supraballs was advanced in the following report toward Janus supraballs with dual photonic band gaps by using combinations of monodisperse PS and SiO_2 colloidal particles as well as PS and superparamagnetic monodispersed Fe_3O_4 colloidal particles in the opposite hemispheres (Fig. 13(b)) [140]. The latter PS/Fe_3O_4 Janus supraballs were used to prepare display panels with dual-color properties, which can respond to strong and weak visible light resulting from optical differentiations between the PS and Fe_3O_4 CPC hemispheres, showing obvious switchable behaviors (Fig. 13(b)). Moreover, thermosensitivity was further introduced to the Janus CPC supraballs, resulting in a temperature-magnetism-optics triple sensitivity of the JPs (Fig. 13(c)) [141]. In this case, Cd^{2+}-loaded PS CPCs, *N*-isopropylacrylamide (NIPAAm) and Fe_3O_4 NPs were used as building blocks in the assembly process. Besides the magnetic response, the Janus supraballs can also display based on temperature response, and an on-demand fluorescent pattern can be quickly displayed by a reaction-induced response process (Fig. 13(c)).

A combined electrically and magnetically dual-driven twisting ball display was fabricated by Komazaki *et al.* utilizing black/white dual-driven JPs produced via microfluidics [134]. The black and white monomers were prepared by dispersing carbon black and TiO_2 particles in an acrylic monomer, whereas superparamagnetic NPs and a negative charge control agent were added into the black monomer for dual-driving. A twisting ball display was prepared from the produced JPs and tested for the electric color control, and handwriting using a magnet. As a result, the display permitted handwriting with a small magnet in addition to conventional color control using an applied voltage of 80 V. The handwriting performance could be improved by increasing the concentration of superparamagnetic NPs. Such a dual-driven twisting ball display can be applicable to low-cost, lightweight, highly visible, and energy-saving e-writers and electronic whiteboards.

10.2.2.4. *Catalysis*

Different strategies can be envisioned concerning the use of JPs as building blocks for active interfaces toward catalytic applications. The high interfacial energy and thus excellent emulsion-stabilizing properties of the JPs can be used to create Pickering emulsions efficiently stabilized by the JPs toward interfacial catalysis. Here, catalytic compounds can be introduced to the different emulsion phases, or to the JPs themselves. In the first case, the JP-loaded interfaces perform as active barriers separating two phases of the emulsion, while in the second case, they additionally perform as catalytically active interfaces due to the functionalization of the JPs.

The first strategy was realized by Cao *et al.* who reported on the use of amphiphilic silica JPs in a Pickering emulsion toward the encapsulation of enzyme molecules for catalytic purposes in organic media [142]. In this way, the submicrometer-sized silica JPs modified with silanes of different wettability at their opposite sides self-assembled into capsules with a shell comprising a monolayer of silica JPs at the oil–water interface (Fig. 14(a)). A model enzyme, lipase from *Candida* sp., was encapsulated within the JP-based capsules during their formation and its catalytic performance was evaluated according to the esterification of 1-hexanol with hexanoic acid. Ultimately, it was found that the specific activity of the encapsulated enzymes was more than 5.6 times higher than that of free enzymes in a biphasic system. The enzyme activity could be further increased by varying the volume ratio of water to oil and the JP loadings. The produced enzyme-loaded capsules demonstrated high stability during the reaction process and good recyclability. Moreover, they created a large interfacial area and had low mass transfer resistance through the monolayer shell. The jellification of agarose in the JP capsules further enhanced their operating stability.

An approach in between the two strategies was employed by Gao *et al.* who generated water-in-oil high internal phase emulsions (HIPEs) with styrene and divinylbenzene as the oil phase, containing Janus polymer NPs that consist of a linear PMMA "tail" and a cross-linked

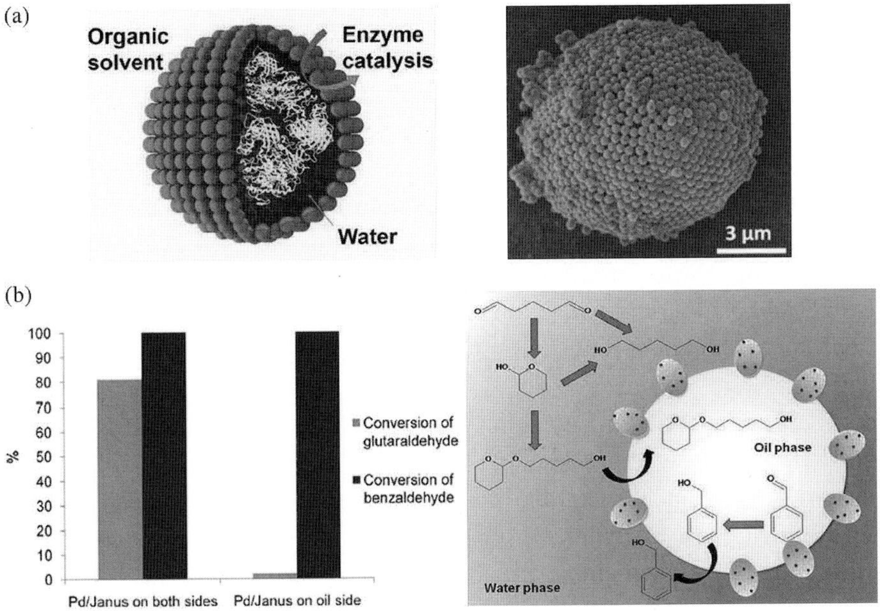

Figure 14. (a) Schematic structure (left) and an SEM image (right) of an enzyme-containing capsule consisting of a monolayer of JPs. (b) Aldehyde conversion (left) in a semi-batch reactor using JPs as catalysts in an emulsion, where benzaldehyde and glutaraldehyde were dissolved in decalin and deionized water, respectively; phase selectivity was demonstrated when the catalyst had Pd selectively deposited on the hydrophobic side; schematic illustration (right) of the hydrogenation reactions taking place at the water/oil interface catalyzed by Pd clusters supported on both sides of the Janus NPs. Reproduced with permission from Refs. [142,145]. Copyright (2010) Wiley-VCH Verlag GmbH & Co. KGaA, (2015) American Chemical Society (Ref. [145] (Wiley) to be requested).

poly(4-vinylpyridine) (P4VP) "head" [143]. The produced water-in-oil HIPEs were stabilized solely by the Janus polymer NPs, and were converted to polyHIPE foams by polymerizing the external continuous oil phase, leaving the Janus NPs embedded in the polyHIPEs. Benefiting from the strong interaction between the P4VP "heads" of the JP emulsifiers and the H_2PdCl_4 precursor, Pd NPs were supported *in situ* within the macrocellular polyHIPEs. The obtained Pd-modified polymer hybrid foams were then applied for the heterogeneous Suzuki–Miyaura carbon–carbon coupling reactions between iodobenzene and benzeneboronic acid, and a high catalytic activity along with a good recyclability could be demonstrated.

The second strategy is based on the prior asymmetric functionalization of the JPs with catalytic compounds and their subsequent use to stabilize fluid–fluid emulsions, while catalyzing reactions at the interface. An important advantage of such Janus-type catalysts is their easy recovery from the reaction mixtures, which makes them recyclable. Moreover, high interfacial area provided by the JPs can facilitate better conversion and mass transfer between the phases. This strategy was realized by Resasco *et al.* who demonstrated the ability of solid hybrid Janus-type catalysts to catalyze reactions at the water–oil interface [144]. The Janus-type catalysts were produced by selectively depositing palladium onto CNTs–inorganic oxide (silica or MgO) hybrid NPs. Herein, the deposition of Pd on the hydrophilic face can be used to catalyze aqueous reactions, whereas its deposition on the hydrophobic face can favor chemistry in the organic solvent. The application of the catalytic nanohybrids in emulsions was demonstrated on the example of biphasic hydrodeoxygenation and condensation catalysis in three substrate classes of interest in biomass refining. In another contribution from the same group, a slightly different Janus type-catalyst was used, comprising silica particles with different wettability at their opposite sides achieved by modifying one side with a silane, whereas Pd was anchored either on the entire JP surface, or only on the hydrophobic silane-modified side (Fig. 14(b)) [145]. The concept of phase selectivity in emulsion systems was demonstrated by choosing two aldehydes with different solubilities, benzaldehyde (oil-soluble) and glutaraldehyde (water-soluble), and performing their hydrogenation in the organic and aqueous phases (Fig. 14(b)). When the catalyst contained Pd on both sides of the JPs, high conversion levels were obtained for both reactants. However, when the catalyst had Pd selectively deposited on the hydrophobic side, the conversion of benzaldehyde was kept at 100%, while the conversion of glutaraldehyde decreased to 2%, demonstrating high phase selectivity.

Another concept regarding JPs for interfacial catalysis was presented by Synytska *et al.* based on the design of an advanced "hairy" hybrid Janus-type catalyst that comprises an inorganic silica core covered with two distinct polymeric shells, hydrophilic polyacrylic acid (PAA), and hydrophobic PS, at its opposite sides, while the catalytic species, in this case silver or gold NPs, were immobilized directly into the hydrophilic stimuli-responsive polymer shell (Fig. 15(a)) [95]. The hybrid hairy JPs

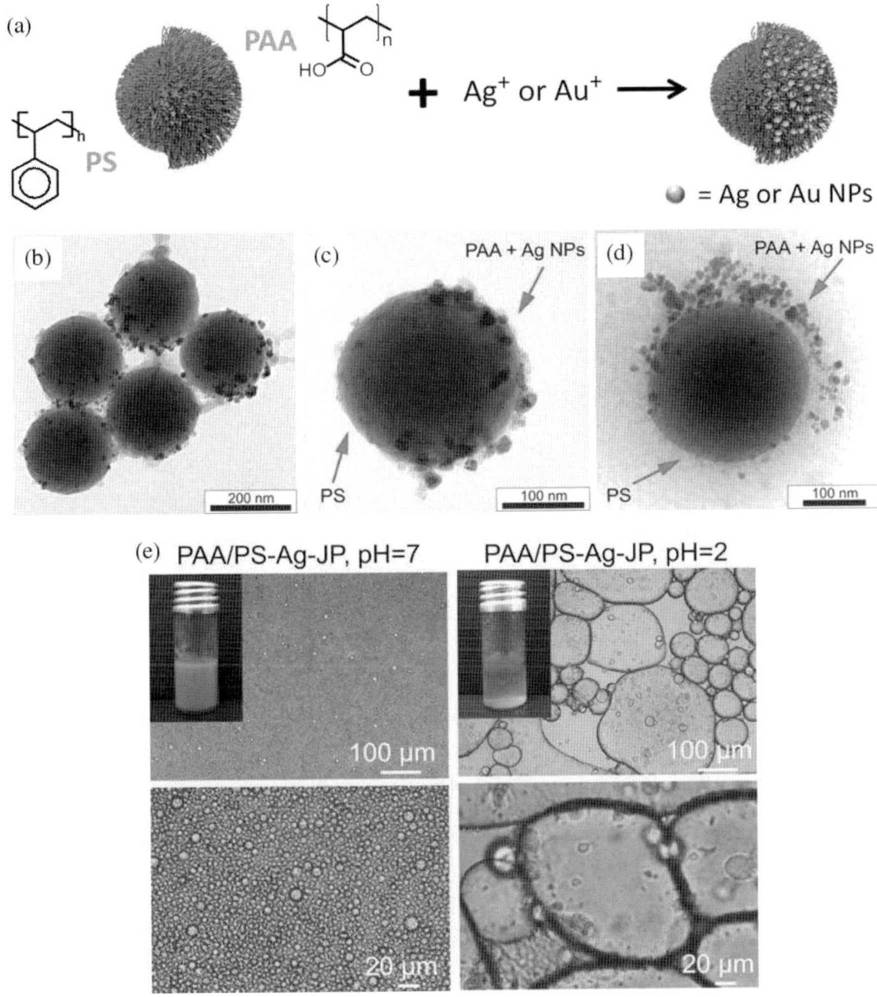

Figure 15. (a) Scheme of the selective NP immobilization onto PAA/PS-JP for catalytic applications; representative TEM (b, c) and cryo-TEM (d) images of the PAA/PS-JP with selectively immobilized Ag NPs into the PAA shell, the PAA chains are swollen and stretched out in (d); (e) photographs (insets) and light microscopy images at different magnifications of the emulsions prepared with PAA/PS-Ag-JP catalyst system at pH 7 (fully stable) and pH 2 (decomposed). Reproduced with permission from Ref. [95]. Copyright (2015) American Chemical Society.

with a 200 nm large core diameter were synthesized through a combination of "grafting from" and "grafting to" approaches, according to the scheme displayed in Fig. 3 and described in detail in Ref. [68]. The role of the polymer is considered very important in the creation of novel hybrid Janus catalysts: (1) polymer shells with a high contrast in hydrophobicity on the opposite sides of the JPs increase their affinity toward different phases (e.g., oil and water) in an emulsion, thus being able to stabilize it to a higher extent than the corresponding bare particles; (2) the "hairy" polymeric architectures and their ability to swell and de-swell provides a better distribution of the catalytic NP species throughout the catalyst surface, thus allowing better substance transport to the catalyst and higher reaction rates; (3) the use of stimuli-responsive polymers as shells would allow a great extent of control over the reaction, for example, inducing or, on the contrary, blocking the catalytic activity upon external stimuli. In the case of the PAA/PS hybrid JPs, the role of the hydrophilic PAA was to ensure a homogeneous distribution of the NPs throughout its hairy chains upon swelling in water, thus providing better substance transport possibilities, while the role of hydrophobic PS was to provide the catalyst with amphiphilicity required for more efficient interfacial catalysis. Successful selective functionalization of the amphiphilic JPs with either Ag or Au NPs could be achieved in a straightforward manner and directly visualized with TEM (Figs. 15(b)–15(d)). The JPs with incorporated metallic NPs demonstrated high interfacial activity and efficient stabilization of oil–water emulsions that can be tuned by pH (Fig. 15(e)). Furthermore, the accessibility of Ag and Au NPs was confirmed by successfully applying convenient benchmark reactions like the reduction of methylene blue, Eosin Y, and 4-nitrophenol. Finally, a combination of the stabilizing and catalytic properties of the fabricated hybrid Janus catalysts was demonstrated by preparing an oil–water emulsion stabilized by PAA/PS-Ag-JP and containing the Eosin Y dye in the water phase (Fig. 16(a)), and performing a successful catalytic reduction of the dye (Figs. 16(b)–16(d)). Moreover, it was shown that the JP catalyst system can be easily recovered after the reaction by simple centrifugation (Fig. 16(e)). Therefore, the developed catalytically active JPs possess a number of advantages: they effectively stabilize emulsions, the emulsion can be destabilized by utilizing the responsive properties of the JPs, and they can easily be recovered after reaction and reused again.

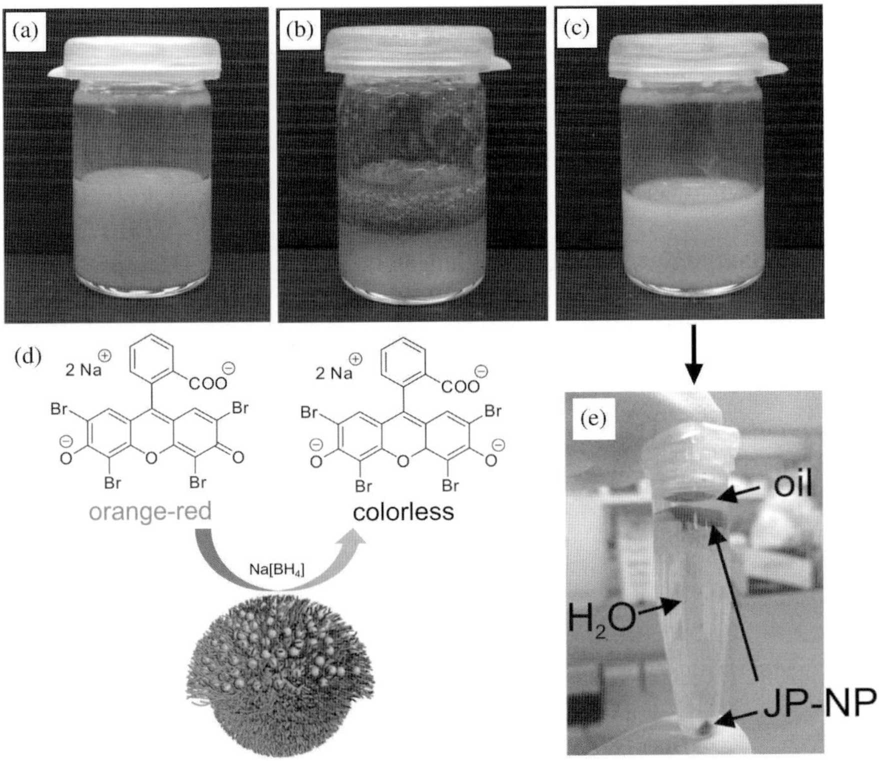

Figure 16. Reduction of Eosin Y in a water–oil emulsion stabilized by the PAA/PS-Ag-JP hybrid catalyst system: photographs of (a) emulsion stabilized by PAA/PS-Ag-JP and containing Eosin Y in the water phase; (b) the same emulsion after the addition of $NaBH_4$ and 10 min of incubation; (c) restabilized emulsion at pH = 6 after the reduction of Eosin Y; (e) emulsion from (c) after centrifugation; (d) schematic illustration of the Eosin Y reduction. Reproduced with permission from Ref. [95]. Copyright (2015) American Chemical Society.

10.3. Active Functional Surfaces

Active functional surfaces are another very promising and appealing field of application for the JPs, where their multifunctionality can lead to various favorable features of the resulting material, and often to synergistic effects if compared to the corresponding mono-functional surfaces. Here, roughly two strategies can be envisioned to create an active JP-based surface. One way is to employ the multifunctional JPs for functionalizing non-planar substrates such as cells in terms of biological and imaging applications, or textiles in terms of technological applications. The second strategy is to fabricate structured continuous surfaces based on JPs toward coatings with specific functions, such as superhydrophobicity, anti-icing, or antifouling properties.

10.3.1. Bioactive Surfaces and Imaging

In terms of biological applications, Lahann *et al.* developed a microstructured biohybrid material based on anisotropic polymer particles with spatially controlled affinity toward human endothelial cells (Fig. 17) [101]. The polymeric bi-compartmental JPs were synthesized via electrohydrodynamic cojetting and selectively modified to achieve two biologically distinguishable hemispheres: one exhibiting high binding affinity for human endothelial cells and the other being essentially resistant toward cell binding. For this purpose, one side of the JPs was modified with biotin tags and each side additionally labeled with green and red fluorescence emitting dyes, respectively (Fig. 17(a)). After streptavidin modification, the modified JPs showed selective binding to human endothelial cells that were labeled with biotinylated antibody against platelet endothelial cell-adhesion molecules (PECAM), which is highly expressed on endothelial cells including human umbilical vein endothelial cells (HUVECs). A control experiment with a biotinylated isotype antibody, which exhibits only low-level, non-specific binding to the cells, showed no association between the HUVEC cells and the JPs (Fig. 17(b)). Additional modification of the particle surface with polyethylene glycol (PEG) significantly decreased the non-specific binding (Fig. 17(b)). As the JPs exhibited a bipolar behavior with one compartment showing binding affinity toward human endothelial cells, while the other compartment was non-binding, the particles not only bound to the top of the cells, but also associated all around the perimeter of the cells forming a single particle lining, thus creating an active JP-based surface (Fig. 17(c)). Notably, the biocompatibility experiments showed that the JPs did not appear to exhibit significant cytotoxic effects that would result in a decrease of cell proliferation and viability.

In terms of imaging applications, Sun *et al.* reported on dumbbell-like Au–Fe_3O_4 Janus NPs, which possess both a magnetic (Fe_3O_4) and an

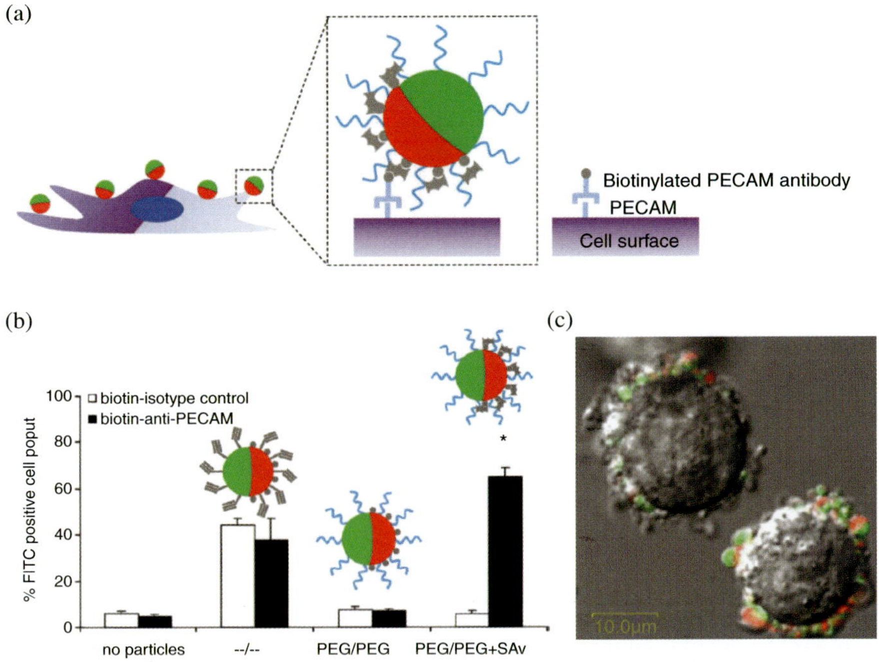

Figure 17. (a) Scheme for the binding of bi-compartmental JPs to cells via streptavidin present on the particle surface in one compartment with biotinylated cell-specific antibodies (in this case PECAM); (b) quantification of the percent of cell population with bound bi-compartmental particles: (from left to right) no particles, unmodified bi-compartmental particles showing non-specific binding to the cells, bi-compartmental particles with PEG on the surface of both compartments showing suppressed non-specific binding, and bi-compartmental particles with PEG on the surface of both compartments and streptavidin (SAv) only on the surface of one compartment showing specific binding; (c) overlay of a confocal laser scanning microscopy (CLSM) image with a differential interference contrast image of human endothelial cells self-assembled with PEG/PEG+SAv particles. Reproduced with permission from Ref. [101]. Copyright (2009) Wiley-VCH Verlag GmbH & Co. KGaA.

optically active plasmonic (Au) unit, and are therefore suitable for simultaneous magnetic and optical detection [102]. The Fe_3O_4 part of the NPs was modified with the epidermal growth factor receptor antibody (EGFRA), which was linked to the surface through PEG and dopamine, while the Au part was protected by the SH-PEG-NH_2 polymer. Further, the dumbbell NPs were shown to be suitable as a probe for A431 (human epithelial carcinoma cell line) cell imaging, which are known to overexpress EGFR. The binding affinity between EGFR in the cells and EGFRA on the NPs enabled the dumbbell NPs to adsorb on the A431 cells. Magnetic resonance imaging analysis showed that cells labeled with NPs shorten the T2 relaxation of the surrounding water molecules, whereas the Au part allowed obtaining reflection images with a scanning confocal microscope. Moreover, the durability of the optical detection was pointed out, as the prepared particles show far less signal loss as compared to standard fluorescent dyes.

A combined imaging and magnetolytic therapy approach was proposed by Hu and Gao [103]. They synthesized JPs (180 nm) based on a fluorescent dye-loaded amphiphilic block copolymer, poly(styrene-*block*-allyl alcohol), in one hemisphere and superparamagnetic NPs assembled in the other one. It was further demonstrated that the orientation of such JPs can be magnetically modulated toward optical imaging applications (Fig. 18(a)). For this purpose, the magneto-optical nanocomposites were placed between two glass coverslips for optical imaging and microscopy. A magnet was placed immediately below the sample to make the magnetic NP phase face down. When placed onto the stage of an inverted microscope where both incident light and fluorescence travel through objectives on the bottom, fluorescence was found to be significantly reduced due to blockage of the excitation and emission. On the contrary, when the magnet was placed on top of the coverslips to switch the orientation of the magneto-optical JPs, strong fluorescence could be observed. Furthermore, the possibility of magnetolytic therapy was realized through magnetic field modulated cell membrane damage (Fig. 18(b)). The nanocomposites attached quickly to the cell surface and the subsequent application of a spinning magnetic field resulted in the majority of the tumor cells being killed due to the mechanical force acting on the cell membranes. Future development of this new class of JPs may include targeted imaging and therapy *in vitro* and *in vivo*.

Another kind of biocompatible JPs suitable for both optical and magnetic detection was developed by Tremel *et al.* [146]. They synthesized monodisperse Au–MnO JPs with different sizes and morphologies via seed-mediated nucleation and growth technique, which allowed a precise control over domain sizes, surface functionalization, and dye labeling.

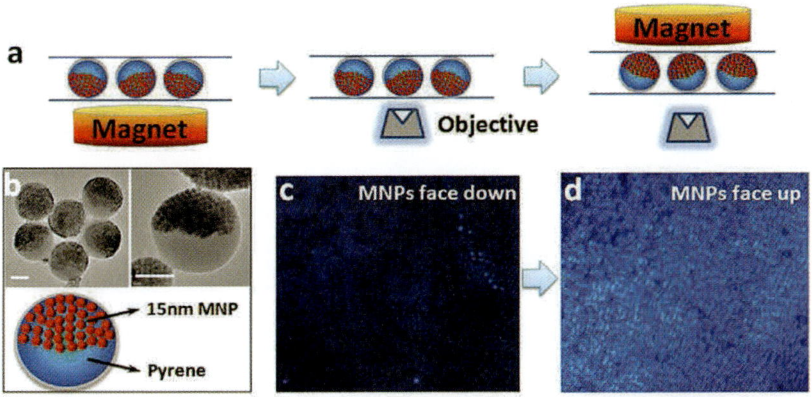

(A)

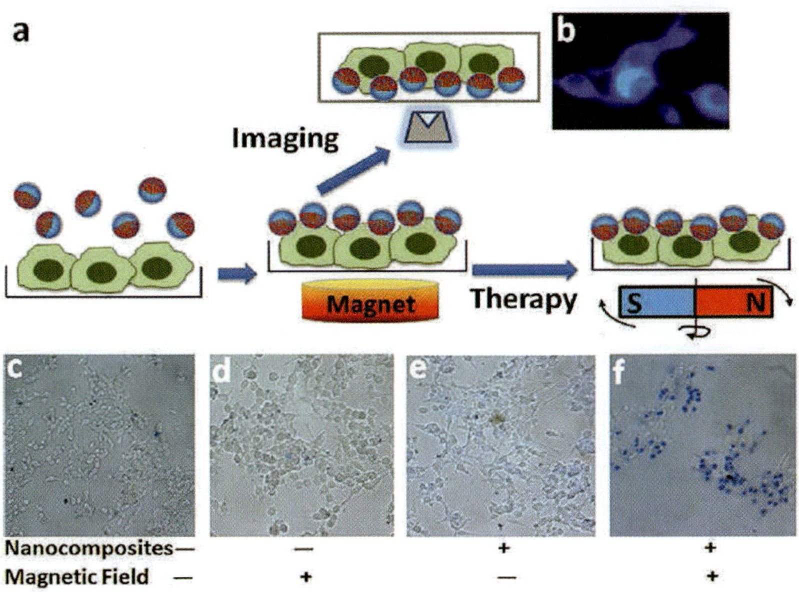

(B)

Figure 18. (A) Schematic of nanocomposite orientation modulation with magnetic fields, (a), (b) TEM images of the nanocomposites, scale bar 100 nm; (c), (d) fluorescence imaging of nanocomposites of opposite orientations. Simultaneous imaging and treatment of cancer cells using biphasic nanocomposites (B) (a) schematic illustration of the experimental conditions and magnetolytic therapy, JPs are brought down to the cell surface by magnetic attraction; (b) fluorescence imaging of cells coated with nanocomposites; (c)–(f) magnetolytic therapy on tumor cells in comparison with the controls missing either magnetic fields or nanocomposites, the viability of cells exposed to both is significantly reduced (dead cells appear blue due to Trypan blue staining). Reproduced with permission from Ref. [103]. Copyright (2010) American Chemical Society.

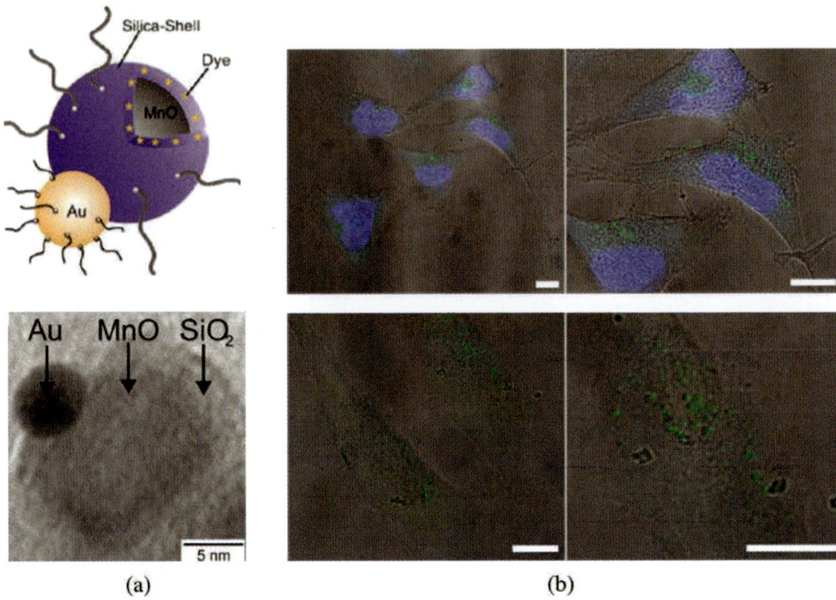

Figure 19. (a) Schematic illustration and a TEM image of a single Au–MnO–SiO_2 JP. (b) Upper row: CLSM images of HeLa cells coincubated with Au–MnO–SiO_2-Atto495 (green) for 24 h at 37°C, the cell nuclei (blue) were stained with DAPI, the sample was excited at 488 nm. Lower row: two-photon images of the same sample, excited with a two-photon laser at 970 nm, 30 mW. Scale bars: 10 μm. Reproduced with permission from Ref. [146]. Copyright (2014) American Chemical Society.

The metal oxide component was subsequently coated with a thin silica layer, leaving the metal domain untouched (Fig. 19(a)). The SiO_2 coating of the oxide domain provides the JPs with water solubility and allows biomolecule conjugation (e.g., antibodies, proteins) in a single step for converting the JPs into multifunctional, efficient vehicles for theranostics. The functionalized NPs are superparamagnetic and two-photon active, and therefore useful for simultaneous magnetic and optical detection. The latter was confirmed by time-resolved fluorescence spectroscopy in combination with CLSM (Fig. 19(b)). Furthermore, cell viability assays carried out to address the biocompatibility and potential biomedical applications revealed the Au–MnO–SiO_2 JPs to be non-cytotoxic.

10.3.2. Water-repellent Textiles

Textile surface modification is a very important issue widely used to improve their durability as well as to provide them with water-repellent properties. An alternative approach for the modification of textiles was proposed by Synytska *et al.* based on the chemical immobilization of amphiphilic (sub)micrometer large JPs [96]. The JPs were synthesized according to the procedure developed by Granick *et al.* [52] based on the preparation of a wax-water Pickering emulsion stabilized by silica particles. The exposed side of the particles was subsequently modified with moderately hydrophilic (3-aminopropyl)triethoxysilane (APTES), and after the dissolution of the wax the opposite side was modified with hydrophobic octadecyltrichlorosilane (OTS), yielding 200 nm or 1 μm large amphiphilic JPs. In order to chemically immobilize the prepared JPs onto the textile surface, the textile was preliminary modified by poly(glycidyl methacrylate) (PGMA), which provided reactive epoxy groups. The JPs were bound to the surface with their amino-modified sides through reaction with the epoxy groups, while their hydrophobic sides were facing the environment (Fig. 20(a)). It was found that the character of the JP distribution on the textile surface depended on their size: larger 1 μm JPs tended to assemble between the individual fibers, whereas smaller 200 nm JPs bound to the whole fiber surface, forming raspberry-like aggregates (Fig. 20(b)). In both cases, the immobilization of JPs resulted in the hydrophobization of the hydrophilic textile surface, indicating on the selective binding of the JPs to the textile surface (Fig. 20(b)). Moreover, higher concentration of JPs in dispersion (up to 20 wt.%) during the deposition led to an increase in the advancing water contact angle (Fig. 21). On the other hand, the tilt angle of a water droplet remained high and almost constant with increasing concentration of JPs in the case of large particles (the water droplet is pinned to the textile surface), while in the case of small JPs, higher concentration of particles led to a very low tilt angle, resulting in the sliding of the droplet off the textile surface (Fig. 21). Thus, water-repellent behavior was observed on the textiles modified with sufficiently high amounts of 200 nm large JPs.

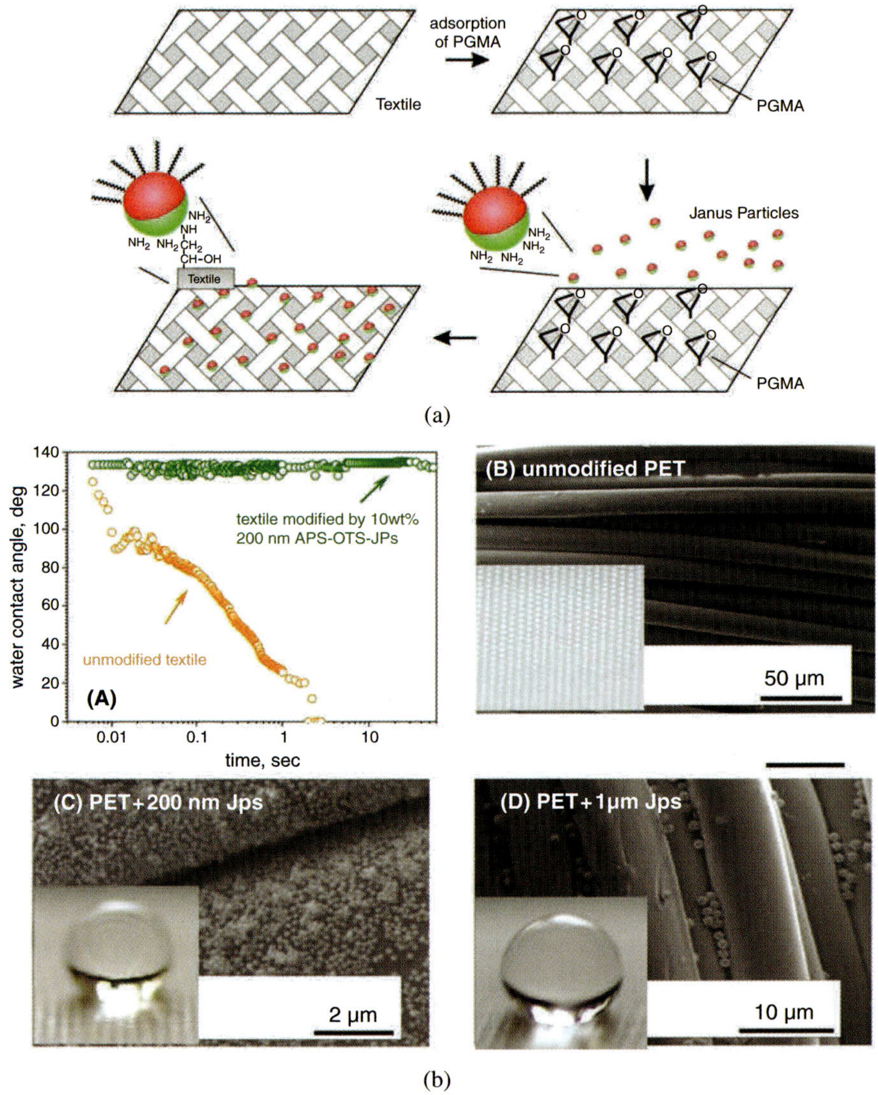

Figure 20. (a) Schematic illustration of the textile surface modification with amphiphilic JPs. (b) Wetting properties of the textiles modified with JPs — (A) time-resolved water contact angle on the native textile (orange) and the textile modified with JPs (green), (B, C, D) SEM morphology images and optical images (insets) of the water droplets on the native textile, textile modified with 200 nm large JPs, and with 1 μm large JPs, respectively. Reproduced with permission from Ref. [96]. Copyright (2011) American Chemical Society.

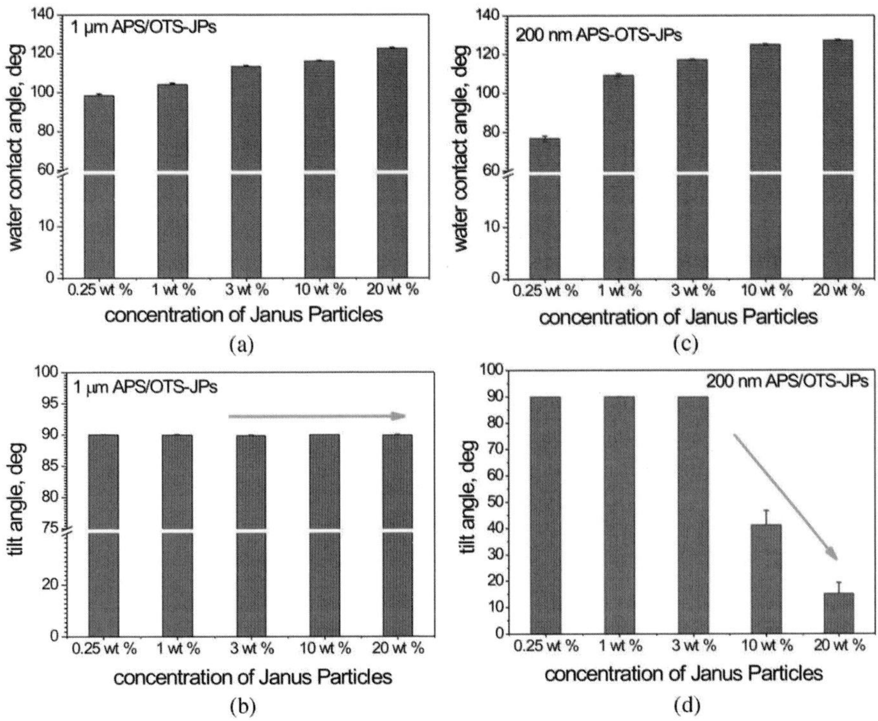

Figure 21. Advancing water contact angles (upper panel) and tilt angles of water droplets (lower panel) on the textiles with grafted 1 μm large (a) and (b), and 200 nm large (c) and (d) JPs. Reproduced with permission from Ref. [96]. Copyright (2011) American Chemical Society.

10.3.3. Photocatalysis

Another approach for the production of an active surface based on functional JPs was realized by Fu *et al.* who generated arrays of Au–TiO_2 Janus NPs toward photocatalytic applications [147]. For this purpose, monolayer organic–inorganic hybrid films were produced by spin coating a mixture of a polystyrene-*block*-poly(ethylene oxide) (PS-*b*-PEO)/ $HAuCl_4$ solution and a titania sol–gel precursor solution, where $HAuCl_4$ and titania were incorporated in the PEO domains and after subsequent removal of the organic matrix through UV irradiation,

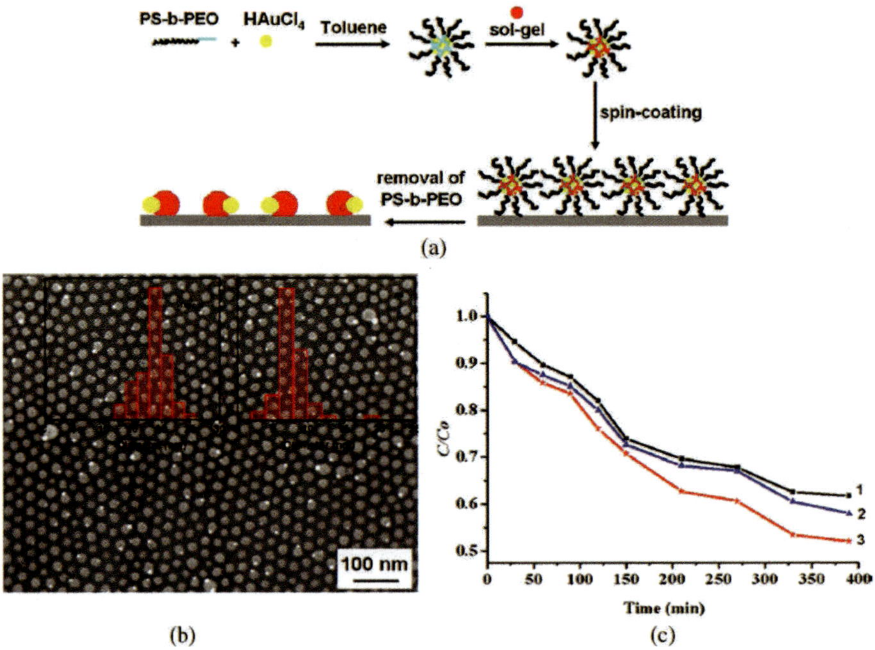

Figure 22. (a) Schematic illustration of the Au–TiO_2 Janus NP array production; (b) field emission scanning electron microscopy (FESEM) image of a Au–TiO_2 Janus NP array, insets: diameter distributions of titania and Au NPs; (c) change in the methylene blue concentration as a function of irradiation time, curves: (1) in the presence of TiO_2 NPs as a catalyst, (2) in the presence of Au–TiO_2 composite NPs as a catalyst; (3) in the presence of Au–TiO_2 Janus NPs as a catalyst. Reproduced with permission from Ref. [147]. Copyright (2011) Elsevier.

arrays of Au–TiO_2 Janus NPs on the substrate surface could be obtained (Figs. 22(a) and 22(b)). Further, the photocatalytic activity of the prepared Janus NPs was tested using the photocatalytic degradation of methylene blue and compared to that of the TiO_2 NPs and Au–TiO_2 composite NPs serving as a reference (Fig. 22(c)). Ultimately, the Janus NP arrays exhibited a better catalytic performance than the references for the decomposition of methylene blue if illuminated by UV light. The increased photocatalytic activity of Au–TiO_2 Janus NPs was attributed to the Au–TiO_2 heterointerfaces.

10.3.4. Optically Active Surfaces

An approach toward the fabrication of colloidal crystals from flower-shaped porphyrin JPs was proposed by Wang *et al.* [148]. The physically anisotropic flower-shaped JPs with orientated assembly were generated from a porphyrin derivative based on the two-step water droplet condensation process, where the first water condensation resulted in the formation of the Janus petal microparticles, while the subsequent second condensation of the droplet induced the integrated aggregation of the Janus petals toward flower-shaped particles. As a result, the flower-shaped particles possess a smooth back side and a concave front side, suggesting the physically anisotropic Janus character of the whole particle (Fig. 23(a)). The orientated arrangement of the synthesized JPs could be assigned to the well-ordered assembly of the droplet template, with its concave side facing the suspension phase and its flat side facing the air phase (Fig. 23(b)). Ultimately, the as-prepared flower-shaped colloidal crystals showed a multiwavelength optical signal produced from the hierarchical structures (Fig. 23(b)), which may imply an important insight into the creation of novel optical materials.

10.3.5. Superhydrophobic Surfaces

Superhydrophobic coatings are highly promising for various types of applications, for instance, for engineering materials with self-cleaning properties. The first approach based on the use of JPs to produce ultrahydrophobic coatings was developed by Synytska and *et al.* [97]. The idea was to use controlled aggregation of mono- and bi-component hybrid hairy JPs into hierarchical structures in dispersions, which then form structured rough layers after deposition on a substrate, displaying ultra-hydrophobic properties (Fig. 24(a)). The mono- and bi-component micrometer-sized JPs were synthesized according to the procedure shown in Fig. 3 and described in detail in Ref. [68], and comprised a silica particle core along with either one grafted polymer shell (PS or poly(*tert*-butyl acrylate), PtBA) on one hemisphere, or two grafted polymer shells (PS-poly(2-vinylpyridine) (P2VP) or PtBA-P2VP) on the opposite hemispheres of the particles. The synthesized hybrid hairy JPs were subsequently deposited onto two kinds of substrates

Figure 23. (a) SEM images of the front side (A) and the back side (B) of as-synthesized flower-shaped JPs (insets — front and back side of an actual flower). (b) Large-area flower-shaped assembly with orientated arrangement and its optical properties — (A) and (B) SEM images of the as-synthesized flower-shaped JPs, inset in part (B) — the reflection spectrum of the sample. Reproduced with permission from Ref. [148]. Copyright (2015) Royal Society of Chemistry.

with chemically grafted polymers, hydrophilic P2VP or hydrophobic PS. The adsorption of polymer-modified mono-functional silica particles was investigated and compared to the behavior of the JPs. It was found that the topography of the layers formed by JPs differs from that formed by the monofunctional ones due to the tendency of JPs to form aggregates, their random orientation on the surface, and the influence of the chemical nature of the substrate on the morphology of the formed JP layers. For instance, the fractal dimension of the JP layers formed on P2VP-modified substrates was higher than that on the PS-modified substrates. In terms of wettability, the particle layers formed on P2VP-modified substrates were more hydrophobic than those formed on the PS-modified ones. Moreover, the wetting properties of the PtBA-P2VP-JP layers approach an almost ultrahydrophobic behavior (both advancing and receding water contact angles were higher than 140°), originating from the multi scale roughness of the layers ranging from the nanometer scale up to the dimension of several hundreds of microns, resulting in three levels of surface roughness (Fig. 24(b)). Finally, bicomponent JPs with a hydrophilic P2VP side and a highly hydrophobic poly(3,3,4,4,5,5,6,6,7,7,8,8,9,9,10,10,10-heptadecafluorodecyl methacrylate) (PHDFMA) side were synthesized and deposited onto P2VP-modified substrates, yielding an ultrahydrophobic surface (Fig. 25, the advancing and receding water contact angles were higher than 150°). Therefore, due to the ability of JPs to self-assemble and form hierarchical structures in dispersions, they are highly promising for the generation of materials with superhydrophobic and self-cleaning properties.

In a later contribution by Liang and Yang *et al.*, cone-like JPs were utilized to fabricate coatings with tunable wettability that could be changed from highly adhesive to water to superhydrophobic by changing the size distribution of the JPs (Fig. 26(a)) [149]. The cone-like JPs were synthesized by emulsion polymerization-induced phase separation. Briefly, a patchy emulsion interface was created after the phase separation of binary surfactants, hydrolyzed styrene–maleic anhydride (HSMA) copolymer, and Tween-80, the cone-like PS particles then protruded from the patchy domains due to Pickering effect, and finally selective modification of the cone surface could be achieved to introduce the Janus character to the particles. Further, the cone-like JPs were employed to fabricate robust coatings based on phenolic epoxy resin F-44 as an intermediate layer to attach the JPs onto the surface via cross-linking (Fig. 26(a)). After spraying a dispersion of the modified PS/titania composite JPs, they self-organized into a layer upon drying, where the amino groups on the flat sides of the particles were covalently bound to the epoxy resin, while the hydrophobic cone sides were facing the air. The wettability of the coatings could be tuned from highly adhesive to water to superhydrophobic by simply changing the roughness of the coating, which was achieved by using different JPs with varied size distribution.

The same group further demonstrated the utilization of another kind of JPs, strawberry-like JPs, for the fabrication of superhydrophobic coatings (Fig. 26(b)) [150]. Strawberry-like JPs were synthesized through a

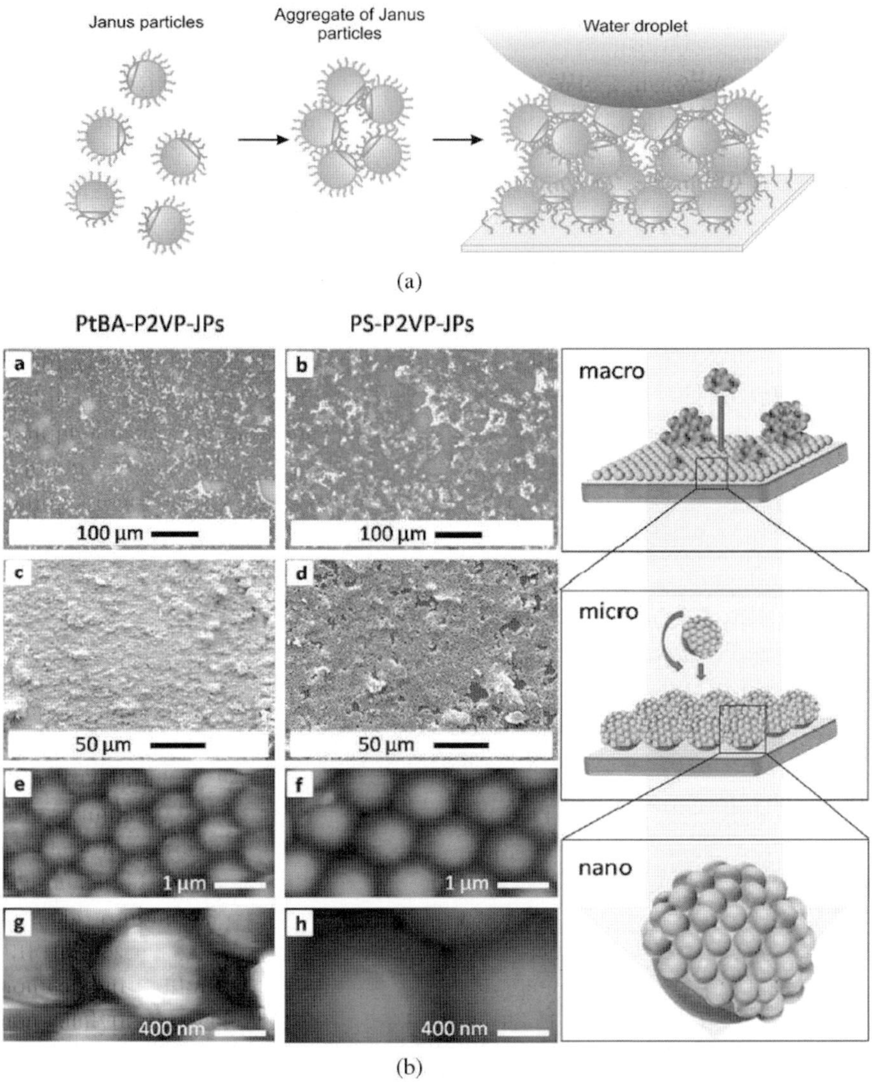

Figure 24. (a) Schematic illustration of the design of functional surfaces using bi-component JPs; JPs form aggregates in solution, which, after deposition on a substrate, possess ultra-hydrophobic properties. (b) Illustration of the multilevel hierarchy of the JP layers — first, second, and third levels of surface roughness are formed by particle agglomerates, spherical shape of the particles, and an intrinsically rough PtBA layer, respectively (cartoons); microscopy images of PtBA-P2VP-JPs (a), (c), (e) and (g) and PS-PVP-JPs (b), (d), (f) and (h), wherein (a) and (b) — optical microscopy images, (c) and (d) — SEM images, (e)–(h) — atomic force microscopy (AFM) images. Reproduced with permission from Ref. [97]. Copyright (2011) Wiley-VCH Verlag GmbH & Co. KGaA.

Figure 25. Topography (representative SEM images) of (a) bi-polymeric PHDFMA-P2VP-JPs on a P2VP substrate as well as snapshots of (b) static, and (c) receding water droplets on the layer formed by PHDFMA-P2VP-JPs. Reproduced with permission from Ref. [97]. Copyright (2011) Wiley-VCH Verlag GmbH & Co. KGaA.

self-organized sol–gel process at a patchy emulsion interface, similarly to the cone-like JPs described above. When tetraethyl orthosilicate (TEOS) and (3-aminopropyl) trimethoxysilane were mixed in the oil phase, corresponding silica JPs formed, their exterior coarse side terminated with amino groups, which allowed its selective modification with 2-bromoisobutyryl bromide and subsequent polymerization of styrene via ATRP. As the flat side of the particles contained silanol groups, the imidazolin group could be introduced through modification with a silane. To produce a JP-based surface, a thin layer of liquid epoxy resin (E51) was previously coated onto the substrates, and after the water evaporated from an aqueous dispersion of the JPs, they self-organized into an ordered layer on the substrate (Fig. 26(b)). The hydrophobic side was facing the air, whereas the hydrophilic side remained emerged within the epoxy resin layer, binding the particles covalently to the surface via cross-linking. As a result, the nanoscale roughness on the hydrophobic sides of the JPs determined the superhydrophobic performance of the coating (Fig. 26(b)).

10.3.6. Anti-icing Surfaces

The development of surfaces that possess anti-icing and de-icing properties is pivotal for the aircraft and automotive industries as well as for the wind turbine technology, potentially resulting in a substantial reduction of

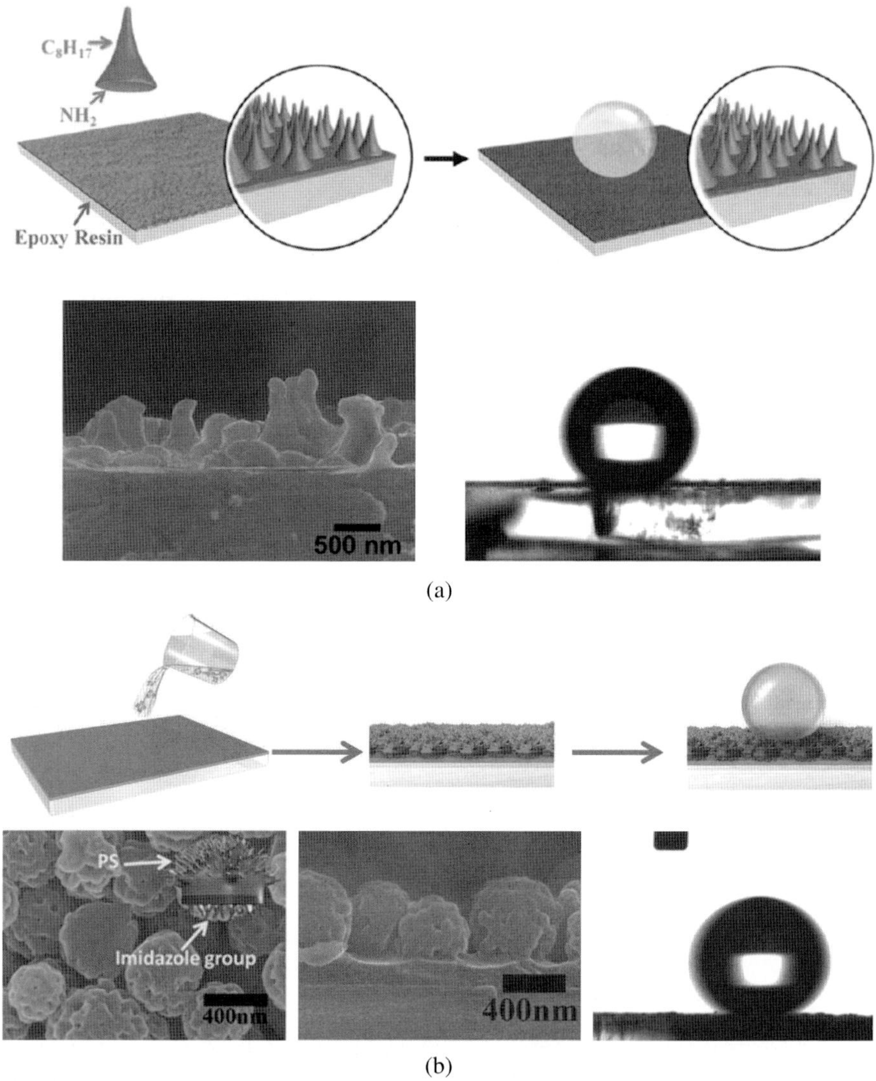

Figure 26. (a) Schematic illustration of a cone-like JP-based coating fabrication (upper panel); side-view SEM image of a superhydrophobic coating from cone-like JPs and an image of a water droplet contour on the coating with a contact angle of 165 ± 2° and a tilt angle ~6° (lower panel). (b) Schematic illustration of a superhydrophobic coating fabrication based on the drying of a strawberry-like JP dispersion (upper panel); lower panel (from left to right) — SEM image of strawberry-like Janus composite particles (inset: structure of a single JP), side-view SEM image of the superhydrophobic coating based on strawberry-like JPs, image of a water droplet contour on the coating [150]. Reproduced with permission from Ref. [149]. Copyright (2015) American Chemical Society. (No copyright needed for Ref. [150], NPG Asia Materials)

costs, and a decreased number of emergency situations. Therefore, the design of passive anti-icing coatings, which can be defined as coatings that prevent icing or reduce the ice accumulation rate, is the most beneficial solution to the problem of icing, allowing the reduction of power consumption necessary to remove the accumulated ice. There are different strategies for the fabrication of anti-icing coatings such as, for instance, the reduction of the ice adhesion strength through the use of hydrophobic or superhydrophobic surfaces, or the inhibition of the ice growth through the utilization of antifreeze proteins or the colligative properties of solutions. Furthermore, a completely new approach was developed by Synytska *et al.* based on the employment of amphiphilic hybrid hairy JPs as building blocks for the design of structured surfaces with controlled ice nucleation and growth, as well as extremely low ice adhesion strength [151].

Robust active surfaces were fabricated, possessing hydrophilic/hydrophobic heterogeneity, based on 1 μm large hybrid hairy JPs composed of a silica core along with poly(poly(ethylene glycol) methyl ether methacrylate) and poly(dimethylsiloxane) polymer shells at the opposite sides of the core (SiO_2-P(PEGMA)/PDMS-JP). The Janus ratio of the employed particles was 2:1 (hydrophilic P(PEGMA):hydrophobic PDMS). The JPs were synthesized through a combination of “grafting from” and “grafting to” approaches, as illustrated in Fig. 3 and described in Ref. [68]. Next, the active structured surfaces were prepared by a simple solvent casting method via depositing and subsequent drying of particle suspensions on silicon wafers premodified with poly(glycidyl methacrylate), serving as an adhesion promoter. The quality of the fabricated layers was checked with optical microscopy and SEM (Fig. 27), revealing mechanically stable monolayers of JPs, which were not removed after the subsequent anti-icing performance experiments. In the SEM, all kinds of JP orientation in the layer could be observed (Figs. 27(d) and 27(e)): JPs turned with their hydrophilic P(PEGMA)-modified or their hydrophobic PDMS-modified side up, as well as JPs lying on their side, thus exposing both polymers in different extents to the surrounding. In order to get statistical insight into the orientation of JPs in the layer, AFM direct force measurements under humid air conditions were performed. Under these conditions, the adhesive properties of the polymers are different, therefore, allowing to classify the recorded force–distance curves as either typical for P(PEGMA),

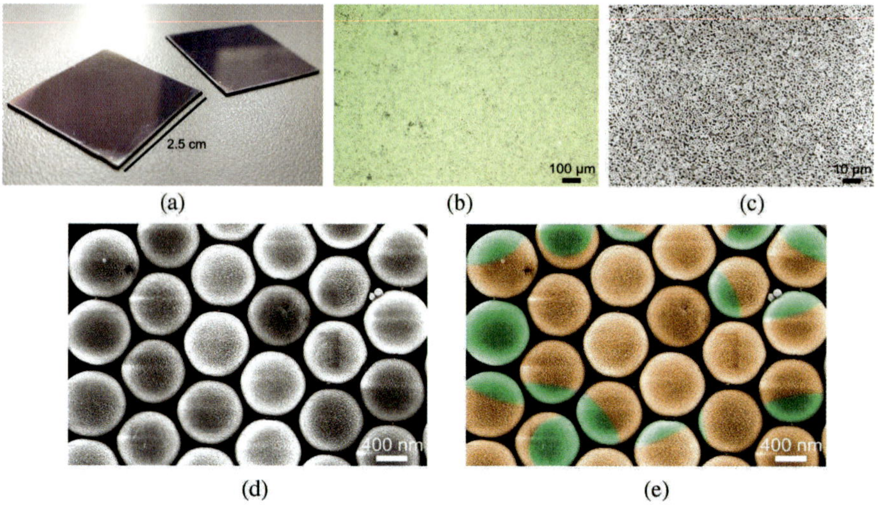

Figure 27. Representative images of the surfaces prepared from spherical hybrid hairy SiO_2-P(PEGMA)/PDMS-JP JPs: (a) optical image, (b) optical microscopy image, (c,d) SEM images, and (e) false color SEM image showing the orientation of JPs (orange: P(PEGMA), green: PDMS) [151].

or for PDMS. As a result, statistical analysis of the force–distance curves revealed a 55%:45% distribution of the particles exposing their hydrophobic or hydrophilic side to the surrounding, respectively.

In terms of the wetting properties, the advancing and receding water contact angle values were found to be intermediate on the surfaces composed of the JPs compared to the surfaces composed of particles homogeneously decorated with either P(PEGMA) or P(PDMSMA)[1] polymers. The wetting properties of the fabricated surfaces were additionally investigated with cryo-SEM by examining condensed and frozen water droplets on the surface of the particle layers (Fig. 28). It was found that in the case of the hydrophilic P(PEGMA)-modified particle surface, the frozen water droplets are very flat in shape, indicating on the spreading of water on the surface (Fig. 28(a)), while in the case of the hydrophobic P(PDMSMA)-modified

[1]The differences in abbreviations are due to the fact that poly(monomethacryloxypropyl terminated polydimethylsiloxane) (P(PDMSMA)) was grafted onto homogeneously decorated particles using the "grafting from" approach, while carboxy terminated PDMS was grafted onto one side of the JPs using the "grafting to" approach.

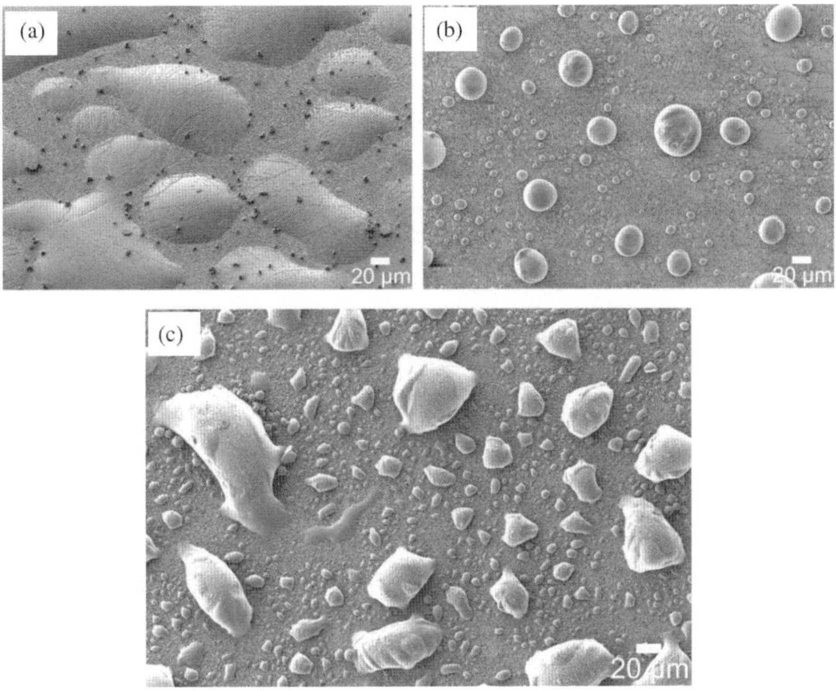

Figure 28. Representative cryo-SEM images of frozen water droplets on the surfaces prepared from 1 μm SiO_2 (a) P(PEGMA)-modified, (b) P(PDMSMA)-modified, and (c) Janus P(PEGMA)/PDMS particles [151].

particle surface, the frozen water droplets have compact spherical shapes, and no spreading of the water in between the particles was observed (Fig. 28(b)). In the case of the heterogeneous amphiphilic JP-based surface (Fig. 28(c)), both flattened and spherical frozen water droplets were observed corresponding to the P(PEGMA) and PDMS polymers on the surface of the particles, respectively, whereas a large fraction of droplets had "intermediate" shapes, which were almost spherical, but simultaneously having clear pinning points to the surface. The analysis of the wetting properties of the JP-based surfaces indicated on the presence of both polymers on the surface of the particle layers, whereas cryo-SEM investigations additionally revealed an entirely different mechanism of water condensation on the JP-based surface compared to that on the corresponding mono-functional surfaces.

Next, the anti-icing performance of the heterogeneous JP-based surfaces was investigated using different experimental techniques. First, the JP-based surface was examined during icing on a Peltier element under an optical microscope (Fig. 29), where during a gradual decrease in the temperature, water droplets condensed from the air (Fig. 29(b)) and subsequently froze. However, instead of forming small ice crystals that slowly grow, as typically observed during icing of homogeneous hydrophilic or hydrophobic surfaces, large crystals were formed on the surface composed of the amphiphilic JPs due to heterogeneous nucleation (Figs. 29(c) and 29(d)). These large ice crystals (crystal dendrites) were found to expand extremely quickly due to the consumption of the neighboring not-yet-frozen water droplets from the surface, resulting in substantial ice-free areas of the surface (Figs. 29(d) and 29(e)), and subsequently also water-free areas during the thawing of ice (Fig. 29 (f)). Additionally, ice growth rates were determined for the model flat systems (silicon wafers covered with the corresponding polymer brushes), homogeneously decorated particle surfaces, and the JP-based particle surfaces, by measuring the ice weight and the thickness of the ice layer after different periods of time (Fig. 30(a)). It was found that the model flat surfaces and the respective homogeneous particle-based surfaces behave very similarly, where the highest and the lowest rates of ice formation correspond to the hydrophilic P(PEGMA) and the hydrophobic P(PDMSMA) polymers on the surface (Figs. 30(a) and 30(b)). The JP-based surfaces, on the other hand, demonstrated an intermediate ice growth rate (Figs. 30(a) and 30(b)). Moreover, ice adhesion measurements were performed on the aforementioned surfaces (Fig. 30(c)). It was found that the adhesion of ice to the heterogeneous amphiphilic JP-based surfaces is lower than that corresponding to the homogeneously decorated hydrophilic or hydrophobic particle-based surfaces (Fig. 30(c)). The reduced ice adhesion to the JP-based surfaces could be explained by a diminished contact area between the ice layer and the surface, which is caused by the evaporation of liquid droplets in the proximity of the large crystals. Therefore, great potential may be seen in the active functional surfaces based on hybrid hairy JPs, allowing controlled nucleation and growth of ice as well as possessing extremely low ice adhesion strength, for the development of advanced and effective anti-icing coatings with synergistic effects.

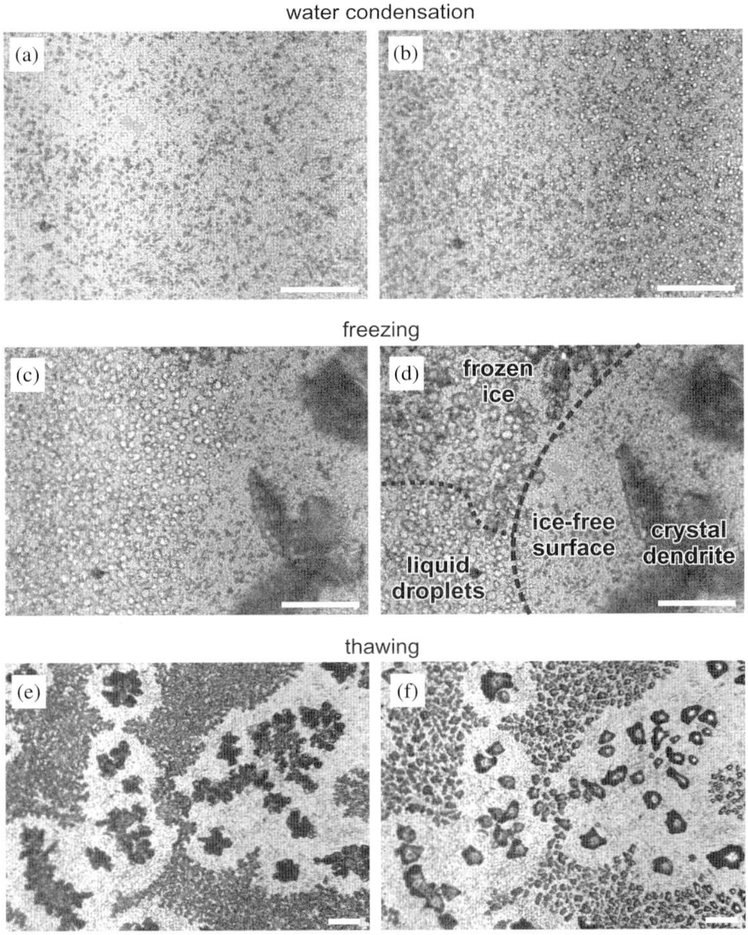

Figure 29. Representative optical microscopy images of a SiO_2-P(PEGMA)/PDMS-JP layer during icing: (a) start, (b) surface after the water droplet condensation (32s), (c) heterogeneous nucleation and growth of large ice crystals (80 s), (d) and (e) freezing of condensed water droplets (different magnifications, 96 s), (f) thawing of ice. Scale bars: 100 μmRef [151].

10.3.7. Antifouling Surfaces

Marine biofouling is the undesired accumulation of microorganisms, algae, or animals on different seawater contacting equipment, such as ship hulls, fishing nets, bridge pillars, etc. It leads to severe consequences,

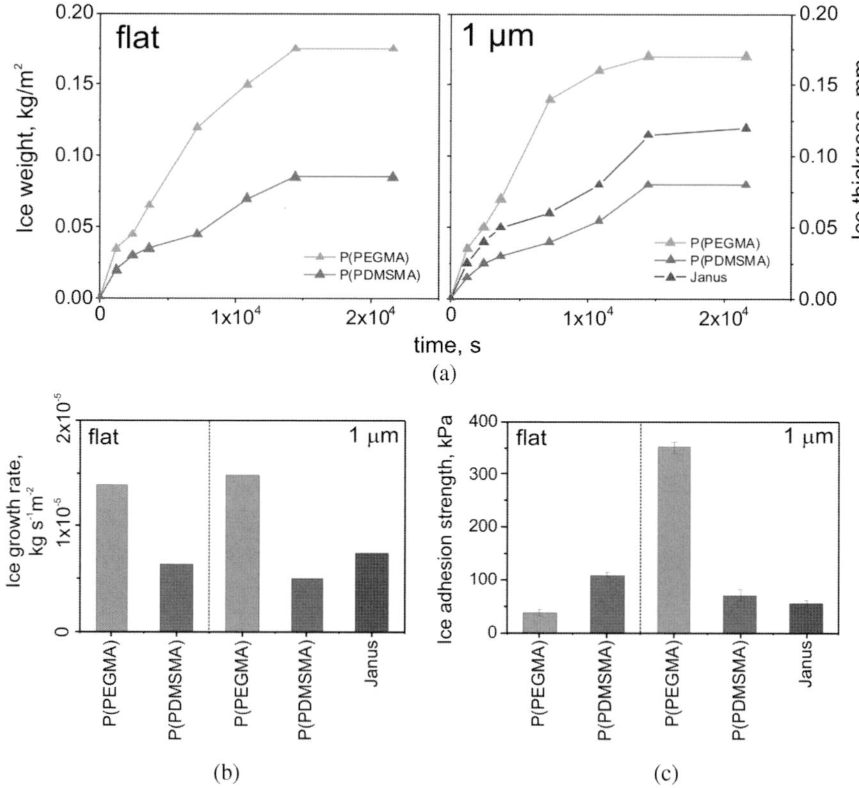

Figure 30. Summary of the (a) ice weight, (b) ice growth rate, and (c) ice adhesion measurement results on flat surfaces, mono-functionalized particle-based surfaces (SiO_2-P(PEGMA), SiO_2-P(PDMSMA)), and bi-functional Janus particle-based surfaces (SiO_2-P(PEGMA)/PDMS-JP) [151].

resulting in, for instance, substantially increased fuel consumption in the case of ship hull biofouling, and the subsequent increase of billions of dollars in the annual worldwide costs. Therefore, there is a strong demand for the development of appropriate effective antifouling strategies. Many of such strategies are focused on the prevention of the initial biofilm formation rather than on the removal of mature biofilms. The initial adhesion of bacteria to a surface is greatly influenced by the properties of the surface, such as its chemistry, energy, mechanical properties, topography, as well as the environmental conditions in general. Consequently, a variety of different antifouling approaches were proposed, based on homogeneous hydrophilic or hydrophobic, heterogeneous (amphiphilic), or 3D microtopographic patterned surfaces. A new strategy for the design of active functional surfaces with antifouling properties based on hybrid hairy JPs as solid amphiphilic block copolymer prototypes was proposed by Synytska *et al.* [152].

The JPs used as building blocks for the production of heterogeneous surfaces were composed of either a spherical silica, or a platelet-like kaolinite core, along with hydrophilic P(PEGMA) and hydrophobic PDMS (or P(PDMSMA)) polymer shells grafted at the opposite sides of the core. The spherical hybrid hairy JPs were synthesized according to the scheme illustrated in Fig. 3 and the procedure described in Ref. [68], whereas the platelet-like JPs were synthesized according to the approach described in Ref. [69], via one-step simultaneous grafting of both polymers from the opposite sides of inorganic kaolinite core particles. Subsequently, robust structured surfaces were prepared with both kinds of JPs via simple solvent casting using wafers pre-modified with PGMA serving as an adhesion promoter. As a result, surfaces with different morphologies were formed (Figs. 31(a), 31(c) and 31(e) — spherical JPs; Figs. 31(b), 31(d) and 31(f) — platelet-like JPs), and their quality was evaluated using optical microscopy (Figs. 31(a) and 31(b)), SEM (Figs. 31(c) and 31(d)), and AFM (Figs. 31(e) and 31(f)). In addition, reference surfaces were prepared with non-functionalized native, and monofunctionalized particles with either P(PEGMA) or P(PDMSMA) polymers on their surface. The negative control was an unmodified silicon wafer.

The orientation of spherical JPs in the layers was random (Figs. 27(d) and 27(e)), and to estimate the statistical distribution of hydrophobic and hydrophilic moieties exposed to the surrounding and accessible to bacteria under water, AFM colloidal probe force measurements were performed. Under these conditions, the hydrophilic swollen P(PEGMA) polymer is completely non-adhesive to the colloidal probe, while the hydrophobic collapsed PDMS (P(PDMSMA)) is adhesive to the colloidal probe, allowing the force–distance curves to be distinguished. Statistical analysis of the force–distance curves recorded on the spherical JP-based surfaces revealed a 52%:48% distribution of the hydrophobic to hydrophilic moieties, respectively. The platelet-like JPs typically expose either their

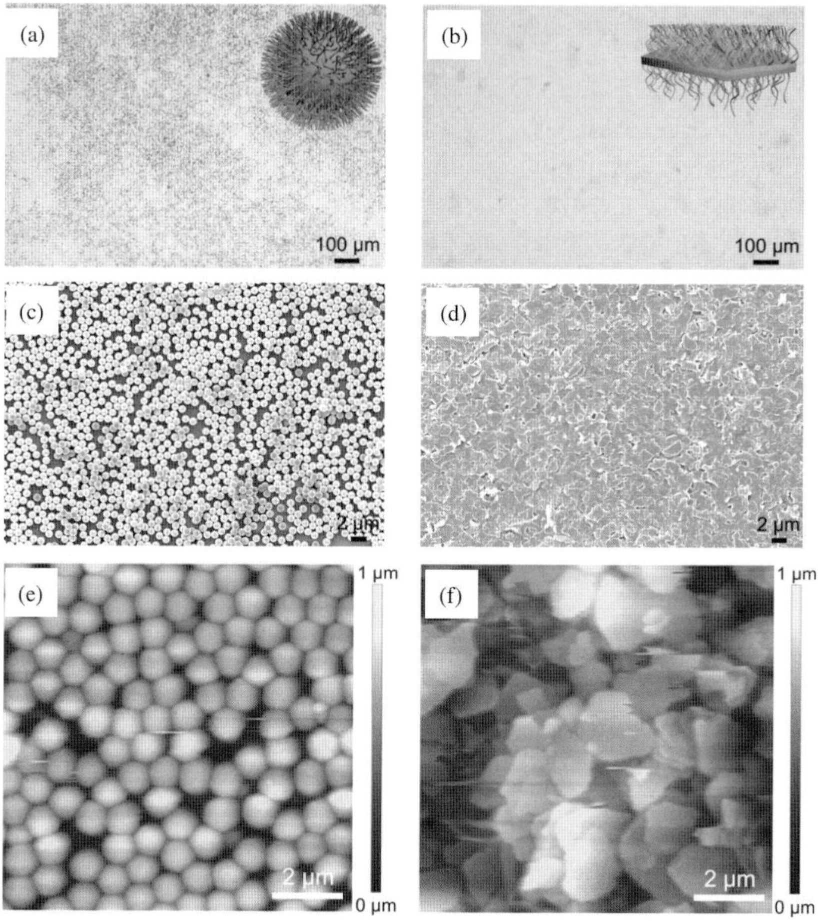

Figure 31. Representative topographical images of the JP-based surfaces: (a), (c) and (e) spherical SiO_2-P(PEGMA)/PDMS-JP surface, (b), (d) and (f) platelet-like K-P(PEGMA)/P(PDMSMA)-JP surface; (a) and (b) optical microscopy images, (c) and (d) SEM images, (e, f) AFM topography images recorded under water [152].

P(PEGMA)- or P(PDMSMA)-modified side to the surrounding due to their flat geometry (Figs. 31(d) and 31(f)). Similarly, the distribution of force–distance curves corresponding to the hydrophobic or the hydrophilic polymers was found to be 44%:56%, respectively.

Furthermore, the electrokinetic behavior of the designed particle-based surfaces was evaluated using streaming potential measurements

(Figs. 32(a) and 32(b)), and it was found that the zeta-potential vs. pH curves recorded for both spherical and platelet-like JP-based surfaces are in between the curves recorded for the respective mono-functionalized particle-based surfaces, indicating the presence of both polymers on the surface. Moreover, the wetting behavior of the surfaces was investigated via water contact angle measurements (Fig. 32(c)). Similarly, the advancing and receding contact angle values measured for both spherical and platelet-like JP-based surfaces were in between the values determined for the surfaces composed of particles mono-functionalized with P(PEGMA) or P(PDMSMA) (Fig. 32(c)), additionally confirming the presence and accessibility of both polymers on the surface of the JP layers.

Additionally, the mechanical robustness of the fabricated JP-based and mono-functionalized particle-based surfaces under water was evaluated using two different kinds of experiments. First, AFM imaging in contact mode was performed under water by sequentially increasing the contact force exerted by the tip on the sample. Second, the samples were incubated for four days in DI water under constant shaking on a shaking device. Ultimately, both types of experiments demonstrated that the particle-based surfaces mono-functionalized with P(PDMSMA) as well as the JP-based surfaces are stable under application-relevant conditions, while the hydrophilic particle-based surfaces mono-functionalized with P(PEGMA) were found to be insufficiently stable.

In order to test the antifouling performance of the designed surfaces, a marine bacteria *Cobetia marina* was seeded in growth medium onto different samples and incubated for 24 h under static, low shear stress, or dynamic, high shear stress conditions. The samples with the adhered bacteria were subsequently investigated with SEM (Fig. 33). It was found that a much higher number of bacteria adhered to the control surfaces (unmodified silicon wafer, native SiO_2 and kaolinite particles) compared to the functionalized particle-based surfaces (Fig. 33). In the case of latter ones, much less bacteria adhered to the surfaces under dynamic conditions, compared to the static ones (Fig. 33). Subsequently, the number of the attached bacteria was quantified using a chemiluminescence assay that quantifies the adhering viable bacteria via determination of the ATP amount present (Fig. 34). Similarly, it was found that less bacteria adhered to all the tested surfaces under dynamic conditions if compared to the static ones, and that

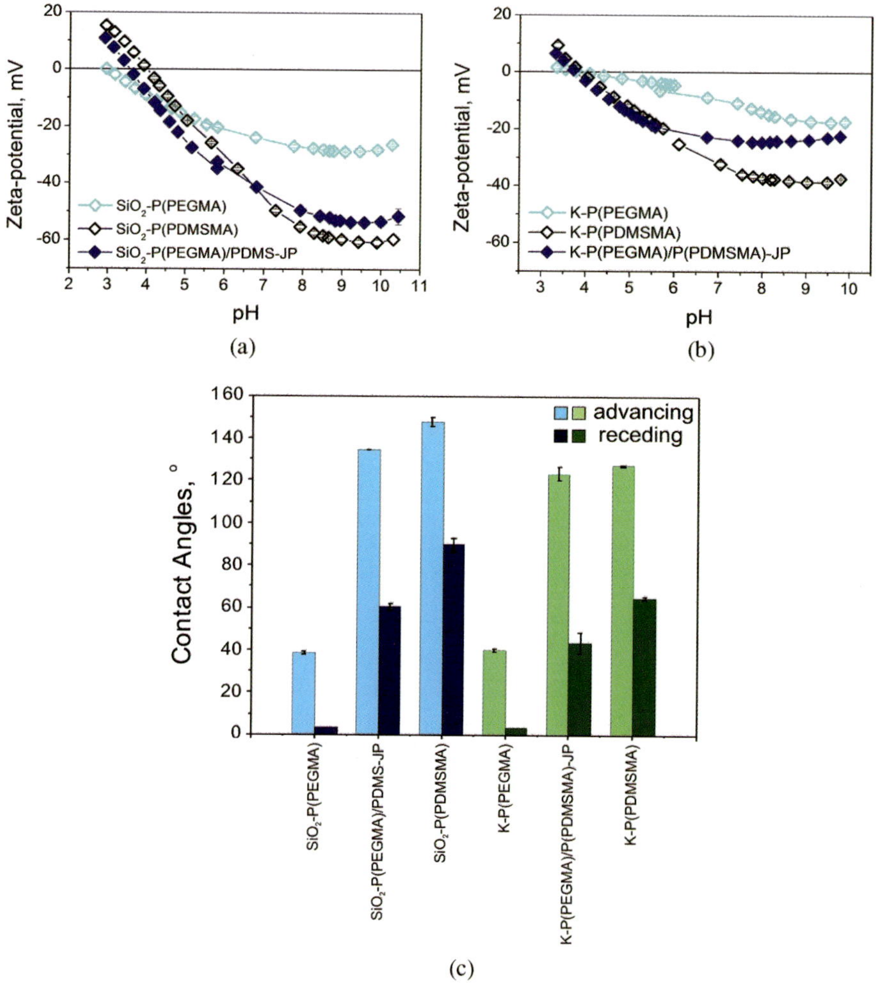

Figure 32. Electrokinetic (a) and (b) and wetting (c) properties of the JP-based surfaces showing an intermediate behavior between the respective mono-functionalized particle-based surfaces: streaming potential measurements of (a) spherical SiO_2-based mono-functionalized (cyan and black lines) and Janus (blue line) particles, and (b) platelet-like kaolinite-based mono-functionalized (cyan and black lines) and Janus (blue line) particles; water contact angle measurements (c) on the surfaces prepared from SiO_2-based (blue columns), and kaolinite-based (green columns) mono-functionalized and JPs [152].

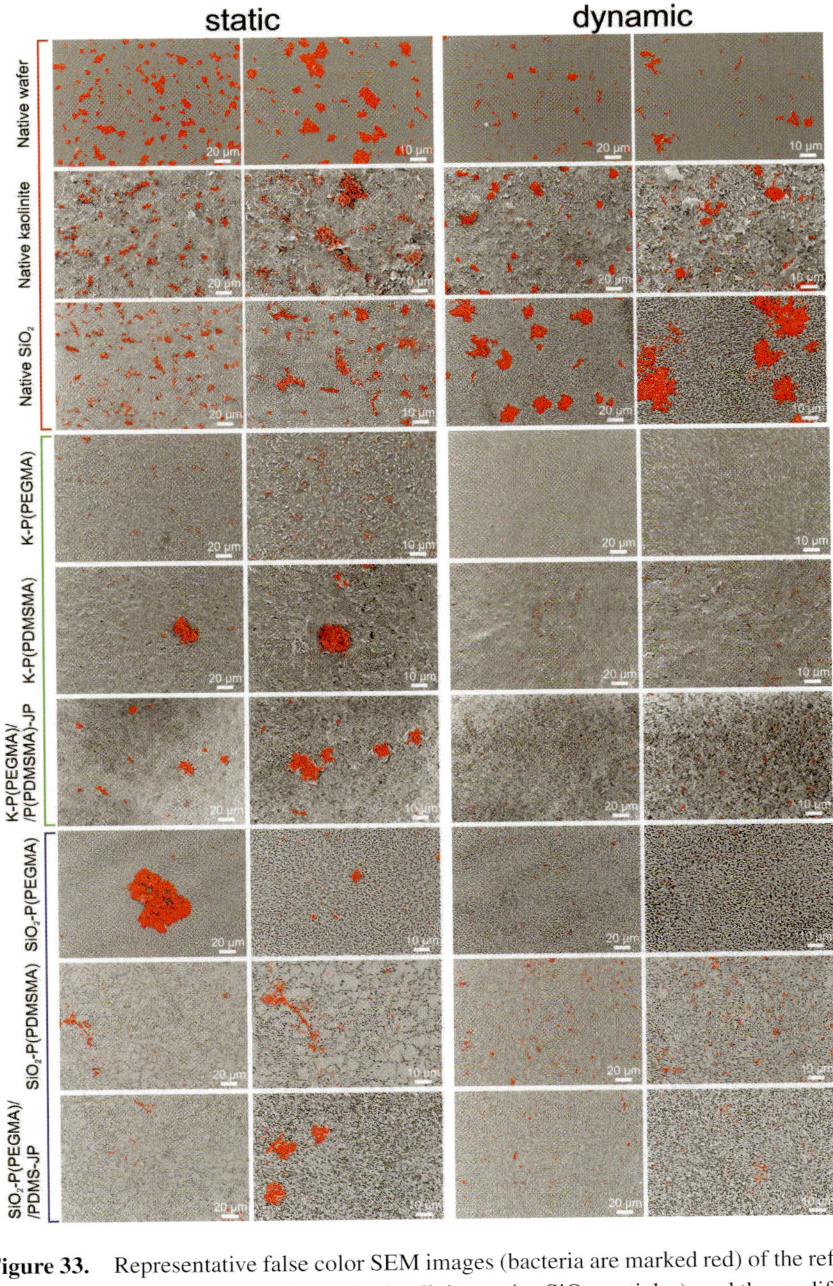

Figure 33. Representative false color SEM images (bacteria are marked red) of the reference native surfaces (native wafer, native kaolinite, native SiO_2 particles), and the modified kaolinite and SiO_2 particle-based surfaces (mono-functionalized and Janus) after the biofilm formation assay under static or dynamic conditions [152].

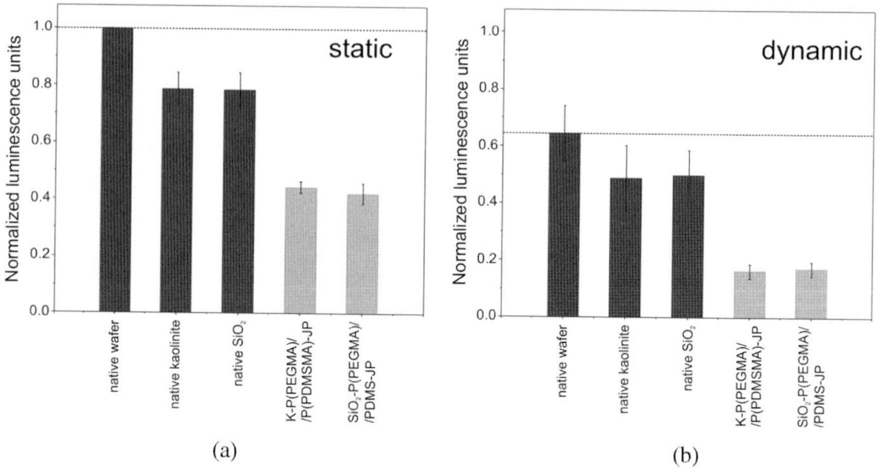

Figure 34. Antifouling and fouling-release performance of the reference native and JP-based surfaces: *Cobetia marina* bacterial cells were incubated for 24 h with the samples under (a) static (45 rpm), or (b) dynamic (90 rpm) conditions, and the number of attached cells was quantified. All values are normalized against the negative control (native wafer, static conditions) [152].

the JP-based surfaces, independently of their geometry, successfully reduced bacterial adhesion under both static and dynamic conditions compared to the control unmodified flat and particle-based surfaces (Fig. 34). As a result, it could be concluded that introducing chemical heterogeneity to the surfaces through the use of JPs allowed a significant lowering of bacterial adhesion compared to the native control surfaces; the JP-based surfaces were proven to be very robust, favoring their further application; furthermore, the kaolinite-based JPs offer a scalable system that may be prepared through a reduced number of steps, allowing the fabrication of large-area coatings. Therefore, the developed active functional surfaces based on hybrid hairy JPs possess great potential for the future fabrication of novel robust antifouling and fouling-release coatings.

10.4. Future Perspectives

Research activities in the field of JPs have flourished in the past decade, indicating the importance of the related investigations and novel kinds of asymmetric building blocks, ready to fulfill the requirement of multi-functionality in the field of materials science. As it was discussed in Chapter 1, the number of contributions regarding the topic of JPs is rapidly growing, indicating a broad interest and great potential of this field of research (Fig. 6(a)). Due to the numerous developed synthetic strategies for the fabrication of JPs, it is now possible to acquire controlled functionalities and architectures with reasonable resources. Nevertheless, there is still scope for advancement concerning the up-scalability of the synthetic procedures, which is a crucial point for the production of JPs on an industrial scale and further applicability. Further, taking advantage of their unique asymmetry, the self-assembly of JPs has been extensively studied in the recent years, revealing a diversity of advanced structures formed through different kinds of interactions between the anisotropic building blocks. The discovered structures have a great potential from the perspective of materials science, although demand for more fundamental investigations still exists in order to translate the mechanisms of the JP assembly toward the production of real materials. On the other hand, arguably the most important part of the topic is the applications of the JPs toward rational material design. Considering the number of contributions published each year on the topic of JPs in general (Fig. 6(a)), the number of contributions regarding their applications is by far smaller (Fig. 6(b)), which almost certainly indicates on the need of a more extensive research in order to apply the synthesized building blocks.

At the same time, there is a substantial need for the development of new progressive materials with specific functions in order to address various technological applications. For instance, surfaces and interfaces with controlled adhesion and wettability in dry, wet, underwater, cold, etc. conditions are essential to fabricate novel materials, coatings, or membranes with antifouling, anti-icing, or self-cleaning properties. Herein, an increasing attention is being drawn to heterogeneous instead of homogeneous surfaces, where a combination of advantageous properties of several components in one surface may lead to synergistic effects and enhanced desired features of the resulting material. Therefore, active multifunctional surfaces and interfaces present great capabilities in technological applications, potentially allowing reducing costs, fuel consumption, and even the number of emergency situations in certain applications.

Furthermore, the robustness and durability of the developed surfaces is pivotal for their future long-term operation.

JPs may present an ideal solution toward the development of multifunctional active surfaces and interfaces (Scheme 1). They can be utilized as advanced building blocks to construct materials with controlled chemical and topographical heterogeneity as well as wettability and adhesion. Especially, combining both chemical and topographical heterogeneity, such materials may produce synergistic effects favorable for specific applications. Some investigations have already been made in this direction, and contributions regarding the use of JPs for active surfaces as well as interfaces are starting to appear in the recent years (Fig. 7). The most promising examples were highlighted in the previous chapters, showing a diversity of advanced properties imparted to the materials by the asymmetric multifunctional JPs (Scheme 1). However, more investigations are needed in order to understand the complex effects of the JP assembly on the resulting surface/interface functions.

Moreover, aside from organic polymer-based or inorganic metal/metal oxide-based JPs, a blend of these physical properties can be

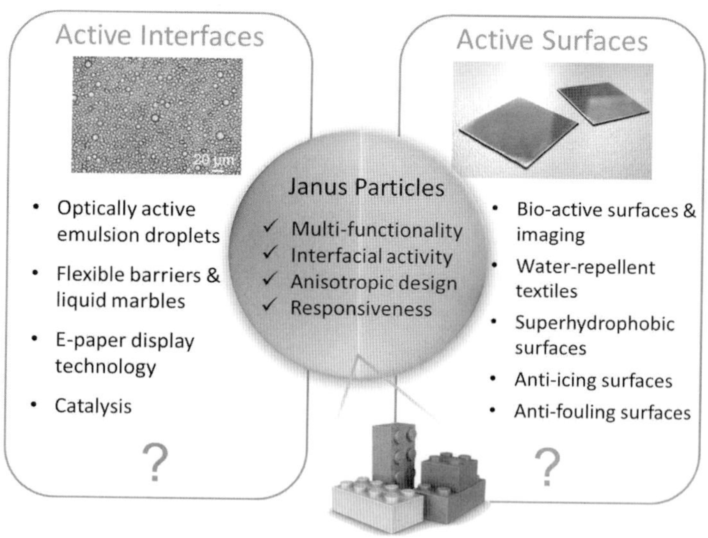

Scheme 1. JPs as building blocks for active interfaces and surfaces: Summary of the state-of-the-art application possibilities for the JPs.

employed to create hybrid JPs, where the mechanical stability would be provided by the inorganic core, while the functionality and stimuli-responsive properties would be introduced by the polymeric shells. By this time, such hybrid hairy JPs were utilized to produce functional active surfaces with controlled wettability toward ultrahydrophobic surfaces, and controlled adhesiveness of ice as well as fouling organisms toward anti-icing or antifouling coatings, respectively. Due to the robust design of the single building blocks, the developed structured surfaces also possess mechanical stability and enhanced robustness, while allowing an easy and scalable production via simple solvent casting or spraying.

Despite the fact that the JPs already offer many feasible solutions for the rational design of active multifunctional materials, there are still several challenges to be overcome for the future industrial applications, such as the production of materials in a reasonably low time and with cost effectiveness as well as the enhancement of their durability toward their long-term operation [153].

References

[1] Walther, A.; Müller, A. H. E. *Chem. Rev.*, **113**, 5194 (**2013**).

[2] Mao, Z.; Xu, H.; Wang, D. *Adv. Funct. Mater.*, **20**, 1053 (**2010**).

[3] Li, F.; Josephson, D. P.; Stein, A. *Angew. Chem.*, **50**, 360 (**2011**).

[4] Glotzer, S. C.; Solomon, M. J. *Nat. Mater.*, **6**, 557 (**2007**).

[5] Yi, G. R.; Pine, D. J.; Sacanna, S. *J. Phys. Condens. Matter.*, **25**, 193101 (**2013**).

[6] Casagrande, C.; Fabre, P.; Raphaël, E.; Veyssié, M. *EPL* (*Europhys. Lett.*), **9**, 251 (**1989**).

[7] de Gennes, P.-G. *Angew. Chem. Int. Ed. in English*, **31**, 842 (**1992**).

[8] Hawker, C. J.; Wooley, K. L.; Frechet, J. M. J. *J. Chem. Soc., Perkin Transactions*, **1**, 1287 (**1993**).

[9] Wooley, K. L.; Hawker, C. J.; Frechet, J. M. J. *J. Am. Chem. Soc.*, **115**, 11496 (**1993**).

[10] Percec, V.; Wilson, D. A.; Leowanawat, P.; Wilson, C. J.; Hughes, A. D.; Kaucher, M. S.; Hammer, D. A.; Levine, D. H.; Kim, A. J.; Bates, F. S.; Davis, K. P.; Lodge, T. P.; Klein, M. L.; DeVane, R. H.; Aqad, E.; Rosen, B. M.; Argintaru, A. O.; Sienkowska, M. J.; Rissanen, K.; Nummelin, S.; Ropponen, J. *Science*, **328**, 1009 (**2010**).

[11] Zhang, S.; Sun, H.-J.; Hughes, A. D.; Moussodia, R.-O.; Bertin, A.; Chen, Y.; Pochan, D. J.; Heiney, P. A.; Klein, M. L.; Percec, V. *Proc. Nat. Acad. Sci.*, **111**, 9058 (**2014**).

[12] Zhang, S.; Moussodia, R.-O.; Sun, H.-J.; Leowanawat, P.; Muncan, A.; Nusbaum, C. D.; Chelling, K. M.; Heiney, P. A.; Klein, M. L.; André, S.; Roy, R.; Gabius, H.-J.; Percec, V. *Angew. Chem. Int. Ed.*, **53**, 10899 (**2014**).

[13] Ge, Z.; Xu, J.; Hu, J.; Zhang, Y.; Liu, S. *Soft Matter*, **5**, 3932 (**2009**).

[14] Bolton, J.; Bailey, T. S.; Rzayev, J. *Nano Lett.*, **11**, 998 (**2011**).

[15] Linder, M. B.; Szilvay, G. R.; Nakari-Setälä, T.; Penttilä, M. E. *FEMS Microbiol. Rev.*, **29**, 877 (**2005**).

[16] Kostiainen, M. A.; Szilvay, G. R.; Smith, D. K.; Linder, M. B.; Ikkala, O. *Angew. Chem. Int. Ed.*, **45**, 3538 (**2006**).

[17] Suci, P. A.; Kang, S.; Young, M.; Douglas, T. *J. Am. Chem. Soc.* **131**, 9164 (**2009**).

[18] Semrau, S.; Schmidt, T. *Soft Matter*, **5**, 3174 (**2009**).

[19] Christian, D. A.; Tian, A.; Ellenbroek, W. G.; Levental, I.; Rajagopal, K.; Janmey, P. A.; Liu, A. J.; Baumgart, T.; Discher, D. E. *Nat. Mater.*, **8**, 843 (**2009**).

[20] Voets, I. K.; de Keizer, A.; de Waard, P.; Frederik, P. M.; Bomans, P. H. H.; Schmalz, H.; Walther, A.; King, S. M.; Leermakers, F. A. M.; Cohen Stuart, M. A. *Angew. Chem. Int. Ed.*, **45**, 6673 (**2006**).

[21] Voets, I. K.; Fokkink, R.; Hellweg, T.; King, S. M.; de Waard, P.; de Keizer, A.; Cohen Stuart, M. A. *Soft Matter*, **5**, 999 (**2009**).

[22] Fang, B.; Walther, A.; Wolf, A.; Xu, Y.; Yuan, J.; Müller, A. H. E. *Angew. Chem. Int. Ed.*, **48**, 2877 (**2009**).

[23] Gröschel, A. H.; Walther, A.; Löbling, T. I.; Schmelz, J.; Hanisch, A.; Schmalz, H.; Müller, A. H. E. *J. Am. Chem. Soc.*, **134**, 13850 (**2012**).

[24] Erhardt, R.; Zhang, M. F.; Böker, A.; Zettl, H.; Abetz, C.; Frederik, P.; Krausch, G.; Abetz, V.; Müller, A. H. E. *J. Am. Chem. Soc.*, **125**, 3260 (**2003**).

[25] Erhardt, R.; Böker, A.; Zettl, H.; Kaya, H.; Pyckhout-Hintzen, W.; Krausch, G.; Abetz, V.; Müller, A. H. E. *Macromolecules*, **34**, 1069 (**2001**).

[26] Saito, R.; Fujita, A.; Ichimura, A.; Ishizu, K. *J. Polym. Sci., Part A: Polym. Chem.*, **38**, 2091 (**2000**).

[27] Liu, Y.; Abetz, V.; Müller, A. H. E. *Macromolecules*, **36**, 7894 (**2003**).

[28] Wolf, A.; Walther, A.; Müller, A. H. E. *Macromolecules*, **44**, 9221 (**2011**).

[29] Walther, A.; Drechsler, M.; Müller, A. H. E. *Soft Matter*, **5**, 385 (**2009**).

[30] Saito, N.; Kagari, Y.; Okubo, M. *Langmuir*, **22**, 9397 (**2006**).
[31] Tanaka, T.; Okayama, M.; Minami, H.; Okubo, M. *Langmuir*, **26**, 11732 (**2010**).
[32] Higuchi, T.; Tajima, A.; Yabu, H.; Shimomura, M. *Soft Matter*, **4**, 1302 (**2008**).
[33] Misra, A.; Urban, M. W. *Macromolecules Rapid Commun.*, **31**, 119 (**2010**).
[34] Park, J. G.; Forster, J. D.; Dufresne, E. R. *J. Am. Chem. Soc.*, **132**, 5960 (**2010**).
[35] Roh, K. H.; Martin, D. C.; Lahann, J. *Nat. Mater.*, **4**, 759 (**2005**).
[36] Bhaskar, S.; Hitt, J.; Chang, S. W. L.; Lahann, J. *Angew. Chem. Int. Ed.*, **48**, 4589 (**2009**).
[37] Nie, Z.; Li, W.; Seo, M.; Xu, S.; Kumacheva, E. *J. Am. Chem. Soc.*, **128**, 9408 (**2006**).
[38] Shah, R. K.; Kim, J. W.; Weitz, D. A. *Adv. Mater.*, **21**, 1949 (**2009**).
[39] Dendukuri, D.; Pregibon, D. C.; Collins, J.; Hatton, T. A.; Doyle, P. S. *Nat. Mater.*, **5**, 365 (**2006**).
[40] Higuchi, T.; Tajima, A.; Motoyoshi, K.; Yabu, H.; Shimomura, M. *Angew. Chem. Int. Ed.*, **47**, 8044 (**2008**).
[41] Bhaskar, S.; Roh, K. H.; Jiang, X.; Baker, G. L.; Lahann, J. *Macromolecules Rapid Commun.*, **29**, 1655 (**2008**).
[42] Pregibon, D. C.; Toner, M.; Doyle, P. S. *Science*, **315**, 1393 (**2007**).
[43] Takei, H.; Shimizu, N. *Langmuir*, **13**, 1865 (**1997**).
[44] Chen, Q.; Bae, S. C.; Granick, S. *Nature*, **469**, 381 (**2011**).
[45] Lin, C. C.; Liao, C. W.; Chao, Y. C.; Kuo, C. *ACS Appl. Mater. Interfaces*, **2**, 3185 (**2010**).
[46] Ye, S.; Carroll, R. L. *ACS Appl. Mater. Interfaces*, **2**, 616 (**2010**).
[47] McConnell, M. D.; Kraeutler, M. J.; Yang, S.; Composto, R. J. *Nano Lett.*, **10**, 603 (**2010**).
[48] Jiang, S.; Granick, S. *Langmuir*, **25**, 8915 (**2009**).
[49] Ling, X. Y.; Phang, I. Y.; Acikgoz, C.; Yilmaz, M. D.; Hempenius, M. A.; Vancso, G. J.; Huskens, J. *Angew. Chem. Int. Ed.*, **48**, 7677 (**2009**).
[50] Kim, S. H.; Lee, S. Y.; Yang, S. M. *Angew. Chem. Int. Ed.*, **49**, 2535 (**2010**).
[51] He, Z.; Kretzschmar, I. *Langmuir*, **29**, 15755 (**2013**).
[52] Hong, L.; Jiang, S.; Granick, S. *Langmuir*, **22**, 9495 (**2006**).
[53] Jiang, S.; Granick, S. *Langmuir*, **24**, 2438 (**2008**).
[54] Jiang, S.; Schultz, M. J.; Chen, Q.; Moore, J. S.; Granick, S. *Langmuir*, **24**, 10073 (**2008**).
[55] Perro, A.; Meunier, F.; Schmitt, V.; Ravaine, S. *Colloids Surf. A*, **332**, 57 (**2009**).

[56] Pradhan, S.; Xu, L.; Chen, S. *Adv. Funct. Mater.*, **17**, 2385 (**2007**).
[57] Sardar, R.; Heap, T. B.; Shumaker-Parry, J. S. *J. Am. Chem. Soc.*, **129**, 5356 (**2007**).
[58] Yu, H.; Chen, M.; Rice, P. M.; Wang, S. X.; White, R. L.; Sun, S. *Nano Lett.*, **5**, 379 (**2005**).
[59] Wang, C.; Xu, C.; Zeng, H.; Sun, S. *Adv. Mat.*, **21**, 3045 (**2009**).
[60] Sacanna, S.; Rossi, L.; Pine, D. J. *J. Am. Chem. Soc.*, **134**, 6112 (**2012**).
[61] Teo, B. M.; Suh, S. K.; Hatton, T. A.; Ashokkumar, M.; Grieser, F. *Langmuir*, **27**, 30 (**2011**).
[62] Tang, C.; Zhang, C.; Liu, J.; Qu, X.; Li, J.; Yang, Z. *Macromolecules*, **43**, 5114 (**2010**).
[63] Zhang, C.; Liu, B.; Tang, C.; Liu, J.; Qu, X.; Li, J.; Yang, Z. *Chem. Commun.*, **46**, 4610 (**2010**).
[64] Shepherd, R. F.; Conrad, J. C.; Rhodes, S. K.; Link, D. R.; Marquez, M.; Weitz, D. A.; Lewis, J. A. *Langmuir*, **22**, 8618 (**2006**).
[65] Anderson, K. D.; Luo, M.; Jakubiak, R.; Naik, R. R.; Bunning, T. J.; Tsukruk, V. V. *Chem. Mater.*, **22**, 3259 (**2010**).
[66] Liu, B.; Zhang, C.; Liu, J.; Qu, X.; Yang, Z. *Chem. Commun.*, 3871 (**2009**).
[67] Liu, B.; Wei, W.; Qu, X.; Yang, Z. *Angew. Chem. Int. Ed.*, **47**, 3973 (**2008**).
[68] Berger, S.; Synytska, A.; Ionov, L.; Eichhorn, K.-J.; Stamm, M. *Macromolecules*, **41**, 9669 (**2008**).
[69] Kirillova, A.; Stoychev, G.; Ionov, L.; Eichhorn, K. J.; Malanin, M.; Synytska, A. *ACS Appl. Mater. Interfaces*, **6**, 13106 (**2014**).
[70] Wang, B.; Li, B.; Zhao, B.; Li, C. Y. *J. Am. Chem. Soc.*, **130**, 11594 (**2008**).
[71] Synytska, A.; Kirillova, A.; Isa, L. *Chem. Plus. Chem.*, **79**, 656 (**2014**).
[72] Synytska, A.; Ionov, L. *Particle & Particle Systems Characterization*, **30**, 922 (**2013**).
[73] Walther, A.; Drechsler, M.; Rosenfeldt, S.; Harnau, L.; Ballauff, M.; Abetz, V.; Müller, A. H. E. *J. Am. Chem. Soc.*, **131**, 4720 (**2009**).
[74] Chen, Q.; Whitmer, J. K.; Jiang, S.; Bae, S. C.; Luijten, E.; Granick, S. *Science*, **331**, 199 (**2011**).
[75] Hong, L.; Cacciuto, A.; Luijten, E.; Granick, S. *Langmuir*, **24**, 621 (**2008**).
[76] Kirillova, A.; Stoychev, G.; Ionov, L.; Synytska, A. *Langmuir*, **30**, 12765 (**2014**).
[77] Cayre, O.; Paunov, V. N.; Velev, O. D. *Chem. Commun.*, **18**, 2296 (**2003**).
[78] Hong, L.; Cacciuto, A.; Luijten, E.; Granick, S. *Nano Letters*, **6**, 2510 (**2006**).
[79] Goyal, A.; Hall, C. K.; Velev, O. D. *Phys. Rev. E*, **77**, 031401 (**2008**).

[80] Goyal, A.; Hall, C. K.; Velev, O. D. *Soft Matter*, **6**, 480 (**2010**).
[81] Gangwal, S.; Cayre, O. J.; Velev, O. D. *Langmuir*, **24**, 13312 (**2008**).
[82] Gangwal, S.; Cayre, O. J.; Bazant, M. Z.; Velev, O. D. *Phys . Rev. Lett.*, **100**, 058302 (**2008**).
[83] Zhang, L.; Zhu, Y. *Langmuir*, **28**, 13201 (**2012**).
[84] Shah, A. A.; Schultz, B.; Zhang, W.; Glotzer, S. C.; Solomon, M. J. *Nat Mater*, **14**, 117 (**2015**).
[85] Smoukov, S. K.; Gangwal, S.; Marquez, M.; Velev, O. D. *Soft Matter*, **5**, 1285 (**2009**).
[86] Isojima, T.; Suh, S. K.; Vander Sande, J. B.; Hatton, T. A. *Langmuir*, **25**, 8292 (**2009**).
[87] Yan, J.; Bloom, M.; Bae, S. C.; Luijten, E.; Granick, S. *Nature*, **491**, 578 (**2012**).
[88] Yan, J.; Chaudhary, K.; Chul Bae, S.; Lewis, J. A.; Granick, S. *Nat Commun*, **4**, 1516 (**2013**).
[89] Yan, J.; Bae, S. C.; Granick, S. *Adv. Mat.*, **27**, 874 (**2015**).
[90] Walther, A.; Hoffmann, M.; Müller, A. H. E. *Angew. Chem. Int. Ed.*, **47**, 711 (**2008**).
[91] Walther, A.; Matussek, K.; Müller, A. H. E. *ACS Nano*, **2**, 1167 (**2008**).
[92] Yan, L. T.; Popp, N.; Ghosh, S. K.; Böker, A. *ACS Nano*, **4**, 913 (**2010**).
[93] Link, J. R.; Sailor, M. J. *Proc. Natl. Acad. Sci. U.S.A.*, **100**, 10607 (**2003**).
[94] Bucaro, M. A.; Kolodner, P. R.; Taylor, J. A.; Sidorenko, A.; Aizenberg, J.; Krupenkin, T. N. *Langmuir*, **25**, 3876 (**2009**).
[95] Kirillova, A.; Schliebe, C.; Stoychev, G.; Jakob, A.; Lang, H.; Synytska, A. *ACS Appl. Mater. & Interfaces*, **7**, 21218 (**2015**).
[96] Synytska, A.; Khanum, R.; Ionov, L.; Cherif, C.; Bellmann, C. *ACS Appl. Mater. Interfaces*, **3**, 1216 (**2011**).
[97] Berger, S.; Ionov, L.; Synytska, A. *Adv. Funct. Mater.*, **21**, 2338 (**2011**).
[98] Nisisako, T.; Torii, T.; Takahashi, T.; Takizawa, Y. *Adv. Mater.*, **18**, 1152 (**2006**).
[99] Sheridon, N. K.; Richley, E. A.; Mikkelsen, J. C.; Tsuda, D.; Crowley, J. M.; Oraha, K. A.; Howard, M. E.; Rodkin, M. A.; Swidler, R.; Sprague, R. *J. Soc. Inf. Disp.*, **7**, 141 (**1999**).
[100] Kim, S. H.; Jeon, S. J.; Jeong, W. C.; Park, H. S.; Yang, S. M. *Adv. Mater.*, **20**, 4129 (**2008**).
[101] Yoshida, M.; Roh, K. H.; Mandal, S.; Bhaskar, S.; Lim, D.; Nandivada, H.; Deng, X.; Lahann, J. *Adv. Mater.*, **21**, 4920 (**2009**).
[102] Xu, C.; Xie, J.; Ho, D.; Wang, C.; Kohler, N.; Walsh, E. G.; Morgan, J. R.; Chin, Y. E.; Sun, S. *Angew. Chem. Int. Ed.*, **47**, 173 (**2007**).
[103] Hu, S. H.; Gao, X. *J. Am. Chem. Soc.*, **132**, 7234 (**2010**).

[104] Anker, J. N.; Kopelman, R. *Appl. Phys. Lett.*, **82**, 1102 (**2003**).
[105] Anker, J. N.; Behrend, C.; Kopelman, R. *J. Appl. Phys.*, **93**, 6698 (**2003**).
[106] Behrend, C. J.; Anker, J. N.; McNaughton, B. H.; Kopelman, R. *J. Magn. Magn. Mater.*, **293**, 663 (**2005**).
[107] Howse, J. R.; Jones, R. A. L.; Ryan, A. J.; Gough, T.; Vafabakhsh, R.; Golestanian, R. *Phys. Rev. Lett.*, **99**, 048102 (**2007**).
[108] Pavlick, R. A.; Sengupta, S.; McFadden, T.; Zhang, H.; Sen, A. *Angew. Chem. Int. Ed.*, **50**, 9374 (**2011**).
[109] Ma, X.; Jannasch, A.; Albrecht, U.-R.; Hahn, K.; Miguel-López, A.; Schäffer, E.; Sánchez, S. *Nano Lett.*, **15**, 7043 (**2015**).
[110] Jurado-Sánchez, B.; Sattayasamitsathit, S.; Gao, W.; Santos, L.; Fedorak, Y.; Singh, V. V.; Orozco, J.; Galarnyk, M.; Wang, J. *Small*, **11**, 499 (**2015**).
[111] Thomson Reuters, webofknowledge.com.
[112] Pickering, S. U. *J. Chem. Soc. Trans.*, **91**, 2001 (**1907**).
[113] Binks, B. P.; Fletcher, P. D. I. *Langmuir*, **17**, 4708 (**2001**).
[114] Fernandez-Rodriguez, M. A.; Rodriguez-Valverde, M. A.; Cabrerizo-Vilchez, M. A.; Hidalgo-Alvarez, R. *Adv. Coll. Int. Sci.*, (**2015**).
[115] Kumar, A.; Park, B. J.; Tu, F.; Lee, D. *Soft Matter*, **9**, 6604 (**2013**).
[116] Park, B. J.; Lee, D. *MRS Bulletin*, **39**, 1089 (**2014**).
[117] Lenis, J.; Razavi, S.; Cao, K. D.; Lin, B.; Lee, K. Y.; Tu, R. S.; Kretzschmar, I. *J Am Chem Soc*, **137**, 15370 (**2015**).
[118] Hirose, Y.; Komura, S.; Nonomura, Y. *J. Chem. Phys.*, **127**, 054707 (**2007**).
[119] Gao, H. M.; Lu, Z. Y.; Liu, H.; Sun, Z. Y.; An, L. J. *J Chem Phys.*, **141**, 134907 (**2014**).
[120] Jiang, S.; Granick, S. *J. Chem.Phys.*, **127**, 161102 (**2007**).
[121] Xie, Q.; Davies, G. B.; Gunther, F.; Harting, J. *Soft Matter*, **11**, 3581 (**2015**).
[122] Dendukuri, D.; Hatton, T. A.; Doyle, P. S. *Langmuir*, **23**, 4669 (**2007**).
[123] Kim, J.-W.; Lee, D.; Shum, H. C.; Weitz, D. A. *Adv. Mater.*, **20**, 3239 (**2008**).
[124] Cao, Z.; Wang, G.; Chen, Y.; Liang, F.; Yang, Z. *Macromolecules*, **48**, 7256 (**2015**).
[125] Passas-Lagos, E.; Schüth, F. *Langmuir*, **31**, 7749 (**2015**).
[126] Zhao, Z.; Liang, F.; Zhang, G.; Ji, X.; Wang, Q.; Qu, X.; Song, X.; Yang, Z. *Macromolecules*, **48**, 3598 (**2015**).
[127] Yamagami, T.; Kitayama, Y.; Okubo, M. *Langmuir*, **30**, 7823 (**2014**).
[128] Bryson, K. C.; Löbling, T. I.; Müller, A. H. E.; Russell, T. P.; Hayward, R. C. *Macromolecules*, **48**, 4220 (**2015**).

[129] Bahrami, R.; Löbling, T. I.; Gröschel, A. H.; Schmalz, H.; Müller, A. H. E.; Altstädt, V. *ACS Nano*, **8**, 10048 (**2014**).
[130] Huang, M.; Guo, H. *Soft Matter*, **9**, 7356 (**2013**).
[131] Kim, J. U.; Matsen, M. W. *Physical Rev. Lett.*, **102**, 078303 (**2009**).
[132] Ou, F. S.; Shaijumon, M. M.; Ajayan, P. M. *Nano Lett.*, **8**, 1853 (**2008**).
[133] Zhao, Y.; Gu, H.; Xie, Z.; Shum, H. C.; Wang, B.; Gu, Z. *J Am Chem Soc.*, **135**, 54 (**2013**).
[134] Komazaki, Y.; Hirama, H.; Torii, T. *J. Appl. Phys.*, **117**, 154506 (**2015**).
[135] Sheridon, N. K.; Xerox Corporation, Stamford, Conn.: Unites States, *U.S. Patent*, 4,126,854 (**1978**).
[136] Crowley, J. M.; Richley, E. A.; Sheridon, N. K.; Xerox Corporation, Stamford, Conn.: United States, *U.S. Patent US***5262098** *A*, (**1993**).
[137] Sheridon, N. K.; Xerox Corporation, Stamford, Conn.: United States, *U.S. Patent US*5344594 *A*, (**1994**).
[138] Yin, S. N.; Wang, C. F.; Yu, Z. Y.; Wang, J.; Liu, S. S.; Chen, S. *Adv. Mater.*, **23**, 2915 (**2011**).
[139] Yu, Z.; Wang, C. F.; Ling, L.; Chen, L.; Chen, S. *Angew. Chem.*, **51**, 2375 (**2012**).
[140] Liu, S.-S.; Wang, C.-F.; Wang, X.-Q.; Zhang, J.; Tian, Y.; Yin, S.-N.; Chen, S. *J. Mater. Chem.*, *C* **2**, 9431 (**2014**).
[141] Wang, H.; Yang, S.; Yin, S. N.; Chen, L.; Chen, S. *ACS Appl. Mater. Interfaces*, **7**, 8827 (**2015**).
[142] Cao, W.; Huang, R.; Qi, W.; Su, R.; He, Z. *ACS Appl. Mater. Interfaces*, **7**, 465 (**2015**).
[143] Yi, F.; Xu, F.; Gao, Y.; Li, H.; Chen, D. *RSC Adv.*, **5**, 40227 (**2015**).
[144] Crossley, S.; Faria, J.; Shen, M.; Resasco, D. E. *Science*, **327**, 68 (**2010**).
[145] Faria, J.; Ruiz, M. P.; Resasco, D. E. *Adv. Synthesis & Catalysis*, **352**, 2359 (**2010**).
[146] Schick, I.; Lorenz, S.; Gehrig, D.; Schilmann, A. M.; Bauer, H.; Panthofer, M.; Fischer, K.; Strand, D.; Laquai, F.; Tremel, W. *J. Am. Chem. Soc.*, **136**, 2473 (**2014**).
[147] Fu, X.; Liu, J.; Yang, H.; Sun, J.; Li, X.; Zhang, X.; Jia, Y. *Mater. Chem. and Phys.*, **130**, 334 (**2011**).
[148] Wang, T.; Chen, S. R.; Jin, F.; Cai, J. H.; Cui, L. Y.; Zheng, Y. M.; Wang, J. X.; Song, Y. L.; Jiang, L. *Chem. Commun. (Camb.)* **51**, 1367 (**2015**).
[149] Zhao, H.; Liang, F.; Qu, X.; Wang, Q.; Yang, Z. *Macromolecules*, **48**, 700 (**2015**).

[150] Yang, H.; Liang, F.; Chen, Y.; Wang, Q.; Qu, X.; Yang, Z. *NPG Asia Mat.*, **7**, e176 (**2015**).

[151] Kirillova, A.; Ionov, L.; Roisman, I. V.; Synytska, A. *Chem. Mater.*, **28**, 6995–7005 (**2016**).

[152] Kirillova, A.; Marschelke, C.; Friedrichs, J.; Werner, C.; Synytska, A. *ACS Appl. Mater. Interfaces*, **8**, 32591–32603 (**2016**).

[153] Synytska, A. "Hairy Janus Particles: From Development and Assembly towards Different Facets of Applications", Habiliration Thesis, TU Dresden (**2017**).

Index